Reactive Oxygen Species in Plants

Reactive Oxygen Species in Plants

Boon Or Bane - Revisiting the Role of ROS

Edited by

Vijay Pratap Singh
Govt. Ramanuj Pratap Singhdev Post Graduate College
Chhattisgarh, India

Samiksha Singh
Ranjan Plant Physiology and Biochemistry Laboratory
Department of Botany
University of Allahabad, Allahabad, India

Durgesh Kumar Tripathi
Centre of Advanced Study in Botany
Banaras Hindu University
Varanasi, India

Sheo Mohan Prasad
Ranjan Plant Physiology and Biochemistry Laboratory
Department of Botany
University of Allahabad
Allahabad, India

Devendra Kumar Chauhan
D.D. Pant Interdisciplinary Research Laboratory
Department of Botany
University of Allahabad, Allahabad, India

WILEY

This edition first published 2018

Registered Offices
John Wiley & Sons, Inc., 111 River Street, Hoboken, NJ 07030, USA
John Wiley & Sons Ltd, The Atrium, Southern Gate, Chichester, West Sussex, PO19 8SQ, UK

Editorial Office
9600 Garsington Road, Oxford, OX4 2DQ, UK

For details of our global editorial offices, customer services, and more information about Wiley products visit us at www.wiley.com.

Wiley also publishes its books in a variety of electronic formats and by print-on-demand. Some content that appears in standard print versions of this book may not be available in other formats.

Library of Congress Cataloging-in-Publication data applied for

ISBN: 9781119287292

Cover Design: Wiley
Cover Image: © amenic181/Gettyimages

Set in 10/12pt Warnock by SPi Global, Pondicherry, India

10 9 8 7 6 5 4 3 2 1

Contents

List of Contributors

Madhoolika Agrawal
Laboratory of Air Pollution and Global Climate Change, Department of Botany, Institute of Science, Banaras Hindu University, Varanasi, India

S.B. Agrawal
Laboratory of Air Pollution and Global Climate Change, Department of Botany, Institute of Science, Banaras Hindu University, Varanasi, India

Neelam Atri
MMV, Banaras Hindu University, Varanasi, India

Aditya Banerjee
Post Graduate Department of Biotechnology, St Xavier's College (Autonomous), Kolkata, India

Nivedita Chaudhary
Laboratory of Air Pollution and Global Climate Change, Department of Botany, Institute of Science, Banaras Hindu University, Varanasi, India

Krishna Kumar Choudhary
Laboratory of Air Pollution and Global Climate Change, Department of Botany, Institute of Science, Banaras Hindu University, Varanasi, India

Institute of Soil, Water and Environmental Sciences, Volcani Center, Agricultural Research Organization, Bet Dagan, Israel

Farah Deeba
Plant Ecology & Environmental Science, CSIR-National Botanical Research Institute, Rana Pratap Marg, Lucknow, India

Gábor Feigl
Department of Plant Biology, University of Szeged, Szeged, Hungary

Sunil K. Gupta
Plant Ecology & Environmental Science; and Academy of Scientific and Innovative Research (AcSIR), CSIR-National Botanical Research Institute, Rana Pratap Marg, Lucknow, India

Gajendra S. Jeena
CSIR-Central Institute of Medicinal and Aromatic Plants, Lucknow, India

Byoung Ryong Jeong
Horticulture Major, Division of Applied Life Science (BK21 Plus), Graduate School; Institute of Agriculture and Life Science; and Research Institute of Life Science, Gyeongsang National University, Jinju, South Korea

Sunita Kataria
School of Life Sciences, Devi Ahilya Vishwavidyalaya, Khandwa Road Campus, Indore, India

Zsuzsanna Kolbert
Department of Plant Biology, University of Szeged, Szeged, Hungary

Chun-Yan Liu
Institute of Root Biology, Yangtze University, Jingzhou, Hubei, China

Abinaya Manivannan
Institute of Agriculture and Life Science, Gyeongsang National University, Jinju, South Korea

Shakti Mehrotra
CSIR-Central Institute of Medicinal and Aromatic Plants, Lucknow, India

Sonal Mishra
University of Jammu, Jammu, India

Vivek Pandey
Plant Ecology & Environmental Science; and Academy of Scientific and Innovative Research (AcSIR), CSIR-National Botanical Research Institute, Rana Pratap Marg, Lucknow, India

Ujjal J. Phukan
CSIR-Central Institute of Medicinal and Aromatic Plants, Lucknow, India

Calogero Pinzino
Research National Council (CNR), Istituto di Chimica dei Composti Organo Metallici (ICCOM), Area della Ricerca del CNR di PISA, Pisa, Italy

Sheo Mohan Prasad
Ranjan Plant Physiology and Biochemistry Lab., Department of Botany, University of Allahabad, Allahabad, India

Syed Naved Quadri
Center for Transgenic Plant Development, Department of Biotechnology, Jamia Hamdard University, New Delhi, India

Mike Frank Quartacci
Department of Agriculture, Food and Environment, University of Pisa, Pisa, Italy

Aryadeep Roychoudhury
Post Graduate Department of Biotechnology, St Xavier's College (Autonomous), Kolkata, India

Cristina Sgherri
Department of Agriculture, Food and Environment, University of Pisa, Pisa, Italy

Marisha Sharma
Plant Ecology & Environmental Science, CSIR-National Botanical Research Institute, Rana Pratap Marg, Lucknow, India

Rakesh K. Shukla
CSIR-Central Institute of Medicinal and Aromatic Plants, Lucknow, India

Anita Singh
Ranjan Plant Physiology and Biochemistry Lab., Department of Botany, University of Allahabad, Allahabad, India

Madhulika Singh
Ranjan Plant Physiology and Biochemistry Lab., Department of Botany, University of Allahabad, Allahabad, India

Prabhakaran Soundararajan
Division of Applied Life Science (BK21 Plus), Graduate School, Gyeongsang National University, Jinju, South Korea

Vikas Srivastava
Department of Plant Sciences, Central University of Jammu, Jammu, India

Dibyendu Talukdar
Department of Botany, R.P.M. College, Hooghly, India

Sanjesh Tiwari
Ranjan Plant Physiology and Biochemistry Lab., Department of Botany, University of Allahabad, Allahabad, India

Santwana Tiwari
Ranjan Plant Physiology and Biochemistry Lab., Department of Botany, University of Allahabad, Allahabad, India

Supriya Tiwari
Department of Botany, Institute of Science, Banaras Hindu University, Varanasi, India

Qiang-Sheng Wu
Institute of Root Biology, Yangtze University, Jingzhou, Hubei, China;
Department of Chemistry, Faculty of Science, University of Hradec Kralove, Hradec Kralove, Czech Republic

Shivam Yadav
Molecular Biology Section, Center of Advanced Study in Botany, Banaras Hindu University, Varanasi, India

Yu-Jie Yang
Institute of Root Biology, Yangtze University, Jingzhou, Hubei, China

De-Jian Zhang
Institute of Root Biology, Yangtze University, Jingzhou, Hubei, China

1

Generation Mechanisms of Reactive Oxygen Species in the Plant Cell: An Overview

Santwana Tiwari, Sanjesh Tiwari, Madhulika Singh, Anita Singh, and Sheo Mohan Prasad

Ranjan Plant Physiology and Biochemistry Lab., Department of Botany, University of Allahabad, Allahabad, India

Introduction

During the course of evolution, life on the Earth started in a reducing environment and about 3.2 billion years ago, the reducing environment changed to an oxidizing one due to the appearance and proliferation of the first oxygen-evolving photosynthetic organisms, that is, cyanobacteria (Schopf *et al.*, 2007). In other words, the cyanobacteria are considered to be the first organisms to release oxygen in the environment by means of an oxygen evolving complex (OEC) (Bekker *et al.*, 2004). The outermost orbital of the dioxygen (O_2) molecule has two unpaired electrons having same spin quantum number, and this enables O_2 to accept electrons one at a time efficiently, and generate the reactive oxygen species (ROS). Out of the total O_2 utilized by plants, 1% is diverted to produce ROS in various cell organelles (del Rio *et al.*, 2002). Reactive oxygen species are essential by-products of all aerobic organisms that are produced during normal metabolic processes as well as under stress conditions. The ROS-producing subcellular organelles are mainly mitochondria, chloroplasts, peroxisomes, cytosol, and plasma membrane (Corpas *et al.*, 2015; Hasan *et al.*, 2016) (Figure 1.1). The roles of ROS are contradictory, they may have negative as well as positive roles depending upon their concentrations in the particular cell organelles. At higher concentrations, ROS cause damaging effects on proteins, DNA/RNA, and lipids by oxidative modification in plant cells (Gill and Tuteja, 2010; Hasan *et al.*, 2016). On the other hand, previous evidence clearly showed that at lower concentrations ROS act as signaling molecules in plants for regulating developmental pathways and control of redox homeostasis and defense responses against pathogens and environmental stress (Wood *et al.*, 2003; Apel and Hirt, 2004). There are well-described mechanisms in prokaryotes where the concentration of ROS directly activates transcription factors that overexpress the genes to combat oxidative stress (Kiley and Storz, 2004). There are several ROS, such as superoxide radical ($O_2^{\cdot-}$), hydroxyl radical ($\cdot OH$), hydroperoxyl radical ($HO_2\cdot$), hydrogen peroxide (H_2O_2), alkoxy radical ($RO\cdot$), peroxy radical ($ROO\cdot$), singlet oxygen ($^1O_2$), and excited carbonyl (RO^*), all of which are cytotoxic to plants at elevated concentrations (Dismukes *et al.*, 2001; Karuppanapandian *et al.*, 2011). In the cell organelles accumulation of superoxide enhances oxidative stress rather than playing a role in redox signaling. However, in some cases it damages certain proteins that activate specific signaling pathways

Reactive Oxygen Species in Plants: Boon Or Bane - Revisiting the Role of ROS, First Edition. Edited by Vijay Pratap Singh, Samiksha Singh, Durgesh Kumar Tripathi, Sheo Mohan Prasad, and Devendra Kumar Chauhan.

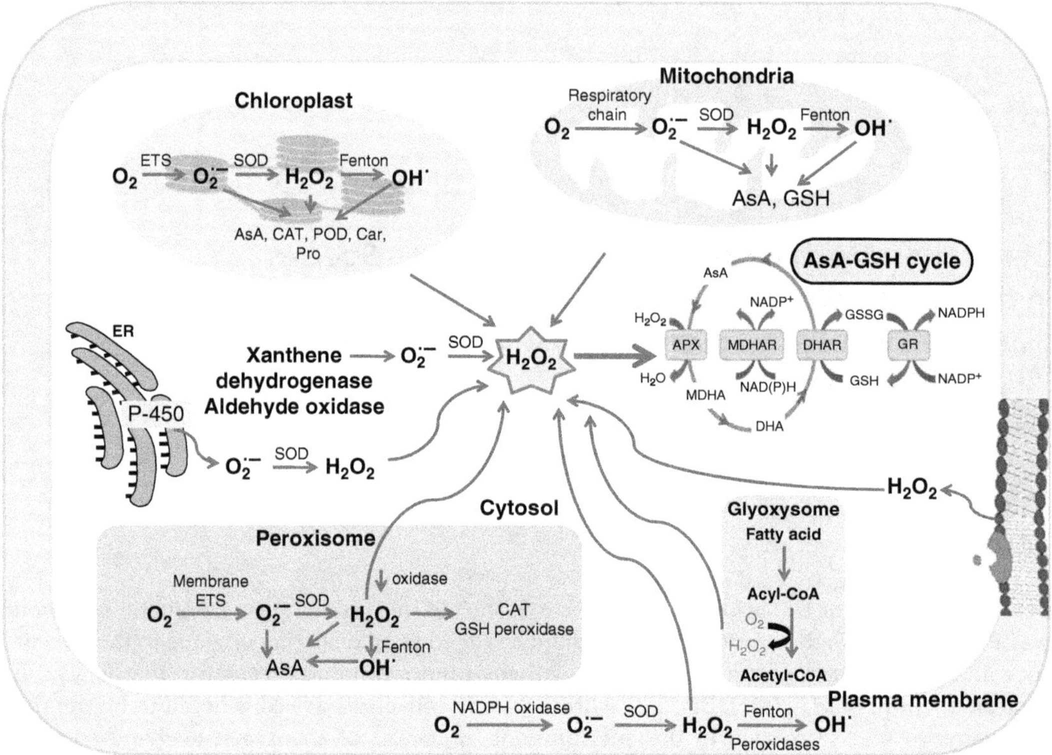

Figure 1.1 Schematic representation of major sites involved in reactive oxygen species (ROS) production and different scavenging mechanisms in plant cells. Abiotic and biotic stresses cause generation of toxic reactive oxygen species (ROS) such as the superoxide radical ($O_2^{\cdot-}$), singlet oxygen ($^1O_2$), hydrogen peroxide (H_2O_2), and hydroxyl radical (·OH). These interact with several essential macromolecules and metabolites causing cellular damage. Moreover, the process of formation of ROS due to the spilling of electrons (e^-) from chloroplasts (ETS or Mehler's reaction), mitochondria (ETS involved in respiratory chain), peroxisomes (ETS involved in photorespiration) and plasma membranes (ETS); these electrons are taken up by molecular O_2 and quickly converted into superoxide radical (SOR). SOR produced during stress conditions is detoxified by superoxide dismutase (SOD) activity, leading to formation of H_2O_2. Detoxification mechanisms involve enzymatic as well non-enzymatic antioxidants to mitigate ROS-induced damage in plants. The ascorbate–glutathione (AsA–GSH) cycle plays an important role in H_2O_2 breakdown. H_2O_2 is reduced to H_2O with the help of ascorbate peroxidase (APX) using ascorbate (AsA) as the specific electron donor. APX is present in different organelles such as chloroplast (chlAPX), mitochondria (mitAPX), peroxisome (mAPX), and cytosol (cAPX). It protects plants from oxidative damage by delivering the electrons as well as minimizing excess production of ROS at these subcellular compartments. APX, ascorbate peroxidase; AsA, ascorbic acid; Car, carotenoids; CAT, catalase; DHA, dehydroascorbate; DHAR, dehydroascorbate reductase; ER, endoplasmic reticulum; ETS, electron transport system; Fenton, breakdown of H_2O_2 to highly reactive ·OH in the presence of iron; GSH, reduced glutathione; GSSG, oxidized glutathione; GR, glutathione reductase; H_2O_2, hydrogen peroxide; MDHA, monodehydroascorbate; MDHAR, monodehydroascorbate reductase; POD, peroxidase; Pro, proline content; SOD, superoxide dismutase. *Source:* Adapted from Jajic *et al.*, 2015.

and consequently leads to death of the particular cell (Chen *et al.*, 2009). Hydrogen peroxide (H_2O_2) acts as a signaling molecule that diffuses across membranes and triggers specific signal transduction pathways (Veal and Day, 2011). The balance between production and elimination of ROS is dependent upon various biotic and abiotic factors such as temperature, heavy metal concentration, drought, salinity, UV radiation, light, nutrient deficiency, and excessive use of pesticides and/or herbicides as well as pathogen attacks. This disturbance ultimately leads to increased concentration of ROS in particular cell organelles. The damaging effects of ROS are ameliorated by different antioxidative defense systems. The antioxidant system consists of enzymatic antioxidants, namely superoxide dismutase (SOD), peroxidase (POD), catalase (CAT), and glutathione *S*-transferase (GST), ascorbate peroxidase (APX) and non-enzymatic antioxidants, including non-protein thiols (NP-SH) ascorbate, tocopherol, caretenoid and cysteine. These act together as a machine to detoxify ROS (Kumar *et al.*, 2015; Singh *et al.*, 2016; Tripathi *et al.*, 2012a,b, 2016a,b) (see Figure 1.1). Among these, SOD is considered as first line of defense; it dismutates the superoxide and subsequently H_2O_2 is generated. Further, H_2O_2 detoxification involves different enzyme systems in distinct cellular compartments. The peroxidase family includes ascorbate peroxidase (APX), glutathione peroxidase (GPX), and peroxidase (POD) for the elimination of H_2O_2. Another enzyme, CAT, is important in the removal of H_2O_2 generated in peroxisomes. GPX also reduces H_2O_2 as well as organic and lipid hydroperoxides by using glutathione (GSH) as substrate. Among the various H_2O_2-detoxifying enzymes, APX plays the most essential role in scavenging ROS. APX is present in thylakoid, glyoxisome, chloroplast stroma, and cytosol, and is involved in the scavenging of H_2O_2 through water-water and ascorbate-glutathione (AsA-GSH) cycles, utilizing AsA as the electron donor. One of the antioxidant enzymes, GST, participates in herbicide detoxification, hormone homeostasis, and regulation of apoptosis and also is involved in plant responses to biotic and abiotic stresses. Non-enzymatic antioxidants include the major cellular redox buffers ascorbate and glutathione, as well as tocopherol, flavonoids, alkaloids, and carotenoids.

With the above context, the present chapter gives an overview of reactive oxygen species, their production sites, and biochemistry as well as the mechanism for their amelioration, particularly in the plant system.

ROS Biochemistry and their Effects

Oxygen is necessary for every aerobic organism. In normal conditions it is involved in several biochemical reactions. The reduction of O_2 to H_2O provides the energy that allows the impressive complexity of higher organisms. However, incomplete reduction of O_2 leads to the production of ROS, which are extremely reactive and can oxidize almost every biological molecule. All ROS can react with DNA, proteins, and lipids (Gill and Tuteja, 2010; Singh *et al.*, 2016). Under these conditions, firstly $^1O_2$ is produced; then H_2O_2 is synthesized via the disproportionation of superoxide catalyzed by SOD, or non-enzymatically in the process of superoxide diffusion with a low yield of the reaction (Quinlan *et al.*, 2013; Singh *et al.*, 2015). H_2O_2 is reduced to water with the involvement of ascorbate peroxidase and ascorbate. Ascorbate is oxidized and then regenerated by the reduced glutathione at the expense of NADPH. Reduction of molecular O_2 proceeds through four steps, thus generating several O_2 radical species (Kalyanaraman *et al.*, 2016). The reaction chain requires initiation at the first step whereas subsequent steps are exothermic and can occur spontaneously, either catalyzed or not. The first step in O_2 reduction produces relatively short-lived ROS that are not readily diffusible: hydroperoxyl ($H_2O_2^-$) and peroxide ($O_2^{\cdot-}$). The second O_2 reduction generates hydrogen peroxide (H_2O_2), which can diffuse upto some distance from its site of production (Quinlan *et al.*, 2013).

Singlet Oxygen Species

Environmental stresses that impact CO_2 fixation (Gul *et al.*, 2016), such as drought and salt stress, ozone, and high or low temperatures, reduce $NADP^+$ regeneration during C-3 cycle, so the photosynthetic electron transport chain is over-reduced, by which singlet oxygen species are produced in the chloroplasts (Wu and Tang, 2004; Bechtold *et al.*, 2005). The chlorophyll pigments associated with the electron transport system (ETS) are the prime source of singlet oxygen ($^1O_2$). The latter may also arise as a by-product of lipoxygenase activity. Like other reactive oxygen species, $^1O_2$ is also highly destructive, and reacts with most biological molecules at near diffusion-controlled rates. This mainly occurs due to the excitement of chlorophyll molecules; although the lifetime of excited chlorophyll is short within these aggregates, its duration varies according to physiological conditions. The excited singlet state of chlorophyll is used for the transfer of energy or electrons. However, there are two other possible routes of de-excitation, radioactive decay (fluorescence) and conversion of the singlet chlorophyll state to the triplet chlorophyll state. The latter interacts with oxygen to produce $^1O_2$.

Superoxide Radical

The half-life for $O_2^{\cdot -}$ is approximately 2–4 ms (Saxena *et al.*, 2016). It is produced at different sites in the cell but the mechanism of its production is almost similar at all sites. Reduction of dioxygen by light in the chloroplasts was first shown by the production of acetaldehyde in the presence of ethanol and catalase, and the photo-reduced product was assumed to be hydrogen peroxide. Under most circumstances, the control of electron flow between photosystems II (PSII) and I (PSI) regulates the reduction state of the acceptor side of PSI. The regulated activation of the C-3 cycle and control of the rate of electron flow are important factors determining the redox state of the ferredoxin pool (Tóth, *et al.* 2007). This is important because ferredoxin and the electron carriers on the reducing side of PSI have sufficiently negative electrochemical potentials to donate electrons to oxygen resulting in the formation of superoxide radical $O_2^{\cdot -}$. The majority of O_2 reduction *in vivo* is thought to proceed via reduced ferredoxin (Fd_{red}), which reduces molecular oxygen to the superoxide radical (Reaction 1). Hydrogen peroxide is then formed through dismutation of O_2^{-} (Reaction 2). The latter occurs spontaneously, but the velocity of the reaction is greatly increased by SOD (Reaction 3):

$$2O_2 + 2Fd_{red} \rightarrow 2O_2^{\cdot -} + 2Fd_{ox} \qquad \text{(Reaction 1)}$$

$$2O_2^{\cdot -} + 2H \rightarrow H_2O_2 + O_2 \qquad \text{(Reaction 2)}$$

$$2O_2^{\cdot -} + 2H \xrightarrow{SOD} H_2O_2 + O_2 \qquad \text{(Reaction 3)}$$

The major site of superoxide formation lies in the electron transfer chain (ETC), of mitochondria especially at the level of Complex I and Complex III. It was shown in animal mitochondria that the flavin mononucleotide (FMN)-containing subunit and an iron-sulfur cluster of the nicotinamide adenine dinucleotide (NADH) dehydrogenase of Complex I are the sites of O_2 generation (Chen *et al.*, 2009), especially when this complex is glutathionylated after oxidative stress (Taylor *et al.*, 2003). This complex could amplify ROS production and participate in the regulation of ROS concentrations in the whole cell. The over-reduction of the ubiquinone pool by Complex I can also lead to a reverse functioning of the chain, and to the formation of large amounts of ROS. In Complex III, the over-reduction state of the ubiquinone pool can lead to a direct electron transfer to molecular oxygen, and to the formation of superoxide anions. Superoxides are known to be produced during NADPH-dependent microsomal electron transport. Two possible loci of O_2^{-} production in microsomes are auto-oxidation of the oxycytochrome-P450 complex that forms during microsomal mixed function oxidase (MFO) reactions, and/or auto-oxidation of cytochrome P450 reductase, a flavoprotein that contains both FAD and FMN.

Hydrogen Peroxide

Hydrogen peroxide is produced by the dismutation of superoxide radicals in a reaction mostly catalyzed by superoxide dismutase (Tripathi *et al.*, 2016). In leaf cells, catalase is exclusively localized in peroxisomes and has not been found in chloroplasts. The hydrogen peroxide in chloroplasts is scavenged by a peroxidase reaction using the photo-reductant produced in the thylakoid as the electron donor. Thus, diffusion of hydrogen peroxide from chloroplasts to peroxisomes and its scavenging by catalase are very unlikely to occur. The electron donor for the peroxidase reaction has been identified as ascorbate.

H_2O_2 is moderately reactive, has a relatively long half-life (1 ms), and can diffuse upto some distance from its site of production. H_2O_2 may inactivate the enzymes by oxidizing their thiol groups. Dismutation and oxidation reactions of superoxide yield hydrogen peroxide. Hydrogen peroxide, although more oxidizing than superoxide, is biologically less toxic: picomolar intracellular levels of superoxide are lethal, whereas micromolar levels of H_2O_2 can be tolerated. H_2O_2 is a potent oxidizer (although not always a fast oxidizer), and is much more diffusible than superoxide, because it is less reactive and is membrane permeable: $O_2^{\cdot-}$ is generally considered membrane impermeable except in its $HO_2\cdot$ form, which is in low abundance at physiological pH. The biological toxicity of H_2O_2 through oxidation of -SH groups has long been known, and it can be enhanced in the presence of metal catalysts through Haber–Weiss or Fenton-type reactions (see Figure 1.1). Fenton showed that the formation of toxic hydroxyl radicals ($\cdot OH$) from hydrogen peroxide (H_2O_2) is catalyzed by iron ions, called the "Fenton reaction."

The Fenton or Haber–Weiss reactions are:

$$O_2\cdot^- + Fe^{3+} \rightarrow O_2 + Fe^{2+}$$

$$Fe^{2+} + H_2O_2 \rightarrow Fe^{3+} + OH^- + \cdot OH$$

Iron ions are required to form toxic $\cdot OH$ radicals. Joseph Weiss and Fritz Haber discovered that $O_2^{\cdot-}$ can be converted into H_2O_2 and further to $\cdot OH$, called the Haber–Weiss reaction. The last species generated by this series of reductions is the hydroxyl radical ($\cdot OH$). It has high reactivity and has half-life of less than 1 ms. As a result, it has a very high affinity for biological molecules to react at its site of production.

Hydroxyl Radical

The generation of $\cdot OH$ from H_2O_2 and $O_2^{\cdot-}$ by the Haber–Weiss process is well known. In this process catalysis is necessary in the presence of a metal since the rate of uncatalyzed reaction is negligible. The hydroxyl radical is highly reactive among ROS. It has a single unpaired electron, and thus can easily bind with oxygen in the triplet ground state. Because cells have no enzymatic mechanism to eliminate $\cdot OH$, its excess production can ultimately lead to cell death (Tripathi *et al.*, 2016). The oxidation of organic substrates by $\cdot OH$ may proceed by two possible reactions: either by addition of $\cdot OH$ to organic molecules or by abstraction of a hydrogen atom. In this context, organic oxygen radicals such as alkoxy, peroxy, semiquinones, reduced hydrogen peroxide, and hydrogen peroxide-electron donor complexes, as well as metallo-oxygen complexes, have been proposed as the ultimate active species besides destructive free $\cdot OH$. These $\cdot OH$ are thought to be largely responsible for mediating oxygen toxicity *in vivo*. The hydroxyl radical can potentially react with all biological molecules, including DNA, proteins, and lipids, and almost any constituent of cells, and due to the absence of any enzymatic mechanism for the elimination of this highly reactive ROS, excess production of $\cdot OH$ ultimately leads to cell death (Table 1.1; see also Figure 1.1).

Table 1.1 Production of reactive oxygen species (ROS) and membrane damage under different stress conditions and their effect on plants.

Serial number	Reactive oxygen species & membrane damage	Stress	Plant	Damaging effect	Antioxidants and other defense systems	Reference
1.	$O_2^{\cdot-}$, H_2O_2, MDA equivalents contents and RNS	Salt and drought	*Ailanthus altissima*	Reduce chlorophyll fluorescence, stomatal conductance, and NR activity	SOD, CAT activity, and Pro content increased for detoxification	Filippou *et al.*, 2014
2.	$O_2^{\cdot-}$, H_2O_2, and MDA equivalents contents	Temperature	*Cucumis sativus* and *Cucurbita ficifolia*	Loss of root cell viability; low root zone changes the mitochondrial electron distribution between the COX and AOX pathway in cucumber root	APX, GPOD, and CAT activity; temperature-mediated production of ROS in cucurbit species may act as signaling molecules, which activate MRR, and subsequently induce expression of genes encoding AOX protein of mitochondria and maintain ROS levels and redox homeostasis	Zhang *et al.*, 2012
3.	Electrolyte leakage (EL)	Salt and zinc	*Vigna radiata*	Reduced plant growth, gas exchange parameters, carbonic anhydrase and nitrate reductase activity	Exogenous epibrassinolide and spermidine application enhances SOD, POD, and CAT activity, which reduces/detoxifies the damaging effect of stress It increases uptake of nutrients, stabilizes the photosynthetic enzyme and pH and enhances carboxylase activity that consequently increases growth of *Vigna* plant	Mir *et al.*, 2015
4.	H_2O_2 and TBARS	Salt	*Brassica juncea*	Reduced photosynthetic rate, stomatal conductance, intercellular CO_2 concentration, quantum yield efficiency of PSII, rubisco activity, and total nitrogen content, and enhanced accumulation of Na^+ and Cl^- ions	Provide salinity tolerance by decreasing Na^+ and Cl^- accumulation and also regulate Pro and ethylene production	Iqbal *et al.*, 2015
5.	$O_2^{\cdot-}$, H_2O_2, MDA equivalents contents, and EL	Arsenic stress	*Solanum melongena*	Reduced growth, photosynthetic pigment, and chlorophyll fluorescence; seedlings accumulate more arsenic (As) content	Reduce the damaging effect of As by upregulating the synthesis of SOD, POD, CAT, GST activity, and Pro as well as Pro synthetic enzyme	Singh *et al.*, 2015

6.	$O_2^{\cdot-}$, H_2O_2, MDA equivalents contents, and EL	Salt	*Solanum melongena*	Reduced growth, K^+ content, photosynthetic pigment, and chlorophyll fluorescence; enhanced accumulation of Na^+	The damaging effect of NaCl is reduced by increased synthesis of enzymatic and non-enzymatic antioxidants	Singh *et al.*, 2016
7.	H_2O_2 and LPO	UV-B radiation	*Ginkgo biloba*	Reduced growth, photosynthetic pigment, and total protein content observed in *Ginkgo biloba*	Increased PAL activity as well as NO enhance accumulation of UV-B filters such as flavonoids	Hao *et al.*, 2009
8.	LPO and MDA equivalent content	UV-B radiation	*Phaseolus vulgaris*	Reduced biomass accumulation, biomass allocation pattern, and physiological and biochemical responses of *P. vulgaris*	UV-B induced enhancement in enzymatic and non-enzymatic antioxidants, ascorbic acid, Pro and TPC, which provide tolerance against UV-B damage. UV-B radiation induces two enzymes, namely chalcone synthase and PAL of the phenylpropanoid pathway, that are essential for synthesis of flavonoids, which act as selective UV-B filters	Raghuvanshi and Sharma, 2016

AOX, alternative oxidase; APX, ascorbate peroxidase; As, arsenic; CAT, catalase; COX, cytochrome *c* oxidase; EL, electrolyte leakage; ER, endoplasmic reticulum; GPOD, guaiacol peroxidase; GSH, reduced glutathione; GST, glutathione *S*-transferase; H_2O_2 hydrogen peroxide; LPO, lipid peroxidation; MDA equivalents contents, malondialdehyde; MRR, mitochondrial retrograde regulation; NR, nitrate reductase activity; $O_2^{\cdot-}$, superoxide radical; PAL, phenylalanine ammonium-lyase activity; POD, peroxidase; Pro, proline content; PSII, photosystem II; RNS, reactive nitrogen species; SOD, superoxide dismutase; TBARS, thiobarbituric acid reactive substances; TPC, total phenolic content; UV-B, ultraviolet B.

In response to all these ROS, peroxidation of lipids is considered as the most harmful process known to occur in every living organism. Membrane injury is sometimes taken as a single parameter to determine the level of lipid demolition under various stresses. ROS are very reactive and damage membranes and various cell components; this results in mobilization of various defense systems to reduce ROS generation and enhance ROS scavenging. This response entails *de novo* synthesis of antioxidant enzymes (i.e., superoxide dismutase, catalase, ascorbate peroxidase, glutathione reductase) and/or activation of their low molecular weight antioxidant precursors (i.e., ascorbate, glutathione, tocopherols, flavonoids) (Minibaeva and Gordon, 2003; Foyer and Noctor, 2005; Hung *et al.*, 2005). Production and removal of ROS must be strictly controlled in order to avoid oxidative stress. When the level of ROS overwhelms the defense mechanisms, a cell is said to be in a state of "oxidative stress." However, the balance between production and scavenging of ROS is disturbed under a number of stressful conditions such as salinity, drought, high light levels, toxicity due to metals, pathogens, and so forth (Table 1.1). The damage caused to biomolecules by enhanced levels of ROS can alter core membrane properties like fluidity and ion transport, lead to loss of enzyme activity, affect protein cross-linking, inhibit protein synthesis, damage DNA, and so forth, ultimately resulting in cell death. When ROS levels exceed a certain threshold, enhanced lipid peroxidation takes place in both the cellular as well as organellar membranes, which, in turn, affects normal cellular functioning. Lipid peroxidation intensifies the oxidative stress through production of lipid-derived radicals that themselves can react with and damage proteins and DNA. The level of lipid peroxidation has been widely used as an indicator of ROS-mediated damage to cell membranes under stressful conditions. Two common sites of ROS attack on the phospholipid molecules are the unsaturated (double) bond between two carbon atoms and the ester linkage between glycerol and the fatty acid. The polyunsaturated fatty acids (PUFAs) present in membrane phospholipids are particularly sensitive to attack by ROS. A single $\cdot$OH can result in peroxidation of many polyunsaturated fatty acids because the reactions involved in this process are part of a cyclic reaction.

Attack by ROS can modify proteins in a variety of ways, some direct and others indirect. Direct modification involves modulation of a protein's activity through nitrosylation, carbonylation, disulfide bond formation, and glutathionylation. Proteins can be modified indirectly by conjugation with breakdown products of fatty acid peroxidation (Yamauchi *et al.*, 2005). Tissues injured by oxidative stress generally contain increased concentrations of carbonylated proteins, the degree of which is widely used as a marker of protein oxidation (Møller and Kristensen, 2004). Greater alteration of proteins has been reported in plants under various stresses (Romero-Puertas *et al.*, 2002; Sharma and Dubey, 2005; Maheshwari and Dubey, 2009; Tanou *et al.*, 2009).

Production Sites of ROS

Oxidative stress occurs when there is a serious imbalance in any cell compartment between production of ROS and antioxidant defense, which leads to cellular damage (Halliwell and Gutteridge, 1999). ROS are a group of free radicals, reactive molecules, and ions that are derived from O_2. They are produced in unstressed and stressed cells in several cell organelles, chiefly chloroplasts, mitochondria, and peroxisomes (see Figure 1.1). There is little contribution from the apoplast, cell wall, or endoplasmic reticulum. ROS are continuously formed by the leakage of electrons to O_2 from the electron transport in different cell organelles. They are also formed in various metabolic pathways as by-products, which are localized in different cellular organelles. Photosynthesizing plants are under threat of oxidative damage, because of their oxygenic conditions and the abundance of the photosensitizer in the chloroplast envelope. It has been reported that two primary processes are mainly involved in the formation of ROS during photosynthesis: (i) direct photoreduction of O_2 to the superoxide radical by reduced electron transport components associated with PSI; and (ii) reactions linked to the photorespiratory cycle, including rubisco (in chloroplasts) and glycolate-oxidase and CAT-peroxidase reactions (in peroxisomes). Of the three major cell organelles, chloroplasts and peroxisomes produce ROS in the presence of light (Foyer and Noctor, 2003) while mitochondria generate ROS in the absence of light. Because ROS can cause damage to proteins, lipids, and DNA, their generation and scavenging must be strictly controlled. To manage this problem, the cell has a survival strategy including mechanisms for scavenging ROS and repairing damage caused by ROS. Excessive ROS reduce the rate of electron transport in the photosynthetic ETC; this leads to activation of alternative pseudocyclic electron transport and photorespiration. Under these conditions, first $^1O_2$ is produced; then H_2O_2 is synthesized in the reaction of disproportionation of superoxide catalyzed by SOD or non-enzymatically in the process of superoxide diffusion with a low yield of the reaction.

Chloroplast

Chloroplasts are considered the most powerful source of ROS generation in plants (Foyer *et al.*, 1994). It is the cellular site of photosynthesis, which proceeds through successive redox reactions during which light energy is transferred to different reaction centers of the two photosystems with the help of the light-harvesting complexes. Oxygen is continuously produced during light-driven photosynthetic electron transport and simultaneously removed from chloroplasts by reduction and assimilation. There are three types of oxygen-consuming processes closely associated with photosynthesis: (i) direct reduction of molecular oxygen by photosystem I (PSI) electron transport; (ii) the oxygenase reaction of ribulose 1,5-bisphosphate carboxylase-oxygenase (RuBisCO); and (iii) chlororespiration within thylakoid membrane (Alric *et al.*, 2010). The electrons released during the process are transferred to an ultimate acceptor via a path called the photosynthetic electron transport chain. The cyclic electron transport chain includes a number of enzymes on the reducing (acceptor) side of PSI: Fe-S centers, reduced thioredoxin, and ferredoxin (Noctor and Foyer, 1998). These electron transport components are auto-oxidizable (Biehler and Fock, 1996). Further generation of ROS is elaborated by the Mehler peroxidase reaction, which explains the transfer of electrons from H_2O to O_2, resulting in production of O_2^- at PSII, $O_2^{\cdot-}$ at PSI, and the trans-thylakoid proton gradient necessary to drive phosphorylation and photochemistry of PSII. It has been suggested that photoreduction of O_2 to water by the Mehler peroxidase pathway in intense light may involve up to 30% of the total electron transport (Oukarroum, 2016). Oxygen reduction sustains significant levels of photosynthetic electron flux, not only through its role in photorespiration but also by its direct reduction through PSI (Asada, 1999). This would suggest that O_2 plays an important role as an alternative electron acceptor in photo-protection. Producing large amounts of ROS is an unavoidable consequence of the photosynthetic reduction of oxygen, and plants have to evolve efficient strategies to deal with the accumulation of these potentially toxic compounds that are integral components of oxygenic photosynthesis.

Mitochondria

Mitochondria, which are considered the cell's "energy hub," are believed to be the foremost sites of ROS production. It has been shown that ROS generated in mitochondria and protein oxidation are contributing factors to the "oxidative stress" syndrome in plants (Sweetlove *et al.*, 2002; Kristensen *et al.*, 2004; Møller and Kristensen 2004). It has also been proven that in the dark or in non-green tissues of plants, mitochondria are a major source of ROS. Firstly, in 1966 it was reported that the respiratory electron transport system produces ROS, and their production can be enhanced in response to various biotic and abiotic stresses. Subsequent innovative work of (Belt *et al.*, 2017) showed that isolated mitochondria produce H_2O_2. The mitochondrial inner membrane is where respiratory electron transport occurs. This mitochondrial ETS harbors electrons with sufficient free energy to directly reduce O_2, which is considered a primary source of ROS generation. It was confirmed later that H_2O_2 arose from the dismutation of superoxide ($O_2^{\cdot-}$) generated within mitochondria. There are two pathways of O_2 consumption, namely: (i) O_2 consumption via cytochrome oxidase to produce H_2O, a process that accounts for more than 95% of O_2 consumption under normal conditions; and (ii) direct reduction of O_2 to $O_2^{\cdot-}$ in the flavoprotein region of the NADH dehydrogenase segment of the respiratory chain (Jezek and Hlavata, 2005). During mitochondrial electron transport, the oxygen radical is markedly enhanced in the presence of antimycin A, which blocks electron flow after ubiquinone. This results in the accumulation of reduced ubiquinone, which may undergo auto-oxidation, resulting in the production of $O_2^{\cdot-}$ (Li *et al.*, 2016). Several observations reveal ubiquinone as a major H_2O_2-generating location of the mitochondrial electron transport chain *in vitro*, and it would appear that $O_2^{\cdot-}$ is a major precursor of H_2O_2 (Winston, 1990). The mitochondrial electron transport chain is comprised of several dehydrogenase complexes that reduce a common pool of ubiquinone (Møller, 1997). The ubiquinone pool is then oxidized by either the cytochrome or the alternative pathway. In general, the main $O_2^{\cdot-}$ generators in the mitochondria are the ubiquinone radical and NADH dehydrogenases (Richter and Schweizer, 1997). Because the ETC harbors electrons with sufficient free energy to directly reduce molecular oxygen, it is considered the unavoidable primary source of mitochondrial ROS production, a necessary accompaniment to aerobic respiration. Production of ROS will increase if the rate of electrons leaving the ETC through the terminal oxidases is slowed and/or the rate of electron input increases in excess of the ability of the two respiratory pathways to process the electrons, leading to an over-reduced ubiquinone pool.

Peroxisomes

Peroxisomes are single membrane-bounded subcellular organelles with an essentially oxidative type of metabolism and a simple morphology that does not reflect the complexity of their enzymatic composition. At the beginning of the 1960s, when peroxisomes were first isolated and characterized from mammalian tissues, their main function was perceived to be the removal of toxic H_2O_2 by catalase. H_2O_2 is typically generated in the peroxisomal respiratory pathway by different flavin oxidases (see Figure 1.1). However, it has become increasingly clear that peroxisomes are involved in a range of important cellular functions in almost all eukaryotic cells. An important property of peroxisomes is their metabolic plasticity, because their enzymatic content can vary depending on the organism, cell or tissue type, and environmental conditions (Jezek and Hlavata, 2005). ROS are also generated by major metabolic pathways, especially those in the peroxisomes, and are used as a weapon against invading pathogens in the oxidative burst. There is another route by which H_2O_2 can be produced during photosynthesis (Elstner, 1982). During carbon assimilation, ribulose 1,5-bisphosphate carboxylase uses CO_2 to carboxylate ribulose 1,5-bisphosphate. However, ribulose 1,5-bisphosphate carboxylase can also use O_2 to oxygenate ribulose 1,5-bisphosphate. Oxygenation yields two glycolates, which are then transported from the chloroplasts to the peroxisomes. Therefore, glycolate oxidation is catalyzed by glycolate oxidase yielding H_2O_2. In addition, the microbodies contain fatty acid beta-oxidase and xanthine oxidase as $H_2O_2^{-}$ and $O_2^{\cdot-}$-producing enzymes, respectively. Peroxisomes are small, usually spherical microbodies bounded by a single lipid bilayer membrane. They are subcellular organelles with an essentially oxidative type of metabolism and are probably the major sites of intracellular ROS production. Like mitochondria and chloroplasts, peroxisomes produce $O_2^{\cdot-}$ radicals as a consequence of their normal metabolism. Two sites of $O_2^{\cdot-}$ generation are established in peroxisomes (del Río *et al.*, 2002). The first is in the organelle matrix, where xanthine oxidase (XOD) catalyzes the oxidation of xanthine and hypoxanthine to uric acid (Corpas *et al.*, 2001). The second site, in the peroxisome membranes, is dependent on NAD(P)H where a small ETC is composed of a flavoprotein NADH and cytochrome *b*; here $O_2^{\cdot-}$ is produced by the peroxisome ETC. Monodehydroascorbate reductase (MDHAR) participates in $O_2^{\cdot-}$ production by peroxisome membranes.

General Mechanisms to Ameliorate the Toxic Effects of ROS

In general the ROS play dual roles, that is, positive as well as negative roles, depending upon their concentrations. In positive ways they can act as signaling molecules to activate the different signaling pathways that participate in development and growth of the plant as well as being involved in defense mechanisms; levels of ROS are maintained via the production of antioxidants in different organelles. However, when the concentration of ROS exceeds than the capacity of antioxidant system, damaging effects of ROS occur. Thus, cells had to evolve sophisticated strategies to keep the concentrations of superoxide radical, hydrogen peroxide, and other reactive oxygen species under tight control (Apel and Hirt, 2004).

Enzymatic ROS Scavenging Mechanisms

Various enzymatic antioxidants are present in certain plant cell organelles to detoxify the ROS; these include SOD, POD, CAT, APX, GST and ascorbate-glutathione cycle enzymes.

Superoxide Dismutase

Superoxide dismutase (SOD) (EC.1.15.1.1) is considered as a first line of defense and is a key enzyme in the plant's defense against oxidative damage induced by various environmental factors (Shi and Zhu, 2008; Mora *et al.*, 2009; Srivastava and Dubey, 2011) (see Table 1.1 and Figure 1.1). The SODs participate in removing the superoxide anion ($O_2^{\cdot-}$) by dismutation whereby one $O_2^{\cdot-}$ is reduced to form H_2O_2 and O_2 is formed by oxidation of another $O_2^{\cdot-}$. It decreases the formation of OH^- due to the absence of O_2^- via a metal-catalyzed Haber–Weiss-type reaction (Abouzari and Fakheri, 2015). In general, several isoforms of SOD are classified according to their subcellular localization on the basis of various metal prosthetic groups. In vascular plants there are three isoforms: Fe-SOD, conserved in chloroplast and cytosol; Mn-SODs, mainly localized in the matrix of mitochondria; and Cu/Zn-SODs, which occur in cytosol, peroxisomes, and plastids (Bowler *et al.*, 1992; Perry *et al.*, 2010). Isomers of SOD show variation in their structure: the prokaryotic Mn-SOD and Fe-SOD, and the eukaryotic Cu/Zn-SOD enzymes are dimers, whereas Mn-SOD of mitochondria is a tetramer. Among these three isozymes, in eukaryotic cells Cu/Zn-SOD comprises approximately 90% of total SOD activity (Liu, 2004). All isoforms of SOD are nuclear encoded, and targeted to their respective subcellular compartments by an amino-terminal targeting sequence. SOD activity under various abiotic stress conditions, such as drought, salinity, extreme temperature, waterlogging, and the presence of heavy metals, suggests that different mechanisms may be involved in various oxidative stress injuries (Babu and Devaraj, 2008; Karuppanapandian *et al.*, 2009; Singh *et al.*, 2011; Singh *et al.*, 2015, 2017; Tripathi *et al.*, 2017a–c).

Catalase

Catalase (CAT) (EC 1.11.1.6) is a tetrameric, heme-containing enzyme found in all aerobic organisms. Catalase activity is largely located in subcellular organelles known as peroxisomes. It converts hydrogen peroxide into water and oxygen (Weydert and Cullen, 2010). Among all antioxidative enzymes, CAT has one of the highest turnover rates: one molecule of CAT can convert around 6 million H_2O_2 molecules to H_2O and O_2 per minute, and stress conditions reduce the rate of protein turnover (Hojati *et al.*, 2010). In general, generation of H_2O_2 occurs through β-oxidation of fatty acids, photorespiration, and purine catabolism during oxidative stress in peroxisomes (Vellosillo *et al.*, 2010).

Among all H_2O_2-degrading enzymes, catalase has the unique property of degrading H_2O_2 without consuming cellular reducing equivalents. Hence, catalase provides the cell with a very energy-efficient mechanism to remove hydrogen peroxide. Therefore, when cells are stressed for energy, cells start to produce H_2O_2 through catabolic processes, and catalase degrades H_2O_2 in an energy-efficient manner (Mallick and Mohn, 2000). This should result in a net gain of reducing equivalents and therefore cellular energy. Like the D1 protein of PSII, the CAT molecule is also highly sensitive to light due to the presence of a heme group that absorbs light quanta. Various researchers have investigated the role of CAT in pathogen defense, by either overexpressing or suppressing CAT in transgenic plants (Vandenabeele *et al.*, 2004). Increase in CAT activity is supposed to be an adaptive trait possibly helping to overcome the damage to tissue metabolism by reducing toxic levels of H_2O_2.

Glutathione Peroxidase (GPX)

Peroxidases are enzymes involved in many physiological and developmental processes of plants along with pathogen infection and countering abiotic stresses. Many other functions like auxin catabolism and biosynthesis of secondary metabolites and ethylene are regulated by peroxidases (Cosio and Dunand, 2009). Glutathione peroxidase (GPX) (EC 1.11.1.9) is the general name for an enzyme family with peroxidase activity whose main biological role is to protect the organism from oxidative damage. The plant glutathione peroxidase (GPX) family consists of multiple isozymes with distinct subcellular locations and different tissue-specific expression patterns and environmental stress responses; they are named AtGPX1 to AtGPX8 in *Arabidopsis* (Passia *et al.*, 2014). The plant GPXs have lower efficiency compared to mammalian GPXs due to the presence of solenocysteine instead of cysteine (Bela *et al.*, 2015), and generally use thioredoxin as a reducing agent rather than glutathione (Navrot *et al.*, 2006). The biochemical function of glutathione peroxidase is to reduce lipid hydroperoxides to their corresponding alcohols and to reduce free hydrogen peroxide to water (Noctor *et al.*, 2002). In the stress response, they actively participate in the maintenance of H_2O_2 homeostasis by the elimination of H_2O_2 and organic hydroperoxides. This enzyme also participates in protein complexes involved in stress defense. Several reports have demonstrated that GPX genes are also regulated by the level of phytohormones in plants (Zhai *et al.*, 2013), indicating their role in plant development. During differentiation of roots the AtGPX2, 3, and 8 are overexpressed while others are repressed. AtGPX4 and 5 are involved in pollen tube growth, as evidenced by the high level of expression of these genes in stamens and pollens. The exact mechanism of plant GPXs is not yet known, but they can be regarded as more than simple antioxidant enzymes. The different expression patterns and intracellular locations of plant GPXs indicate that individual isoforms have particular functions.

The Ascorbate-glutathione Cycle Enzymes

The AsA-GSH cycle of mitochondria, chloroplasts, cytosol, and other cell organelles is one of the major antioxidant protection systems for detoxifying H_2O_2 to water at the expense of AsA. The AsA-GSH cycle comprises enzymatic as well as non-enzymatic antioxidants (Bashri and Prasad, 2016). Enzymatic antioxidants include APX, MDHAR, dehydroascorbate reductase

(DHAR), and glutathione reductase (GR), while non-enzymatic antioxidants include ascorbate and glutathione, which act as substrates for AsA-GSH cycle enzymes (Foyer and Noctor, 2011) (see Figure 1.1). APX is involved in the primary reaction, reducing H_2O_2 to water using ascorbate as the electron donor. Several isoforms of APX are present including thylakoid (tAPX) and glyoxisome membrane forms (gmAPX), as well as a chloroplast stromal soluble form (sAPX) and a cytosolic form (cAPX) (Noctor and Foyer, 1998). This enzyme has a greater affinity towards H_2O_2 detoxification than CAT and POD, hence it plays a crucial role in maintaining the ROS level inside the cell. The ascorbate has to be regenerated to maintain the activity of APX, and this is performed by MDHAR (Locato *et al.*, 2008), which yields oxidized ascorbate (monodehydroascorbate). MDHAR is present in two isoforms, chloroplastic and cytosolic, and has FAD as a cofactor. MDHAR exhibits a high specificity for monodehydroascorbate (MDHA) as the electron acceptor, preferring NADH rather than NADPH as the electron donor. Along with APX, MDHAR also scavenges H_2O_2 in mitochondria and peroxisomes (del Rio *et al.*, 2002). Being a radical, if monodehydroascorbate is not rapidly reduced, it disproportionates into ascorbate and dehydroascorbate. After this, due to involvement of DHAR, ascorbate is regenerated via the reduction of dehydroascorbate at the expense of GSH yielding oxidized glutathione (GSSG), which is crucial for tolerance to various abiotic stresses that cause the production of ROS. At the end of the AsA-GSH cycle, reduced glutathione (GSH) is formed from oxidized glutathione (GSSG) via involvement of the enzyme glutathione reductase (GR) using NADPH as electron donor. GSH is the most abundant non-protein -SH-containing metabolite and takes part in the regeneration of AsA (Foyer and Noctor, 2005). It is present in both prokaryotes and eukaryotes (Romero-Puertas *et al.*, 2006) and is a flavoprotein oxidoreductase localized in mitochondria and cytosol. It is a potential enzyme of the AsA-GSH cycle and plays an essential role in defense against ROS by sustaining the reduced status of GSH. It is predominantly localized. Thus it is concluded that non-enzymatic antioxidants: ascorbate and glutathione, are not consumed and net electron flow is from NADPH to H_2O_2. Recent studies have reported that reduction of dehydroascorbate (DHA) may be non-enzymatic or catalyzed by proteins with dehydroascorbate reductase (DHAR) activity, such as GST omega 1 or glutaredoxins (Wood *et al.*, 2003). Moreover, ascorbate and glutathione are associated with the cellular redox balance, and the ratios of AsA:DHA and GSH:GSSG may function as signals for the regulation of antioxidant mechanisms (Mittler, 2002).

Glutathione *S*-transferase

Detoxification of the xenobiotic compounds that are produced from oxidative stress, like secondary metabolites, as well as human-derived chemicals such as herbicides involve a three-phase detoxification system (Neuefeind *et al.*, 1997). The first phase reactions (oxidation, reduction, or hydrolysis) are catalyzed by cytochrome P450 monooxygenases and result in the exposure of a functional group. After this, with the help of sugars or tripeptide glutathione (GSH), these metabolites are conjugated via glutathione *S*-transferases (GSTs). GSTs are dimeric multifunctional enzymes that catalyze the conjugation of GSH to a variety of electrophilic, hydrophobic, and often toxic substrates thereby reducing their toxicity; they are present in both plant and animal cells (Dixon *et al.*, 1998). In addition to this, GSTs may also exhibit glutathione peroxidase (GPX) or isomerase activities, or function as binding proteins known as ligandins (Edwards *et al.*, 2000). It has also been found that GST overexpression also enhances plant tolerance to various abiotic stresses.

Non-enzymatic Antioxidants

Non-enzymatic antioxidative defense systems include the major cellular redox buffers like ascorbate (AsA) and glutathione (γ-glutamyl-cysteinyl-glycine, GSH) as well as tocopherols and carotenoids. They interact with many cellular components and in addition to their essential roles in defense and as enzyme cofactors, these antioxidants affect plant growth and development by controlling processes ranging from mitosis and cell elongation to senescence and ultimately to cell death (Pinto and Gara, 2004) (see Figure 1.1). Mutants with decreased non-enzymatic antioxidants have been shown to be hypersensitive to stress (Gao and Zhang, 2008; Semchuk *et al.*, 2009).

Ascorbic Acid

Ascorbic acid is the most abundant, influential, and water-soluble antioxidant acting to prevent or reduce the damage caused by ROS in plants (Yabuta *et al.*, 2002; Wang *et al.*, 2005). It is present in almost all plant cell types, organelles, and the apoplast (Horemans *et al.*, 2000; Smirnoff, 2000). Under physiological conditions, it exists mostly in the reduced form (90% of the ascorbate pool) in chloroplasts (Smirnoff, 2000). Although ubiquitous in plant tissues, it is usually most abundant in photosynthetic cells and meristems (and some fruits). Its concentration is reported to be highest in mature leaves with fully developed chloroplasts and highest chlorophyll concentrations. It has also been reported that ascorbic acid is mostly available in reduced form in leaves and chloroplasts under normal physiological conditions. The ability of ascorbic acid to donate electrons in a wide range of enzymatic and non-enzymatic reactions makes it the main ROS-detoxifying compound in the aqueous phase. It can directly scavenge $O_2^{\cdot-}$, $\cdot OH$, and $^1O_2$, and can reduce H_2O_2 to H_2O via the APX reaction. The majority of the AsA pool in plants is contributed by D-mannose/L-galactose through what is commonly called the Smirnoff–Wheeler pathway, which proceeds via GDP-D-mannose, GDP-L-galactose, L-galactose, and L-galactono-1,4-lactone. It is also synthesized via uronic acid intermediates, such as D-galacturonic acid. In this pathway D-galacturonic acid is reduced to L-galactonic acid by galacturonic acid reductase, which is subsequently converted to L-galactono-1,4-lactone. The L-galactono-1,4-lactone is further oxidized to ascorbic acid by L-galactono-1,4-lactone dehydrogenase (GALDH). In mitochondria it is synthesized by L-galactono-γ-lactone dehydrogenase. Then it is transported to the other cell components by a proton-electrochemical gradient or through facilitated diffusion. The level of ascorbic acid under various environmental stresses depends on the balance between the rate and capacity of ascorbic acid biosynthesis and turnover related to antioxidant demand (Chaves *et al.*, 2002). Overexpression of enzymes involved in AsA biosynthesis confers abiotic stress tolerance in plants. GDP-mannose 3,5-epimerase (GME) catalyzes the conversion of GDP-D-mannose to GDP-L-galactose, an important step in the Smirnoff–Wheeler pathway of AsA biosynthesis in higher plants. Overexpression of two members of the *GME* gene family resulted in increased accumulation of ascorbate and improved tolerance to abiotic stresses in tomato plants (Zhang *et al.*, 2011). Ascorbic acid present in apoplast is believed to represent the first line of defense against external oxidants, protecting critical macromolecules from oxidative damage. It regenerates tocoperoxyl (TOC) from its radical (TOC·), which provides protection to membranes (Horemans *et al.*, 2000; Smirnoff, 2000). Thus, elevated levels of endogenous ascorbic acid in plants are necessary to combat oxidative stress in addition to regulate other plant metabolic process (Smirnoff, 2000). Plant mitochondria not only synthesize AsA by L-galactono-γ-lactone dehydrogenase but also take part in the regeneration of AsA from its oxidized forms (Szarka *et al.*, 2007). The regeneration of AsA is extremely important because fully oxidized dehydroascorbic acid has a short half-life and would be lost unless it is reduced back. In addition to the importance of AsA in the AsA-GSH cycle, it also plays an important role in preserving the activities of enzymes that contain prosthetic transition metal ions (Noctor *et al.*, 1998). The AsA redox system consists of L-ascorbic acid, MDHA, and DHA. Both oxidized forms of AsA are relatively unstable in aqueous environments, while DHA can be chemically reduced by GSH to AsA (Foyer and Halliwell, 1976).

Tocopherols

Tocopherols are lipid-soluble antioxidants and are considered as potential scavengers of ROS and lipid radicals (Holländer-Czytko, 2005). Tocopherols are considered to be major antioxidants in biomembranes, where they have both antioxidant and non-antioxidant functions. Tocopherols are considered as general antioxidants that help on the protection of membrane stability, including quenching or scavenging of ROS like $^1O_2$. In plants tocopherols are localized in the thylakoid membrane of plant chloroplasts. Relative antioxidant activity of the tocopherol isomers *in vivo* is $\alpha > \beta > \gamma > \delta$, which is due to the methylation pattern and the amount of methyl groups attached to the phenolic ring of the polar head structure. Hence, α-tocopherol with its three methyl substituents has the highest antioxidant activity. Tocopherols (TOCs) are synthesized only by photosynthetic organisms and are present only in green parts of plants. In higher plants, chloroplast membranes containing TOCs were also known to protect lipids and other membrane components by physically quenching and chemically reacting with O_2 in chloroplasts, thus protecting the PSII structure and function (Igamberdiev *et al.*, 2004). α-TOC is a chain-breaking antioxidant, that is, it is able to repair oxidizing radicals directly and thereby prevent the chain propagation step during lipid auto-oxidation. α-TOC reacts with RO·, ROO·, and RO* derived from polyunsaturated fatty acid (PUFA) oxidation. The reaction between α-TOC and lipid radicals occurs at the membrane-water interface, where α-TOC donates hydrogen atoms to lipid radicals, with the consequent formation of TOH·, which can be recycled back to the corresponding α-TOC by reacting with AsA or other antioxidants (Igamberdiev *et al.*, 2004). The tocopherol biosynthetic pathway utilizes two compounds as precursors, homogentisic acid (HGA) and phytyl diphosphate (PDP). At least five enzymes, 4-hydroxyphenylpyruvate dioxygenase (HPPD), homogentisate phytyl transferases (VTE2), 2-methyl-6-phytylbenzoquinol methyltransferase (VTE3), tocopherol cyclase (VTE1), and γ-tocopherol methyltransferase (VTE4), are involved in the biosynthesis of tocopherols, exclusive of the bypass pathway of phytyl-tail synthesis and utilization (Ahmad *et al.*, 2008). A high level of α-tocopherol has been found in the leaves of many plant species, including *Arabidopsis*. Nitration of α-tocopherol is considered to be an important mechanism for the regulation and detoxification of NOx in animal tissues. In plants also *in vivo* 5-nitro-γ-tocopherol (5-NgT) was identified in leaves of an *Arabidopsis* mutant line (vte4). Germinating seeds of *Brassica napus, Nicotiana tabacum* and *A. thaliana* also showed the presence of 5-NgT. It can be said that γ-tocopherol or 5-NgT prolongs early development by reducing NOx concentration (Desel *et al.*, 2007). Tocopherol has been shown to prevent the chain propagation step in lipid auto-oxidation, which makes it an effective free radical trap. Additionally, it has been estimated that one molecule of α-tocopherol can scavenge up to 120 $^1O_2$ molecules by resonance energy transfer (Munné-Bosch, 2005). Recently, it has been found that oxidative stress activates the expression of genes responsible for the synthesis of tocopherols in higher plants. Regeneration of the oxidized tocopherol back to its reduced form can be achieved by AsA, GSH, or coenzyme Q (Kagan, 2000). Accumulation of α-tocopherol has been shown to induce tolerance to chilling, water deficit, and salinity in different plant species (Guo *et al.*, 2004).

Carotenoids (CARs)

Plants have evolved several mechanisms to get rid of excess energy present in photosynthetic membranes. In all photosynthetic organisms, the carotenoids β-carotene and zeaxanthin and tocopherols play an important photoprotective role, either by dissipating excess excitation energy as heat or by scavenging ROS and suppressing lipid peroxidation (LPO). Carotenoids are pigments found in plants and microorganisms, and they exhibit different forms in nature. CARs are lipophilic organic compounds located in the plastids of both photosynthetic and non-photosynthetic plant tissues. CARs have a multitude of functions in plant metabolism including a role in oxidative stress tolerance. They are also referred as antenna molecules because they absorb light in the region 450–570 nm of the visible spectrum and transfer the captured energy to the chlorophyll. In chloroplasts, CARs function as accessory pigments in light harvesting; however, perhaps a more important role is their ability to detoxify various forms of ROS. CARs can exist in a ground state or in one of two excited states after the absorption of light energy. In terms of their antioxidant properties, CARs can protect photosystems in one of four ways: (i) by reacting with lipid peroxidation (LP) products to terminate chain reactions; (ii) by scavenging $^1O_2$ and dissipating the energy as heat; (iii) by reacting with triplet chlorophyll (3Chl*) or excited chlorophyll (Chl*) molecules to prevent the formation of $^1O_2$; and (iv) by dissipating excess excitation energy through the xanthophyll cycle. The main protective role of β-carotene in photosynthetic tissue may be accomplished via direct quenching of 3Chl*, which prevents $^1O_2$ generation and thereby inhibits oxidative damage (Collins, 2001). During quenching of 3Chl*, energy is transferred from Chl to CAR, which subsequently dissipates the energy in a non-radiative form (i.e., heat). Thus, CARs act as competitive inhibitors of $^1O_2$ formation, and this is aided by their proximity to Chl in the light-harvesting complex. This method of protection is especially critical when light intensity increases above saturating levels (Collins, 2001). Another form of CAR, zeaxanthin, has been implicated in the dissipation of thermal energy, but the precise mechanism underlying this dissipation has not been resolved. Zeaxanthin appears to facilitate the conversion of 3Chl* to 1Chl* more efficiently than does β-carotene (Mortensen *et al.*, 2001).

Conclusions and Future Perspectives

Environmental stresses such as heavy metals, salinity, drought, radiation, temperature, pathogens, and so forth are the major factors that limit plant growth, development, and productivity. A consequence of the generation of ROS in plants is the loss of agricultural productivity due to impairment in the functioning of the plant's metabolism. ROS are produced due to leakage of electrons during general metabolic processes and lead to oxidative modification in nucleic acids, lipids, and proteins. On the other hand, cells can deploy several counteracting mechanisms involving enzymatic and non-enzymatic antioxidant defense systems to combat the damaging effects of ROS. Several plant cell organelles, such as chloroplasts, mitochondria, peroxisomes, and glyoxysomes, have antioxidant defense systems to protect themselves against ROS. Thus, plants have the capability to scavenge or reduce the level of ROS, and hence tolerate harsh environmental stresses/conditions.

References

Abouzari A, Fakheri BA. (2015) Reactive oxygen species: Generation, oxidative damage, and signal transduction. *Int J Life Sci* 5: 3–17.

Ahmad P, Sarwat M, Sharma S. (2008) Reactive oxygen species, antioxidants and signaling in plants. *J Plant Biol* 51: 167–173.

Alric J, Lavergne J, Rappaport F. (2010) Redox and ATP control of photosynthetic cyclic electron flow in *Chlamydomonas reinhardtii* (I) aerobic conditions. *Biochim Biophys Acta Bioenerg* 1797: 44–51.

Aono M, Kubo A, Saji H, Tanaka K, Kondo N. (1993) Enhanced tolerance to photooxidative stress of transgenic *Nicotiana tabacum* with high chloroplastic glutathione reductase activity. *Plant Cell Physiol* 1: 129–135.
Apel K, Hirt H. (2004) Reactive oxygen species: Metabolism, oxidative stress, and signal transduction. *Annu Rev Plant Biol* 55: 373–399.
Asada K. (1999) The water–water cycle in chloroplasts: scavenging of active oxygen and dissipation of excess photons. *Annu Rev Plant Physiol Plant Mol Biol* 50: 601–639.
Bashri G, Prasad SM. (2016) Exogenous IAA differentially affects growth, oxidative stress and antioxidants system in Cd stressed *Trigonella foenum-graceum* L. seedlings: Toxicity alleviation by up-regulation of ascorbate-glutathione cycle. *Ecotoxicol Environ Saf* 132: 329–338.
Bechtold U, Karpinski S, Mullineaux PM. (2005) The influence of the light environment and photosynthesis on oxidative signaling responses in plant-biotrophic pathogen interactions. *Plant Cell Environ* 28: 1046–1055.
Bekker A, Holland HD, Wang PL, *et al.* (2004) Dating the rise of atmospheric oxygen. *Nature* 427: 417–420.
Bela K, Horvath G, Galle A, Szabados L, Tari I, Casizar J. (2015) Plant glutathione peroxidases; emerging role of the antioxidant enzymes in plant development and stress response. *J Plant Physiol* 176: 192–201.
Belt K, Huang S, Thatcher LF, Casarotto H, Singh K, Van Aken O, Millar AH. (2017) Salicylic acid-dependent plant stress signalling via mitochondrial succinate dehydrogenase. *Plant Physiol* 16: 00060.
Biehler K, Fock H. (1996) Evidence for the contribution of the Mehler-peroxidase reaction in dissipating excess electrons in drought-stressed wheat. *Plant Physiol* 112: 265–272.
Bowler C, Van Montagu M, Inze D. (1992) Superoxide dismutase and stress tolerance. *Annu Rev Plant Physiol* 43: 83–116.
Chaves MM, Pereira JS, Maroco J, *et al.* (2002) How plants cope with water stress in the field. Photosynthesis and growth. *Ann Bot* 7: 907–916.
Chen Y, Azad MB, Gibson SB. (2009) Superoxide is the major reactive oxygen species regulating autophagy. *Cell Death Differ* 16: 1040–1052.
Collins A. (2001) Carotenoids and genomic stability. *Mutat Res* 475: 1–28.
Corpas FJ, Barroso JB, del Río LA. (2001) Peroxisomes as a source of reactive oxygen species and nitric oxide signal molecules in plant cells. *Trends Plant Sci* 6: 145–150.
Corpas FJ, Gupta DK, Palma JM. (2015) Production sites of reactive oxygen species (ROS) in organelles from plant cells. In: Gupta DK, Palma JM, Corpas FJ (eds) *Reactive Oxygen Species and Oxidative Damage in Plants Under Stress*. Springer International Publishing, pp. 1–22.
Cosio C, Dunand C (2009) Specific functions of individual class III peroxidase genes. *J Exp Bot* 60: 391–408.
del Río LA, Corpas FJ, Sandalio LM, Palma JM, Gomez M, Barroso JB. (2002) Reactive oxygen species, antioxidant systems and nitric oxide in peroxisomes. *J Exp Bot* 53: 1255–1272.
Desel C, Hubbermann EM, Schwarz K, Krupinska K. (2007) Nitration of γ-tocopherol in plant tissues. *Planta* 5: 1311–1322.
Dismukes GC, Klimov VV, Baranov SV, Kozlov YN, Das Gupta J, Tyryshkin A. (2001) The origin of atmospheric oxygen on Earth: The innovation of oxygenic photosynthesis. *Proc Natl Acad Sci U. S. A.* 98: 2170–2175.
Dixon DP, Cole DJ, Edwards R. (1998) Purification, regulation and cloning of a glutathione transferase (GST) from maize resembling the auxin inducible type III GSTs. *Plant Mol Biol* 36: 75–87.
Edwards R, Dixon DP, Walbot V. (2000) Plant glutathione-*S*-transferase enzymes with multiple functions in sickness and in health. *Trends Plant Sci* 5: 193–198.

Elstner EF. (1982) Oxygen activation and oxygen toxicity. *Annu Rev Plant Physiol* 33: 73–96.
Filippou P, Bouchagier P, Skotti E, Fotopoulos V. (2014) Proline and reactive oxygen/nitrogen species metabolism is involved in the tolerant response of the invasive plant species *Ailanthus altissima* to drought and salinity. *Environ Exp Bot* 97: 1–10.
Foyer CH, Halliwell B. (1976) The presence of glutathione and glutathione reductase in chloroplasts: a proposed role in ascorbic acid metabolism. *Planta* 133: 21–25.
Foyer CH, Noctor G. (2003) Redox sensing and signaling associated with reactive oxygen in chloroplasts, peroxisomes and mitochondria. *Physiol Plant* 119: 355–364.
Foyer CH, Noctor G. (2005) Oxidant and antioxidant signalling in plants: a re-evaluation of the concept to oxidative stress in a physiological context. *Plant Cell Environ* 28: 1056–1071.
Foyer CH, Noctor G. (2011) Ascorbate and glutathione: the heart of the redox hub. *Plant Physiol* 155: 2–18.
Foyer CH, Lelandais M, Kunert KJ. (1994) Photooxidative stress in plants. *Physiol Plant* 92: 696–717.
Gao F, Zhang CT. (2008) Ori-Finder: A web-based system for finding ori Cs in unannotated bacterial genomes. *BMC Informatics* 9: 79.
Gill SS, Tuteja N. (2010) Reactive oxygen species and antioxidant machinery in abiotic stress tolerance in crop plants. *Plant Physiol Biochem* 48: 909–930.
Gul A, Ahad A, Akhtar S, Ahmad Z, Rashid B, Husnain T. (2016) Microarray: gateway to unravel the mystery of abiotic stresses in plants. *Biotechnol Lett* 38: 527–543.
Guo T, Zhang G, Zhou M, WuF, Chen J. (2004) Effects of aluminum and cadmium toxicity on growth and antioxidant enzyme activities of two barley genotypes with different Al resistance. *Plant Soil* 258: 241–248.
Halliwell B, Gutteridge JMC. (1999) *Free Radicals in Biology and Medicine*. Oxford: Oxford University Press.
Hao GX, Du F, Zhao R, Wang J. (2009) Role of nitric oxide in UV-B-induced activation of PAL and stimulation of flavonoid biosynthesis in *Ginkgo biloba* callus. *Plant Cell Tiss Org* 97: 175–185.
Hasan MR, Ghosh A, Kaur, C, Pareek, A, Singla-Pareek SL. (2016) Glyoxalase pathway and drought stress tolerance in plants. In: *Drought Stress Tolerance in Plants*, Vol. 1. Springer International Publishing, pp. 379–399.
Hojati M, Modarres Sanavy SAM, Karimi M, Ghanati F. (2010) Responses of growth and antioxidant systems in *Carthamus tinctorius L.* under water deficit stress. *Acta Physiol Plant* 33: 105–112.
Holländer-Czytko H, Grabowski J, Sandorf I, Weckermann K, Weiler EW. (2005) Tocopherol content and activities of tyrosine amino transferase and cystine lyase in *Arabidopsis* under stress conditions. *J Plant Physiol* 162: 767–770.
Horemans N, Foyer CH, Potters G, Asard H. (2000) Ascorbate function and associated transport systems in plants. *Plant Physiol Biochem* 38: 531–540.
Hung SH, Yu CW, Lin CH. (2005) Hydrogen peroxide functions as a stress signal in plants. *Bot Bull Acad Sinica* 46: 1–10.
Igamberdiev AU, Seregelyes C, Manach N, Hill RD. (2004) NADH-dependent metabolism of nitric oxide in alfalfa root cultures expressing barley haemoglobin. *Planta* 219: 95–102.
Iqbal N, Umara S, Khan NA. (2015) Nitrogen availability regulates proline and ethylene production and alleviates salinity stress in mustard (*Brassica juncea*). *J Plant Physiol* 178: 84–91.
Jajic I, Sarna T, Strzalka K. (2015) Senescence, stress, and reactive oxygen species. *Plants* 4: 393–411.
Jezek P, Hlavata L. (2005) Mitochondria in homeostasis of reactive oxygen species in cell, tissues, and organism. *Int J Biochem Cell Biol* 37: 2478–2503.

Kagan HM. (2000) Intra- and extracellular enzymes of collagen biosynthesis as biological and chemical targets in the control of fibrosis. *Acta Tropica* 77: 147–152.

Kalyanaraman B, Hardy M, Zielonka J. (2016) A critical review of methodologies to detect reactive oxygen and nitrogen species stimulated by NADPH oxidase enzymes: Implications in pesticide toxicity. *Cur. Pharmacol Rep* 2: 193–201.

Karuppanapandian T, Sinha PB, Kamarul Haniya A, Manoharan K. (2009) Chromium-induced accumulation of peroxide content, stimulation of antioxidative enzymes and lipid peroxidation in green gram (*Vigna radiata* L. cv. Wilczek) leaves. *Afr J Biotechnol* 8: 475–479.

Karuppanapandian1 T, Moon JC, Kim C, Manoharan K, Kim W. (2011) Reactive oxygen species in plants: their generation, signal transduction, and scavenging mechanisms. *Aust J Crop Sci* 6: 709–725.

Kiley PJ, Storz G. (2004) Exploiting thiol modifications. *PLoS Biol* 2: 400.

Kristensen BK, Askerlund P, Bykova NV, Egsgaard H, Møller IM. (2004) Identification of oxidised proteins in the matrix of rice leaf mitochondria by immune precipitation and two-dimensional liquid chromatography-tandem mass spectrometry. *Phytochem* 65: 1839–1851.

Krumova K, Cosa G. (2016) Overview of reactive oxygen species. In: Nonell S, Flors C (eds), *Singlet Oxygen: Applications in Biosciences and Nanoscience.* Royal Society of Chemistry, pp. 1–21.

Kumar J, Singh VP, Prasad SM. (2015) NaCl-induced physiological and biochemical changes in two cyanobacteria *Nostoc muscorum* and *Phormidium foveolarum* acclimatized to different photosynthetically active radiation. *J Photochem Photobio.* 151: 221–232.

Li C, Chen Q, Gao X, *et al.* (2005) AtHsfA2 modulates expression of stress responsive genes and enhances tolerance to heat and oxidative stress in *Arabidopsis. China Ser C–Life Sci* 48: 540.

Li Y, Zhu H, Kuppusamy P, Zweier JL, Trush MA. (2016) Mitochondrial electron transport chain-derived superoxide exits macrophages: implications for mononuclear cell-mediated pathophysiological processes. *Reactive Oxygen Species* 1: 81–98.

Liu J. (2004) Redox regulation of pancreatic cancer cell growth: role of glutathione peroxidase in the suppression of the malignant phenotype. *Hum Gene Ther* 15: 239–250.

Locato V, Gadaleta C, De Gara L, de Pinto MC. (2008) Production of reactive species and modulation of antioxidant network in response to heat shock: a critical balance for cell fate. *Plant Cell Environ* 31: 1608–1619.

Maheshwari R, Dubey RS. (2009) Nickel-induced oxidative stress and the role of antioxidant defense in rice seedlings. *Plant Growth Regulation* 59: 37–49.

Mallick N, Mohn FH. (2000) Reactive oxygen species: response of algal cells. *J Plant Physiol* 157: 183–193.

Minibaeva FV, Gordon LKh. (2003) Superoxide production and the activity of extracellular peroxidase in Plant Tissues under stress conditions. *Russ J Plant Physiol* 50: 411–416.

Mir BA, Khan TA, Fariduddin Q. (2015) 24-epibrassinolide and spermidine modulate photosynthesis and antioxidant systems in *Vigna radiata* under salt and zinc stress. *Int J Adv Res* 3: 592–608.

Mittler R. (2002) Oxidative stress, antioxidants and stress tolerance. *Trends Plant Sci* 7: 405–410.

Møller IM. (1997) The oxidation of cytosolic NAD(P)H by external NAD(P)H dehydrogenases in the respiratory chain of plant mitochondria. *Physiol Plant* 100: 85–90.

Møller IM, Kristensen BK. (2004) Protein oxidation in plant mitochondria as a stress indicator. *Photochem Photobiol Sci* 3: 730–735.

Mora ML, Rosas A, Ribera A, Rengel Z. (2009) Differential tolerance to Mn toxicity in perennial ryegrass genotypes: involvement of antioxidative enzymes and root exudation of carboxylates. *Plant Soil* 320: 79–89.

Mortensen A, Skibsted LH, Truscott TG. (2001) The interaction of dietary carotenoids with radical species. *Arch Biochem Biophys* 385: 13–19.

Munné-Bosch S. (2005) The role of α-tocopherol in plant stress tolerance. *J Plant Physiol* 162: 743–748.

Nagesh Babu R, Devaraj VR. (2008) High temperature and salt stress response in French bean (*Phaseolus vulgaris*). *Aust J Crop Sci* 2: 40–48.

Navrot N, Collin V, Gualberto E, Hirasawa M, Rey P. (2006) Plant glutathione peroxidases are functional peroxiredoxins distributed in several subcellular compartments and regulated during biotic and abiotic stresses. *Plant Physiol* 142: 1364–1379.

Neuefeind T, Reinemer P, Bieseler B. (1997) Plant glutathione-*S*-transferases and herbicide detoxification. *Bio Chem* 378: 199–205.

Noctor G, Foyer CH. (1998) Ascorbate and glutathione: keeping active oxygen under control. *Annu Rev Plant Physiol Plant Mol Biol* 49: 249–279.

Noctor G, Arisi ACM, Jouanin L, Kunert KJ, Rennenberg H, Foyer CH. (1998) Glutathione: biosynthesis, metabolism and relationship to stress tolerance explored in transformed plants. *J Exp Bot* 49: 623–647.

Noctor G, Gomez, L, Vanacker, H, Foyer CH. (2002) Interactions between biosynthesis, compartmentation, and transport in the control of glutathione homeostasis and signalling. *J Exp Bot* 53: 1283–1304.

Oukarroum A. (2016) Change in photosystem II photochemistry during algal growth phases of *Chlorella vulgaris* and *Scenedesmus obliquus*. *Curr Microbiol* 72: 692.

Passia G, Queval G, Bai J, Margis-Pinherio M, Foyer CH. (2014) The effects of redox controls mediated by glutathione peroxidases on root architecture in *Arabidopsis thaliana*. *J Exp Bot* 65: 1403–1413.

Perry JJ, Shin DS, Getzoff ED, Tainer JA. (2010) The structural biochemistry of the superoxide dismutases. *Biochim Biophys Acta* 2: 245–262.

Pinto MC, Gara LD. (2004) Changes in the ascorbate metabolism of apoplastic and symplastic spaces are associated with cell differentiation. *J Exp Bot* 55: 2559–2569.

Quinlan CL, Perevoshchikova IV, Hey-Mogensen M, Orr AL, Brand MD. (2013) Sites of reactive oxygen species generation by mitochondria oxidizing different substrates. *Redox Biol* 1: 304–312.

Raghuvanshi R, Sharma RK. (2016) Response of two cultivars of *Phaseolus vulgaris* L. (French beans) plants exposed to enhanced UV-B radiation under mountain ecosystem. *Environ Sci Pollut Res* 23: 831–842.

Richter C, Schweizer M. (1997) Oxidative stress in mitochondria. In: Scandalios JG (ed.), *Oxidative Stress and the Molecular Biology of Antioxidant Defenses*. Cold Spring Harbor Laboratory Press, pp. 169–200.

Romero-Puertas MC, Palma JM, Gomez M, Del Rio LA, Sandalio LM. (2002) Cadmium causes the oxidative modification of proteins in pea plants. *Plant, Cell and Environment* 25: 677–686.

Romero-Puertas MC, Corpas FJ, Sandalio LM, *et al.* (2006) Glutathione reductase from pea leaves: response to abiotic stress and characterization of the peroxisomal isozyme. *New Phytol* 170: 43–52.

Saxena A, Gautam S, Arya KR, Singh RK. (2016) Comparative study of phytochemicals, antioxidative potential and activity of enzymatic antioxidants of *Eclipta alba* and *Plumbago zeylanica* by in vitro assays. *Free Radicals and Antioxidants* 2: 139–144.

Schopf JW, Kudryavtsev AB, Czaja AD, Tripathi AB. (2007) Evidence of Archean life: Stromatolites and microfossils. *Precambrian Res* 158: 141–155.

Semchuk NM, Lushchak OV, Falk J, Krupinska K, Lushchak VI. (2009) Inactivation of genes encoding tocopherol biosynthetic pathway enzymes results in oxidative stress in outdoor grown *Arabidopsis thaliana*. *Plant Physiol Biochem* 47: 384–390.

Sharma P, Dubey RS. (2005) Drought induces oxidative stress and enhances the activities of antioxidant enzymes in growing rice seedlings. *Plant Growth Regul* 46: 209–221.

Shi Q, Zhu Z. (2008) Effects of exogenous salicylic acid on manganese toxicity, element contents and antioxidative system in cucumber. *Environ Exp Bot* 63: 317–326.

Singh M, Singh VP, Dubey G, Prasad SM. (2015) Exogenous proline application ameliorates toxic effects of arsenate in *Solanum melongena* L. seedlings. *Ecotoxicol Environ Saf* 117: 164–173.

Singh M, Singh VP, Prasad SM. (2016) Nitrogen modifies NaCl toxicity in eggplant seedlings: Assessment of chlorophyll *a* fluorescence, antioxidative response and proline metabolism. *Biocat Agri Biotechnol* 7: 76–86.

Singh R, Singh S, Parihar P, *et al.* (2016) Reactive oxygen species (ROS): beneficial companions of plants' developmental processes. *Front Plant Sci* 7: 1299.

Singh S, Srivastava PK, Kumar D, Tripathi DK, Chauhan DK, Prasad SM. (2015) Morpho-anatomical and biochemical adapting strategies of maize (Zea mays L.) seedlings against lead and chromium stresses. *Biocatal Agric Biotechnol* 4: 286–295.

Singh S, Tripathi DK, Singh S, *et al.* (2017) Toxicity of aluminium on various levels of plant cells and organism: a review. *Environ Exp Bot* 137: 177–193.

Singh VP, Tripathi DK, Kumar D, Chauhan DK. (2011) Influence of exogenous silicon addition on aluminium tolerance in rice seedlings. *Biol Trace Elem Res* 144: 1260–1274.

Smirnoff N. (2000) Ascorbic acid: metabolism and functions of a multi-faceted molecule. *Curr Opin Plant Biol* 3: 229–235.

Srivastava S, Dubey RS. (2011) Manganese-excess induces oxidative stress, lowers the pool of antioxidants and elevates activities of key antioxidative enzymes in rice seedlings. *Plant Growth Regul* 64: 1–16.

Sweetlove L, Heazlewood JL, Herald V, Holtzapffil R, Day DA, Leaver CJ, Millar AH. (2002) The impact of oxidative stress on *Arabidopsis* mitochondria. *Plant J* 32: 891–904.

Szarka A, Horemans N, Kovacs Z, Grof P. (2007) Dehydroascorbate reduction in plant mitochondria is coupled to the respiratory electron transfer chain. *Physiol Plant* 129: 225–232.

Tanou G, Molassiotis A, Diamantidis G. (2009) Induction of reactive oxygen species and necrotic death-like destruction in strawberry leaves by salinity. *Environ Exp Bot* 65: 270–281.

Taylor ER, Hurrell F, Shannon RJ, Lin T-K, Hirst J, Murphy MP. (2003) Reversible glutathionylation of complex I increases mitochondrial superoxide formation. *J Biol Chem* 30: 278.

Tóth SZ, Schansker G, Strasser RJ. (2007) A non-invasive assay of the plastoquinone pool redox state based on the OJIP-transient. *Photosynth Res* 93: 193–203.

Tripathi DK, Singh VP, Kumar D, Chauhan DK. (2012a) Impact of exogenous silicon addition on chromium uptake, growth, mineral elements, oxidative stress, antioxidant capacity, and leaf and root structures in rice seedlings exposed to hexavalent chromium. *Acta Physiol Plant* 34: 279–289.

Tripathi DK, Singh VP, Kumar D, Chauhan DK. (2012b) Rice seedlings under cadmium stress: effect of silicon on growth, cadmium uptake, oxidative stress, antioxidant capacity and root and leaf structures. *Chem Ecol* 28: 281–291.

Tripathi DK, Singh S, Singh VP, Prasad SM, Chauhan DK, Dubey NK. (2016a) Silicon nanoparticles more efficiently alleviate arsenate toxicity than silicon in maize cultivar and hybrid differing in arsenate tolerance. *Front Environ Sci* 4: 46.

Tripathi DK, Singh VP, Prasad SM, Dubey NK, Chauhan DK, Rai AK. (2016b) LIB spectroscopic and biochemical analysis to characterize lead toxicity alleviative nature of silicon in wheat (*Triticum aestivum* L.) seedlings. *J Photochem Photobiol B: Biol* 154: 89–98.

Tripathi DK, Singh S, Singh VP, Prasad SM, Dubey NK, Chauhan DK. (2017a) Silicon nanoparticles more effectively alleviated UV-B stress than silicon in wheat (*Triticum aestivum*) seedlings. *Plant Physiol Biochem* 110: 70–81.

Tripathi DK, Singh S, Singh S, *et al.* (2017b) Nitric oxide alleviates silver nanoparticles (AgNps)-induced phytotoxicity in Pisum sativum seedlings. *Plant Physiol Biochem* 110: 167–177.

Tripathi DK, Mishra RK, Singh S, *et al.* (2017c) Nitric oxide ameliorates zinc oxide nanoparticles phytotoxicity in wheat seedlings: implication of the ascorbate–glutathione cycle. *Front Plant Sci* 8: 1.

Vandenabeele S, Vanderauwera S, Vuylsteke M, *et al.* (2004) Catalase deficiency drastically affects gene expression induced by high light in *Arabidopsis thaliana. Plant J* 39: 45–58.

Veal EA, Day AM. (2011) Hydrogen peroxide as a signalling molecule. *Antioxid Redox Signal* 15: 147–151.

Vellosillo T, Vicente J, Kulasekaran S, Hamberg M, Castresana C. (2010) Emerging complexity in reactive oxygen species production and signaling during the response of plants to pathogens. *Plant Physiol* 154: 444–448.

Wang Y, Chang CF, Chou J, *et al.* (2005) Dietary supplementation with blueberries, spinach, or *Spirulina* reduces ischemic brain damage. *Exp Neurol* 1: 75–84.

Weydert CJ, Cullen JJ. (2010) Measurement of superoxide dismutase, catalase, and glutathione peroxidase in cultured cells and tissue. *Nat Protoc* 1: 51–66.

Winston GW. (1990) Free radicals in cells. In: Alscher R, Hess JL (eds), *Stress Responses in Plants: Adaptation and Acclimation Mechanisms.* Wiley-Liss, Inc., pp. 57–86.

Wood ZA, Poole LB, Karplus PA. (2003) Peroxiredoxin evolution and the regulation of hydrogen peroxide signalling. *Science* 300: 650–653.

Wu YS, Tang KX. (2004) MAP kinase cascades responding to environmental stress in plants. *Acta Bot Sin* 2: 127–136.

Yabuta Y, Motoki T, Yoshimura K, Takeda T, Ishikawa T, Shigeoka S. (2002) Thylakoid-membrane bound ascorbate peroxidase is a limiting factor of antioxidative systems under photo-oxidative stress. *Plant J* 32: 915–926.

Yamauchi Y, Furutera A, Seki K, Toyoda Y, Tanaka K, Sugimoto Y. (2005) Malondialdehyde generated from peroxidized linolenic acid causes protein modification in heat-stressed plants. *Plant Physiol Biochem* 46: 786–793.

Zhai CZ, Zhao L, Yin LJ, Chen M, Wang QY, Li LC. (2013) Two wheat glutathione peroxidase genes whose products are located in chloroplasts improve salt and H_2O_2 tolerance in *Arabidopsis. PLoS ONE* 8: 73989.

Zhang C, Liu J, Zhang Y, *et al.* (2011) Over expression of SlGMEs leads to ascorbate accumulation with enhanced oxidative stress, cold, and salt tolerance in tomato. *Plant Cell Rep* 30: 389–398.

Zhang Y-P, Qiao Y-X, Yu J-Q, *et al.* (2012) Temperature effects on the reactive oxygen species formation and antioxidant defence in roots of two cucurbit species with contrasting root zone temperature optima. *Acta Physiol Plant* 34: 713–720.

2

Abiotic Stress, Generation of Reactive Oxygen Species, and Their Consequences: An Overview

Aditya Banerjee and Aryadeep Roychoudhury

St Xavier's College (Autonomous), Kolkata, India

Introduction

Abiotic stresses are actually environmental adversities to which crop plants are mostly exposed. This exposure almost always leads to decreased productivity resulting in huge economic losses. Such stresses mainly include salinity, drought, low or high light intensity, high temperature, freezing or cold stress, and heavy metal toxicity. Plants, being sessile organisms, have evolved regulated, complex mechanisms that confer protection and hence aid survival under such critical conditions. In order to activate or upregulate these complex signaling cascades, it is essential for the plant system to successfully sense the stress factors and trigger an appropriate response. Hence, it is essential primarily to focus on the point where the multitudes of signaling pathways converge, rather than to emphasize the myriad branching and diverging pathways. Reactive oxygen species (ROS) are universally considered as the common focal points (the common factor in almost all abiotic stresses that triggers downstream responses) in abiotic stress signaling (Mittler *et al.*, 2011; V.P. Singh *et al.*, 2011; Tripathi *et al.*, 2012a,b, 2016a,b; 2017a–c; Noctor *et al.*, 2014; S. Singh *et al.*, 2015, 2017; Xia *et al.*, 2015; R. Singh *et al.*, 2016).

About 2.7 billion years ago, molecular oxygen was introduced to the Earth's reducing atmosphere by O_2-evolving photosynthetic organisms. This event heralded the advent of ROS as unwanted by-products (Halliwell, 2006). ROS are continuously generated as a by-product of aerobic metabolism. They are generally confined to the plant organelles like chloroplasts, mitochondria, and peroxisomes. The apoplast could also be a potential site for the generation of ROS (Roychoudhury and Basu, 2012). Estimates show that only 1–2% of the O_2 utilized by plant tissues leads to the formation of ROS. Control conditions favoring the usual growth and development of plants also constitutively induce the production of ROS. However, this induction is at a basal level and the ROS produced are scavenged via various antioxidant mechanisms without causing any major physiological damage (Foyer and Noctor, 2005). Homeostasis of the ROS and the levels of ROS scavengers are maintained via a delicate balance, which gets tipped toward uncontrolled ROS generation as the plants are subjected to any abiotic stress mentioned above. The ROS scavengers in the sensitive (usually high-yielding) crop varieties cannot efficiently scavenge this sudden ROS burst and the system effectively succumbs to death. However, the crop varieties tolerant for the particular stressor have a high constitutive expression of the ROS scavenging genes even under control conditions and can survive the oxidative stress caused by the sudden ROS burst. Hence, to guard against such oxidative stress, a plant

Reactive Oxygen Species in Plants: Boon Or Bane - Revisiting the Role of ROS, First Edition. Edited by Vijay Pratap Singh, Samiksha Singh, Durgesh Kumar Tripathi, Sheo Mohan Prasad, and Devendra Kumar Chauhan.

has to modify its growth and developmental physiology and also has to swiftly adapt to the changing energy equation (Miller *et al.*, 2010).

The ROS comprise singlet oxygen ($^1O_2$), hydrogen peroxide (H_2O_2), superoxide radical ($O_2^{\cdot -}$), and hydroxyl radical (·OH), which cause deleterious and lethal damage to proteins, DNA, and lipids (Foyer and Noctor, 2005). Antioxidants involved in scavenging of ROS can be enzymes like superoxide dismutase (SOD), ascorbate peroxidase (APX), guaiacol peroxidase (GPX), glutathione *S*-transferase (GST), and catalase (CAT) or low molecular weight compounds like ascorbic acid (AsA), reduced glutathione (GSH), α-tocopherol, carotenoids, phenolics, flavonoids, and proline (Pro) (Gill *et al.*, 2011). Apart from highlighting the individual role of each of these antioxidants, it is also essential to review the signaling cascade and epigenetic phenomena involving ROS as the central player. In this chapter, we have tried to provide a holistic approach of ROS production, their damaging effects, signaling, and the way they interconnect various pathways to generate an integrated stress response.

Types of ROS

Superoxide Radical ($O_2^{\cdot -}$)

The main site of the formation of superoxide radical is photosystem I (PSI), localized in the thylakoids. It is generated by the non-cyclic electron transport chain (ETC). Interaction of O_2 with cytochrome *c* oxidase produces water under normal conditions. However, superoxide radicals are produced as O_2 reacts with different ETC components. The Fenton reaction illustrates the chemistry of the superoxide radical, thus:

$$O_2^{\cdot -} + Fe^{3+} \rightarrow {}^1O_2 + Fe^{2+}$$

$$2O_2^{\cdot -} + 2H^+ \rightarrow O_2 + H_2O_2Fe^{3+}$$

$$Fe^{2+} + H_2O_2 + Fe^{3+} \rightarrow Fe^{3+} + OH^- + OH^{\cdot}$$

Superoxide radical, having a short half-life of 2–4 μs, does not itself cause much damage. However, it undergoes transformation to hydroxyl radicals, which cause massive lipid peroxidation and membrane disintegration (Halliwell, 2006).

Singlet Oxygen ($^1O_2$)

Singlet oxygen is usually generated by the reaction of chlorophyll (Chl) triplet state with O_2 in the antenna system of the chloroplasts:

$$Chl \rightarrow {}^3Chl$$

$${}^3Chl + {}^3O_2 \rightarrow Chl + {}^1O_2$$

In spite of having a short half-life of about 3 μs, singlet oxygen diffuses to a distance of about 100 nm, causing widespread oxidation of target molecules including proteins, pigments, lipids, and nucleic acids (Krieger-Liszkay *et al.*, 2008). Singlet oxygen is the major ROS produced as a result of photo-oxidative stress, causing light-induced loss of PSII. It can also react with the D1 protein of PSII. The essential scavengers of this ROS are tocopherol, plastoquinone, and β-carotene (Krieger-Liszkay *et al.*, 2008). Singlet oxygen produced in the chloroplasts can also induce a genetic program, via the Executor1 and Executor2 pathways, leading to growth inhibition in plants (Lee *et al.*, 2007).

Hydrogen Peroxide (H_2O_2)

Hydrogen peroxide is formed via univalent reduction and protonation of the superoxide radical, as has been highlighted earlier in the Fenton reaction. Generation of hydrogen peroxide may also occur due to the action of superoxide dismutase (SOD):

$$2O_2^{\cdot-} + 2H^+ \rightarrow H_2O_2 + O_2$$

Different abiotic stresses induce the accumulation of a large amount of ROS. H_2O_2 is produced from multiple sources in a plant cell. These include the over-energization of the electron transport chain (ETC) or the redox reactions in the chloroplasts or mitochondria, lipid oxidation, and photorespiration (Miller *et al.*, 2010). Oxidation of glycolate in the peroxisome during the photosynthetic carbon oxidation cycle is also a major source of H_2O_2 production. The class of cell membrane-bound NADPH-dependent oxidases, similar to the respiratory burst oxidase homologs (RBOH), are also significant H_2O_2 generators (Agrawal *et al.*, 2003). The cell wall-associated peroxidases and alternative oxidases have also been found to regulate H_2O_2 generation and determine cell survival under stress (Kapoor *et al.*, 2015).

The common antioxidants (both enzymatic and non-enzymatic), including peroxyredoxin (Prx), are essential scavengers of H_2O_2 (Hossain *et al.*, 2015). According to Bhattacharjee (2012), a paradox of H_2O_2 physiology exists with opposing activities. H_2O_2 at higher concentrations causes damage to important cellular macromolecules, but at lower concentrations, it induces abiotic stress responsive signaling. Hence, recent priming experiments with H_2O_2 have resulted in tolerance to multiple stresses via a relatively less characterized molecular mechanism. It is possible that such stress tolerance is generated through the modulation of multiple physiological processes like photosynthesis and upregulation of the abscisic acid (ABA)-dependent and ABA-independent cascades, as well as of those like ROS and methylglyoxal (MG) detoxification pathways (Hossain *et al.*, 2015).

Hydroxyl Radical (·OH)

Transition metals like Fe^{2+} and Fe^{3+} catalyze the production of hydroxyl radicals from the Fenton reaction at neutral pH. The hydroxyl radical is the most toxic ROS known so far. Hydroxyl radicals blatantly damage different cellular compartments by lipid peroxidation of the organelle membranes. They also induce catastrophic protein and DNA damage. The most crucial aspect is that there is no scavenging mechanism for this particular ROS. Hence, the excess accumulation of hydroxyl radicals, brought about indirectly by the higher accumulation of superoxide radicals, causes acute cytotoxicity, resulting in cell death (Pinto *et al.*, 2003).

Major Players in ROS Signaling

In spite of several developments in this field, a major gap still exists in our understanding of how synchronized and globally coordinated gene expression can be induced by ROS. An interwoven feedback and feed-forward regulation exists in ROS signaling. ROS-dependent signaling is mediated by a number of secondary messengers like G-proteins, mitogen activated protein kinases (MAPKs), Ca^{2+}, jasmonic acid (JA), salicylic acid (SA), and abscisic acid (ABA). Intuitive minds have also wondered whether ROS, produced passively during different stresses through their damaging effects on cellular structures, induce signaling events that differ from those signaling events produced actively through activation of cell membrane-bound enzymes like NADPH oxidases (Sewelam *et al.*, 2016).

Photosynthesis

Photosynthesis is a major source of ROS formation. During this process, light energy is captured by photosynthetic pigments via intersystem electron transport. The captured energy gets transformed to chemical energy, which channels reductants like NADPH and ATP into the Calvin cycle for dark reactions to occur (Kromer, 1995). Apart from photosynthesis, the chloroplasts also synthesize and store carbohydrates, amino acids, fatty acids, and plant hormones. ROS production possibly creates an imbalance in overall homeostasis, which ultimately affects the entire plant. Godoy Herz *et al.* (2014) recently reported the ability of chloroplasts to sense light conditions and develop a modulation of nuclear gene expression with changing environment. The redox state of the photosynthetic components can sense the changing environment and induce appropriate gene expression. Xiao *et al.* (2012) showed that the plastidial metabolite, methylerythritol cyclodiphosphate (MEcPP), regulated stress responsive-gene expression through retrograde signaling from the chloroplast to the nucleus.

Excess of photosynthetically active light induces ROS (mainly singlet oxygen and hydroxyl radical) generation in the chloroplast. These might target the thylakoid and plastidial membranes and even the light energy harvesting PSII. Sewelam *et al.* (2014) showed that the receptors or sensors at the cell surface or membranes sense the stress stimuli indirectly through chloroplast stress signals. This was because H_2O_2 production in the chloroplasts induced several membrane-bound receptor and signaling component genes. Chloroplastidic ROS activated the membrane-associated NADPH oxidases in intercellular signaling in ozone-treated *Arabidopsis* leaves. Accordingly, it was proposed that the chloroplast is responsible for ROS generation under ozone stress (Serrato *et al.*, 2013).

Membranes

The plasma membrane acts as a selective barrier between the protoplast and the environment and is also responsible for primary perception of any kind of stress stimulus. As a result, H_2O_2-induced changes in membrane fluidity trigger rapid accumulation of small heat-shock proteins (sHSPs) (Königshofer *et al.*, 2008). Ion channels, which act as hydrophilic permeation pores for selective passage of specific charged ions, also perceive multiple stress stimuli. Dynowski *et al.* (2008) induced expression of the *PIP2.1* gene for *Arabidopsis* plasma membrane intrinsic protein in yeast and detected higher H_2O_2-induced toxicity in the transformed yeast cells. The phospholipids of the membrane backbone serve as precursors for the generation of second messengers like phosphatidylinositol (PI), inositol 1,4,5-trisphosphate (IP_3), diacylglycerol (DAG), and jasmonates. DAG and IP_3 are critical regulators of downstream calcium signaling pathways (Guo *et al.*, 2002).

Another group of essential regulators of ROS generation are the membrane-bound NADPH oxidases, which catalyze the production of superoxide radicals ($O_2^{\cdot -}$) in plants. These NADPH oxidases in plants are known as respiratory burst oxidase homologs (Rboh) (Torres and Dangl, 2005). The Rboh enzymes regulate various downstream cascades including calcium signaling and protein phosphorylation along with ROS production (Kadota *et al.*, 2014). Miller *et al.* (2009) reported the involvement of RbohD in ROS-induced rapid systemic signal transduction in *Arabidopsis* subject to various environmental cues like heat, cold, high light, and salinity. The accumulation of the universal stress phytohormone, ABA, was also regulated by RbohD and RbohF in several ROS-ABA signaling cascades (Xue and Seifert, 2015). The correlation can be validated from previous microarray experiments where ABA treatment downregulated *RbohD* expression whereas H_2O_2 treatment upregulated the same gene in *Arabidopsis* (Barraud *et al.*, 2006). ROS production in wild-type plants under stress is not always disadvantageous. This was proved in *Arabidopsis* plants exposed to salt stress, where ROS produced by RbohD and RbohF acted as signals for regulating Na^+/K^+ homeostasis. In contrast, the double mutants, *rbohD2/F2* accumulated less ROS during stress, yielding a salt-sensitive phenotype (L. Ma *et al.*, 2012).

The G-protein coupled receptors (GPCRs) are membrane-bound seven-pass transmembrane receptors, which are controversially simpler in plants than in animals when considered in terms of structural significance. The *Arabidopsis* genome encodes single Gα and Gβ subunits, two Gγ subunits, only one GPCR protein, and just a single regulator of G-protein signaling (Assmann, 2004). Joo *et al.* (2005) highlighted the fact that during ozone stress, the extracellular ROS activated the G-protein either directly or indirectly. This was concluded from the observation that on exposure to ozone, the first biphasic oxidative burst was short-lived in leaves of *Arabidopsis* mutants lacking Gα or Gβ proteins. This obviously suggested that the ROS accumulating in the apoplast did not enter the cells to trigger the intracellular ROS-generating machinery. RbohD and RbohF have also been suggested to receive and relay the initial signals from G-proteins in order to stimulate the ozone responses in *Arabidopsis* stomatal guard cells (Suharsono *et al.*, 2002). Zhang *et al.* (2011) also highlighted the importance of G-proteins in Ca^{2+} signaling in plants as the *gpa1 Arabidopsis* mutants deficient in the Gα protein had inconsistent ABA–ROS interplay, resulting ultimately in the inhibition of Ca^{2+} channel activation and further downstream signaling required for triggering positive stress responses. Further research regarding the role of G-proteins in oxidative stress responses in plants is required to unravel the potential interplay between the upstream elicitors, G-proteins, and the downstream signals triggered during abiotic stress.

Calcium Signaling as an Oxidative Stress Response

Calcium (Ca^{2+}), a major secondary messenger in plants, is stored in the vacuole, endoplasmic reticulum (ER), and the apoplast in plant cells. As a secondary messenger, calcium perceives multiple stress stimuli and connects them with the downstream cellular responses. The major calcium sensors, which can detect even transient alterations in calcium homeostasis, are calmodulin (CAM), calmodulin-like protein (CML), calcium-dependent protein kinase (CDPK), calcineurin B-like protein (CBL), and CBL-interacting protein kinases (CIPKs) (Asai *et al.*, 2013). A direct interconnection between CBL-CIPK-mediated calcium and ROS signaling in plants was reported as a proof for a synergistic activation of RbohF by direct calcium binding to its EF hands (Sewelam *et al.*, 2016). Such calcium-mediated CBL-CIPK networks in tonoplasts have also been reported to regulate magnesium homeostasis in *Arabidopsis* (Tang *et al.*, 2015). Choi *et al.* (2014) also showed that pulsatile bursts of calcium can regulate long-distance root-to-shoot signaling.

Exogenous treatment with both ABA and H_2O_2 increased the cytosolic calcium concentration and expression of *CAM1*. However, seed pre-treatment with CAM inhibitors prohibited the upregulation of multiple antioxidant genes (Hu *et al.*, 2007). This obviously points to the fact that CAM functions downstream to the calcium signal and relays the stimulation to further downstream candidates. Several instances have been reported showing that a calcium burst occurs upstream to ROS production, whereas some others indicate that ROS production induces the calcium bursts (Bowler and Fluhr, 2000). Pre-treatment of *Arabidopsis* with lanthanum chloride ($LaCl_3$: a calcium channel blocker) reduced the H_2O_2-induced activation of *ERF6*, clearly indicating that calcium operates downstream to ROS during induction of this transcription factor (Sewelam *et al.*, 2013). Inhibitors of calcium fluxes blocked both ROS production and calcium signaling, whereas inhibitors of Rboh prohibited the occurrence of only the oxidative burst (Abuharbeid *et al.*, 2004). These experiments developed contrasting views on the spatial expansions of calcium as a secondary messenger in the oxidative stress signaling cascade.

Tobacco seedlings exhibited cytosolic calcium bursts in response to oxidative stress, whereas the allelopathic toxin, catechin, rapidly produced ROS, thus triggering the ROS-induced calcium bursts in *Centaurea diffusa* and *Arabidopsis* roots (Bais *et al.*, 2003; Bhattacharjee, 2005). The ROS-calcium network in oxidative stress still requires exhaustive research to identify the diverse sources of ROS production and their interlinking with the calcium bursts and the anticipated feedback mechanisms essential to control stress responses.

Nitric Oxide (NO) and Oxidative Stress Responses

Nitric oxide is a water- and lipid-soluble, free radical gas acting as an important signaling molecule during plant growth, development, and defense responses (Moreau *et al.*, 2010). Correa-Aragunde *et al.* (2015) reported that the toxic or protective roles of NO within the cell are dependent on its concentration along with its subcellular localization and interaction with ROS. H_2O_2-induced activation of MAPK6 was responsible for the NO biosynthesis in *Arabidopsis* (Wang *et al.*, 2010). Overlap between H_2O_2 and NO was reported in case of salinity stress in citrus plants using a proteome-based approach (Tanou *et al.*, 2010). Drought stress triggered the ABA biosynthetic pathway, via ROS–NO interactions, to induce ABA-mediated stomatal closure to prohibit unwanted water loss. NO modulates protein activities by nitrosylation and tyrosine nitration and also behaves as a Ca-mobilizing messenger (Besson-Bard *et al.*, 2008). In this regard, NO negatively regulates ABA signaling in stomatal guard cells, via S-nitrosylation-mediated inhibition of the open stomata 1 (OST1)/sucrose non-fermenting 1 (SNF1)-related protein kinase 2.6 (SnRK2.6) (Wang *et al.*, 2015). In contrast, several antioxidative roles of NO have been reported during oxidative stress. NO modulated the formation of superoxide radicals and prevented lipid peroxidation. Pre-treatment of *Sorghum* embryos with sodium nitroprusside and diethylenetriamine NONOate as sources of exogenous NO alleviated oxidative damages. In spite of these positive regulatory roles of NO, the homeostasis of cellular ROS and NO is crucial, as excess NO negatively regulates plant growth due to development of nitrosative stress (Sewelam *et al.*, 2016).

Protein Kinases and Phosphatases

Crucial roles have been reported for MAPKs in ROS signaling. Studies in *Arabidopsis* have shown that ROS not only trigger MAPK signaling, but also are the consequence of activation of the same (Pitzschke *et al.*, 2009). MKK1 and MKK5 (MAP kinase kinases) from *Gossypium hirsutum* (cotton) have been reported to be involved in maintaining ROS homeostasis. Transgenic tobacco overexpressing *GhMKK1* showed improved salt and drought stress tolerance (Lu *et al.*, 2013). However, overexpression of *GhMKK5* in tobacco reduced the plants' tolerance to salinity and drought stresses due to the occurrence of hypersensitive response (HR)-like cell death and over-accumulation of H_2O_2 (Zhang *et al.*, 2012). The rice *drought-hypersensitive mutant1* (*dsm1*) lacking a putative MAPKK encoding gene was sensitive to oxidative stress due to downregulated expression of two *peroxidase* (*POD*) genes (Ning *et al.*, 2010).

The CDPKs also play important regulatory roles in mediating ROS-induced signaling. OsCPK12 was reported to enhance salt stress tolerance in rice by reducing ROS accumulation. The *OsCPK12*-overexpressing plants had upregulated levels of *OsAPX2* and *OsAPX8*, whereas *OsRboh1* was downregulated. High ROS accumulation was noted in the *oscpk12* mutants, which were more sensitive to high salinity in comparison to the wild-type plants (Asano *et al.*, 2012). Campo *et al.* (2014) reported lower levels of ROS and lipid peroxidation in transgenic rice plants overexpressing *OsCPK4*. The plants also exhibited higher salt stress tolerance. A high K^+/Na^+ ratio and calcium levels were maintained in transgenic tobacco seedlings overexpressing wheat *CIPK29*. These plants also exhibited improved salt tolerance due to upregulation of ROS-scavenging enzymes (Deng *et al.*, 2013). Hu *et al.* (2015) reported that overexpression of a CIPK gene from apple, *MdSOS2L1*, conferred stress tolerance in tomato and apple. Li *et al.* (2014) reported that a rice lectin receptor-like kinase, salt intolerance 1 (SIT1), could mediate salt sensitivity by regulating ROS and ethylene signaling. SIT1 could sense the salt stress stimulus in order to phosphorylate MPK3 and MPK6. SIT1 also promoted MPK3, MPK6, and ethylene signaling-dependent ROS accumulation and plant cell death during stress, thus acting as a negative stress regulator.

Dephosphorylation and termination of the kinase signaling cascade is mediated in plants by protein phosphatases. The *stress responsive NAC1* gene (SNAC1) regulated downstream gene *protein phosphatase 18* (*OsPP18*)-mediated drought stress tolerance by regulating ROS homeostasis, via an ABA-dependent pathway. The *ospp18* mutant exhibited sensitivity toward drought and oxidative stresses, with downregulated levels of ROS-scavenging enzymes (You *et al.*, 2014).

Roles of Phytohormones

Abscisic Acid (ABA)

ABA has been unanimously regarded as an universal stress hormone in plants due to its accumulation under most kinds of stresses, leading to activation of downstream stress responses such as stomatal conductance and overall plant development (Banerjee and Roychoudhury, 2016a). In rice, a Ca^{2+} /calmodulin (CaM)-dependent protein kinase (CCaMK), *does not make infections 3* (*OsDMI3*), mediates the ABA-induced expression of *SOD* and *CAT* genes. OsDMI3 acts upstream to OsMPK1 to regulate the activity of antioxidants by stimulating ROS production via upregulation of the *Rboh* genes (Shi *et al.*, 2012, 2014). Wen *et al.* (2015) recently showed that rice histidine kinase OsHK3 functions upstream to OsDMI3. ABA-induced H_2O_2 production was found to be involved in the expression of C_2H_2-type *Zinc finger protein 36* (*ZFP36*) and *ZFP182*, both regulating ABA-induced antioxidative responses (Zhang *et al.*, 2014). In maize, an interesting cross-talk between a MAPK and a CDPK in the ABA signaling cascade was recently reported. Ding *et al.* (2013) showed that ZmCPK11 acted upstream of ZmMPK5 to essentially upregulate the activities of SOD and APX in maize leaves. F. Ma *et al.* (2012) reported the necessary presence of ZmCCaMK to induce the ABA-dependent activation of antioxidants and the H_2O_2-induced NO production.

Brassinosteroids (BR)

Brassinosteroids (BRs) are a group of plant steroid hormones that can activate antioxidant defenses to improve stress tolerance in crop plants. MPK5 plays an important role in regulating the NADPH oxidase-dependent self-propagation of ROS in BR-induced antioxidant defense in maize. ZmMPK5 phosphorylates and activates a 65 kDa microtubule-associated protein (MAP65) and positively regulates the BR-induced antioxidant defense (Zhu *et al.*, 2013). A Ca^{2+}/calmodulin (CaM) dependent-protein kinase (CCaMK) in maize was also recently demonstrated to be required for BR-induced antioxidant defence (Yan *et al.*, 2015). H_2O_2 has been reported to mediate the cross-talk between ABA and BR, via a signaling cascade, where BR induces a rapid and transient H_2O_2 production by NADPH oxidase. This results in greater accumulation of ABA and H_2O_2 leading to enhanced stress tolerance in tomato plants (Zhou *et al.*, 2014).

Salicylic Acid (SA)

Salicylic acid (SA) is a versatile metabolite that regulates systemic acquired resistance (SAR) and local disease resistance and also attenuates several abiotic stresses (Vlot *et al.*, 2008; Roychoudhury *et al.*, 2016). Khokon *et al.* (2011) reported that extracellular ROS production stimulated the accumulation of SA leading to stomatal closure. Such closure and drought tolerance are achieved by the ROS accumulation in the *Arabidopsis* guard cells (Miura *et al.*, 2013). Increased H_2O_2 levels triggered high levels of SA through changes in the glutathione (GSH) pool in *Arabidopsis* (Han *et al.*, 2013a).

Jasmonic Acid (JA)

Jasmonic acid (JA) or methyl jasmonate (MeJA) are natural plant products that are strongly correlated with ROS signaling. Pre-treatment of *Arabidopsis* seeds with MeJA resulted in enhanced ozone stress tolerance in the seedlings, with reduced H_2O_2 accumulation and cell death (Rao *et al.*, 2000). Maruta *et al.* (2011) identified MYC2, a transcription factor activated by ROS generated from RbohD and RbohF. MYC2 was found to be involved in JA-mediated stress responses. Han *et al.* (2013b) observed that the intracellular ROS production in *cat2* mutant of *Arabidopsis* led to the activation of the JA pathway and its related genes with GSH accumulation.

Ethylene

Little is known about the roles of the gaseous hormone ethylene in regulating ROS during abiotic stresses. However, it has been documented that ethylene is a major player in programmed cell death (PCD), senescence, and ozone-induced oxidative stress (Sewelam *et al.*, 2016). de Jong *et al.* (2002) suggested the role of ethylene in H_2O_2 production during PCD in tomato suspension cells. Desikan *et al.* (2005) inferred that the ethylene receptor ETR1 regulated H_2O_2 signaling in *Arabidopsis* guard cells, indicating a potential cross-talk between ethylene and ROS in plant cell signaling.

Transcription Factors (TFs)

Transcription factors (TFs) play a pivotal role in plant abiotic stress signaling. These TFs associate with the RNA polymerase complex and regulate the expression pattern of downstream genes. The major TFs that have been characterized in abiotic stress responses are apetala2/ethylene response factor (AP2/ERF), basic leucine zipper (bZIP), zinc finger, WRKY, and NAC (NAM, ATAF, and CUC) families (Yamaguchi-Shinozaki and Shinozaki, 2006; Banerjee and Roychoudhury, 2015, 2016a).

Several zinc finger TFs have been reported to regulate ROS-related defense genes. *ZAT7, ZAT10*, and *ZAT12* were upregulated during oxidative stress in *APX* knockout plants of *Arabidopsis* (Miller *et al.*, 2008). Map-based cloning identified a C_2H_2-type zinc finger TF named drought and salt tolerance (DST), which negatively regulated stomatal closure by directly inducing H_2O_2 accumulation. Thus stomatal closure occurred via DST-mediated H_2O_2 induction (Huang *et al.*, 2009). Sun *et al.* (2010) reported that the transgenic rice plants overexpressing *ZFP179* had elevated levels of ROS-scavenging enzymes, yielding a salt-tolerant phenotype. Jan *et al.* (2013) overexpressed a CCCH-tandem zinc finger TF, *OsTZF1*, in rice which led to enhanced expression of redox homeostasis genes and ROS-scavenging genes.

WRKY TFs are diverse proteins that mediate several signal transduction pathways during multiple abiotic stress responses. These TFs have one or two conserved WRKYGQK heptapeptides at the N-terminus and a zinc finger-like motif at the C-terminus. The WRKY TFs regulate expression of the downstream genes by binding to the cognate W-box in the promoter region (Banerjee and Roychoudhury, 2015). GmWRKY27 was found to reduce ROS levels and enhance salinity and drought tolerance in transgenic soybean hairy roots. F. Wang *et al.* (2015) also reported that GmWRKY27 interacted with GmMYB174, and this association negatively regulated the expression of *GmNAC29*. NAC29 in soybean enhances the levels of ROS and hence increases oxidative stress sensitivity in the plants. Yan *et al.* (2014) overexpressed *GhWRKY17* in tobacco to find lower cellular levels of ROS leading to enhanced drought and salt tolerance. The *Brachypodium distachyon* WRKY36 was reported as a positive regulator of the drought stress response via the regulated control of ROS homeostasis in the cell (Sun *et al.*, 2015).

The dehydration responsive element binding/cold-binding factors (DREB/CBFs) regulate multiple abiotic stress responses and are the major group of the AP2/ERF family of TFs. Schmidt *et al.* (2013) reported the critical role of rice salt and H_2O_2-responsive ERF, namely SERF1, in mediating the H_2O_2-induced signaling in response to salt stress in rice. It has been found that SERF1 regulates salt tolerance in roots, itself being activated by the phosphorylation of MPK5. On activation, SERF1 functions in a positive regulatory loop in which it activates salt-responsive *MAPK* genes like *MPK5* and *MPKK6*, and also the salt-responsive TFs like *ZFP179* and *DREB2A*, through direct interaction with the corresponding promoters. Another *ERF* that gets upregulated during submergence is *SUB1A*. It inhibits ethylene production and gibberellin responsiveness under anoxic conditions, and economically restricts unwanted use of carbohydrates. In the post-submergence phase, when the plant gets re-exposed to atmospheric oxygen and experiences severe post-anoxic oxidative injury, SUB1A promotes ROS scavenging and enhances stress tolerance (Fukao and Xiong, 2013). SUB1A also guarded against dehydration after desubmergence and desiccation during drought by triggering the ABA-dependent, stress-responsive pathways (Fukao *et al.*, 2011). Wu *et al.* (2008) highlighted an interesting cross-talk between ethylene and JA, mediated via an ERF, JERF3. This TF was reported to be responsive to both ethylene and JA. JERF3 recognizes the cognate osmotic and oxidative responsive *cis* elements and regulates the expression of downstream genes to enhance abiotic stress tolerance in tobacco. Regulation of the ROS-responsive TF, *AtERF6*, was found to be regulated by AtRbohD (Sewelam *et al.*, 2013).

ABA stress ripening (ASR) TFs have been considered as important regulators of plant responses to different abiotic stresses. Recently, Golan *et al.* (2014) reported the roles of ASR1 and ASR4 in enhancing drought tolerance in tomato in an ABA-dependent fashion. A wheat *ASR* gene, *TaASR1*, was found to be a positive regulator of osmotic stress due to its ability to activate the antioxidant system and upregulate the downstream stress-responsive genes (Hu *et al.*, 2013). NAC2 from soybean was portrayed as a negative regulator of oxidative stress (Jin *et al.*, 2013). Overexpression of the *EcNAC1* gene from finger millet in tobacco increased the ROS scavenging activity in the transgenic plants due to upregulation of several stress-responsive genes (Ramegowda *et al.*, 2012). Fang *et al.* (2015) identified SNAC3 as a positive regulator of high temperature and drought stress. SNAC3 modulated H_2O_2 homeostasis by controlling the expression of ROS-associated enzymes. An interesting development was noted when Hou *et al.* (2009) correlated the transcriptional coregulator and spliceosome factor, OsSKIPa (rice homolog of human Ski-interacting protein, SKIP), with drought stress tolerance. The transgenic rice plants overexpressing *OsSKIPa* exhibited enhanced tolerance toward drought at both seedling and reproductive stages due to efficient and activated ROS scavenging machinery (Hou *et al.*, 2009).

SRO Proteins

Similar to radical-induced death 1 (RCD1), the SRO protein family has been regarded as a group of plant specific proteins characterized by the presence of domains like poly (ADP-ribose) polymerase catalytic (PARP) and C-terminal RCD1-SRO-TAF4 (RST) domains. Some SRO proteins also have an additional N-terminal WWE domain (You and Chan, 2015).

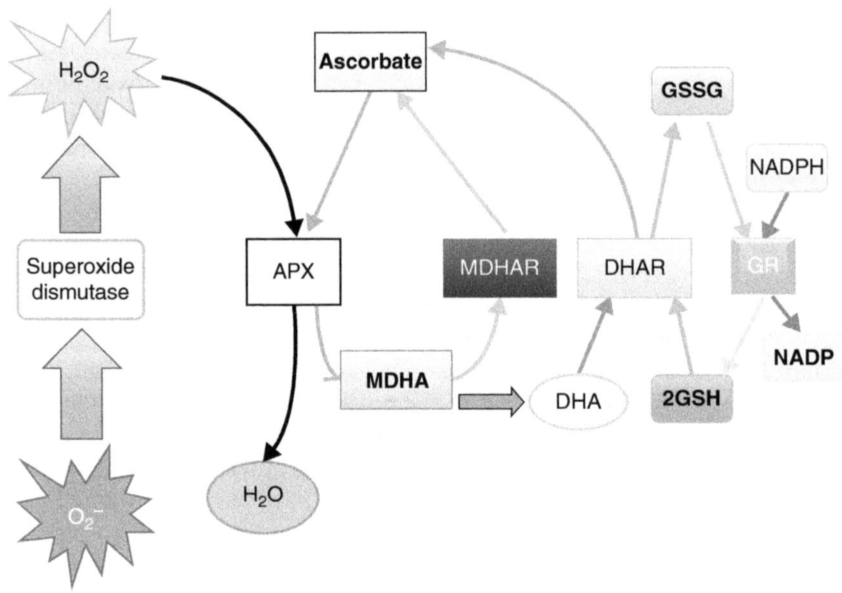

Figure 2.1 The ascorbate-glutathione (AsA-GSH) cycle is often regarded as the redox-regulating hub of the cell. Here, hydrogen peroxide produced on detoxification of superoxide radical by superoxide dismutase (SOD) is immediately scavenged by ascorbate peroxidase (APX) to produce a water molecule. During this reduction, AsA acts as the reducing equivalent and itself gets oxidized into monodehydroascorbate (MDHA). MDHA can be recycled back to AsA by the action of MDHA reductase (MDHAR). The reducing equivalent in this reaction is NADPH, which gets oxidized by MDHAR into NADP (not shown in figure). Usually, MDHA produces dehydroascorbate (DHA) which also gets recycled to form ascorbate catalyzed by DHA reductase (DHAR). For recycling of DHA to AsA, DHAR utilizes the cellular reducing equivalent, that is, reduced glutathione (GSH), and oxidizes it to oxidized glutathione (GSSG). GSH is regenerated from GSSG by the catalytic reaction of glutathione reductase (GR) via the oxidation of NADPH to NADP. Thus, the AsA-GSH cycle regulates the GSH:GSSG ratio in the cell to scavenge the ROS and minimize widespread oxidative damage.

RCD1 in *Arabidopsis* interacts with salt overly sensitive 1 (SOS1) and a number of other TFs. Recent studies also demonstrate the possibility of RCD1 regulating the quantitative changes in gene expression in response to ROS-mediated oxidative stress (Brosche *et al.*, 2014). The drought stress-related TF, SNAC1, acted as a direct regulator of *OsSRO1c* expression in rice. It was also confirmed that OsSRO1c exhibits dual roles in drought and oxidative stresses in rice by promoting stomatal closure and H_2O_2 accumulation following a signaling pathway regulated by SNAC1 and DST TFs (You *et al.*, 2013). Recently a *SRO* gene (*TaSRO1*) in wheat cultivar Shanrong No. 3 was correlated with salinity stress tolerance, via modulation of ROS homeostasis within the cell. Unlike the *Arabidopsis* SRO proteins, wheat SROs had PARP activity. Overexpression of *TaSRO1* in both wheat and *Arabidopsis* led to an improved salt tolerance phenotype via the increased efficiency of the ascorbate-glutathione (AsA-GSH) cycle (Figure 2.1) and the GPX cycle enzymes, which efficiently scavenged the ROS (Liu *et al.*, 2014).

miRNA-mediated Regulation in Oxidative Stress

The microRNAs (miRNAs) are a group of 22 non-coding RNAs that act at the post-transcriptional level to modulate target gene expression. Several *in silico* and high-throughput deep sequencing technologies have been developed to identify miRNAs mediating multiple abiotic stress tolerance in plants. Li *et al.* (2011) reported that H_2O_2 stress mediated differential expression of seven miRNA families. The targets of these miRNAs were post-transcriptional gene silencing (PTGS), nutrient transport, and programmed cell death (PCD). It was found that during heat stress, miR398 inhibits the upregulation of some isoforms of superoxide dismutase (SOD), so that the ROS level is not drastically brought down to a basal level. As a result, more heat-shock factors and hence more heat-shock proteins are produced via an autoregulatory cascade, yielding a more efficient and integrated heat-shock response. This finding highlighted the fact that downregulation of miR398 mediated post-transcriptional induction of two Cu/Zn-SOD isoforms in *Arabidopsis* during oxidative stress (Sunkar *et al.*, 2006).

Consequences of ROS Generation During Abiotic Stresses

Damaging Cellular Lipids

The plasma membrane is the most crucial mechanical barrier of the cell, enveloping it and promoting adaptation to the changing environment. Lipids form the major portion of the plasma membrane. Increased accumulation of ROS results in lipid peroxidation (LPO) during stress. Such LPO becomes so damaging that it is often considered as the single parameter to gauge lipid destruction. The double bond between the C-atoms and the ester linkage between glycerol and fatty acids are the two main targets of ROS in membrane phospholipids. The poly-unsaturated fatty acids (PUFAs) of the plasma membrane, like linoleic and linolenic acids, are highly susceptible to oxidation by singlet oxygen and hydroxyl radicals (Anjum *et al.*, 2015). Smirnoff (2000) divided the entire process of LPO intro three distinct phases: initiation, progression, and termination. The rate-limiting step of the entire process is the initiation. It energizes the O_2 to yield superoxide and hydroxyl radicals. These ROS then react with the methylene groups of the PUFA, yielding conjugated dienes, lipid peroxyl radicals, and hydroperoxides:

$$\text{PUFA-H} + \text{OH}\bullet \rightarrow \text{PUFA}\bullet\,(\text{PUFA alkyl radical}) + H_2O$$
$$\text{PUFA}\bullet + O_2 \rightarrow \text{PUFA-OO}\bullet\,(\text{Peroxyl radical})$$

Upon formation of the PUFA peroxyl radical, it extracts one H atom from the adjoining PUFA side chains and propagates further LPO:

$$\text{PUFA-OO}\bullet + \text{PUFA-H} \rightarrow \text{PUFA-OOH} + \text{PUFA}\bullet$$

The lipid hydroperoxide (PUFA-OOH) formed is cleaved on reaction with reduced metals like Fe^{2+}:

$$\text{PUFA-OOH} + Fe^{2+} \rightarrow \text{PUFA-O}\bullet + Fe^{3+}$$

The lipid hydroperoxides can also undergo decomposition to form different reactive species such as lipid alkoxyl radicals, aldehydes, alkanes, lipid epoxides, and alcohols. LPO terminates with the formation of different lipid dimers caused by different lipid-derived radicals:

$$\text{PUFA}\bullet + \text{PUFA}\bullet \rightarrow \text{PUFA} + \text{PUFA}\,(\text{Fatty acid dimer})$$
$$\text{PUFA}\bullet + \text{PUFA-OO}\bullet \rightarrow \text{PUFA-OO-PUFA}\,(\text{Peroxide bridged dimer})$$
$$\text{PUFA-OO}\bullet + \text{PUFA-OO}\bullet \rightarrow \text{PUFA-OO-PUFA} + O_2\,(\text{Peroxide bridged dimer})$$

Membrane fluidity is altered due to LPO. This causes the membrane to be leaky to substances that otherwise enter the cell through special channels, damaging the membrane proteins and deactivating the membrane receptors, membrane-localized enzymes, and ion-channels (Anjum *et al.*, 2015).

Damaging Cellular Proteins

Proteins can undergo direct and indirect modifications after translation. Protein carbonylation is the marker for evaluating protein oxidation. Such post-translational modification is direct. Products of LPO interact with proteins and facilitate their indirect modifications. Such modifications are specifically targeted to residues like Arg, Lys, Pro, Thr, and Trp, resulting in higher susceptibility of the modified proteins toward proteolytic degradation (Møller *et al.*, 2007). The singlet oxygen and hydroxyl radical inevitably attack modified residues like Cys and Met containing thiol and sulfur groups. Superoxide radical irreversibly inactivates the enzymes containing Fe-S centers. The oxidized proteins thus become better targets for proteolytic digestion by getting primed for ubiquitination-mediated proteosomal degradation.

Damaging Nucleoid DNA

Plants are eukaryotic organisms with the nuclear DNA well packaged and protected by histones and associated proteins. However, the DNA in the prokaryotic organelles like chloroplasts and mitochondria is arranged in a nucleoid structure without proper packaging and protection. This nucleoid DNA bears the brunt of the ROS attacks. The ROS-induced DNA damage includes oxidation of the deoxyribose sugars, modification of the nucleotide bases, abstraction of nucleotides, DNA strand breaks, and DNA-protein cross-linking. The hydroxyl radical also reacts with double bonds of the purine and pyrimidine bases (Choudhury *et al.*, 2013). The ROS abstracts the C-4H-atom of the deoxyribose sugar and forms a deoxyribose radical, which reacts further to form single-strand breaks (SSBs) in the DNA. The damaged products of base oxidation include 8-hydroxyquinine and other less common ones like hydroxyl methyl urea, dehydro-2′-deoxyguanosine, thymine glycol, and thymine and adenine ring opened. These cross-links are not easily reparable and may be lethal to the plant cell, if they are not repaired in a timely manner before commencement of the next round of critical cellular processes like replication or transcription, which require DNA denaturation and unfolding (Das and Roychoudhury, 2014).

Plant Cellular Defenses Against ROS Activities

Enzymatic Antioxidants

The important enzymes involved in alleviating oxidative stress caused by the above ROS are SOD, CAT, APX, monodehydroascorbate reductase (MDHAR), dehydroascorbate reductase (DHAR), glutathione reductase (GR), and GPX.

Superoxide Dismutase (SOD)

Superoxide dismutase (E.C. 1.15.1.1) is a metalloenzyme found in all aerobic organism, and it forms the first line of defense against ROS-induced damage. It catalyzes the dismutation of a superoxide radical into O_2 and H_2O_2, thus removing the possibility of hydroxyl radical formation by the Haber–Weiss reaction. Based on the bound metal ion, there are three isozymes of SOD. These are Mn-SOD (in mitochondria), Fe-SOD (in chloroplasts), and Cu/Zn-SOD (in cytoplasm, peroxisomes, and chloroplasts) (Das and Roychoudhury, 2014). Abiotic stresses strongly upregulate SOD isozymes, which can be visualized by in-gel isozyme analyses (Boguszewska *et al.*, 2010). The reaction catalyzed by SOD is as follows:

$$O_2^{\bullet -} + O_2^{\bullet -} + 2H^+ \rightarrow 2H_2O_2 + O_2$$

Catalase (CAT)

Catalase (E.C.1.11.1.6) is a tetrameric heme-containing enzyme that catalyzes the dismutation of H_2O_2 into H_2O and O_2. This enzyme has high affinity for H_2O_2. However, it has lesser specificity for organic peroxides (R-O-O-R). The uniqueness of this enzyme lies in the fact that it does not require a reducing equivalent. The high turnover rate of this enzyme (6×10^6 molecules of H_2O_2 to H_2O and O_2 min^{-1}) is essential to tackle the high amounts of H_2O_2 produced especially in the peroxisomes during photorespiration, purine catabolism, and β-oxidation of fatty acids (Roychoudhury and Banerjee, 2016). Mhamdi *et al.* (2010) also highlighted the presence of CAT isozymes in the cytoplasm, chloroplast, and mitochondria. Three *CAT* genes have been reported in angiosperms. Of these, *CAT1* is expressed in the pollen and seed (in peroxisomes and cytosol), *CAT2* is predominantly expressed in the photosynthetic tissues and to some extent in the roots and seeds (in peroxisomes and cytosol), and *CAT3* is localized in the mitochondria of leaves and vascular tissues. These isozymes remove H_2O_2 produced under stress in an energy efficient manner.

$$H_2O_2 \rightarrow H_2O + \tfrac{1}{2}O_2$$

Ascorbate Peroxidase (APX)

Ascorbate peroxidase (E.C. 1.1.11.1) is an integral component of the ascorbate-glutathione (AsA-GSH) cycle. CAT isozymes, though found in the cytosol and chloroplasts, do not perform as efficiently there as in peroxisomes. Hence, the same function of CAT is performed by APX in the cytosol and chloroplast. Unlike CAT, which requires no reducing equivalent, APX uses ascorbic acid (AA) as the reducing agent to reduce H_2O_2 to water and dehydroascorbate (DHA) in the following fashion:

$$H_2O_2 + AA \rightarrow 2H_2O + DHA$$

Five isoforms have been detected for APX, based on differences in the primary structure and localization (cytosol, mitochondria, peroxisome, and chloroplast) of these proteins. APX is a more efficient H_2O_2 scavenger during stress due to its better distribution across cell organelles and higher affinity for H_2O_2 than CAT (Roychoudhury and Basu, 2012).

Monodehydroascorbate Reductase (MDHAR)

Monodehydroascorbate reductase (E.C. 1.6.5.4) uses NADPH as a reducing agent for regenerating AA from MDHA (with a short half-life). This aids in replenishing the AA pool of the cell. MDHAR is colocalized with APX within the peroxisomes and mitochondria. Several isozymes of MDHAR have been identified in chloroplast, mitochondria, peroxisomes, cytosol, and glyoxysomes (Roychoudhury and Basu, 2012).

$$MDHA + NADPH \rightarrow AA + NADP^+$$

Dehydroascorbate Reductase (DHAR)

Dehydroascorbate reductase (E.C. 1.8.5.1) uses GSH as an electron donor to reduce DHA to AA. Thus, along with MDHAR, it also replenishes the cellular AA pool and especially regulates the AA pool size in the symplast and apoplast. This maintains the redox homeostasis in the cell. The DHAR isozymes are abundantly found in seeds, roots, and both green and etiolated shoots (Eltayeb *et al.*, 2007).

$$DHA + 2GSH \rightarrow AA + GSSG \text{ (oxidized glutathione)}$$

Glutathione Reductase (GR)

Glutathione reductase (E.C. 1.6.4.2), a flavoprotein oxidoreductase, reduces GSSG to GSH, using NADPH as a reductant. As a crucial enzyme of the AsA-GSH cycle, GR introduces a disulfide bond in glutathione disulfide in order to maintain a high GSH:GSSG ratio in the cell. GSH, the reduced form of glutathione, is a low molecular weight (MW) compound playing the role of a reductant. GSH prevents the ROS-induced oxidation of thiol groups in proteins by reducing ROS members like singlet oxygen and hydroxyl radicals. GR isozymes have been detected mainly in chloroplasts and with limited activity in the mitochondria and cytosol (Roychoudhury and Basu, 2012).

$$GSSG + NADPH \rightarrow 2GSH + NADP^+$$

Glutathione Peroxidase (GPX)

Glutathione peroxidase (E.C. 1.11.1.7), a heme-containing enzyme formed of 40–50-kDa monomers, plays key roles in eliminating H_2O_2 during stress. It is the most important enzyme in terms of H_2O_2 detoxification due to its high efficiency intracellularly (cytosol, vacuole), in the cell wall, and also extracellularly. GPX has affinity toward aromatic compounds like guaiacol and pyrogallol as electron donors. GPX also plays a vital role in lignin biosynthesis and degradation of indole acetic acid during biotic stress (Roychoudhury *et al.*, 2016).

$$H_2O_2 + GSH \rightarrow H_2O + GSSG$$

Non-enzymatic Antioxidants

The other arm of the antioxidant machinery present in plants comprises AA, GSH, α-tocopherol, carotenoids, flavonoids, and amino acids like Pro. Apart from their ROS-scavenging activities, these antioxidants also participate in plant growth and development by regulating mitosis, cell elongation, senescence, and apoptosis (Das and Roychoudhury, 2014).

Ascorbic Acid (AA)

Ascorbic acid is a powerful antioxidant due to its capability to donate electrons to a wide range of enzymatic and non-enzymatic reactions. Hence, it is the most abundant antioxidant found in cells during stress. The majority (90%) of the AA pool is distributed in the cytosol as well as the apoplast, thus tagging it as the first line of defense against ROS-mediated damage. In plants, the Smirnoff–Wheeler pathway yields the most AA in the mitochondria by virtue of the enzyme L-galactano-γ-lactone dehydrogenase. The remaining AA is produced from D-galacturonic acid (Barnes *et al.*, 2002). Oxidation of AA is performed in two successive steps. Initially, MDHA is produced, which if not reduced immediately to ascorbate, disproportionates to DHA and AA. Apart from scavenging H_2O_2, ·OH, and singlet oxygen, AA also regenerates α-tocopherol from tocopheroxyl radical. These together protect the membranes from oxidative damages. Reduced AA is the cofactor of violaxanthine de-epoxidase and maintains the dissipation of excess excitation energy. Prevention of photo-oxidation via a pH-regulated modulation of PSII activity along with zeaxanthine formation has been credited to AA as its other major role (Smirnoff, 2000).

Reduced Glutathione (GSH)

GSH, a low MW thiol tripeptide (γ-glutamyl-cysteinyl-glycine), is omnipresent in almost all cell organelles like endoplasmic reticulum (ER), mitochondria, chloroplasts, vacuoles, peroxisomes, and even in the cytosol and apoplast. The essential roles of GSH involve regulation of cell differentiation, cell growth or division, cell death and senescence, regulation of sulfate

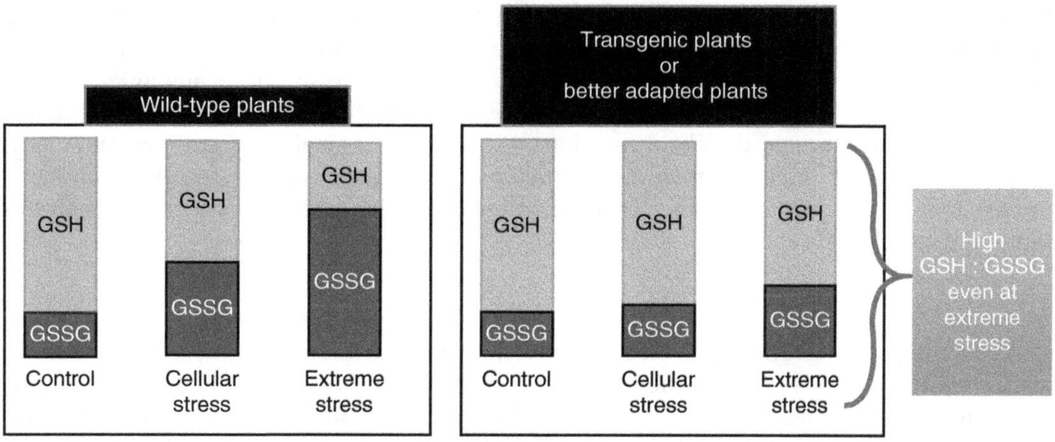

Figure 2.2 Under control conditions, cells maintain a very high ratio of reduced glutathione to oxidized glutathione (GSH:GSSG), which rapidly reduces during extreme oxidative stress due to higher utilization of GSH for scavenging ROS. Transgenic plants or better adapted plants capable of tolerating such oxidative stresses can maintain a high GSH:GSSG ratio even during suboptimal conditions due to a more efficient ascorbate-glutathione (AsA-GSH)cycle.

transport, detoxification of xenobiotics, conjugation of metabolites, regulation of enzymatic activity, synthesis of proteins and nucleotides, synthesis of phytochelatins, and finally expression of stress responsive genes (Mullineaux and Rausch, 2005). The high and effective reductive potential of GSH can be accredited to the central Cys residue with efficient nucleophilic power. GSH scavenges all ROS species and shields the biomacromolecules by forming glutathiolated adducts or reducing them in presence of ROS, itself forming oxidized GSSG. GSSG formed by this process or during the replenishment of the AA pool is ultimately converted to GSH by de novo synthesis or enzymatically by glutathione reductase (GR). Production of effective heavy metal chelators like phytochelatins is also stimulated by GSH via the action of phytochelatin synthase (PS). The delicate balance of GSH/GSSG necessarily maintains the redox homeostasis in the plant cell (Roychoudhury and Chakraborty, 2013) (Figure 2.2).

α-Tocopherol

This antioxidant is lipophilic in nature and is an efficient scavenger of ROS and lipid radicals. Thus, it protects the essential components of membranes. Among the four isomers of tocopherol (α-, β-, γ-, δ-), α-tocopherol has the highest degree of antioxidant capacity (Kiffin *et al.*, 2006). It is produced by photosynthetic organisms and hence is harbored only in the green tissues of the plant. The γ-tocopherol-methyl-transferase (TMT encoded by *VTE4*) catalyzes the synthesis of α-tocopherol from γ-tocopherol. Tocopherols in the chloroplasts protect the PSII structurally and functionally by reacting with O_2 in order to quench its excess energy. Tocopherol inhibits the chain propagation step of the LPO cycle and traps the free lipid radicals like RO·, ROO·, and RO· at the membrane–water interface. Here, α-tocopherol reduces the free radicals and itself metamorphoses into TOH·, which interacts with GSH and AA to recycle back into the reduced active form (Igamberdiev *et al.*, 2004).

Carotenoids

These are a group of lipophilic antioxidants localized in the plastids of both photosynthetic and non-photosynthetic tissues in plants. The antennae complex in chloroplasts harbor these molecules in order to absorb light in the wavelength range 450–570 nm and also to transfer the energy to the chlorophyll molecules. Hence, the major antioxidative function of the carotenoids is rendered in the chloroplasts exposed to rapid photo-oxidation and ROS-induced degradation during stress. The four mechanisms of action of carotenoids include: (i) reaction with LPO products to end the chain reactions; (ii) scavenging singlet oxygen and generating heat as a by-product; (iii) preventing the accumulation of singlet oxygen by reacting with $^3Chl^*$ and excited chlorophyll (Chl*); and (iv) dissipating the excess excitation energy through the xanthophyll cycle (Das and Roychoudhury, 2014).

Flavonoids

Flavonoids in plants have been detected in leaves, floral organs, and pollens. Based on structural variations, these have been classified as flavonols, isoflavones, flavones, and anthocyanins. These molecules are pigmented and have multifarious roles in flower and fruit pigmentation, seed fertility, and pollen germination. These pigmented molecules exhibit their antioxidative roles by preventing ROS-induced damage to the photosynthetic apparatus (Roychoudhury *et al.*, 2016). Flavonoids are also known to scavenge singlet oxygen and protect the outer envelope of the plastidic membrane (Agati *et al.*, 2012).

Proline (Pro)

Proline is a free amino acid acting as a compatible solute as well as a powerful antioxidant in plant species. Due to its low molecular weight and highly soluble organic nature, Pro is non-toxic for the cell at high concentrations. High levels of Pro have been detected in plant species under stresses like high soil salinity, dehydration and water scarcity, chilling temperatures, ROS-mediated oxidative stresses, heavy metal toxicity, and ultraviolet (UV) ray exposure. Accumulation of Pro under abiotic stresses depends on the plant species and can be up to 80% of the cellular amino acid pool (Roychoudhury *et al.*, 2015). Pro is also a stabilizer of the plasma membrane and subcellular structures, which are necessary for the viability of the stressed cell. In such cases, Pro often acts as a protein-compatible hydrotope, alleviating cytoplasmic acidosis. The metabolic $NADP^+$/NADPH ratio, cellular pH, and cellular redox status are also maintained by Pro under stress conditions (Hayat *et al.*, 2012).

The major enzymes catalyzing Pro formation from glutamic acid are δ^1-pyrroline-5-carboxylate synthetase (P5C) and pyrroline-5-carboxylate reductase (P5CR). This accumulation of Pro occurs via the intermediate, pyrroline-5-carboxylate (P5C) (Roychoudhury *et al.*, 2015).

Melatonin

A new inclusion in the list of non-enzymatic antioxidants is melatonin (*N*-acetyl-5-methoxytryptamine), which is present in much higher concentrations in plants than in animals. In plants (rice), *N*-acetyl serotonin is directly produced from tryptamine, and *N*-acetyltryptamine serves as an intermediate product. This pathway is catalyzed by serotonin *N*-acetyltransferase (SNAT) and tryptophan 5-hydroxylase. Melatonin can even be directly produced from serotonin with 5-methoxytryptamine serving as intermediate product. This reaction is catalyzed by hydroxyindole-*O*-methyltransferase (HIOMT) (also referred to as acetyl serotonin methyltransferase: ASMT) and SNAT (Byeon *et al.*, 2014). Melatonin has been depicted to behave as indole-3-acetic acid (IAA) due to its regulation in physiological growth and development of the plant. During stress, melatonin decreases oxidative damage by directly scavenging H_2O_2 and upregulating the other antioxidant enzymes like APX, CAT, and so forth (Nawaz *et al.*, 2016).

Regulation of the Antioxidants Governs Abiotic Stress Tolerance in Plants

The *SOD* genes in various cultivars of *Phaseolus vulgaris* and *Oryza sativa* were strongly upregulated under drought stress (Zlatev *et al.*, 2006). Similar upregulation was also reported in the leaves of white clover (*Trifolium repens*) exposed to drought stress (Chang-Quan and Rui-Chang, 2008). Salinity stress also induced the expression of three isoforms of SOD in chickpea (Eyidogan and Öz, 2007). Overexpression of *Mn-SOD* in *Arabidopsis* led to enhanced salt tolerance in the transgenic lines (Wang *et al.*, 2004). UV-B rays are potentially harmful agents present in sunlight. Exposure to such photo-damaging rays increased SOD activity in wheat, pea, *Arabidopsis*, and rice, but not in barley and soybean. Supplemented UV-B rays also increased SOD activity in wheat and mungbean, whereas a differential response was recorded among soybean cultivars. APX activity was increased in *Arabidopsis* exposed to UV-B rays (Rao *et al.*, 1996).

Metallothioneins (MTs) are a group of low MW proteins with high Cys content like phytochelatins. They are important heavy metal chelators, and their high Cys content suggests their roles in ROS scavenging (Anjum *et al.*, 2015). Overexpression of the MT gene, *OsMT1a*, in rice resulted in higher activities of CAT, APX, and peroxidase (POD) (You and Chan, 2015). Transgenic tobacco overexpressing the halophyte MT gene, *SbMT-2*, showed efficient ROS scavenging and tolerance toward salinity and osmotic stresses (Chaturvedi *et al.*, 2014). Positive correlation between the activity of APX and Pb treatment was seen in *Eichhornia crassipes* seedlings (Malar *et al.*, 2014). Roychoudhury *et al.* (2012) showed that the activities of APX and GPX increased both in IR-29 (salt-sensitive) and Nonabokra (salt-tolerant) rice varieties on exposure to Cd(II) stress. The peroxidase and CAT activities increased progressively with the increase in Cd(II) concentration in *Vigna radiata* (Roychoudhury and Ghosh, 2013). Roychoudhury *et al.* (2016) also reported high expression of the CAT, POD, APX, and other antioxidant parameters in Cd(II)-stressed rice varieties pre-treated with salicylic acid. Isozyme analyses identified stress-inducible isoforms for SOD, CAT, APX, and GPX. *Vaccinium myrtillus* L. is regarded as a species that is a successful colonist of acid- and heavy metal-contaminated soil. Upon analysis of the antioxidant response of this plant from heavily polluted sites (immediate vicinity of zinc smelter, iron smelter, and power plant), it was found that the contents of GSH, non-protein thiols, Pro, and activity of GPX were elevated. The GPX activity seemed to be universal and correlated well with heavy metal stress (Kandziora-Ciupa *et al.*, 2013; Das and Roychoudhury, 2014). Increased CAT activity under Cd(II) stress has been also reported in *Phaseolus aureus*, *Pisum sativum*, *Lemna minor*, barley, and sunflower (Sreedevi and Krishnan, 2012). Arsenic (III) significantly decreased the GSH content in rice roots due to its conversion to phytochelatins. The GSH supplementation resulted in partial protection against arsenic stress, reducing the MDA content and restoring the seedling growth of arsenic (V) exposed seedlings. Under low UV-B doses, increases in AA and GSH pools, as well as AA regeneration ability functioned to keep the balance of cellular H_2O_2 (Roychoudhury and Basu, 2012).

Sairam *et al.* (1998) first compared the antioxidant profiles of drought-tolerant (C306) and drought-sensitive (HD2329) genotypes of wheat. Higher APX and CAT activities, along with higher AA content were detected in the drought-tolerant genotype. Automatically, this genotype exhibited low ROS and malondialdehyde (MDA) accumulation. Overexpression of *MDHAR* and *DHAR* in *Arabidopsis* resulted in improved salt tolerance (Eltayeb *et al.*, 2007). Sharma and Dubey (2005) reported that the rice seedlings exposed to salt stress exhibited high expression of *MDHAR*, *DHAR*, and *GR*, thus replenishing the AA pool. Salt-tolerant cultivars of potato showed marked increase in APX and GR activities during salinity stress (Aghaei *et al.*, 2009). Malik and Ashraf (2012) reported higher chlorophyll content, net photosynthesis, and growth in the wheat plants treated with AA and exposed to drought stress. The adverse effects of drought stress were ameliorated in *Carthamus tinctorius* seeds primed with AA (Razaji *et al.*, 2012). Salt sensitivity was accredited in IR-29 and Pusa Basmati due to a decreased GSH/GSSH ratio compared to the salt-tolerant genotype, Pokkali (Basu *et al.*, 2010). Wang *et al.* (2014) reported lowered oxidative damage due to increased activity of GSH in the chloroplasts of rice plants exposed to salt stress. A lowered level of ROS was detected in the heat-acclimated cool season turfgrass genotype as a result of enhanced activity of the enzymes involved in GSH accumulation and a higher GSH/GSSG ratio (Hasanuzzaman *et al.*, 2013). Transgenic tobacco overexpressing the tocopherol biosynthetic gene, *AtVTE1*, from *Arabidopsis* exhibited better adaptation to drought stress. These plants had lower LPO, electrolytic leakage, and H_2O_2 accumulation (Liu *et al.*, 2008). Another role of α-tocopherol is to maintain the Na^+/K^+ homeostasis and hormonal balance. Hence, the *Arabidopsis* mutants for *vte1* and *vte4* showed salt hypersensitivity (Ellouzi *et al.*, 2013). Jain *et al.* (2003) showed that the α-tocopherol level decreased in cucumber cotyledons on exposure to UV-B rays. This correlates with the fact that the α-tocopherol gets utilized while scavenging the accumulated lipid radicals formed during stress.

The effects of UV-B rays on the flavonoid content were highlighted by Xu *et al.* (2008). The two isolines of soybean cv. Clark, the normal line with moderate levels of flavonoids, and the magenta line with reduced flavonoid levels, were grown in the field with or without natural levels of UV-B. Solar UV-B radiation caused oxidative stress in both the lines and altered ROS metabolism, primarily by decreasing SOD activity and increasing the activities of APX, CAT, and GR. This resulted in decreased AA content and increased DHA content. The magenta line had greater oxidative stress than the normal line, in spite of its enhanced oxidative defense capacity as compared to the normal line even under UV-B exclusion. These results indicate enhanced sensitivity in the magenta line, especially under UV-B exclusion, which was likely due to the absence of flavonoid epidermal screening compounds and a resultant increased penetration of solar UV radiation into the leaf (Xu *et al.*, 2008; Das and Roychoudhury, 2014). Luo *et al.* (2016) overexpressed the *anthocyanidin reductase* (*ANR*) gene (responsible for biosynthesis of proanthocyanidins) from rose in tobacco. The transgenic tobacco lines exhibited tolerance toward oxidative stress via increased accumulation of ABA, polyamines, and other antioxidants. Antioxidants like anthocyanin and Pro showed accumulation in the salt-tolerant cv. Nonabokra compared to the salt-sensitive cv. M-1-48 and aromatic rice Gobindobhog, all exposed to 200 mM salt stress (Roychoudhury *et al.*, 2008). Roychoudhury *et al.* (2015) have shown that Pro accumulation confers multiple abiotic stress tolerance in a variety of crop species and model plants.

Wang *et al.* (2016) overexpressed the *Cu/Zn-SOD* (*TaSOD2*) in wheat and *Arabidopsis* to develop salt tolerance. Such tolerance developed due to better activities of H_2O_2 metabolic enzymes. However, the accumulation of superoxide radical was accelerated resulting from high NADPH oxidase (NOX) activity. The inhibition of NOX activity and the mutation of NOX-encoding genes both abolished the salt resistance of *TaSOD2* overexpression lines. These data indicated that Cu/Zn-SOD enhanced salt tolerance, which was accomplished via modulation of the redox homeostasis by promoting NOX activity. Ramabulana *et al.* (2016) reported oxidative stress tolerance in *Moringa oleifera* exposed to gamma radiation, via upregulation of polyphenolic compounds like flavonoids. Though some polyphenolic compounds like hydroxycinnamoyl derivatives were downregulated following gamma irradiation, a compensatory protection was provided by the other upregulated flavonoids.

de Abreu Neto and Frei (2016) recently conducted a meta-analysis (MA) of microarray experiments in rice by mining the transcriptomic data from public databases. This represented 36 different experiments and investigations in multiple abiotic stresses. Inherent artefacts of different MA methods were removed by processing the data in Fisher, rOP, REM, and product of rank (GeneSelector). The genes identified were regarded as shared differentially expressed genes (DEGs). Quite interestingly, in all tissues and stress conditions, the priority of gene clusters overrepresented among the DEGs was ROS-related. These genes represented 4% of all expressed genes, but 7–10% of the DEGs. The main utility of such a MA approach is that though this methodology may exclude many important genes, its stringency always removes all false positives that could be obtained as artefacts of different statistical methods (de Abreu Neto and Frei, 2016). A very recent development on melatonin in scavenging ROS was reported by Ma *et al.* (2016) in cassava. Exogenous melatonin treatment reduced the post-harvest physiological deterioration (PPD) in cassava by reducing H_2O_2 accumulation and increasing the activities of SOD, CAT, and GR, but interestingly not APX. Transcript analyses also validated higher expression of *Cu/Zn-SOD, CAT1, glutathione peroxidase, POD3,* and *glutathione-S-transferase* (*GST*) in cassava roots treated with melatonin. This proves that it is via the direct or indirect maintenance of ROS homeostasis and preventing root deterioration that melatonin delays cassava PPD.

Conclusion

The discussion presented in this chapter clearly depicts the complicated signaling cascades and the regulatory components present in plants to tackle ROS, produced universally in any kind of abiotic stresses. The antioxidative pathways discussed also point to the fact that all types of abiotic stress responses are cross-linked via these regulations and hence are all correlated with one another. ROS participate in any kind of stresses as central players, acting upstream or downstream to other crucial players like Rbohs, membrane associated proteins and ion channels, MAPKs, CDPKs, and, most importantly, the phytohormones, which ultimately dictate plant viability, growth, and development. Though several components have come to our knowledge, a full proof blueprint of ROS signaling in abiotic stress is yet to be constructed.

A future goal in the field of ROS biology is to create a clear roadmap that would better aid researchers to develop multiple stress-tolerant cultivars, via controlling toxic levels of ROS accumulated within the system under any form of abiotic stresses. Bioinformatic software, strategies for data mining, and so forth would serve as powerful tools to accelerate crop development worldwide.

Acknowledgments

Financial support from the Science and Engineering Research Board (SERB), Department of Science and Technology, Government of India in the form of research Grant (SR/FT/LS-65/2010), and from the Council of Scientific and Industrial Research (CSIR), Government of India, through the Project [38(1387)/14/EMR-II] to Dr Aryadeep Roychoudhury, is gratefully acknowledged.

References

Abuharbeid S, Apel J, Sander M, *et al.* (2004) Cytotoxicity of the novel anti-cancer drug Viscumin depends on HER-2 levels in SKOV-3 cells. *Biochem Biophys Res Commun* 321: 403–412.

Agati G, Azzarello E, Pollastri S, Tattini M. (2012) Flavonoids as antioxidants in plants: location and functional significance. *Plant Sci* 196: 67–76.

Aghaei K, Ehsanpour AA, Komatsu S. (2009) Potato responds to salt stress by increased activity of antioxidant enzymes. *J Integr Plant Biol* 51: 1095–1103.

Agrawal GK, Iwahashi H, Rakwal R. (2003) Small GTPase 'Rop': molecular switch for plant defense responses. *FEBS Lett* 546: 173–180.

Anjum NA, Sofo A, Scopa A, *et al.* (2015) Lipids and proteins—major targets of oxidative modifications in abiotic stressed plants. *Environ Sci Pollut Res* 22: 4099–4121.

Asai S, Ichikawa T, Nomura H, *et al.* (2013) The variable domain of a plant calcium-dependent protein kinase (CDPK) confers subcellular localization and substrate recognition for NADPH oxidase. *J Biol Chem* 288: 14332–14340.

Asano T, Hayashi N, Kobayashi M, *et al.* (2012) A rice calcium-dependent protein kinase OsCPK12 oppositely modulates salt-stress tolerance and blast disease resistance. *Plant J* 69: 26–36.

Assmann SM (2004) Plant G proteins, phytohormones, and plasticity: three questions and a speculation. *Sci STKE* 2004: re20.

Bais HP, Vepachedu R, Gilroy S, Callaway RM, Vivanco JM. (2003) Allelopathy and exotic plant invasion: from molecules and genes to species interactions. *Science* 301: 1377–1380.

Banerjee A, Roychoudhury A. (2015) WRKY proteins: signaling and regulation of expression during abiotic stress responses. *Sci World J* 2015: 807560.

Banerjee A, RoychoudhuryA. (2016) Abscisic-acid-dependent basic leucine zipper (bZIP) transcription factors in plant abiotic stress. *Protoplasma* doi: 10.1007/s00709-015-0920-4.

Banerjee A, Roychoudhury A. (2016) Group II late embryogenesis abundant (LEA) proteins: structural and functional aspects in plant abiotic stress. *Plant Growth Regul* 79: 1–17.

Barnes J, Zheng Y, Lyons T. (2002) Plant resistance to ozone: the role of ascorbate. In: Osama K, Saji H, Youssefian S, Kondo N (eds), *Air Pollution and Plant Biotechnology – Prospects for Phytomonitoring and Phytoremediation*. Springer-Verlag, Tokyo, pp. 235–252.

Barraud N, Hassett DJ, Hwang SH, Rice SA, Kjelleberg S, Webb JS. (2006) Involvement of nitric oxide in biofilm dispersal of *Pseudomonas aeruginosa*. *J Bacteriol* 188: 7344–7353.

Basu S, Roychoudhury A, Saha PP, Sengupta DN. (2010) Comparative analysis of some biochemical responses of three indica rice varieties during polyethylene glycol-mediated water stress exhibits distinct varietal differences. *Acta Physiol Plant* 32: 551–563.

Besson-Bard A, Pugin A, Wendehenne D. (2008) New insights into nitric oxide signaling in plants. *Annu Rev Plant Biol* 59: 21–39.

Bhattacharjee S. (2005) Reactive oxygen species and oxidative burst: roles in stress, senescence and signal transduction in plants. *Curr Sci* 89: 1113–1121.

Bhattacharjee S. (2012) An inductive pulse of hydrogen peroxide pretreatment restores redox-homeostasis and mitigates oxidative membrane damage under extremes of temperature in two rice cultivars (*Oryza sativa* L., cultivars Ratna and SR26B). *Plant Growth Regul* 68: 395–410.

Boguszewska D, Grudkowska M, Zagdańska B. (2010) Drought-responsive antioxidant enzymes in potato (*Solanum tuberosum* L.). *Potato Res* 53: 373–382.

Bowler C, Fluhr R. (2000) The role of calcium and activated oxygen as signals for controlling cross-tolerance. *Trends Plant Sci* 5: 241–246.

Brosche M, Blomster T, Salojarvi J, *et al.* (2014) Transcriptomics and functional genomics of ROS-induced cell death regulation by RADICAL-INDUCED CELL DEATH1. *PLoS Genet* 10: e1004112.

Byeon Y, Lee HY, Lee K, Park S, Back K. (2014) Cellular localization and kinetics of the rice melatonin biosynthetic enzymes SNAT and ASMT. *J Pineal Res* 56: 107–114.

Campo S, Baldrich P, Messeguer J, Lalanne E, Coca M, SanSegundo B. (2014) Overexpression of a calcium-dependent protein kinase confers salt and drought tolerance in rice by preventing membrane lipid peroxidation. *Plant Physiol* 165: 688–704.

Chang-Quan W, Rui-Chang L. (2008) Enhancement of superoxide dismutase activity in the leaves of white clover (*Trifolium repens* L.) in response to polyethylene glycol-induced water stress. *Acta Physiol Plant* 30: 841–847.

Chaturvedi AK, Patel MK, Mishra A, Tiwari V, Jha B. (2014) The *SbMT2* gene from a halophyte confers abiotic stress tolerance and modulates ROS scavenging in transgenic tobacco. *PLoS ONE* 9: e111379.

Choi WG, Toyota M, Kim SH, Hilleary R, Gilroy S. (2014) Salt stress-induced Ca2C waves are associated with rapid, long-distance root-to-shoot signaling in plants. *Proc Natl Acad Sci USA* 111: 6497–6502.

Choudhury S, Panda P, Sahoo L, Panda SK. (2013) Reactive oxygen species signaling in plants under abiotic stress. *Plant Signal Behav* 8: e23681.

Correa-Aragunde N, Foresi N, Lamattina L. (2015) Nitric oxide is a ubiquitous signal for maintaining redox balance in plant cells: regulation of ascorbate peroxidase as a case study. *J Exp Bot* 66: 2913–2921.

Das K, Roychoudhury A. (2014) Reactive oxygen species (ROS) and response of antioxidants as ROS-scavengers during environmental stress in plants *Front Environ Sci* 2: 53.

de Abreu Neto JB, Frei M. (2016) Microarray meta-analysis focused on the response of genes involved in redox homeostasis to diverse abiotic stresses in rice. *Front Plant Sci* 6: 1260.

de Jong A, Yakimova E, Kapchina V, Woltering E (2002) A critical role for ethylene in hydrogen peroxide release during programmed cell death in tomato suspension cells. *Planta* 214: 537–545.

Deng X, Hu W, Wei S, *et al.* (2013) TaCIPK29, a CBL-interacting protein kinase gene from wheat, confers salt stress tolerance in transgenic tobacco. *PLoS ONE* 8: e69881.

Desikan R, Hancock JT, Bright J, *et al.* (2005) A role for ETR1 in hydrogen peroxide signaling in stomatal guard cells. *Plant Physiol* 137: 831–834.

Ding Y, Cao J, Ni L, *et al.* (2013) ZmCPK11 is involved in abscisic acid-induced antioxidant defence and functions upstream of ZmMPK5 in abscisic acid signalling in maize. *J Exp Bot* 64: 871–884.

Dynowski M, Schaaf G, Loque D, Moran O, Ludewig U. (2008) Plant plasma membrane water channels conduct the signaling molecule H_2O_2. *Biochem J* 414: 53–61.

Ellouzi H, Hamed K, Cela J, Müller M, Abdelly C, Munné-Bosch S. (2013) Increased sensitivity to salt stress in tocopherol-deficient *Arabidopsis* mutants growing in a hydroponic system. *Plant Signal Behav* 8: e23136.

Eltayeb AE, Kawano N, Badawi GH, *et al.* (2007) Overexpression of *monodehydroascorbate reductase* in transgenic tobacco confers enhanced tolerance to ozone, salt and polyethylene glycol stresses. *Planta* 225: 1255–1264.

Eyidogan F, Öz MT. (2007) Effect of salinity on antioxidant responses of chickpea seedlings. *Acta Physiol Plant* 29: 485–493.

Fang Y, Liao K, Du H, *et al.* (2015) A stress-responsive NAC transcription factor SNAC3 confers heat and drought tolerance through modulation of reactive oxygen species in rice. *J Exp Bot* 66: 6803.

Foyer CH, Noctor G. (2005) Redox homeostasis and antioxidant signaling: a metabolic interface between stress perception and physiological responses. *Plant Cell* 17: 1866–1875.

Fukao T, Xiong L. (2013) Genetic mechanisms conferring adaptation to submergence and drought in rice: simple or complex? *Curr Opin Plant Biol* 16: 196–204.

Fukao T, Yeung E, Bailey-Serres J. (2011) The submergence tolerance regulator SUB1A mediates crosstalk between submergence and drought tolerance in rice. *Plant Cell* 23: 412–427.

Gill SS, Khan NA, Anjum NA, Tuteja N. (2011) Amelioration of cadmium stress in crop plants by nutrients management: morphological, physiological and biochemical aspects. In: Anjum NA, Lopez-Lauri F (eds), *Plant Nutrition and Abiotic Stress Tolerance III, Plant Stress 5 (Special Issue 1).* Global Science Books Ltd, pp. 1–23.

Godoy Herz MA, Kornblihtt AR, Barta A, Kalyna M, Petrillo E. (2014) Shedding light on the chloroplast as a remote control of nuclear gene expression. *Plant Signal Behav* 9: e976150.

Golan I, Dominguez PG, Konrad Z, *et al.* (2014) Tomato *abscisic acid stress ripening* (*ASR*) gene family revisited. *PLoS ONE* e: 107117.

Guo Y, Xiong L, Ishitani M, Zhu JK. (2002) An *Arabidopsis* mutation in translation elongation factor 2 causes superinduction of CBF/DREB1 transcription factor genes but blocks the induction of their downstream targets under low temperatures. *Proc Natl Acad Sci USA* 99: 7786–7791.

Halliwell B. (2006) Reactive species and antioxidants. Redox biology is a fundamental theme of aerobic life. *Plant Physiol* 141: 312–322.

Han Y, Chaouch S, Mhamdi A, Queval G, Zechmann B, Noctor G. (2013a) Functional analysis of *Arabidopsis* mutants points to novel roles for glutathione in coupling H_2O_2 to activation of salicylic acid accumulation and signaling. *Antioxid Redox Signal* 18: 2106–2121.

Han Y, Mhamdi A, Chaouch S, Noctor G. (2013b) Regulation of basal and oxidative stress-triggered jasmonic acid-related gene expression by glutathione. *Plant Cell Environ* 36: 1135–1146.

Hasanuzzaman M, Nahar K, Alam MM, Roychowdhury R, Fujita M. (2013) Physiological, biochemical, and molecular mechanisms of heat stress tolerance in plants. *Int J Mol Sci* 14: 9643–9684.

Hayat S, Hayat Q, Alyemeni MN, Wani AS, Pichtel J, Ahmad A. (2012) Role of proline under changing environments. *Plant Signal Behav* 7: 1456–1466.

Hossain MA, Bhattacharjee S, Armin S-M, *et al.* (2015) Hydrogen peroxide priming modulates abiotic oxidative stress tolerance: Insights from ROS detoxification and scavenging. *Front Plant Sci* 6: 420.

Hou X, Xie K, Yao J, Qi Z, Xiong L. (2009) A homolog of human ski-interacting protein in rice positively regulates cell viability and stress tolerance. *Proc Natl Acad Sci USA* 106: 6410–6415.

Hu DG, Ma QJ, Sun CH, Sun MH, You CX, Hao YJ. (2015) Overexpression of *MdSOS2L1*, a CIPK protein kinase, increases the antioxidant metabolites to enhance salt tolerance in apple and tomato. *Physiol Plant* doi: 10.1111/ppl.1235.

Hu W, Huang C, Deng X, *et al.* (2013) TaASR1, a transcription factor gene in wheat, confers drought stress tolerance in transgenic tobacco. *Plant Cell Environ* 36: 1449–1464.

Hu X, Jiang M, Zhang J, Zhang A, Lin F, Tan M. (2007) Calcium-calmodulin is required for abscisic acid-induced antioxidant defense and functions both upstream and downstream of H_2O_2 production in leaves of maize (*Zea mays*) plants. *New Phytol* 173: 27–38.

Huang XY, Chao DY, Gao JP, Zhu MZ, Shi M, Lin HX. (2009) A previously unknown zinc finger protein, DST, regulates drought and salt tolerance in rice via stomatal aperture control. *Genes Dev* 23: 1805–1817.

Igamberdiev AU, Seregelyes C, Manac N, Hill RD. (2004) NADH-dependent metabolism of nitric oxide in alfalfa root cultures expressing barley hemoglobin. *Planta* 219: 95–102.

Jain K, Kataria S, Guruprasad KN. (2003) Changes in antioxidant defenses of cucumber cotyledons in response to UV-B and to the free radicals generating compound AAPH. *Plant Sci* 165: 551–557.

Jan A, Maruyama K, Todaka D, *et al.* (2013) OsTZF1, a CCCH-tandem zinc finger protein, confers delayed senescence and stress tolerance in rice by regulating stress-related genes. *Plant Physiol* 161: 1202–1216.

Jin HX, Huang F, Cheng H, Song HN, Yu DY. (2013) Overexpression of the GmNAC2 gene, an NAC transcription factor, reduces abiotic stress tolerance in tobacco. *Plant Mol Biol Rep* 31: 435–442.

Joo JH, Wang S, Chen JG, Jones AM, Fedoroff NV. (2005) Different signaling and cell death roles of heterotrimeric G protein a and b subunits in the *Arabidopsis* oxidative stress response to ozone. *Plant Cell* 17: 957–970.

Kadota Y, Sklenar J, Derbyshire P, *et al.* (2014) Direct regulation of the NADPH oxidase RBOHD by the PRR-associated kinase BIK1 during plant immunity. *Mol Cell* 54: 43–55.
Kandziora-Ciupa M, Ciepal R, Nadgórska-Socha A, Barczyk G. (2013) A comparative study of heavy metal accumulation and antioxidant responses in *Vaccinium myrtillus* L. leaves in polluted and non-polluted areas. *Environ Sci Pollut Res* 20: 4920–4932.
Kapoor D, Sharma R, Handa N, *et al.* (2015) Redox homeostasis in plants under abiotic stress: role of electron carriers, energy metabolism mediators and proteinaceous thiols. *Front Environ Sci* 3:13.
Khokon AR, Okuma E, Hossain MA, *et al.* (2011) Involvement of extracellular oxidative burst in salicylic acid-induced stomatal closure in *Arabidopsis*. *Plant Cell Environ* 34: 434–443.
Kiffin R, Bandyopadhyay U, Cuervo AM. (2006) Oxidative stress and autophagy. *Antioxid Redox Signal* 8: 152–162.
Königshofer H, Tromballa HW, Loppert HG. (2008) Early events in signaling high-temperature stress in tobacco BY2 cells involve alterations in membrane fluidity and enhanced hydrogen peroxide production. *Plant Cell Environ* 31: 1771–1780.
Krieger-Liszkay A, Fufezan C, Trebst A. (2008) Singlet oxygen production in photosystem II and related protection mechanism. *Photosyn Res* 98: 551–564.
Kromer S (1995) Respiration during photosynthesis. *Annu Rev Plant Physiol Plant Mol Biol* 46: 45–70.
Lee KP, Kim C, Landgraf F, Apel K. (2007) EXECUTER1-and EXECUTER2-dependent transfer of stress-related signals from the plastid to the nucleus of *Arabidopsis thaliana*. *Proc Natl Acad Sci USA* 104: 10270–10275.
Li CH, Wang G, Zhao JL, *et al.* (2014) The receptor-like kinase SIT1 mediates salt sensitivity by activating MAPK3/6 and regulating ethylene homeostasis in rice. *Plant Cell* 26: 2538–2553.
Li T, Li H, Zhang YX, Liu JY. (2011) Identification and analysis of seven H_2O_2-responsive miRNAs and 32 new miRNAs in the seedlings of rice (*Oryza sativa* L. ssp. *indica*). *Nucleic Acids Res* 39: 2821–2833.
Liu X, Hua X, Guo J, *et al.* (2008) Enhanced tolerance to drought stress in transgenic tobacco plants overexpressing *VTE1* for increased tocopherol production from *Arabidopsis thaliana*. *Biotechnol Lett* 30: 1275–1280.
Liu S, Wang M, Wei T, Meng C, Xia G. (2014) A wheat *SIMILAR TO RCD-ONE* gene enhances seedling growth and abiotic stress resistance by modulating redox homeostasis and maintaining genomic integrity. *Plant Cell* 26: 164–180.
Lu W, Chu X, Li Y, Wang C, Guo X. (2013) Cotton GhMKK1 induces the tolerance of salt and drought stress, and mediates defence responses to pathogen infection in transgenic *Nicotiana benthamiana*. *PLoS ONE* 8: e68503.
Luo P, Shen Y, Jin S, *et al.* (2016) Overexpression of *Rosa rugosa anthocyanidin reductase* enhances tobacco tolerance to abiotic stress through increased ROS scavenging and modulation of ABA signaling. *Plant Sci* 245: 35–49.
Ma F, Lu R, Liu H, *et al.* (2012) Nitric oxide-activated calcium/calmodulin-dependent protein kinase regulates the abscisic acid-induced antioxidant defence in maize. *J Exp Bot* 63: 4835–4847.
Ma L, Zhang H, Sun L, *et al.* (2012) NADPH oxidase AtrbohD and AtrbohF function in ROS-dependent regulation of NaC/KC homeostasis in *Arabidopsis* under salt stress. *J Exp Bot* 63: 305–317.
Ma Q, Zhang T, Zhang P, Wang ZY. (2016) Melatonin attenuates postharvest physiological deterioration of cassava storage roots. *J Pineal Res* doi: 10.1111/jpi.12325.
Malar S, Vikram SS, Favas PJC, Perumal V. (2014) Lead heavy metal toxicity induced changes on growth and antioxidative enzymes level in water hyacinths [*Eichhornia crassipes* (Mart.)]. *Bot Stud* 55: 54.
Malik S, Ashraf M. (2012) Exogenous application of ascorbic acid stimulates growth and photosynthesis of wheat (*Triticum aestivum* L.) under drought. *Soil Environ* 31: 72–77.

Maruta T, Inoue T, Tamoi M, *et al.* (2011) *Arabidopsis* NADPH oxidases, AtrbohD and AtrbohF, are essential for jasmonic acid-induced expression of genes regulated by MYC2 transcription factor. *Plant Sci* 180: 655–660.
McCord JM, Fridovich I. (1969) Superoxide dismutase. An enzymic function for erythrocuprein (hemocuprein). *J Biol Chem* 244: 6049–6055.
Mhamdi A, Queval G, Chaouch S, Vanderauwera S, VanBreusegem F, Noctor G. (2010) Catalase function in plants: a focus on *Arabidopsis* mutants as stress-mimic models. *J Exp Bot* 61: 4197–4220.
Miller G, Shulaev V, Mittler R. (2008) Reactive oxygen signaling and abiotic stress. *Physiol Plant* 133: 481–489.
Miller G, Schlauch K, Tam R, *et al.* (2009) The plant NADPH oxidase RBOHD mediates rapid systemic signaling in response to diverse stimuli. *Sci Signal* 2: ra45.
Miller G, Suzuki N, Ciftci-Yilmaz S, Mittler R. (2010) Reactive oxygen species homeostasis and signalling during drought and salinity stresses. *Plant Cell Environ* 33: 453–467.
Mittler R, Vanderauwera S, Suzuki N, *et al.* (2011) ROS signaling: the new wave? *Trends Plant Sci* 16: 300–309.
Miura K, Okamoto H, Okuma E, *et al.* (2013) SIZ1 deficiency causes reduced stomatal aperture and enhanced drought tolerance via controlling salicylic acid-induced accumulation of reactive oxygen species in *Arabidopsis*. *Plant J* 73: 91–104.
Møller IM, Jensen PE, Hansson A. (2007) Oxidative modifications to cellular components in plants. *Annu Rev Plant Biol* 58: 459–481.
Moreau M, Lindermayr C, Durner J, Klessig DF. (2010) NO synthesis and signaling in plants – where do we stand? *Physiol Plant* 138: 372–383.
Mullineaux PM, Rausch T. (2005) Glutathione, photosynthesis and the redox regulation of stress-responsive gene expression. *Photosyn Res* 86: 459–474.
Nawaz MA, Huang Y, Bie Z, *et al.* (2016) Melatonin: Current status and future perspectives in plant science. *Front Plant Sci* 6: 1230.
Ning J, Li X, Hicks LM, Xiong L. (2010) ARaf-like MAPKKK gene *DSM1* mediates drought resistance through reactive oxygen species scavenging in rice. *Plant Physiol* 152: 876–890.
Noctor G, Mhamdi A, Foyer CH. (2014) The roles of reactive oxygen metabolism in drought: not so cut and dried. *Plant Physiol* 164: 1636–1648.
Pinto E, Sigaud-kutner T, Leitao MA, Okamoto OK, Morse D, Colepicolo P. (2003) Heavy metal-induced oxidative stress in algae. *J Phycol* 39: 1008–1018.
Pitzschke A, Djamei A, Bitton F, Hirt H. (2009) A major role of the MEKK1-MKK1/2-MPK4 pathway in ROS signalling. *Mol Plant* 2: 120–137.
Ramabulana T, Mavunda RD, Steenkamp PA, Piater LA, Dubery IA, Madala NE. (2016) Perturbation of pharmacologically relevant polyphenolic compounds in *Moringa oleifera* against photo-oxidative damages imposed by gamma radiation. *J Photochem Photobiol* 156: 79–86.
Ramegowda V, Senthil-Kumar M, Nataraja KN, Reddy MK, Mysore KS, Udayakumar M. (2012) Expression of a finger millet transcription factor, *EcNAC1*, in tobacco confers abiotic stress-tolerance. *PLoS ONE* 7: e40397.
Rao MV, Paliyath G, Ormrod DP. (1996) Ultraviolet-B and ozone-induced biochemical changes in antioxidant enzymes of *Arabidopsis thaliana*. *Plant Physiol* 110: 125–136.
Rao MV, Lee H, Creelman RA, Mullet JE, Davis KR. (2000) Jasmonic acid signaling modulates ozone-induced hypersensitive cell death. *Plant Cell* 12: 1633–1646.
Razaji A, Asli DE, Farzanian M. (2012) The effects of seed priming with ascorbic acid on drought tolerance and some morphological and physiological characteristics of safflower (*Carthamus tinctorius* L.). *Ann Biol Res* 3: 3984–3989.

Roychoudhury A, Banerjee A. (2016) Endogenous glycine betaine accumulation mediates abiotic stress tolerance in plants. *Trop Plant Res* 3: 105–111.

Roychoudhury A, Basu S. (2012) Ascorbate-glutathione and plant tolerance to various abiotic stresses. In: Anjum NA, Umar S, Ahmad A (eds), *Oxidative Stress in Plants: Causes, Consequences and Tolerance.* IK International Publishers, New Delhi, pp. 177–258.

Roychoudhury A, Chakraborty M. (2013) Biochemical and molecular basis of varietal difference in plant salt tolerance. *Annu Rev Res Biol* 3: 422–454.

Roychoudhury A, Ghosh S. (2013) Physiological and biochemical responses of mungbean (*Vigna radiata* L. Wilczek) to varying concentrations of cadmium chloride or sodium chloride. *Unique J Pharm Biol Sci* 1: 11–21.

Roychoudhury A, Basu S, Sarkar SN, Sengupta DN. (2008) Comparative physiological and molecular responses of a common aromatic indica rice cultivar to high salinity with non-aromatic indica rice cultivars. *Plant Cell Rep* 27: 1395–1410.

Roychoudhury A, Basu S, Sengupta DN. (2012) Antioxidants and stress-related metabolites in the seedlings of two indica rice varieties exposed to cadmium chloride toxicity. *Acta Physiol Plant* 34: 835–847.

Roychoudhury A, Banerjee A, Lahiri V. (2015) Metabolic and molecular-genetic regulation of proline signaling and its cross-talk with major effectors mediates abiotic stress tolerance in plants. *Turk J Bot* 39: 887–910.

Roychoudhury A, Ghosh S, Paul S, Mazumdar S, Das G, Das S. (2016) Pre-treatment of seeds with salicylic acid attenuates cadmium chloride-induced oxidative damages in the seedlings of mungbean (*Vigna radiata* L. Wilczek). *Acta Physiol Plant* 38: 11.

Sairam RK, Deshmukh PS, Saxena DC. (1998) Role of antioxidant systems in wheat genotypes tolerance to water stress. *Biol Plant* 41: 387–394.

Schmidt R, Mieulet D, Hubberten HM, *et al.* (2013) Salt-responsive ERF1 regulates reactive oxygen species-dependent signaling during the initial response to salt stress in rice. *Plant Cell* 25: 2115–2131.

Serrato AJ, Fernández-Trijueque J, Barajas-López JD, Chueca A, Sahrawy M. (2013) Plastid thioredoxins: a "one-for-all" redox-signaling system in plants. *Front Plant Sci* 21: 463.

Sewelam N, Kazan K, Thomas-Hall S, Kidd BN, Manners JM, Schenk PM. (2013) Ethylene response factor 6 is a regulator of reactive oxygen species signaling in *Arabidopsis*. *PLoS ONE* 8: e70289.

Sewelam N, Jaspert N, Kelen KVD, *et al.* (2014) Spatial H_2O_2 signaling specificity: H_2O_2 from chloroplasts and peroxisomes differentially modulates the plant transcriptome. *Mol Plant* 7: 1191–1210.

Sewelam N, Kazan K, Schenk PM. (2016) Global plant stress signaling: reactive oxygen species at the cross-road. *Front Plant Sci* 7: 187.

Sharma P, Dubey RS. (2005) Drought induces oxidative stress and enhances the activities of antioxidant enzymes in growing rice seedlings. *Plant Growth Regul* 46: 209–221.

Shi B, Ni L, Zhang A, *et al.* (2012) OsDMI3 is a novel component of abscisic acid signaling in the induction of antioxidant defense in leaves of rice. *Mol Plant* 5: 1359–1374.

Shi B, Ni L, Liu Y, Zhang A, Tan M, Jiang M. (2014) OsDMI3-mediated activation of OsMPK1 regulates the activities of antioxidant enzymes in abscisic acid signalling in rice. *Plant Cell Environ* 37: 341–352.

Singh R, Singh S, Parihar P, *et al.* (2016) Reactive oxygen species (ROS): beneficial companions of plants' developmental processes. *Front Plant Sci* 7: 1299.

Singh S, Srivastava PK, Kumar D, Tripathi DK, Chauhan DK, Prasad SM. (2015) Morpho-anatomical and biochemical adapting strategies of maize (*Zea mays* L.) seedlings against lead and chromium stresses. *Biocatal Agric Biotechnol* 4: 286–295.

Singh S, Tripathi DK, Singh S, *et al.* (2017) Toxicity of aluminium on various levels of plant cells and organism: a review. *Environ Exp Bot* 137: 177–193.
Singh VP, Tripathi DK, Kumar D, Chauhan DK. (2011) Influence of exogenous silicon addition on aluminium tolerance in rice seedlings. *Biol Trace Elem Res* 144: 1260–1274.
Smirnoff N. (2000) Ascorbic acid: metabolism and functions of a multi-faceted molecule. *Curr Opin Plant Biol* 3: 229–235.
Sreedevi S, Krishnan PN. (2012) Detoxification of heavy metal induced oxidative stress in plants, enzymatic and non-enzymatic machanisms – a mini review. In: Anjum NA, Umar S, Ahmad A (eds), *Oxidative Stress in Plants: Causes, Consequences and Tolerance.* IK International Publishing House, New Delhi, pp. 291–307.
Suharsono U, Fujisawa Y, Kawasaki T, Iwasaki Y, Satoh H, Shimamoto K. (2002) The heterotrimeric G protein alpha subunit acts upstream of the small GTPase Rac in disease resistance of rice. *Proc Natl Acad Sci USA* 99: 13307–13312.
Sun J, Hu W, Zhou R, *et al.* (2015) The *Brachypodium distachyon BdWRKY36* gene confers tolerance to drought stress in transgenic tobacco plants. *Plant Cell Rep* 34: 23–35.
Sun SJ, Guo SQ, Yang X, *et al.* (2010) Functional analysis of a novel Cys2/His2-type zinc finger protein involved in salt tolerance in rice. *J Exp Bot* 61: 2807–2818.
Sunkar R, Kapoor A, Zhu JK. (2006) Post transcriptional induction of two *Cu/Zn superoxide dismutase* genes in *Arabidopsis* is mediated by downregulation of miR398 and important for oxidative stress tolerance. *Plant Cell* 18: 2051–2065.
Tang RJ, Zhao FG, Garcia VJ, *et al.* (2015) Tonoplast CBL-CIPK calcium signaling network regulates magnesium homeostasis in *Arabidopsis*. *Proc Natl Acad Sci USA* 112: 3134–3139.
Tanou G, Job C, Belghazi M, Molassiotis A, Diamantidis G, Job D. (2010) Proteomic signatures uncover hydrogen peroxide and nitric oxide cross-talk signaling network in citrus plants. *J Proteome Res* 9: 5994–6006.
Torres MA, Dangl JL. (2005) Functions of the respiratory burst oxidase in biotic interactions, abiotic stress and development. *Curr Opin Plant Biol* 8: 397–403.
Tripathi DK, Singh VP, Kumar D, Chauhan DK. (2012a) Impact of exogenous silicon addition on chromium uptake, growth, mineral elements, oxidative stress, antioxidant capacity, and leaf and root structures in rice seedlings exposed to hexavalent chromium. *Acta Physiol Plant* 34: 279–289.
Tripathi DK, Singh VP, Kumar D, Chauhan DK. (2012b) Rice seedlings under cadmium stress: effect of silicon on growth, cadmium uptake, oxidative stress, antioxidant capacity and root and leaf structures. *Chem Ecol* 28: 281–291.
Tripathi DK, Singh S, Singh VP, Prasad SM, Chauhan DK, Dubey NK. (2016a) Silicon nanoparticles more efficiently alleviate arsenate toxicity than silicon in maize cultivar and hybrid differing in arsenate tolerance. *Front Environ Sci* 4: 46.
Tripathi DK, Singh VP, Prasad SM, Dubey NK, Chauhan DK, Rai AK. (2016b) LIB spectroscopic and biochemical analysis to characterize lead toxicity alleviative nature of silicon in wheat (Triticum aestivum L.) seedlings. *J Photochem Photobiol B: Biol* 154: 89–98.
Tripathi DK, Singh S, Singh VP, Prasad SM, Dubey NK, Chauhan DK. (2017a) Silicon nanoparticles more effectively alleviated UV-B stress than silicon in wheat (Triticum aestivum) seedlings. *Plant Physiol Biochem* 110: 70–81.
Tripathi DK, Singh S, Singh S, *et al.* (2017b) Nitric oxide alleviates silver nanoparticles (AgNps)-induced phytotoxicity in Pisum sativum seedlings. *Plant Physiol Biochem* 110: 167–177.
Tripathi DK, Mishra RK, Singh S, *et al.* (2017c) Nitric oxide ameliorates zinc oxide nanoparticles phytotoxicity in wheat seedlings: implication of the ascorbate–glutathione cycle. *Front Plant Sci* 8: 1.

Vlot AC, Liu P-P, Cameron RK, *et al.* (2008) Identification of likely orthologs of tobacco salicylic acid-binding protein 2 and their role in systemic acquired resistance in *Arabidopsis thaliana. Plant J* 56: 445–456.

Wang F, Chen HW, Li QT, *et al.* (2015) GmWRKY27 interacts with GmMYB174 to reduce expression of GmNAC29 for stress tolerance in soybean plants. *Plant J* 83: 224–236.

Wang M, Zhao X, Xiao Z, Yin X, Xing T, Xia G. (2016) A wheat superoxide dismutase gene *TaSOD2* enhances salt resistance through modulating redox homeostasis by promoting NADPH oxidase activity. *Plant Mol Biol* 91: 115–130.

Wang P, Du Y, Li Y, Ren D, Song CP. (2010) Hydrogen peroxide-mediated activation of MAP kinase 6 modulates nitric oxide biosynthesis and signal transduction in *Arabidopsis. Plant Cell* 22: 2981–2998.

Wang P, Du Y, Hou YJ, *et al.* (2015) Nitric oxide negatively regulates abscisic-acid signaling in guard cells by S-nitrosylation of OST1. *Proc Natl Acad Sci USA* 112: 613–618.

Wang R, Liu S, Zhou F, Ding C, Hua C. (2014) Exogenous ascorbic acid and glutathione alleviate oxidative stress induced by salt stress in the chloroplasts of *Oryza sativa* L. *Z Naturforsch C* 69: 226–236.

Wang Y, Ying Y, Chen J, Wang X. (2004) Transgenic *Arabidopsis* overexpressing Mn-SOD enhanced salt-tolerance. *Plant Sci* 167: 671–677.

Wen F, Qin T, Wang Y, *et al.* (2015) OsHK3 is a crucial regulator of abscisic acid signaling involved in antioxidant defense in rice. *J Integr Plant Biol* 57: 213–228.

Wu L, Zhang Z, Zhang H, Wang XC, Huang R. (2008) Transcriptional modulation of ethylene response factor protein JERF3 in the oxidative stress response enhances tolerance of tobacco seedlings to salt, drought, and freezing. *Plant Physiol* 148: 1953–1963.

Xia XJ, Zhou YH, Shi K, Zhou J, Foyer CH, Yu JQ. (2015) Interplay between reactive oxygen species and hormones in the control of plant development and stress tolerance. *J Exp Bot* 66: 2839–2856.

Xiao Y, Savchenko T, Baidoo EE, *et al.* (2012) Retrograde signaling by the plastidial metabolite MEcPP regulates expression of nuclear stress-response genes. *Cell* 149: 1525–1535.

Xu C, Sullivan JH, Garrett WM, Caperna TJ, Natarajan S. (2008) Impact of solar ultraviolet-B on proteome in soybean lines differing in flavonoid contents. *Phytochemistry* 69: 38–48.

Xue H, Seifert GJ. (2015) Fasciclin like arabinogalactan protein 4 and respiratory burst oxidase homolog D and F independently modulate abscisic acid signaling. *Plant Signal Behav* 10: e989064.

Yamaguchi-Shinozaki K, Shinozaki K. (2006) Transcriptional regulatory networks in cellular responses and tolerance to dehydration and cold stresses. *Annu Rev Plant Biol* 57: 781–803.

Yan H, Jia H, Chen X, Hao L, An H, Guo X. (2014) The cotton WRKY transcription factor GhWRKY17 functions in drought and salt stress in transgenic *Nicotiana benthamiana* through ABA signaling and the modulation of reactive oxygen species production. *Plant Cell Physiol* 55: 2060–2076.

Yan J, Guan L, Sun Y, *et al.* (2015) Calcium and ZmCCaMK are involved in brassinosteroid-induced antioxidant defense in maize leaves. *Plant Cell Physiol* 56: 883–896.

You J, Chan Z. (2015) ROS regulation during abiotic stress responses in crop plants. *Front Plant Sci* 6: 1092.

You J, Zong W, Li X, *et al.* (2013) The SNAC1- targeted gene *OsSRO1c* modulates stomatal closure and oxidative stress tolerance by regulating hydrogen peroxide in rice. *J Exp Bot* 64: 569–583.

You J, Zong W, Hu H, Li X, Xiao J, Xiong L (2014) A stress-responsive NaCl-regulated protein phosphatase gene *rice protein phosphatase 18* modulates drought and oxidative stress tolerance through abscisic acid-independent reactive oxygen species scavenging in rice. *Plant Physiol* 166: 2100–2114.

Zhang H, Liu Y, Wen F, *et al.* (2014) A novel rice C2H2-type zinc finger protein, ZFP36, is a key player involved in abscisic acid-induced antioxidant defence and oxidative stress tolerance in rice. *J Exp Bot* 65: 5795–5809.

Zhang L, Li Y, Lu W, Meng F, Wu CA, Guo X. (2012) Cotton GhMKK5 affects disease resistance, induces HR-like cell death, and reduces the tolerance to salt and drought stress in transgenic *Nicotiana benthamiana*. *J Exp Bot* 63: 3935–3951.

Zhang W, Jeon BW, Assmann SM. (2011) Heterotrimeric G-protein regulation of ROS signaling and calcium currents in *Arabidopsis* guard cells. *J Exp Bot* 62: 2371–2379.

Zhou J, Wang J, Li X, *et al.* (2014) H_2O_2 mediates the crosstalk of brassinosteroid and abscisic acid in tomato responses to heat and oxidative stresses. *J Exp Bot* 65: 4371–4383.

Zhu Y, Zuo M, Liang Y, *et al.* (2013) MAP65-1a positively regulates H_2O_2 amplification and enhances brassinosteroid-induced antioxidant defence in maize. *J Exp Bot* 64: 3787–3802.

Zlatev ZS, Lidon FC, Ramalho JC, Yordanov IT. (2006) Comparison of resistance to drought of three bean cultivars. *Biol Plant* 50: 389–394.

3

Balancing Roles of Reactive Oxygen Species in Plants' Response to Metalloid Exposure

Dibyendu Talukdar

R.P.M. College, Hooghly, India

Introduction

Since the advent of photosynthetic organisms on our planet, the reducing environment of the atmosphere has changed to an aerobic state, since then the generation of reactive oxygen species (ROS) has become an integral part of our living system. Being an autotrophic organism, a plant produces its own food through photosynthesis, which determines numerous events related to plant growth and development. ROS are continuously generated as unavoidable consequences of these aerobic events. Superoxide anion ($O_2^{\cdot-}$), hydrogen peroxide (H_2O_2), singlet oxygen ($^1O_2$), and the hydroxyl radical ($\cdot OH$) are some of the abundant ROS molecules that plants generate continually through photosynthesis, respiration, and other cellular and metabolic events. The versatility of ROS biology in plants' life involves different cellular organelles, compartments, and enzymatic as well as non-enzymatic events. While superoxide anion and singlet oxygen are mainly generated through leakage of electrons to O_2 in an over-energized electron transport chain operating in chloroplasts and mitochondria and by plasma membrane-bound NADPH oxidase during reduction of apoplastic O_2, H_2O_2 is produced by the dismutation of superoxides, activity of cell wall-anchored class III peroxidases, and photorespiratory activity of the flavin-containing enzyme glycolate oxidase and β-oxidation of fatty acids (by the flavin-containing enzyme acetyl-CoA oxidase) in peroxisomes (del Río, 2015; Gupta and Igamberdiev, 2015). H_2O_2 can give rise to highly toxic $\cdot$OH radicals through the Fenton and Haber–Weiss reactions with the help of free redox-active metal ions. These reactive intermediates and their by-products generated through different physiological processes have strong oxidizing potential toward cellular DNA, RNA, proteins, and lipids but are not equally reactive. While superoxides and H_2O_2 are selective in their actions and targets, hydroxyl radicals randomly attack surrounding molecules (Halliwell, 2006; Møller *et al.*, 2007). In steady-state biological conditions, the ROS level is delicately balanced between ROS production and scavenging, for which plants have evolved robust but tightly controlled antioxidant defense machinery in different cellular compartments (Foyer and Noctor, 2009, 2011, 2012). Any disturbances in this process can lead to increases in different ROS levels, which can tilt ROS balance in favor of oxidative imbalance and trigger a series of cascading reactions toward oxidative damage (Apel and Hirt, 2004). The cues to these disturbances are often appropriately and coordinately responded to by plants through a low titer of ROS molecules, which then

Reactive Oxygen Species in Plants: Boon Or Bane - Revisiting the Role of ROS, First Edition. Edited by Vijay Pratap Singh, Samiksha Singh, Durgesh Kumar Tripathi, Sheo Mohan Prasad, and Devendra Kumar Chauhan.

plays balancing roles as signaling molecules in acclimation and tolerance of external stress factors (Petrov and Van Breusegem, 2012; Hossain *et al.*, 2015; Cuypers *et al.*, 2016).

Plants are sessile organisms, and thus they have to continually confront diverse types of abiotic and biotic stresses. Throughout the world, the negative impacts on human health of soil and water contamination by heavy metals like lead (Pb), cadmium (Cd), mercury (Hg), chromium (Cr), copper (Cu), nickel (Ni), aluminum (Al), and toxic metalloids like arsenic (As) have now entered a dangerous stage due to their toxic accumulations in edible plant parts through the soil-water-plant system, jeopardizing plant productivity (both quality and quantity of edible parts) and accelerating aura of "hidden hungers" (Nagajyoti *et al.*, 2010; V.P. Singh *et al.*, 2011; Tripathi *et al.*, 2012ab, 2016a,b, 2017a–c; Sethy and Ghosh, 2013; S. Singh *et al.*, 2015, 2017; R. Singh *et al.*, 2016). Metalloid toxicity is one of the severest abiotic stress factors, which undergoes bioenrichment in the food chain and alters cellular events in plants' life. The term "metalloid" in this chapter will be used to refer to heavy metals + metalloid (arsenic). Extensive literatures are available regarding metalloid-induced toxicity symptoms in different aspects of plant growth and development (Nagajyoti *et al.*, 2010; Anjum *et al.*, 2014). However, new areas are being unveiled in ROS biology, showing it to be fundamental in metalloid-induced events (Cuypers *et al.*, 2016). This chapter will mainly focus on the diversity of roles played by ROS during plants' metalloid exposure against the background of the possible dual roles of ROS in stress perception, signaling, and adaptation.

ROS: Oxidizing Agents During Metalloid Exposure

Excess ROS is the first cellular signal in response to toxic metalloid exposure. Heavy metals/metalloids can induce ROS generation in the apoplast, mitochondria, and in illuminated chloroplasts (Gill and Tuteja, 2010). However, based on different chemical properties of metalloids, two modes of action have been proposed: (i) under physiological conditions, redox active metals such as Fe and Cu, which exist in different oxidation states, can directly participate in the Fenton and Haber–Weiss reactions and ultimately produce highly reactive hydroxyl radicals from H_2O_2; (ii) non-redox active metals like Cd, Hg, and Zn indirectly induce ROS generation through alterations in antioxidant defense components (Sharma and Dietz, 2009). Metalloids can negatively interfere with the redox balance and energy equilibrium of the cell and destroy the ROS homeostasis. Metalloid-induced oxidative stress is usually manifest as poor seed germination, reduced photosynthesis, decreased plant growth, low yield at a morpho-physiological level, compromised capacity of antioxidants and enzymes, altered electron transport chain, and high levels of membrane lipid peroxidation, protein carbonylation, and membrane leakage at a biochemical level (Anjum *et al.*, 2010, 2011; Nagajyoti *et al.*, 2010 and references therein; Talukdar, 2011). Modulations of morpho-physiological and biochemical traits are actually governed by regulation of expression of the involved genes during metalloid exposure in plants (Finnegan and Chen, 2012). In all these cases, ROS levels play fundamental roles in modulation of plants' cellular responses. Membrane lipid peroxidation yields malondealdehyde, a cytotoxic aldehyde, which is often found positively correlated with H_2O_2 accumulation in metalloid-exposed plant organs (Srivastava *et al.*, 2009, 2014). In addition, protein carbonylation, which represents the oxidative transformation of -SH rich protein moieties, is another symptoms of metalloid-induced oxidative damage. Carbonylated protein levels were found to be significantly increased in Pb-exposed rice seedlings (Srivastava *et al.*, 2014), Al-exposed *Cucumis sativus* and *Allium cepa* roots (Achary *et al.*, 2008; Pereira *et al.*, 2010), and in Cd-exposed lettuce plants (Monteiro *et al.*, 2012). Among the early events responding to As exposure, photosynthesis is the most sensitive event to As-induced ROS accumulation (del Río, 2015). Excess ROS generation was attributed to low pigment status, reduced chlorophyll *a*/*b* ratio, and decreased rate of photosynthesis in several crop plants (Mishra *et al.*, 2016). However, in a recent investigation at physiologically relevant concentration of As, Mishra *et al.* (2014) opined that oxidative stress was not the reason for inhibition of photosynthesis and concomitant reduced plant growth. Rather, the study pointed out that a decrease in photosynthetic pigment level was the earliest event at low As concentrations, followed by inhibition of photosynthetic linear electron transport and the photosystem II reaction center at higher concentrations, which ultimately led to oxidative stress (Mishra *et al.*, 2014).

Although ROS generation has been considered as a general feature of plants experiencing metalloid-toxicity, it seems likely that excess ROS accumulation in metal-exposed plants may be due to (i) reduced capability of plants' thiol-antioxidant defense, and (ii) involvement of metal detoxification processes itself, including the reduction of arsenate [As(V)] to arsenite [As(III)] and the induction of phytochelatins (PCs) synthesis (Anjum *et al.*, 2012; Gupta *et al.*, 2013a, 2013b). Over-accumulation of H_2O_2 and consequent oxidative damage has been reported in *Arabidopsis thaliana*, rice, wheat, maize, oats, vegetables, pulses, red clover, and in several other plant species under conditions of metalloid toxicity (Gupta *et al.*, 2004, 2015; Anjum *et al.*, 2012). To mitigate oxidative damage, plants are equipped with multiple levels of ROS-detoxification and scavenging mechanisms (Foyer and Noctor, 2011). Reduced glutathione (GSH) and GSH-dependent antioxidant defense, comprising enzymes of the ascorbate (AsA)-GSH cycle, such as ascorbate peroxidase (APX), dehydroascorbate reductase (DHAR), and GSH-reductase (GR), three redox couples of this cycle like AsA-DHA (dehydroascorbate), GSH-GSSG (oxidized glutathione), and $NADP^+$/NADPH, and GSH peroxidase (GPX) as well as glutathione-*S*-transferases (GSTs) outside this cycle, create powerful shields against metalloid-induced excess ROS generation. GSH is central in this process, and its redox pool is critical in maintaining redox homeostasis in plants under metalloid exposure (Anjum *et al.*, 2011, 2012; Noctor *et al.*, 2012). Following an influx of ROS, GSH rapidly absorbs electrons and is spontaneously converted to the oxidized form, GSSG, to prevent irreversible modifications of critical plant cell machinery. Besides the detoxification process, both AsA and GSH play key roles as enzyme cofactors, and GSH is also a storage and transport form of cysteine (Cys) (Noctor *et al.*, 2012). Overproduction of superoxide radicals is the first distinct cellular symptom of metalloid-induced toxicity, in which superoxide dismutase (SOD) constitutes the first line of defense (Foyer and Noctor, 2011, 2012; Cuypers *et al.*, 2016). SOD in different cellular compartments dismutates highly reactive superoxide radicals to H_2O_2, which is then scavenged by APX by using AsA as exclusive cofactor. The resultant DHA is then recycled back to AsA by the activity of DHAR, which uses GSH as reductant. This leads to oxidation of GSH to GSSG, which is then recycled back to GSH by a NADPH-dependent reaction catalyzed by GR in the AsA-GSH cycle (Noctor *et al.*, 2012). The cycle is tightly regulated and different isoforms of SOD, APX, GR, and DHAR take part in the process. Outside this cycle, catalases (CAT) can decompose bulk H_2O_2 without using any reducing power and thus save reducing power to be oxidized during ROS metabolism (Anjum *et al.*, 2012; Hossain *et al.*, 2015). GSTs are mainly involved in detoxification of both endogenous and xenobiotic compounds with electrophilic centers by the nucleophilic addition of GSH (Finnegan and Chen, 2012). GST isoforms sometimes additionally function as GPXs in the presence of ROS (Finnegan and Chen, 2012). Plant GPXs play pivotal roles in metalloid detoxification and scavenging; besides using GSH, they can use thioredoxins (Trxs) as their electron donor (Bela *et al.*, 2015). Although GST-mediated As-GSH conjugates have not yet been detected in plants, upregulation of Tau-class GSTs genes is found in As-induced *Arabidopsis* (Abercrombie *et al.*, 2008; Fu *et al.*, 2014). This whole process is well coordinated and integrated, and any alterations in any of the components during ROS metabolism can seriously compromise plants' tolerance mechanisms to metalloid exposure.

Fu *et al.* (2014) studied transcriptional profiling of genes and pathways involved in As toxicity in As(V)-sensitive and tolerant clones of *A. thaliana* and found a number of oxidative stress-related genes were differentially expressed. Notable among these were genes regulating ROS metabolism like alternative oxidase (AOX), DHAR, glutaredoxins (Grxs), Trxs, and class III peroxidases. A number of genes controlling Trxs were found to be induced in tolerant clones, whereas 14 genes governing expression of GSTs were elevated in sensitive clones. Nine GST genes were induced in As-tolerant genotypes. The majority of As-responsive GST genes belong to the 40-member tau class of GSTs, with the remaining genes belonging to phi and lambda subfamilies (Abercrombie *et al.*, 2008; Fu *et al.*, 2014).

In a significant development, Mishra *et al.* (2016) reported the presence of As in the cell nucleus. Using the X-ray source PETRA III, this group beamed X-rays through leaves of the aquatic metal accumulator *Ceratophyllum demersum* exposed to As concentrations between 1 and 5 $\mu mol\,L^{-1}$. By studying the X-ray tomogram they revealed that As was first accumulated in the epidermis and with increasing concentration it gradually entered into the mesophyll of leaves. Significantly, As was found in cell nuclei, and at a concentration of 5 $\mu mol\,L^{-1}$, its presence was revealed in the vacuole and randomly throughout the cell (Mishra *et al.*, 2016). The findings have immense significance in view of the assumption that As can damage the DNA within the cell nucleus by replacing its phosphate, breaking phosphodiester bonds, and thereby causing destabilization of the whole DNA structure and its biological function. So far, no proof has been found regarding As-DNA binding in plant cells; however, As can bind to other karyological ligands and indirectly trigger As-induced genotoxic events (Mishra *et al.*, 2016). This study also pointed out that As-induced inhibition of plant photosynthesis was not due to removal of pigments but to inhibition of their biosynthetic enzymes, particularly 5-aminolevulinic acid dehydratase by As (Mishra *et al.*, 2016).

ROS: Signaling Roles During Metalloid Exposure

The concept of ROS as having only damaging roles in plant cells, as manifested by oxidative stress, has shifted in recent years toward a more balanced view of ROS as signaling molecules in a broad range of plant growth and developmental processes (Foyer and Noctor, 2009; Cuypers *et al.*, 2016). Usually, raised levels of ROS, and H_2O_2 in particular, may lead to cell death (Gechev and Hille, 2004; Petrov and Van Breusegem, 2012; Petrov *et al.*, 2015). This event is crucial during the hypersensitive reaction and senescence in the leaf, and may be an avoidance tactic of plants in response to oxidative stress. However, at low concentrations ROS have the ability to perform a balancing act between inducing oxidative damage and inducing antioxidant defense against their own toxic titers. This act is generally performed through shifting the cellular redox balance to a more oxidized state, alterations in activity of target molecules that can act as primary signals, and oxidation of any biomolecules that have the potential to perform as second messengers during signal transduction (Apel and Hirt, 2004; Petrov and Van Breusegem, 2012; Cuypers *et al.*, 2016). Emerging evidence is indicating that ROS, and H_2O_2 in particular, play dual roles in the metal/metalloid-exposed cell, which employs low levels of ROS as signaling molecules to effectively respond to developmental and environmental cues (Petrov and Van Breusegem, 2012; Cuypers *et al.*, 2016; Farnese *et al.*, 2016).

Mechanistic Versatility in Duality of ROS Action

Diverse manifestations of the roles of ROS molecules during metalloid exposure have been unveiled in plant cells. To perform the signaling roles, the different pools of ROS, which are generated in several cellular compartments, communicate with each other and regulate plant metabolism (Suzuki *et al.*, 2012). The language of this communication, however, is modulated in the plant cell facing oxidative challenge. For example, the apoplastic ROS burst often generated in response to an external stressor is transduced to chloroplast, which upon illumination can trigger a second wave of ROS production (Shapiguzov *et al.*, 2012). Superoxides produced in the apoplast can be converted to H_2O_2, which then can traverse through chloroplastic lipid bilayers. Chloroplastic ROS can send retrograde signaling to the nucleus, which modulates expression of a number of genes related to photosynthesis (de Dios Barajas-López *et al.*, 2013). Among the ROS molecules, H_2O_2 has drawn huge attention as the predominant signaling component due to its longer half-life (1 ms), abundant presence in the cellular environment (up to the low millimolar range), capability to diffuse across membranes, and relatively low toxicity in comparison to superoxides, hydroxyl radicals, and singlet oxygen (Neill *et al.*, 2002; Petrov and Van Breusegem, 2012; Bienert and Chaumont, 2014; Cuypers *et al.*, 2016). With its relatively long half-life, H_2O_2 can travel cellular distances of up to 1 μM, which brings it close to target molecules to perform as a primary messenger (Cuypers *et al.*, 2016). A number of studies indicate dual roles of ROS/H_2O_2 in the regulation of plant stress responses, and most of the primary evidence in this regard comes from H_2O_2 priming experiments. However, the process is extremely delicate as ROS beyond a certain threshold limit can induce oxidative imbalance while at appropriate concentrations it can initiate a signaling cascade to modulate numerous upstream and downstream cellular events, responsive to stressors. Often, it is an uphill task to pinpoint the concentration at which it functions as a signaling molecule due to varied experimental set-ups and lack of a precision technology to measure its real-time concentration at different cellular compartments under stressful environments. Priming of H_2O_2 at low concentrations (100–500 μM) has proved beneficial to minimize metalloid-induced growth inhibition, low biomass production, lipid peroxidation, and programmed cell death (PCD) in different plant species (Hu *et al.*, 2009; Bai *et al.*, 2011). In rice, priming with H_2O_2 alleviated Cd toxicity by reducing metal uptake, upward translocation of metals, and oxidative damage (Chao *et al.*, 2009; Bai *et al.*, 2011). Similarly, pretreatment with H_2O_2 significantly ameliorated Al-induced oxidative damage in wheat seedlings (Xu *et al.*, 2011) and Cu-induced deleterious effects on maize leaves by restoring plant dry matter production, mineral contents, and osmolyte levels (Guzel and Terzi, 2013). Although ultimate manifestations of the outcome of these investigations are mainly related to alleviation of metalloid stress, there are huge mechanistic differences in how exogenous H_2O_2 induced metal tolerance. For example, Al tolerance in H_2O_2-primed wheat seedlings was not only related to stimulated levels of thiol metabolites like GSH but also related to increased AsA biosynthesis (Xu *et al.*, 2011). This effect was, however, not conspicuous in Al-tolerant genotypes. Similarly, the responses of antioxidant enzymes, particularly SOD, APX, GPX, GST, and CAT, are not similar in all H_2O_2-primed plants (Chao *et al.*, 2009; Hu *et al.*, 2009; Bai *et al.*, 2011). Besides reduced metal uptake and translocation and induction of antioxidant defense activities, H_2O_2 priming has been found to be associated with (i) enhanced metal scavenging by PCs and GSH and/or by GSTs; (ii) reduction of ROS levels from toxic cellular concentrations; (iii) restoration of relative water content, water potential, and minerals, sugar, and protein contents; and (iv) maintenance of osmotic adjustment by proline (Guzel and Terzi, 2013). Besides its role as an osmoregulator during metalloid exposure, proline has been envisaged as a metal-chelator and ROS-scavenger (Sharma and Dietz, 2009). High ROS levels have significant effects on the functioning of different antioxidant enzymes. ROS and H_2O_2 in particular at high intracellular concentrations can oxidize two of the sulfur-containing amino acids of almost every protein, like Cu/Zn-SOD and Fe-SOD (Das and Roychoudhury, 2014). Similarly, an abnormal H_2O_2 titer toward the oxidative state can kinetically inhibit the functioning of several APX isoforms (Hiner *et al.*, 2000). The disruption of enzymes' functioning due to ROS-induced irreversible damage, such as protein carbonylation and phosphorylation, as postulated recently, is another way to perceive stress signals and relay them to the plant cell (Hardin *et al.*, 2009; Møller and Sweetlove, 2010; Petrov and Van Breusegem, 2012; Chmielowska-Bąk *et al.*, 2015). However, the identity of H_2O_2 receptors in plant cells is still elusive, and it is our assumption that H_2O_2 can oxidize redox-sensitive transcription factors, which can then directly activate signaling molecules downstream (Mittler *et al.*, 2011; Dietz, 2014). The production of heat-shock factors in different metal-exposed plants suggested the potential involvement of these proteins in perceiving H_2O_2 signals (Cuypers *et al.*, 2011; Mittler *et al.*, 2011). According to a model proposed by Miller *et al.* (2008) different heat-shock proteins may function as "H_2O_2 sensors" upstream of different transcription factors belonging to the WRKY and ZAT families, some members of which, like WRKY25 and ZAT12, were found induced in *Arabidopsis* plants exposed to metals (Opdenakker *et al.*, 2012a). This differential manifestation of a defense response suggests versatility of ROS as signaling molecules, which are intimately associated with almost every cellular process and homeostasis; but it poses important questions about whether the oxidation of cellular proteins and lipids is an inseparable part of the signaling process (Møller and Sweetlove, 2010). Interestingly, most of the beneficial effects of H_2O_2 priming were not conspicuous in the absence of metalloid exposure. This indicates that H_2O_2 at low titers can perceive the stress signals through a "mild oxidative challenge" generated during metal exposure, and is either percolating the effects downstream and upstream or is itself inducing a number of basal antioxidant defense components (Cuypers *et al.*, 2012; Hossain *et al.*, 2015). This reveals a unique cellular phenomenon whereby an exogenous ROS at appropriate concentration can intrinsically regulate its own uncontrolled production through modulation of a number of defense circuits in a highly integrated way. Conversely, this raises the intriguing question of why endogenous ROS, when it passes through this concentration, fails to stabilize its own level and trigger signaling events that can prime plants intrinsically to modulate their defense response against oxidative stress. Comparisons between sensitive and metal-tolerant genotypes in *Arabidopsis* and crop plants indicated (i) a constitutive upregulation of thiol-regenerating enzymes and ROS-scavenging enzymes, and (ii) lower H_2O_2 levels in tolerant genotypes compared to sensitive ones. This suggests that the effect generated by H_2O_2 is modified by an inherent property of plants' metal tolerance, and a genotype-stressor interaction in relation to this modification cannot be ruled out.

ROS and Cellular Signaling Through Premature Senescence

Leaf senescence is a normal physiological process that is tightly regulated in fine tune with other cellular and metabolic events to mobilize the nutrients to growing tissues. Senescence is usually accompanied by degradation of cellular macromolecules like chlorophyll, proteins, lipids, and nucleic acids. Metalloid exposure breaks this tune, accelerating degradation of cellular macromolecules and inducing premature leaf senescence through generation of excess ROS. The regulation of cellular signals leading to leaf senescence, however, involves several factors such as the ROS pool, balancing of plant hormone pools, and regulation of gene expression of several transcription factors responsive to senescence. Increased levels of H_2O_2 and superoxide radical beyond certain limits under metalloid exposure is accompanied by an enhanced ratio of senescence-promoting hormones such as abscisic acid (ABA), ethylene, salicylic acid (SA), and jasmonic acid (JA) over those delaying senescence such as auxins, cytokinins, and gibberellic acids. This process is accompanied by an enhanced rate of lipid peroxidation and membrane leakage, which triggers premature leaf senescence, as was reported in Cd-exposed pea plants, Pb-treated *C. demersum*, Zn-exposed *Hydrilla verticillata*, and plants subjected to As exposure (Rodriguez-Serrano *et al.*, 2009; Gupta *et al.*, 2013b, 2015; Xu *et al.*, 2013; Cuypers *et al.*, 2016 and references therein). Enhanced lipid peroxidation may also be orchestrated through ROS-induced decrease in plants' antioxidant defense under metalloid treatment. This was evident from repressed expression of *CAT2* and an upregulation of *CAT3* isoforms in Cd-exposed *Arabidopsis* seedlings (Cuypers *et al.*, 2011), pointing to possible metal-induced acceleration of premature leaf senescence.

Leaf senescence is the ultimate stage of plants' PCD and is transcriptionally regulated. While several genes encoding photosynthetic proteins are downregulated, genes promoting senescence-associated factors/pathways are upregulated during senescence. These so-called "senescence associated genes," or SAGs, encode proteins that are involved in the breakdown of cellular molecules like proteins, nucleic acids, and cell wall components and in the remobilization of nutrients to developing plant tissues (Zhou *et al.*, 2009). Several transcription factors like ORESARA1 SISTER1 (ORS1) and WRKY53, belonging to the NAC and WRKY families, respectively, are overexpressed during senescence in *Arabidopsis* and are important constituents of SAGs. *Arabidopsis vtc1-1* mutant, having a constitutive deficiency in the total AsA pool and a low AsA redox balance, exhibited enhanced expression of certain SAGs, which was accompanied by an earlier appearance of senescence symptoms, as compared to wild-type seedlings (Barth *et al.*, 2004). Apart from regulating antioxidant defense, the SAG transcription factors, particularly WRKY53, are phosphorylated by MAP3K-MEKK1 in the mitogen activated protein kinase (MAPK) cascade, and their upregulated expression along with MPK6 and MKK9 was strongly confirmed as a factor accelerating senescence in plants under stress (Zhou *et al.*, 2009; Cuypers *et al.*, 2016).

ROS Signaling Through PCD Under Metalloid Exposure

The versatile capability of ROS as signaling molecules is further evidenced in relation to their role in plant cell death (Van Breusegem and Dat, 2006). Accumulating evidence indicates roles for ROS in PCD in plants subject to stressful environments (Gechev and Hille, 2004; Petrov and Van Breusegem, 2012; Petrov *et al.*, 2015). Metalloid-induced increases in ROS levels in roots and photosynthetic organs of different plants have been widely studied and reviewed (Cuypers *et al.*, 2016). However, the roles of ROS in regulating plants' PCD during metal exposure have been recognized only very recently. *Arabidopsis* cell suspension culture and *Nicotiana tabacum* cells exposed to Cd triggered excess H_2O_2 production and subsequent cell death (Garnier *et al.*, 2006; De Michele *et al.*, 2009). The Cd-dependent ROS-mediated PCD in *N. tabacum* bright yellow-2 cells occurs through an endoplasmic reticulum (ER)-dependent mechanism, in which unfolded proteins accumulate (Xu *et al.*, 2013). Upon Cd exposure, several markers of ER stress, like *NtBLP4* and *NtPDI*, were upregulated whereas expression of the *Arabidopsis AtBiP2* gene can suppress PCD-related genes such as *NtHsr203J*. Interestingly, the volatile hormone ethylene, which is modulated by metal-induced ROS production, can also participate in PCD of plants (Petrov *et al.*, 2015). The ROS burst, mitochondrial dysfunction, and onset of PCD all participate in ROS-mediated signaling events during Al exposure (Li and Xing, 2010). Huang *et al.* (2014) summarized some of the ROS-induced PCD programs during Al toxicity: (i) a sharp elevation in gene expression of respiratory burst oxidase homolog (*Rboh*) and RBPH activity; (ii) "opening of the mitochondrial permeability pore (PTP)"; (iii) reduced mitochondrial inner membrane potential; (iv) increase in cytosolic cytochrome *c* level; and (v) induction of the caspase-like 3 protease activity. Reversal of this process in the presence of H_2O_2-scavenger CAT, AsA and its precursor molecules, and NADPH oxidase inhibitors suggests involvement of ROS in the PCD program in the presence of Al (Petrov *et al.*, 2015). Besides the RBOH/NADPH oxidase, Liu *et al.* (2014) recently demonstrated that mitochondrial AOX, which prevents over-reduction of the mitochondrial electron transport chain, is upregulated (*AOX1a*) in the presence of Al-derived superoxide radicals. The AOX1a in turn maintains mitochondrial functions properly and thus, can mitigate the following PCD during metal exposure (Liu *et al.*, 2014). In leaves of Cd-stressed *A. thaliana*, AOX1a regulates oxidative challenge through modulation of the ROS level and associated cellular functions related to PCD (Keunen *et al.*, 2015).

Similar to Cd, the toxicity of tungsten is also mediated through ER stress, but the metal also inhibits plant growth by binding with cellular molybdoenzymes and frequent switching over of the cell to PCD (Brychkova *et al.*, 2008). Most facts in this regard have been unraveled through study of the *Arabidopsis xdh1* mutant (Brychkova *et al.*, 2008). The mutant, deficient for the molybdoenzyme xanthine dehydrogenase (XDH), accumulated xanthine and triggered premature cell death with a high ROS level (Brychkova *et al.*, 2008). A ROS-mediated PCD process is also linked with NO, as was evident in the roots of *Solanum nigrum* under Zn deficiency. Zn is an essential element in plant growth and a constituent of the ROS metabolic enzyme SOD, but high level of Zn can be deleterious for plant growth (Petrov *et al.*, 2015). It is reported that Zn exposure initially provokes NO generation, which in turn negatively regulates antioxidant defense components, thus elevating ROS production and subsequent onset of PCD. The opposite scenario was observed when NO production was inhibited, linking Zn-induced ROS-mediated PCD via NO (Xu *et al.*, 2010). The relationship between metal toxicity, oxidative stress, and cell death has also been confirmed by other studies that strongly suggest that ROS can mediate the signaling process during cell death (Gadjev *et al.*, 2008). This again revealed the intricate relationship between oxidative death and cellular ROS signaling in metal-exposed plants (Cuypers *et al.*, 2016).

ROS in Cross-talk with NO During Metalloid Exposure

The mode of action of ROS and NO and their interaction as signaling molecules during metalloid exposure is now the subject of intense study (Silveira *et al.*, 2015; Thao *et al.*, 2015). During metalloid exposure, the redox state of plant cells is rapidly changed due to rising levels of ROS and NO, which in turn triggers a series of cascading reactions to induce antioxidant defense (Shi *et al.*, 2014a,b; del Río, 2015). ROS and NO have several distinct functions during stress responses, and the enzymes involved in their metabolism are distributed in several cellular compartments. The spatial and temporal differentiation of these enzymes is one of the determining factors of their signaling roles (Groß *et al.*, 2013). Several reports indicate that ROS/H_2O_2 can induce NO generation/degradation and vice versa, although the signaling process is not fully understood. NO may induce H_2O_2 production (Liao *et al.*, 2012; Niu and Liao, 2016) but equally has the ability to reduce H_2O_2 levels during abiotic stress responses (Correa-Aragunde *et al.*, 2015). The NO level was reportedly elevated by stimulated nitrate reductase (NR) activity in Al-exposed wheat seedlings, which modulated the number of antioxidant events under Al exposure (Sun *et al.*, 2014). However, H_2O_2 acted upstream of NO, which consequently mediated antioxidant defense in heat-stressed *Arabidopsis* (Zhang *et al.*, 2007; Wang *et al.*, 2015). Under metal(loid) exposure, ROS can induce NO generation whereas NO can limit ROS accumulation through diverse mechanisms. NO-induced metalloid tolerance mainly involved (i) modulation of antioxidant defense for enhanced ROS detoxification and prevention of consequent oxidative damage; (ii) prevention of metalloid entry into plant cells; (iii) their vacuolar sequestration/compartmentalization; and (iv) post-translational changes in activity of ROS-generating enzymes like glycolate oxidase and NADPH oxidase (Singh *et al.*, 2013; Andrade *et al.*, 2016). Addition of NO increased the activity of SOD up to 110%, which coupled with enhanced AsA-GSH cycle enzymes and CAT, conferred tolerance to sorghum plants exposed to As (Saxena and Shekhawat, 2013). In *Pogonatherum crinitum* root cells, Yu *et al.* (2012) reported increased activity of NR cytosolic isoforms and enhanced tolerance to Pb. On the other hand, in *A. thaliana*, As stimulated both NO production (via NR) and *S*-nitrosoglutathione reductase (GSNOR) activity, indicating a balance between NO generation and its modification as a signaling molecule during As exposure (Leterrier *et al.*, 2012).

Available reports indicate that the effect of NO is percolated through chemical changes of proteins such as *S*-nitrosylation, metal nitrosylation, and tyrosine nitration (Farnese *et al.*, 2016). The first event involves (i) the binding of the NO molecule to cysteine residues in target proteins, and (ii) transferring NO from an *S*-nitrosylated residue to another thiol group through a *trans*-nitrosylation reaction by GSNOR (Lamotte *et al.*, 2015). This is a post-translational modification that can alter structure and function of other biomolecules, thus regulating diverse arrays of cellular functions during plants' metal stress response. *S*-nitrosoglutathione (GSNO) is a significant NO reservoir in plant cells and is formed by *S*-nitrosylation of GSH by NO. GSNO has the capacity to be transported in the phloem, a property that can contribute to the long-distance transport of this signaling molecule, with significant implications for plant stress tolerance. GSNO also regulates the intracellular concentration of NO by influencing the nitrate assimilation pathway. Interestingly, the GSNOR, which controls the cellular GSNO pool through deamination, generates oxidized glutathione (GSSG). The GSSG level is extremely important for thiol redox status as its increasing concentration relative to that of GSH in metalloid-exposed plants can tilt the redox balance in favor of oxidative stress during metalloid exposure (Noctor *et al.*, 2012; Lamotte *et al.*, 2015). Regulation of GSNOR activity is thus an important event during signaling cross-talk between ROS and NO in the presence of stress (Wang *et al.*, 2015). Besides changing cellular protein dynamics, *S*-nitrosylation can also alter gene expression in the presence of ROS through nitrosylation of target proteins like transcription factors, which can modify their affinity with DNA. Similarly, *S*-nitrosylated histone deacetylases, which are responsible for chromatin condensation through the removal of acetyl groups on histones, may alter gene expression related to the nuclear events of the cell (Mengel *et al.*, 2013). However, the homeostasis of *S*-nitrosothiols depends on the balance between nitrosylation and denitrosylation, which besides GSNO/GSNOR is maintained by Trxs/thioredoxin reductase (Trx/TR) systems (Lamotte *et al.*, 2015). The dithiol moiety of Trx reduces nitrosothiols and gets itself oxidized; the process is reversed by TR and NADPH. This indicates an intense interaction between NO and thiol groups, which is involved in regulation of the redox status of thiol groups during stress signaling process (Benhar, 2015). Metal nitrosylation is also a post-translational modification of proteins, but in this case NO has to bind with transition metals in metalloproteins. However, addition of a nitro group from peroxynitrite – which is a product of a reaction between NO and superoxides – to tyrosine residues initiates tyrosine nitration. Nitration of tyrosine residues was initially considered as a stress indicator but recent studies have revealed its signaling role during stress acclimation and tolerance (Mengel *et al.*, 2013).

Like NO, ROS can also transmit stress signals via post-translational modifications of proteins, and the sulfur-rich cysteine moiety is the prime target. ROS-induced oxidative post-translational modifications involve *S*-glutathionylation, disulfide bond formation, and sulfhydration (Akter *et al.*, 2015). The glutathionylation process includes reverse oxidation of cysteine residues and consequent reaction with free protein thiol, which results in the formation of disulfide bonds and covalent attachment with a low molecular weight thiol buffer like GSH. This process was reversed by Grxs and Trx and disrupts the signaling process initiated by glutathionylation (Waszczak *et al.*, 2015). Compared to animals and bacteria, many events of *S*-glutathionylation are still not known in plant systems undergoing stress conditions.

A complex interaction among ROS, NO, Ca^{2+}, and plant hormones has been reported in pea plants subjected to Cd toxicity (Rodriguez-Serrano *et al.*, 2009). Depletion of cytosolic Ca^{2+} in the presence of Cd led to reduced ROS-scavenging capacity on the one hand, and impeded NO production on the other, which cumulatively increased oxidative load in the plant cell.

The ROS accumulation, however, turns on plant hormones like jasmonic acid (JA) and an ethylene-dependent signaling pathway against Cd toxicity (Rodriguez-Serrano *et al.*, 2009).

ROS/H_2O_2 in Dialogue with Hydrogen Sulfide (H_2S) during Metalloid Tolerance

One of the remarkable discoveries in the history of plant biology is the potential role of H_2S as a gasso-signaling molecule in plant growth, development, and abiotic stress tolerance. In plants, apart from the reductive sulfate assimilatory pathway, in which ferredoxin-dependent sulfite reductase converts sulfite to sulfide, H_2S in plant cells may also be produced from cysteine by reverse reaction of *O*-acetylserine thiol lyase, during cyanide detoxification by β-cyanoalanine synthase, and through cysteine desulfhydration by L-cysteine and D-cysteine desulfhydrase activity (Calderwood and Kopriva, 2014). Due to its high degree of permeability through biological membranes, H_2S can freely traverse lipid membrane with positive effects at low concentrations, indicating its role as a signaling molecule. Treatment with H_2S-releasing compounds conferred protection against Cd, Al, and As toxicity (Zhang *et al.*, 2010; Sun *et al.*, 2013; Singh *et al.*, 2015). In most of these cases, exogenous H_2S donor improved the endogenous H_2S pool, which in turn induced antioxidant defense through a stimulated AsA and GSH redox pool, prevented metal uptake, maintained metal homeostasis, ionic balance, and photosynthesis, and finally conferred metal tolerance and achieved acclimation (Mostofa *et al.*, 2015). H_2S may also act as an upstream signaling molecule to stimulate NO production. The available NO pool in turn counterbalances the ROS level and ROS-induced oxidative damage (Li *et al.*, 2013). In contrast, NO activates H_2S production in Cd-stressed *Cynodon dactylon*. This upstream induction of H_2S generation modulated antioxidant defense to reduce ROS levels and lipid peroxidation and conferred metal tolerance (Shi *et al.*, 2014b). The results indicated that cross-talk between NO and H_2S is bidirectional during metal exposure. However, in the absence of any significant "H_2S burst" it is still not clear whether these reported positive effects are from H_2S alone or relayed through downstream antioxidant defense components. In a significant development, Li *et al.* (2013) reported that use of sodium nitroprusside (a NO donor) elevated the endogenous H_2S level by stimulating the L-cysteine desulfhydrase activity in heat-stressed maize seedlings. Enhanced H_2S in turn prevented ROS over-accumulation and membrane leakage, suggesting that H_2S acted downstream of NO in the process. Although, cross-talk among H_2O_2/ROS, NO, and H_2S is likely during plant stress tolerance, the mechanistic details of both upstream and downstream signaling events are not clear. Fang *et al.* (2014) studied the cross-talk between H_2S and Ca^{2+} signaling in Cr-stressed *Setaria italica* plants and observed that Cr^{6+} stress activated endogenous H_2S synthesis as well as Ca^{2+} signaling. The Ca^{2+} signaling in turn induces a complex physiological process including the activation of metal chelators like PC synthase and metallothionein MT3A to confer Cr tolerance. Hancock and Whiteman (2016) observed that this molecular dialogue may occur at various levels, such as influencing the generation of ROS molecules, interacting synergistically, or competing for control of downstream signaling events such as thiol-antioxidant defense and/or NO-generating systems or other back-up defense circuits. However, further study is needed to reveal the complete sequence of interaction between ROS/H_2O_2 and H_2S during metal tolerance of plants.

ROS and the Kinase Cascade in Plant Cells

Kinases may act as signals during metalloid exposure and produce stress-related proteins and/or secondary plant metabolites that can perform either damage-causing roles or stress-countering activities. *Arabidopsis* roots under As(V) exposure exhibited regulated expression of mitogen-activated protein kinase (MAPK), SNF1-related kinases (SnRKs), and leucine-rich repeat receptor-like kinase VIII (LRR-RLK VIII) pathways with higher magnitude in a tolerant genotype than an As-sensitive one (Fu *et al.*, 2014). Indeed, activation of the MAPK cascade has been considered as one of the early signaling events in response to stress stimuli (Baxter *et al.*, 2014). Another protein kinase cascade involving serine/threonine protein kinase oxidative signal-inducible 1 (OXI1) is directly induced by H_2O_2 and constitutes an important component in the signal transduction pathway leading from ROS production to downstream stress responses (Rentel *et al.*, 2004). Increased OXI1 transcription was reported in Cd- or Cu-induced *Arabidopsis* plants (Opdenakker *et al.*, 2012a). Recently, Smeets *et al.* (2013) reported pivotal roles of the OXI1 factor in the *oxi1* mutant of *Arabidopsis* subjected to Cu stress. This kinase cascade is linked to lipid signaling in a phosphoinositide-dependent kinase (PDK1)-related manner and further downstream to MAPKs. Activation of MAPKs comprising of MAP3K, MAP2K, and MAPK takes place through phosphorylation and activating each others' modules during As, Cd, Cu, Pb, Hg, and Zn exposure, and ROS, particularly H_2O_2, play key roles in this activation process (Rentel *et al.*, 2004; Opdenakker *et al.*, 2012b; Islam *et al.*, 2015). Protein phosphorylation is an important early event during the signaling process and occurs quite quickly in response to stress stimuli (Baxter *et al.*, 2014). In rice, As exposure induced expression of the MAPK cascade components OsMPK3, OsMPK4, and OsMKK4 (Rao *et al.*, 2011). Transcript analysis of the MAPK family and immunokinase assay revealed that OsMPK3 transcripts were significantly induced in leaves, and OsMPK3 and OsMPK4 transcripts in roots. Among the MAPK kinase (MKK) gene family, OsMKK4 transcripts were upregulated in As(III)-treated rice leaves and roots (Rao *et al.*, 2011). In the Japonica rice cultivar "Nipponbare," elevated concentrations of Cu, Cd, and Hg induced activation of a novel MAPK gene *OsMSRMK2* (Agrawal *et al.*, 2002). Similarly, Yeh *et al.* (2004) reported activation of a MAPK gene and MBP kinase upon Cd exposure. In another study, Cu treatment increased mRNA levels of OsMAPK2, whose activity was prevented in the presence of GSH (Yeh *et al.*, 2003). In an interesting study, Jonak *et al.* (2004) found differential kinetics of the MAPK cascade in *Medicago sativa* seedlings under Cd and Cu exposure. Rapid activation of MPK3 and MPK6 was observed in the presence of Cu and Cd (Jonak *et al.*, 2004). Gupta *et al.* (2009) reported the activation of MAPK in two varieties of Indian mustard under As(III) treatment, indicating a role for the MAPK cascade in transducing As(III)-mediated signals. Heavy metal-induced activation of the MAPK cascade was also demonstrated in maize seedlings (Wang *et al.*, 2010). *Arabidopsis* seedlings subjected to Cd or Cu stress exhibited H_2O_2-mediated activation of MPK3 and MPK6 isoforms of MAPK in the presence of OXI1 kinase but failed to do so in plants lacking OXI1 (Opdenakker *et al.*, 2012a). Besides OXI1 kinase, H_2O_2 can also activate the MAP3K *Arabidopsis* NPK1-like protein kinase 1 (ANP1) by inducing MPK3- and MPK6-mediated phosphorylation. This was observed in Cu-induced *A. thaliana* roots after 6 and 24 h of treatment (Opdenakker *et al.*, 2012a). Several lines of evidence suggest that MAPK acts both upstream and downstream of ROS, and that MAPK activated by H_2O_2 can trigger an H_2O_2-mediated oxidative burst (Petrov and Van Breusegem, 2012). This was observed in *Nicotiana* MEK2, which is implicated in ROS production upstream of the respiratory burst homolog (RBOH) gene upon fungal infection (Yoshioka *et al.*, 2003). RBOH with its multigene families has been shown to be the most powerful ROS-signal generator; it triggers a ROS wave and can initiate rapid cell-to-cell systemic signaling (Mittler *et al.*, 2011). However, RBOH actions are tightly controlled by different cellular factors under abiotic stressors, and the temporal and spatial regulation of the ROS wave during metal/metalloid exposure in plants is still not known.

Besides OXI1, MPK3, and MPK6, information regarding the regulation of other transcription factors and signaling pathways by H_2O_2 during metal/metalloid exposure is rather scanty. Available evidence indicates that activation of different kinase cascades by H_2O_2 transforms the ROS signals to activate different downstream antioxidant defense components. MAPK-regulated ZAT12 resulted in increased expression of the cytosolic APX isoform (*APX1*) and

CAT1 genes in response to H_2O_2 in *A. thaliana* plants exposed to Cu, Cd, or Zn (Remans *et al.*, 2010; Cuypers *et al.*, 2011). These two enzymes are vital in cellular ROS homeostasis and thus, are the prime target of MAPK-mediated regulation. Although the exact sequence of events is still not known, it is postulated that MAPK-regulated *RBOH* expression plays vital roles in amplification of the ROS signal, which is strong enough to be sensed by different signaling cascades in metal-exposed *Arabidopsis* plants (Remans *et al.*, 2010; Cuypers *et al.*, 2011, 2012).

The calcium-dependent protein kinase (CDPK) families play pivotal roles in signaling during diverse events of plant growth and development (Boudsocq and Sheen, 2013; Schulz *et al.*, 2013). A "reciprocal relationship" between Ca^{2+} and H_2O_2 signaling has been implicated in multiple stress conditions (Petrov and Van Breusegem, 2012; Cuypers *et al.*, 2016), and increased flux of Ca^{2+} into cytosol is considered one of the early signaling events under stress conditions (Baxter *et al.*, 2014). ROS have been found to modulate cytosolic Ca^{2+} levels via activation of plasma membrane-bound Ca^{2+} channels. However, H_2O_2 production by plasma membrane NADPH oxidase reversely depends on Ca^{2+} (Mazars *et al.*, 2010). H_2O_2 mediated an increase in cytosolic Ca^{2+}, which in turn stimulated expression of H_2O_2-sensitive *GST1* in Cd-exposed *Arabidopsis* and *Nicotiana* bright yellow-2 seedlings (Rentel and Knight, 2004; Garnier *et al.*, 2006). The study pointed out the close cross-talk between Ca^{2+} and the H_2O_2-generating NADPH oxidase system. Ca^{2+} can directly bind to elongation factor (EF)-hand motifs in the N-terminus domain and phosphorylate this domain by activating CDPK, one isoform of which, *CDPK1*, was found induced in Cd-exposed *A. thaliana* plants (Smeets *et al.*, 2013). Similarly, several CDPK isoforms were found induced by H_2O_2 in wheat seedlings, and a relationship within ROS/H_2O_2-CDPK-ROS scavengers and with NO has been revealed in Cd-exposed pea plants (Rodríguez-Serrano *et al.*, 2009). González *et al.* (2012) revealed that in the green alga *Ulva compressa*, Cu exposure led to H_2O_2 accumulation, which in turn triggered a series of signaling events involving NO production and activation of gene expression related to calmodulins and the CDPK cascade. Recent work of Baliardini *et al.* (2015) unraveled a positive correlation between Cd tolerance and the expression of a gene encoding a Ca^{2+}/H^+ exchanger (CAX1) in *Arabidopsis*. A Ca^{2+} signaling cascade depicting cross-talk with ROS and NO for metal tolerance is presented in Figure 3.1.

ROS in Cross-talk with Plant Hormone Signaling

Plant hormones can diffuse and thus transmit stress signals under varying ROS levels (Yoo *et al.*, 2009). The volatile plant hormone ethylene participates in signaling cascades, regulating both development and responses to stress (Yoo *et al.*, 2009; Iqbal *et al.*, 2013). Coupled with enhanced biosynthesis, induction of the ethylene signaling cascade was found in plants exposed to heavy metals like Cd and Cu (Arteca and Arteca, 2007; DalCorso *et al.*, 2008). Cd-induced enhanced expression of genes related to ACC (1-amino cyclopropane-1-carboxylic acid) synthase and ACC oxidase was observed in *Arabidopsis* plants. Moreover, genes involved in ethylene responsive factors, ERF2 and ERF5, were induced by Cd exposure (Sandalio *et al.*, 2009). Microarray data revealed elevation of genes coding for ethylene-signaling components in roots of As-tolerant *Arabidopsis* seedlings that exhibited elevation of genes involved in ROS metabolism and homeostasis (Fu *et al.*, 2014). A transcription factor gene regulating ERF was found associated with As-tolerance mechanisms. Interestingly, regulatory interaction between ethylene and the reductive sulfur assimilation pathway was found in association with ROS balance. In the presence of high levels of ROS, sulfur assimilatory genes were downregulated, whereas a weak response was manifested by ERF in an As-sensitive genotype. However, the opposite reactions were found in a tolerant genotype (Fu *et al.*, 2014).

Metalloid-induced phytohormone signaling has extensive cross-talk with the MAPK cascade. One of the important constituents in phytohormone signaling is gaseous ethylene, which in

Metalloid exposure (Cu, Zn, Cd, As)

↓

Activation of calmodulins/CDPK/ calcineurin B-like protein kinase

↓

Acquisition of signals of Ca^{2+}

↓ Calmodulins

Activation of mitochondrial NADH-synthesizing enzymes (isocitrate dehydrogenase, 2-oxoglutarate dehydrogenase) in Krebs cycle and pyruvate dehydrogenase

↓

Increase in NADH concentration

↓

Enhancement of mitochondrial electron transport

↓

Increased H_2O_2 and superoxide levels

↓

Enhanced entry of Ca^{2+} via activated Ca-channel/phosphorylation via CDPK

↕ Calmodulins

Metalloid-----Cu, Cd, Al, As, Zn → H_2O_2 production → Increased NO production via NO synthase like activity ↔ Ca^{2+} release → Activation of Ca-channel by H_2O_2 and NO

↓

Calmodulins and CDPK-dependent activation of antioxidant enzymes (SOD, GR, GST, PRX, GRX) by upregulation of gene expression

↓

Metalloid tolerance and acclimation

Figure 3.1 Cascades of cross-talk among reactive oxygen species (ROS), Ca^{2+}, and nitric oxide (NO) involving the Embden–Meyerhof–Parnas (EMP) pathway and Ca-dependent kinase cascade during metalloid exposure of plants. CDPK, calcium-dependent protein kinase; GR, glutathione reductase; GRX, glutaredoxin; GST, glutathione *S*-transferase; PRX, peroxiredoxin; SOD, superoxide dismutase.

recent years has been credited with significant roles in metal tolerance (reviewed by Keunen *et al.*, 2016). The activated MPK3 and MPK6 of the MAPK cascade can phosphorylate isoforms 2 and 6 of 1-aminocyclopropane-1-carboxylate synthase (ACS) and enhance ethylene production by increasing isoform half-life (Joo *et al.*, 2008). Studies indicate that ethylene signaling can be promoted by increasing the transcription of both ACS isoforms via the WRKY33 transcription factor, and by phosphorylating ethylene-insensitive 3 (EIN3), the nuclear transcription factor in metal-exposed *A. thaliana* via MPK3/6 and MKK9-MPK3/6 (Yoo *et al.*, 2008; Li *et al.*, 2012; Schellingen *et al.*, 2015). The cascade of events leading to enhanced ethylene production during metal exposure is depicted in Figure 3.2.

Cadmium

↓

Activation of NADPH oxidase

↓

Oxidative burst through ROS

↓

Stress signals picked up by OXI1

↓ Phosphorylation of MAPK cascade

Activation of MPK3 and MPK6

↓

Transcriptional upregulation of ACS2 and ACS6 isoforms in ethylene biosynthetic pathway

↓

Increased ethylene production

Figure 3.2 A simplified representation of the ethylene production cascade under cadmium exposure. ACS, 1-aminocyclopropane-1-carboxylate synthase; MAPK, mitogen-activated protein kinase; MPK, mitogen-activated kinase; OXI1, serine/threonine protein kinase oxidative signal-inducible 1.

Besides ethylene, altered production of other hormones like abscisic acid (ABA), auxins, JA, and SA was observed in different plant species during metalloid exposure (reviewed by Cuypers *et al.*, 2016). An increased level of JA was reported in Cd- or Cu-exposed *A. thaliana* and *Phaseolus coccineus* plants, suggesting a role for JA signaling in metal stress responses (Maksymiec, 2007). MeJA (methyl JA), a volatile derivative of JA, was found to upregulate the genes involved in GSH synthesis and induced H_2O_2 generation, which was also elevated in Cd- or Cu-exposed *A. thaliana*. SA-induced signaling during plants' exposure to metals has been studied but the interaction between ROS/H_2O_2 and SA is not clear. A high SA level often induces elevated H_2O_2 concentration in SA-accumulating plants, whereas the opposite phenomenon was observed in SA-deficient plants. SA alleviates Cd-induced growth inhibition and oxidative damage by inducing H_2O_2 signaling, which: (i) involves repair processes or Cd binding and compartmentation; and (ii) increases levels of thiol antioxidants like GSH through the activation of serine acetyltransferase (SAT) in the cysteine synthase complex. Similar mechanisms have been proposed in Cd-tolerance of rice, barley, and soybean and Ni-tolerance of *Thalaspi* (Sandalio *et al.*, 2009). Further, links between JA/ABA and the ROS signaling pathway have also been established, but conflicting outcomes from different experimental conditions prevents the drawing of a clear-cut relationship between them. A transient increase in the H_2O_2 level by ABA treatment induces tolerance to oxidative stress. Acquisition of systemic acquired acclimation of stressed plants was correlated with a ROS wave and ABA accumulation in *Arabidopsis* (Suzuki *et al.*, 2013). Zhang *et al.* (2011) established that maize ABP9 protein can enhance multiple stress tolerance in engineered *Arabidopsis* plants by modulating ROS accumulation and ABA signaling. It is now evidently clear that phytohormone signaling can go through redox switches that determine metal tolerance/sensitivity via changes in both ROS and redox homeostasis (Petrov and Van Breusegem, 2012).

The interaction between ROS and phytohormone during metalloid exposure is also exhibited by activation of a different biosynthetic pathway leading to production of oxylipins in a signaling system mediated by JA and MeJA. Metal-induced ROS can induce oxylipin production via oxidation of polyunsaturated fatty acids (PUFAs) through enzymatic activity of lipooxygenase (LOX) and a non-enzymatic reaction (Dave and Graham, 2012). Accumulating evidence indicates that ROS-induced non-enzymatically produced oxylipins, such as hydroxy fatty acids and phytoprostanes, play important roles as signaling molecules through activation of a number of genes including CORONATINE-INSENSITIVE1 (COI1) during oxidative stress (Eckardt, 2008). The reactive electrophile species (RES) possessing a reactive α,β-unsaturated carbonyl structure have recently been classified as a subgroup of oxylipins and have the capacity to modify cellular proteins directly by binding to free thiol groups. This property is not common in weakly electrophilic compounds such as JA, indicating a differential pathway of oxylipin and JA signaling in plant cells (Eckardt, 2008). LOX transcripts and corresponding enzyme activity were found in Cd- or Cu-induced barley and *Arabidopsis* (Tamás *et al.*, 2009; Remans *et al.*, 2010). In *Taxus chinensis*, exogenous MeJA induced NO production, which in turn stimulated H_2O_2 production, LOX activity, and lipid peroxidation (Wang and Wu, 2005). The results, however, indicated ROS-induced oxylipin signaling in an interaction with JA signaling events during the stress response of plants (Dave and Graham, 2012).

Challenges and Future Prospects

With the advances of precision technology in plant ROS biology, the mechanism of the dual nature of ROS functions have gradually been revealed. The complexity of functional biology of ROS is due to its versatile source of generation and mode of actions in different cellular compartments. The origin-function relationship of ROS actions is complicated during metalloid exposure due to diverse chemical properties and different modes of action of metals in plant cells. While excess metal-induced damaging effects are attributed to toxic concentrations of ROS, its quick "burst" or "wave" can initiate a complex network of signaling events that can ultimately confer metal acclimation and tolerance. Although studies using different mutants with different inhibitors unravel significant links between ROS and other cellular signals mediated by NO, hormones, H_2S, and Ca^{2+} via cellular and molecular events, detailed interactions among these players are still not fully known. Furthermore, with emerging concepts and evidence of sugar and microRNA (miRNA) signaling in plant abiotic stress tolerance, future research is needed to unearth the signaling dialogue between ROS and plant sugars as well as miRNA during metalloid exposure. Soluble sugars are an easy source of energy, which is urgently required to meet requirements during oxidative challenge. But the information in this regard is scanty.

The temporal and spatial regulation of ROS and cell-to-cell communication during relay of stress signals also poses serious challenges. While forward signaling events are gradually becoming clearer, retrograde signaling events from cell organelles to nucleus are a topic of intense debate. Thus, a holistic approach should be made to dissect the fundamentals in duality of ROS actions during the metalloid stress response of plants.

References

Abercrombie JM, Halfhill MD, Ranjan P, *et al.* (2008) Transcriptional responses of *Arabidopsis thaliana* plants to As (V) stress. *BMC Plant Biol* 8: 87.

Achary VMM, Jena S, Panda KK, Panda BB. (2008) Aluminium induced oxidative stress and DNA damage in root cells of *Allium cepa* L. *Ecotoxicol Environ Saf* 70: 300–310.

Agrawal GK, Rakwal R, Iwahashi H. (2002) Isolation of novel rice (*Oryza sativa* L.) multiple stress responsive MAP kinase gene, *OsMSRMK2*, whose mRNA accumulates rapidly in response to environmental cues. *Biochem Biophys Res Commun* 294:1009–1016.

Akter S, Huang J, Waszczak C, *et al.* (2015) Cysteines under ROS attack in plants: a proteomics view. *J Exp Bot* 66: 2935–2944.

Apel K, Hirt H. (2004) Reactive oxygen species: metabolism, oxidative stress, and signal transduction. *Annu Rev Plant Biol* 55: 373–399.

Andrade HM, Oliveira JA, Farnese FS, *et al.* (2016) Arsenic toxicity: cell signalling and the attenuating effect of nitric oxide in *Eichhornia crassipes. Biol Plant* 60: 173–180.

Anjum NA, Umar S, Chan M-T. (2010) *Ascorbate-Glutathione Pathway and Stress Tolerance in Plants*. Springer, Dordrecht, The Netherlands.

Anjum NA, Umar S, Iqbal M, Khan NA. (2011) Cadmium causes oxidative stress in mungbean [*Vigna radiata* (L.) Wilczek] by affecting antioxidant enzyme systems and ascorbate-glutathione cycle metabolism. *Russ J Plant Physiol* 58: 92–99.

Anjum NA, Ahmad I, Mohmood I, *et al.* (2012) Modulation of glutathione and its related enzymes in plants' responses to toxic metals and metalloids—a review. *Environ Exp Bot* 75: 307–324.

Anjum NA, Umar S, Iqbal M. (2014) Assessment of cadmium accumulation, toxicity, and tolerance in Brassicaceae and Fabaceae plants: implications for phytoremediation. *Environ Sci Pollut Res* 21: 10286–10293.

Arteca RN, Arteca JM. (2007) Heavy-metal-induced ethylene production in *Arabidopsis thaliana. J Plant Physiol* 164: 1480–1488.

Bai X-J, Liu L-J, Zhang C-H, Ge Y, Cheng W-D. (2011) Effect of H_2O_2 pretreatment on Cd tolerance of different rice cultivars. *Rice Sci* 18: 29–35.

Baliardini C, Meyer C-L, Salis P, Saumitou-Laprade P, Verbruggen N. (2015) CATION EXCHANGER1 cosegregates with cadmium tolerance in the metal hyperaccumulator *Arabidopsis halleri* and plays a role in limiting oxidative stress in *Arabidopsis* spp. *Plant Physiol* 169: 549–559.

Barth C, Moeder W, Klessig DF, Conklin PL. (2004) The timing of senescence and response to pathogens is altered in the ascorbate deficient *Arabidopsis* mutant *vitamin c-1. Plant Physiol* 134: 1784–1792.

Baxter A, Mittler R, Suzuki N. (2014) ROS as key players in plant stress signalling. *J Exp Bot* 65: 1229–1240.

Bela K, Horváth E, Gallé Á, Szabados L, Tari I, Csiszár J. (2015) Plant glutathione peroxidases: Emerging role of the antioxidant enzymes in plant development and stress responses. *J Plant Physiol* 176: 192–201.

Benhar M. (2015) Nitric oxide and the thioredoxin system: a complex interplay in redox regulation. *Biochim Biophys Acta* 1850: 2476–2484.

Bienert GP, Chaumont F. (2014) Aquaporin-facilitated transmembrane diffusion of hydrogen peroxide. *Biochim Biophys Acta* 1840: 1596–1604.

Boudsocq M, Sheen J. (2013) CDPKs in immune and stress signaling. *Trends Plant Sci* 18: 30–40.

Brychkova G, Alikulov Z, Fluhr R, Sagi M. (2008) A critical role for ureides in dark- and senescence-induced purine remobilization is unmasked in the Atxdh1 *Arabidopsis* mutant. *Plant J* 54: 496–509.

Calderwood A, Kopriva S. (2014) Hydrogen sulfide in plants: From dissipation of excess sulfur to signaling molecule. *Nitric Oxide* 41: 72–78.

Chao YY, Hsu YT, Kao CH. (2009) Involvement of glutathione in heat shock- and hydrogen peroxide-induced cadmium tolerance of rice (*Oryza sativa* L.) seedlings. *Plant Soil* 318: 37–45.

Chmielowska-Bąk J, Izbiańska K, Deckert J. (2015) Products of lipid, protein and RNA oxidation as signals and regulators of gene expression in plants. *Front Plant Sci* 6: 405.

Correa-Aragunde N, Foresi N, Lamattina L. (2015) Nitric oxide is a ubiquitous signal for maintaining redox balance in plant cells: regulation of ascorbate peroxidase as a case study. *J Exp Bot* 66: 2913–2921.

Cuypers A, Smeets K, Ruytinx J, *et al.* (2011) The cellular redox state as a modulator in cadmium and copper responses in *Arabidopsis thaliana* seedlings. *J Plant Physiol* 168: 309–316.
Cuypers A, Keunen E, Bohler S, *et al.* (2012) Cadmium and copper stress induce a cellular oxidative challenge leading to damage versus signalling. In: Gupta DK, Sandalio LM (eds), *Metal Toxicity in Plants: Perception, Signaling and Remediation*. Springer-Verlag GmbH, Berlin, Heidelberg, pp. 65–90.
Cuypers A, Hendrix S, dos Reis R, *et al.* (2016). Hydrogen peroxide, signaling in disguise during metal phytotoxicity. *Front Plant Sci* 7: 470.
DalCorso G, Farinati S, Maistri S, Furini A. (2008) How plants cope with cadmium: Staking all on metabolism and gene expression. *J Integr Plant Biol* 50: 1268–1280.
Das K, Roychoudhury A. (2014) Reactive oxygen species (ROS) and response of antioxidants as ROS-scavengers during environmental stress in plants. *Front Environ Sci* 2: 53.
Dave A, Graham IA. (2012) Oxylipin signaling: a distinct role for the jasmonic acid precursor cis-(+)-12-oxo-phytodienoic acid (cis-OPDA). *Front Plant Sci* 3: 42.
de Dios Barajas-López J, Blanco NE, Strand Å. (2013) Plastid-to-nucleus communication: signals controlling the running of the plant cell. *Biochim Biophys Acta* 1833: 425–437.
del Río LA. (2015) ROS and RNS in plant physiology: an overview. *J Exp Bot* 66: 2827–2837.
De Michele R, Vurro E, Rigo C, *et al.* (2009) Nitric oxide is involved in cadmium-induced programmed cell death in *Arabidopsis* suspension cultures. *Plant Physiol* 150: 217–228.
Dietz K-J. (2014) Redox regulation of transcription factors in plant stress acclimation and development. *Antioxid Redox Signal* 21: 1356–1372.
Eckardt NA. (2008) Oxylipin signaling in plant stress responses. *Plant Cell* 20: 495–497.
Fang H, Jing T, Liu Z, Zhang L, Jin Z, Pei Y. (2014) Hydrogen sulfide interacts with calcium signaling to enhance the chromium tolerance in *Setaria italica*. *Cell Calcium* 56: 472–481.
Farnese FS, Menezes-Silva PE, Gusman GS, Oliveira JA. (2016) When bad guys become good ones: The key role of reactive oxygen species and nitric oxide in the plant responses to abiotic stress. *Front Plant Sci* 7: 471.
Finnegan PM, Chen W. (2012) Arsenic toxicity: the effects on plant metabolism. *Front Physio* 3: 182.
Foyer CH, Noctor G. (2009) Redox regulation in photosynthetic organisms: signaling, acclimation, and practical implications. *Antiox Redox Signal* 11: 861–905.
Foyer CH, Noctor G. (2011) Ascorbate and glutathione: the heart of the redox hub. *Plant Physiol* 155: 2–18.
Foyer CH, Noctor G. (2012) Managing the cellular redox hub in photosynthetic organisms. *Plant Cell Environ* 35: 199–201.
Fu S-F, Chen P-Y, Nguyen QTT, *et al.* (2014) Transcriptome profiling of genes and pathways associated with arsenic toxicity and tolerance in *Arabidopsis*. *BMC Plant Biol* 14: 94.
Gadjev I, Stone JM, Gechev TS. (2008) Programmed cell death in plants: new insights into redox regulation and the role of hydrogen peroxide. *Int Rev Cell Mol Biol* 270: 87–144.
Garnier L, Simon-Plas F, Thuleau P, *et al.* (2006) Cadmium affects tobacco cells by a series of three waves of reactive oxygen species that contribute to cytotoxicity. *Plant Cell Environ* 29: 1956–1969.
Gechev TS, Hille J. (2004) Hydrogen peroxide as a signal controlling plant programmed cell death. *J Cell Biol* 168: 17–20.
Gill SS, Tuteja N. (2010) Reactive oxygen species and antioxidant machinery in abiotic stress tolerance in crop plants. *Plant Physiol Biochem* 48: 909–930.
González A, de los Ángeles Cabrera M, *et al.* (2012) Cross talk among calcium, hydrogen peroxide, and nitric oxide and activation of gene expression involving calmodulins and calcium-dependent protein kinases in *Ulva compressa* exposed to copper excess. *Plant Physiol* 158: 1451–1462.

Groß F, Dumer J, Gaupels F. (2013) Nitric oxide, antioxidants and prooxidants in plant defence responses. *Front Plant Sci* 29: 419.

Gupta DK, Tohoyama H, Joho M, Inouhe M. (2004) Changes in the levels of phytochelatins and related metal-binding peptides in chickpea seedlings exposed to arsenic and different heavy metal ions. *J Plant Res* 117: 253–256.

Gupta DK, Inouhe M, Rodríguez-Serrano M, Romero-Puerta MC, Sandalio LM. (2013a) Oxidative stress and arsenic toxicity: Role of NADPH oxidases. *Chemosphere* 90: 1987–1996.

Gupta DK, Huang HG, Nicoloso FT, *et al.* (2013b) Effect of Hg, As and Pb on biomass production, photosynthetic rate, nutrients uptake and phytochelatin induction in *Pfaffia glomerata*. *Ecotoxicology* 22: 1403–1412.

Gupta DK, Palma JM, Corpas FJ. (2015) In: *Reactive Oxygen Species and Oxidative Damage in Plants Under Stress*. Springer-Verlag, Germany, pp 1–22.

Gupta KL, Igamberdiev AU. (2015) Compartmentalization of reactive oxygen species and nitric oxide production in plant cells: an overview. In: Gupta KJ, Igamberdiev AU (eds), *Reactive Oxygen and Nitrogen Species Signaling and Communication in Plants*. Springer International Publishing, pp. 215–237.

Gupta M, Sharma P, Sarin NB, Sinha AK. (2009) Differential response of arsenic stress in two varieties of *Brassica juncea* L. *Chemosphere* 74: 1201–1208.

Guzel S, Terzi R. (2013) Exogenous hydrogen peroxide increases dry matter production, mineral content and level of osmotic solutes in young maize leaves and alleviates deleterious effects of copper stress. *Bot Stud* 54: 10.

Halliwell B. (2006) Reactive species and antioxidants. Redox biology is a fundamental theme of aerobic life. *Plant Physiol* 141: 312–322.

Hancock JT, Whiteman M. (2016) Hydrogen sulfide signaling: interactions with nitric oxide and reactive oxygen species. *Ann NY Acad Sci* 1356: 5–14.

Hardin SC, Larue CT, Oh M-H, Jain V, Huber SC. (2009) Coupling oxidative signals to protein phosphorylation via methionine oxidation in Arabidopsis. *Biochem J* 422: 305–312.

Hiner ANP, Rodríguez-López JN, Arnao MB, Raven EL, García-Cánovas F, Acosta M. (2000) Kinetic study of the inactivation of ascorbate peroxidase by hydrogen peroxide. *Biochem J* 348: 321–328.

Hossain MA, Bhattacharjee S, Armin SM, *et al.* (2015) Hydrogen peroxide priming modulates abiotic oxidative stress tolerance: insights from ROS detoxification and scavenging. *Front Plant Sci* 6: 420.

Hu YL, Ge Y, Zhang CH, Ju T, Cheng WD. (2009) Cadmium toxicity and translocation in rice seedlings are reduced by hydrogen peroxide pretreatment. *Plant Growth Regul* 59: 51–61.

Huang W, Yang X, Yao S, *et al.* (2014) Reactive oxygen species burst induced by aluminum stress triggers mitochondria-dependent programmed cell death in peanut root tip cells. *Plant Physiol Biochem* 82: 76–84.

Iqbal NMA, Khan MIR, Asgher M, Fatma M, Khan NA. (2013) Cross-talk between sulfur assimilation and ethylene signaling in plants. *Plant Signal Behav* 8: 1–9.

Islam E, Khan MT, Irem S. (2015) Biochemical mechanisms of signaling: Perspectives in plants under arsenic stress. *Ecotox Environ Saf* 114: 126–133.

Jonak C, Nakagami H, Hirt H. (2004) Heavy metal stress. Activation of distinct mitogen-activated protein kinase pathways by copper and cadmium. *Plant Physiol* 136: 3276–3283.

Joo S, Liu Y, Lueth A, Zhang S. (2008) MAPK phosphorylation induced stabilization of ACS6 protein is mediated by the non-catalytic C terminal domain, which also contains the cis-determinant for rapid degradation by the 26S proteasome pathway. *Plant J* 54: 129–140.

Keunen E, Schellingen K, Van Der Straeten D, *et al.* (2015) ALTERNATIVE OXIDASE1a modulates the oxidative challenge during moderate Cd exposure in *Arabidopsis thaliana* leaves. *J Exp Bot* 66: 2967–2977.

Keunen E, Schellingen K, Vangronsveld J, Cuypers A. (2016) Ethylene and metal stress: small molecule, big impact. *Front Plant Sci* 7: 23.

Lamotte O, Bertoldo JB, Besson-Bard A, *et al.* (2015) Protein S-nitrosylation: specificity and identification strategies in plants. *Front Chem* 2: 114.

Leterrier M, Airaki M, Palma JM, Chaki M, Barroso JB, Corpas FJ. (2012) Arsenic triggers the nitric oxide (NO) and S-nitrosoglutathione (GSNO) metabolism in *Arabidopsis. Environ Pollut* 166: 136–143.

Li G, Meng X, Wang R, *et al.* (2012) Dual level regulation of ACC synthase activity by MPK3/MPK6 cascade and its downstream WRKY transcription factor during ethylene induction in *Arabidopsis. PLoS Genet* 8: e1002767.

Li Z, Xing D. (2010) Mitochondrial pathway leading to programmed cell death induced by aluminum phytotoxicity in *Arabidopsis. Plant Signal Behav* 5: 1660–1662.

Li Z-G, Yang S-Z, Long W-B, Yang G-X, Shen Z-Z. (2013) Hydrogen sulphide may be a novel downstream signal molecule in nitric oxide-induced heat tolerance of maize (*Zea mays* L.) seedlings. *Plant Cell Environ* 36: 1564–1572.

Liao WB, Huang GB, Yu JH, Zhang ML. (2012) Nitric oxide and hydrogen peroxide alleviate drought stress in marigold explants and promote its adventitious root development. *Plant Physiol Biochem* 58: 6–15.

Liu J, Li Z, Wang Y, Xing D. (2014) Overexpression of ALTERNATIVE OXIDASE1a alleviates mitochondria-dependent programmed cell death induced by aluminium phytotoxicity in *Arabidopsis. J Exp Bot* 65: 4465–4478.

Maksymiec W. (2007) Signaling responses in plants to heavy metal stress. *Acta Physiol Plant* 29: 177–187.

Mazars C, Thuleau P, Lamotte O, Bourque S. (2010) Cross-talk between ROS and calcium in regulation of nuclear activities. *Mol Plant* 3: 706–718.

Mengel A, Chaki M, Shekariesfahlan A, Lidermayr C. (2013) Effect of nitric oxide on gene transcription – S-nitrosylation of nuclear proteins. *Front Plant Sci* 4: 293.

Miller G, Shulaev V, Mittler R (2008) Reactive oxygen signaling and abiotic stress. *Physiol Plantarum* 133: 481–489.

Mishra S, Stärk H-J, Küpper H. (2014) A different sequence of events than previously reported leads to arsenic-induced damage in *Ceratophyllum demersum* L. *Metallomics* 6: 444–454.

Mishra S, Alfeld M, Sobotka R, Andresen E, Falkenberg G, Küpper H. (2016) Analysis of sub-lethal arsenic toxicity to *Ceratophyllum demersum*: Subcellular distribution of arsenic and inhibition of chlorophyll biosynthesis. *J Exp Bot* 67: 4639–4646.

Mittler R, Vanderauwera S, Suzuki N, *et al.* (2011) ROS signaling: the new wave? *Trends Plant Sci* 16: 300–309.

Møller IM, Sweetlove LJ. (2010) ROS signalling – specificity is required. *Trends Plant Sci* 15: 370–374.

Møller IM, Jensen PE, Hansson A. (2007) Oxidative modifications to cellular components in plants. *Annu Rev Plant Biol* 58: 459–481.

Monteiro C, Santos C, Pinho S, Oliveira H, Pedrosa T, Dias MC. (2012). Cadmium-induced cyto- and genotoxicity are organ-dependent in lettuce. *Chem Res Toxicol* 25: 1423–1434.

Mostofa MG, Rahman A, Ansary Md.MU, Watanabe A, Fujita M, Tran L-SP. (2015) Hydrogen sulfide modulates cadmium-induced physiological and biochemical responses to alleviate cadmium toxicity in rice. *Sci Rep* 5: 14078.

Nagajyoti PC, Lee KD, Sreekanth KVM. (2010) Heavy metals, occurrence and toxicity for plants: a review. *Environ Chem Lett* 8: 199–216.

Neill SJ, Desikan R, Clark A, Hurst RD, Hancock JT. (2002) Hydrogen peroxide and nitric oxide as signaling molecules in plants. *J Exp Bot* 53: 1237–1247.

Niu L, Liao W. (2016) Hydrogen peroxide signaling in plant development and abiotic responses: Crosstalk with nitric oxide and calcium. *Front Plant Sci* 7: 230.

Noctor G, Mhamdi A, Chaouch S, *et al.* (2012) Glutathione in plants: an integrated overview. *Plant Cell Environ* 35: 454– 484.

Opdenakker K, Remans T, Keunen E, Vangronsveld J, Cuypers A. (2012a) Exposure of *Arabidopsis thaliana* to Cd or Cu excess leads to oxidative stress mediated alterations in MAPKinase transcript levels. *Environ Exp Bot* 83: 53–61.

Opdenakker K, Remans T, Vangronsveld J, Cuypers A. (2012b) Mitogen activated protein (MAP) kinases in plant metal stress: regulation and responses in comparison to other biotic and abiotic stresses. *Int J Mol Sci* 13: 7828–7853.

Pereira LB, Mazzanti CM, Gonçalves JF, *et al.* (2010). Aluminum-induced oxidative stress in cucumber. *Plant Physiol Biochem* 48: 683–689.

Petrov VD, Van Breusegem F. (2012) Hydrogen peroxide – a central hub for information flow in plant cells. *AoB Plants* 2012: pls014.

Petrov V, Hille J, Mueller-Roeber B, Gechev TS. (2015) ROS-mediated abiotic stress-induced programmed cell death in plants. *Front Plant Sci* 6: 69.

Rao KP, Vani G, Kumar K, *et al.* (2011) Arsenic stress activates MAP kinase in rice roots and leaves. *Arch Biochem Biophys* 506: 73–82.

Remans T, Opdenakker K, Smeets K, Mathijsen D, Vangronsveld J, Cuypers A. (2010) Metal-specific and NADPH oxidase dependent changes in lipoxygenase and NADPH oxidase gene expression in *Arabidopsis thaliana* exposed to cadmium or excess copper. *Funct Plant Biol* 37: 532–544.

Rentel MC, Knight MR. (2004) Oxidative stress-induced calcium signaling in *Arabidopsis*. *Plant Physiol* 135: 1471–1479.

Rentel MC, Lecourieux D, Ouaked F, *et al.* (2004) OXI1 kinase is necessary for oxidative burst-mediated signalling in *Arabidopsis*. *Nature* 427: 858–861.

Rodríguez-Serrano M, Romero-Puertas MC, Pazmiño DM, *et al.* (2009) Cellular response of pea plants to cadmium toxicity: cross talk between reactive oxygen species, nitric oxide, and calcium. *Plant Physiol* 150: 229–243.

Sandalio LM, Rodríguez-Serrano M, del Río LA, Romero-Puertas MC. (2009) Reactive oxygen species and signaling in cadmium toxicity. In: del Río LA, Puppo A (eds), *Reactive Oxygen Species in Plant Signaling*. Springer-Verlag, Berlin, Heidelberg, pp. 182–183.

Saxena I, Shekhawat GS. (2013) Nitric oxide (NO) in alleviation of heavy metal induced phytotoxicity and its role in protein nitration. *Nitric Oxide* 32: 13–20.

Schellingen K, Van Der Straeten D, Remans T, Vangronsveld J, Keunen E, Cuypers A. (2015) Ethylene signalling is mediating the early cadmium induced oxidative challenge in *Arabidopsis thaliana*. *Plant Sci* 239: 137–146.

Schulz P, Herde M, Romeis T. (2013) Calcium-dependent protein kinases: hubs in plant stress signaling and development. *Plant Physiol* 163: 523–530.

Sethy SK, Ghosh S. (2013) Effect of heavy metals on germination of seeds. *J Nat Sci Biol Med* 4: 272–275.

Shapiguzov A, Vainonen JP, Wrzaczek M, Kangasjärvi J. (2012) ROS-talk – how the apoplast, the chloroplast, and the nucleus get the message through. *Front Plant Sci* 3: 292.

Sharma SS, Dietz K-J. (2009) The relationship between metal toxicity and cellular redox imbalance. *Trends Plant Sci* 14: 43–50.

Shi H, Ye T, Zhu J-K, Chan Z. (2014a) Constitutive production of nitric oxide leads to enhanced drought stress resistance and extensive transcriptional reprogramming in *Arabidopsis*. *J Exp Bot* 65: 4119–4131.

Shi H, Ye T, Chan Z. (2014b) Nitric oxide-activated hydrogen sulfide is essential for cadmium stress response in bermudagrass (*Cynodon dactylon* (L). Pers.). *Plant Physiol Biochem* 74: 99–107.

Silveira NM, Oliveira JA, Ribeiro C, Canatto RA, Siman L, Farnese F. (2015) Nitric oxide attenuates oxidative stress induced by arsenic in lettuce (*Lactuca sativa*) leaves. *Water Air Soil Pollut* 226: 379.

Singh R, Singh S, Parihar P, *et al.* (2016) Reactive oxygen species (ROS): beneficial companions of plants' developmental processes. *Front Plant Sci* 7: 1299.

Singh S, Srivastava PK, Kumar D, Tripathi DK, Chauhan DK, Prasad SM. (2015) Morpho-anatomical and biochemical adapting strategies of maize (*Zea mays* L.) seedlings against lead and chromium stresses. *Biocatal Agric Biotechnol* 4: 286–295.

Singh S, Tripathi DK, Singh S, *et al.* (2017) Toxicity of aluminium on various levels of plant cells and organism: a review. *Environ Exp Bot* 137: 177–193.

Singh VP, Tripathi DK, Kumar D, Chauhan DK. (2011) Influence of exogenous silicon addition on aluminium tolerance in rice seedlings. *Biol Trace Elem Res* 144: 1260–1274.

Singh VP, Srivastava PK, Prasad SM. (2013) Nitric oxide alleviates arsenic-induced toxic effects in ridged *Luffa* seedlings. *Plant Physiol Biochem* 71: 155–163.

Singh VP, Singh S, Kumar J, Prasad SM. (2015) Hydrogen sulfide alleviates toxic effects of arsenate in pea seedlings through up-regulation of the ascorbate-glutathione cycle: Possible involvement of nitric oxide. *J Plant Physiol* 181: 20–29.

Smeets K, Opdenakker K, Remans T, *et al.* (2013) The role of the kinase OXI1 in cadmium- and copper-induced molecular responses in *Arabidopsis thaliana*. *Plant Cell Environ* 36: 1228–1238.

Srivastava RK, Pandey P, Rajpoot R, Rani A, Dubey RS. (2014) Cadmium and lead interactive effects on oxidative stress and antioxidative responses in rice seedlings. *Protoplasma* 251: 1047–1065.

Srivastava S, Srivastava AK, Suprasanna P, D'Souza SF (2009) Comparative biochemical and transcriptional profiling of two contrasting varieties of *Brassica juncea* L. in response to arsenic exposure reveals mechanisms of stress perception and tolerance. *J Exp Bot* 60: 3419–3431.

Suzuki N, Koussevitzky S, Mittler R, Miller G. (2012) ROS and redox signalling in the response of plants to abiotic stress. *Plant Cell Environ* 35: 259–270.

Suzuki N, Miller G, Salazar C, *et al.* (2013) Temporal-spatial interaction between ROS and ABA controls rapid systemic acclimation in plants. *Plant Cell* 25: 3553–3569.

Sun C, Lu L, Liu L, *et al.* (2014) Nitrate reductase-mediated early nitric oxide burst alleviates oxidative damage induced by aluminum through enhancement of antioxidant defenses in roots of wheat (*Triticum aestivum*). *New Phytol* 201: 1240–1250.

Sun J, Wang R, Zhang X, *et al.* (2013) Hydrogen sulfide alleviates cadmium toxicity through regulations of cadmium transport across the plasma and vacuolar membranes in *Populus euphratica* cells. *Plant Physiol Biochem* 65: 67–74.

Talukdar D. (2011) Effect of arsenic-induced toxicity on morphological traits of *Trigonella foenum-graecum* L. and *Lathyrus sativus* L. during germination and early seedling growth. *CurrRes J Biol Sci* 3: 116–123.

Tamás L, Dudíková J, Durceková K, Halusková L, Huttová J, Mistrík I. (2009) Effect of cadmium and temperature on the lipoxygenase activity in barley root tip. *Protoplasma* 235: 17–25.

Thao NP, Khan MIR, Thu NBA, *et al.* (2015) Role of ethylene and its cross talk with other signaling molecules in plant responses to heavy metal stress. *Plant Physiol* 169: 73–84.

Tripathi DK, Singh VP, Kumar D, Chauhan DK. (2012a) Impact of exogenous silicon addition on chromium uptake, growth, mineral elements, oxidative stress, antioxidant capacity, and leaf and root structures in rice seedlings exposed to hexavalent chromium. *Acta Physiol Plant* 34: 279–289.

Tripathi DK, Singh VP, Kumar D, Chauhan DK. (2012b) Rice seedlings under cadmium stress: effect of silicon on growth, cadmium uptake, oxidative stress, antioxidant capacity and root and leaf structures. *Chem Ecol* 28: 281–291.

Tripathi DK, Singh S, Singh VP, Prasad SM, Chauhan DK, Dubey NK. (2016a) Silicon nanoparticles more efficiently alleviate arsenate toxicity than silicon in maize cultivar and hybrid differing in arsenate tolerance. *Front Environ Sci* 4: 46.

Tripathi DK, Singh VP, Prasad SM, Dubey NK, Chauhan DK, Rai AK. (2016b) LIB spectroscopic and biochemical analysis to characterize lead toxicity alleviative nature of silicon in wheat (Triticum aestivum L.) seedlings. *J Photochem Photobiol B: Biol* 154: 89–98.

Tripathi DK, Singh S, Singh VP, Prasad SM, Dubey NK, Chauhan DK. (2017a) Silicon nanoparticles more effectively alleviated UV-B stress than silicon in wheat (Triticum aestivum) seedlings. *Plant Physiol Biochem* 110: 70–81.

Tripathi DK, Singh S, Singh S, *et al.* (2017b) Nitric oxide alleviates silver nanoparticles (AgNps)-induced phytotoxicity in Pisum sativum seedlings. *Plant Physiol Biochem* 110: 167–177.

Tripathi DK, Mishra RK, Singh S, *et al.* (2017c) Nitric oxide ameliorates zinc oxide nanoparticles phytotoxicity in wheat seedlings: implication of the ascorbate–glutathione cycle. *Front Plant Sci* 8: 1.

Van Breusegem F, Dat JF. (2006) Reactive oxygen species in plant cell death. *Plant Physiol* 141: 384–390.

Wang D, Liu Y, Tan X, *et al.* (2015) Effect of exogenous nitric oxide on antioxidative system and S-nitrosylation in leaves of *Boehmeria nivea* (L.) Gaud under cadmium stress. *Environ Sci Pollut Res* 22: 3489–3497.

Wang J, Ding H, Zhang A, Ma F, Cao J, Jiang M. (2010) A novel mitogen-activated protein kinase gene in maize (*Zea mays*), ZmMPK3, is involved in response to diverse environmental cues. *J Integr Plant Biol* 52: 442–452.

Wang JW, Wu JY. (2005) Nitric oxide is involved in methyl jasmonate induced defense responses and secondary metabolism activities of *Taxus* cells. *Plant Cell Physiol* 46: 923–930.

Waszczak C, Akter S, Jacques S, Huang J, Messens J, Van Breusegem F. (2015) Oxidative post–translational modifications of cysteine residues in plant signal transduction. *J Exp Bot* 66: 2923–2934.

Xu FJ, Jin CW, Liu WJ, Zhang YS, Lin XY. (2011) Pretreatment with H_2O_2 alleviates aluminum-induced oxidative stress in wheat seedlings. *J Integr Plant Biol* 53: 44–53.

Xu H, Xu W, Xi H, Ma W, He Z, Ma M. (2013) The ER luminal binding protein (BiP) alleviates Cd^{2+}-induced programmed cell death through endoplasmic reticulum stress-cell death signaling pathway into bacco cells. *J Plant Physiol* 170: 1434–1441.

Xu J, Yin H, Li Y, Liu X. (2010) Nitric oxide is associated with long-term zinc tolerance in *Solanum nigrum*. *Plant Physiol* 154: 1319–1334.

Yeh CM, Hsiao LJ, Huang HJ. (2003) Copper treatment activates mitogen-activated protein kinase signalling in rice. *Physiol Plantarum* 119: 392–399.

Yeh CM, Hsiao LJ, Huang HJ. (2004) Cadmium activates a mitogen-activated protein kinase gene and MBP kinases in rice. *Plant Cell Physiol* 45: 1306–1312.

Yoo SD, Cho YH, Tena G, Xiong Y, Sheen J. (2008) Dual control of nuclear EIN3 by bifurcate MAPK cascades in C2H4 signalling. *Nature* 451: 789–795.

Yoo SD, Cho Y, Sheen J. (2009) Emerging connections in the ethylene signaling network. *Trends Plant Sci* 14: 270–279.

Yoshioka H, Numata N, Nakajima K, *et al.* (2003) *Nicotiana benthamiana* gp91^{phox} homologs *NbrbohA* and *NbrbohB* participate in H_2O_2 accumulation and resistance to *Phytophthora infestans*. *Plant Cell* 15: 706–718.

Yu Q, Sun L, Jin H, Chen Q, Chen Z, Xu M. (2012) Lead-induced nitric oxide generation plays a critical role in lead uptake by *Pogonatherum crinitum* root cells. *Plant Cell Physiol* 53: 1728–1736.

Zhang A, Jiang M, Zhang J, *et al.* (2007) Nitric oxide induced by hydrogen peroxide mediates abscisic acid-induced activation of the mitogen-activated protein kinase cascade involved in antioxidant defense in maize leaves. *New Phytol* 175: 36–50.

Zhang H, Tan ZQ, Hu LY, Wang SH, Luo JP, Jones RL. (2010) Hydrogen sulfide alleviates aluminum toxicity in germinating wheat seedlings. *J Integr Plant Biol* 52: 556–567.

Zhang X, Wang L, Meng H, Wen H, Fan Y, Zhao J. (2011) Maize ABP9 enhances tolerance to multiple stresses in transgenic *Arabidopsis* by modulating ABA signaling and cellular levels of reactive oxygen species. *Plant Mol Biol* 75: 365–378.

Zhou C, Cai Z, Guo Y, Gan S (2009) An *Arabidopsis* mitogen-activated protein kinase cascade, MKK9-MPK6, plays a role in leaf senescence. *Plant Physiol* 150: 167–177.

4

Role of Reactive Oxygen Species in Magnetoprimed Induced Acceleration of Germination and Early Growth Characteristics of Seeds

Sunita Kataria

School of Life Science, Devi Ahilya Vishwavidyalaya, Khandwa Road Campus, Indore, India

Introduction

Seed germination, a crucial stage in a plant's life, is complicated by several factors, including plant hormones and environmental factors. Seed vigor includes "those seed properties, which determine the potential for rapid, uniform emergence and development of normal seedling under a wide range of field conditions" (AOSA, 1983). Magnetobiology is a new synthetic discipline encompassing the principles and techniques of many sciences including engineering, physics, chemistry, and biology and centered on biophysics. It studies the biological effects of oscillating or static and low-frequency magnetic fields on tissues without causing heating of tissues. Magnetic and electromagnetic treatments are being used in agriculture for seed priming as a non-invasive technique to improve the germination and vigor of seeds and ultimately yield.

Pre-sowing exposure of seeds of different crops to a magnetic field (MF), called "magnetopriming," is a non-destructive, dry seed priming treatment that has been reported to increase the percentage of germination, rate of germination, and seedling vigor of many crops (Martinez *et al.*, 2009; Shine *et al.*, 2011; Bhardwaj *et al.*, 2012; Kataria *et al.*, 2017a,b). All living processes are highly dependent on energy exchange between cells and their environment. MF, being a part of the environment and a source of energy, thereby affects normal metabolism (Aladjadjiyan, 2007) and has an impact on meristem cell division (Belyavskaya, 2001). MFs may serve in agriculture as a physical pre-treatment for seeds, in a bid to improve germination and seedling emergence, increasing production without harmful effects on the environment (Vasilevski, 2003). The biostimulative impact usually depends on the following factors: genotype; the frequency of alternating fields; magnetic flux density; seed exposure time; absolute exposure dose; and polarity (north or south). Earlier studies indicated that roots were more predisposed to being affected by MFs than shoots (Bhatnagar and Deb, 1977; Kavi, 1983), hence increasing and improving nutrient assimilation.

MF has positive effects on plant characteristics such as seed germination, shoot development, fresh weight and plant length, fruit yield per plant, and average fruit weight (Aladjadjıyan, 2002; Rochalska and Orzeszko-Rywka, 2005; DeSouza *et al.*, 2006; Baghel *et al.*, 2016; Kataria *et al.*, 2017a,b). In addition MF affects water absorption, preservation, and ionization (Taia *et al.*, 2007). It is believed that the MF influences the structure of cell membranes and increases their permeability and ion transport through the ion channels, which then affects metabolic activity (Reina and Pascual, 2001). A higher content of hydrolyzing enzymes after pre-sowing

Reactive Oxygen Species in Plants: Boon Or Bane - Revisiting the Role of ROS, First Edition. Edited by Vijay Pratap Singh, Samiksha Singh, Durgesh Kumar Tripathi, Sheo Mohan Prasad, and Devendra Kumar Chauhan.

treatment with MFs has been reported in *Helianthus annuus* (Vashisth and Nagarajan, 2010). Increased levels of free radicals in seeds after treatment with laser light and MFs, accompanied by enhanced dynamics of seed germination, was also reported in faba bean seeds and pea seeds (Podleśny *et al.*, 2001, 2005). The beneficial effects of a static magnetic field (SMF) pre-sowing on seedling growth are not only the result of stimulation of reserve mobilization and respiration initiated in germination but also of metabolic changes involving protein biosynthesis from mRNAs, gene transcription, and cellular repairs (Racuciu *et al.*, 2008; Shabrangi and Majd, 2009).

Recently, several reviews have summarized the impacts of MFs on many biological processes in plants, such as growth, development, and metabolism (Maffei, 2014; Wolff *et al.*, 2014). Magneto-priming is becoming the most popular technique amongst modern agronomic practices to improve seed vigor, growth, and yield of plants as it is an efficient and eco-friendly production technology based on physical treatment of seeds. The present chapter provides useful data about the beneficial physiological and biochemical changes induced by the latest available MF treatment on the germination of the seeds and also the role of reactive oxygen species in magnetoprimed induced acceleration of germination and early growth characteristics of seeds.

Germination and Seedling Vigor After Magnetopriming

Seed priming advances seed metabolism leading to an improvement of germination, enhancement of seedling performance, and tolerance of plants to stress (McDonald, 2000). However, priming decreases the longevity of high-vigor seeds so post-priming treatments are needed (Varier *et al.*, 2010) but physical methods (laser, MF, ultrasound, microwave) avoid this disadvantage. In fact, these methods introduce energy into cells, inducing molecular transformations and improvement in yield for suitable treatments. A comparison of different physical methods in different seed species concluded that beneficial effects were more pronounced after MF treatments (Aladjadjiyan, 2007). Magnetopriming is reported to affect seed germination and helps in crop establishment. The first report on the effect of MF on growth of seedlings dates back to the work by Savostin (1930) who observed an increase in the rate of wheat seedling elongation under magnetic conditions. Later, Pittman and Ormrod (1970), Bhatnagar and Dev (1977), and Pietruszewski (1999) recorded an increase in rate of germination of cereal seeds exposed to MF. The MF pre-sowing effect would appear to be organ dependent, with roots seemingly much more susceptible than shoots (Kato *et al.*, 1989). Muraji *et al.* (1992) demonstrated an enhancement in root growth of maize (*Zea mays*) by exposing the maize seedling to 5 mT MFs at alternating frequencies of 40–160 Hz.

Alexander and Doijode (1995) noted that the application of an external MF as a pre-germination treatment improved the germination and seedling vigor of low-viability rice and onion seeds. Stationary MF treatment of lettuce seeds led to faster germination compared to untreated seeds, possibly due to an increase in water uptake rate (Garcia and Arze, 2001). Celestino *et al.* (2000) reported enhanced germination and growth of *Quercus suber* seedlings exposed to an electromagnetic field. Germination increased by 1.1–2.8 times in tomato seeds exposed to a suitable MF (Moon and Chung, 2000). Wheat and barley seeds treated with MF showed early sprouting (Martinez *et al.*, 2000, 2002). The germination characteristics of maize seeds exposed to MFs have been enhanced in many studies (Aladjadjiyan, 2002; Flórez *et al.*, 2007; Kataria et *al.*, 2017b). Aladjadjiyan (2002) observed that the MF stimulated the shoot development of maize and led to an increase in germinating energy, germination, fresh weight, and shoot length.

Studies conducted by Rochalska (2002) on the germination and growth of wheat, triticale, maize, and soybean, also indicated that MF can be used as a method of seed vigor improvement. Under continuous exposure, a weak electromagnetic field (EMF) induces an inhibitory effect on flax, pea, and lentil roots for 0.5 and 2 nT intensities, whereas no effect is observed for 1 nT (Belyavskaya, 2004). Exposure of seeds to MF was found to accelerate sprouting and growth of the seedlings of cucumber (Yinan *et al.*, 2005). Germination energy and germination rate of tobacco seeds and growth of germinated *Vicia faba* seedlings was enhanced by the application of power frequency MFs (100 μT), supported by an increased mitotic index and ^{3}H-thymidine uptake (Rajendra *et al.*, 2005). Podleśny *et al.* (2004, 2005) confirmed the positive effect of the magnetic treatment on the germination and emergence of both broad bean and pea cultivars. Soltani *et al.* (2006) have reported the positive effect of MF on *Asparagus officinalis* and *Ocimum basilicum* seed germination and seedling growth. Vashisth and Nagarajan (2007, 2008) reported significant increases in germination, seedling vigor, and shoot/root growth of 1-month-old maize plants and chickpea seeds exposed to static MFs. Alternating MF treatments at low frequency (60 Hz) with combinations of three magnetic flux densities (20, 60, and 100 mT) and three exposure times (7.5, 15, and 30 minutes) were used as pre-sowing seed treatments in three maize (*Zea mays* L.) genotypes to increase seedling emergence rate, seedling dry weight, and emergence in soil (Aguilar *et al.*, 2009). In tomato, treating seed with a stationary MF improved germination percentage and rate of germination (Martínez *et al.*, 2009). Mean germination time (MGT) increased when the duration of exposure of seeds to MF was increased in wheat (Gholami and Sharafi, 2010). Irradiating maize seeds at magnetic induction levels of 160 and 560 mT also produced an increment in seed emergence rate and seedling emergence (Dominguez *et al.*, 2010).

Seedling growth responses to MF pre-sowing include stimulation or inhibition, depending on treatment. The use of MFs could be useful in the alleviation of abiotic stress *in vitro* or in the field, as evidenced by the ability of MFs to allow the continued growth of wheat and common bean seedlings under salt or osmotic stress after exposure for 7 days at 7 mT (Cakmak *et al.*, 2010); their stress-protecting effect under water, salt, heavy-metal, or heat stress (Xi *et al.*, 1994; Ružič and Jerman, 2002; Chen *et al.*, 2011; Selim and El-Nady, 2011; Baghel *et al.*, 2016, 2017) and the enhancement of organogenesis (Radhakrishnan and Kumari, 2013) and carbon and nitrogen metabolism (Baghel *et al.*, 2016) of soybean under salt stress. Reina *et al.* (2001) have also found increased water uptake after SMF pre-treatment, which might be responsible for enhanced germination of lettuce seeds. Previous studies suggested that speed of germination and seedling growth increase by pre-germination exposure of the dry seed to an introduced MF in many plants, like chickpea (Mahajan and Pandey, 2014); soybean (Shine *et al.*, 2011); capsicum (Ahamed *et al.*, 2013); carrot (Dorna *et al.*, 2010); cucumber (Yinan *et al.*, 2005); maize (Flórez *et al.*, 2004); radish (Krawiec *et al.*, 2013); sunflower (Vashisth and Nagarajan, 2010); tomato (Moon and Chung, 2000); *Satureja hortensis* (Pourakbar and Hatami, 2012). Several studies have been performed since 1993 to evaluate plant responses after exposure to different MF strengths, from near-null (0–40 nT) to low (up to 40 mT), to extremely high values (up to 30 T) (Teixeira da Silva and Dobránszki, 2015). The reported results showed a variety of plant responses at the biochemical, molecular, cellular, and whole-plant levels. In most studied species, SMF treatment pre-sowing induces positive effects on germination (Shine *et al.*, 2011; Shine and Guruprasad, 2012; Kataria *et al.*, 2015) through improving its rate and speed. Magnetoprimed seeds of soybean and maize showed improved germination rates, vigor, and seedling biomass or root development (Figure 4.1).

The results obtained by magnetopriming indicate that MF acts as a biostimulant, which can be considered as an alternative to chemical and biological methods for the pre-sowing treatment of the seeds (Vakharia *et al.*,1991; Alexander and Doijode, 1995). Several reports also show that the MF exposure increases germination of low-viability seeds and improves their quality and sprouting rate even under abiotic stress conditions (Carbonell *et al.*, 2008; Thomas *et al.*, 2013; Kataria *et al.*, 2017b). These effects may be because of interaction of MF with ionic currents in the plant embryo cell membrane, which induces changes in both osmotic pressure and ionic concentrations on both sides of the membrane (Yaycili and Alikamanoglu, 2005). Changes in

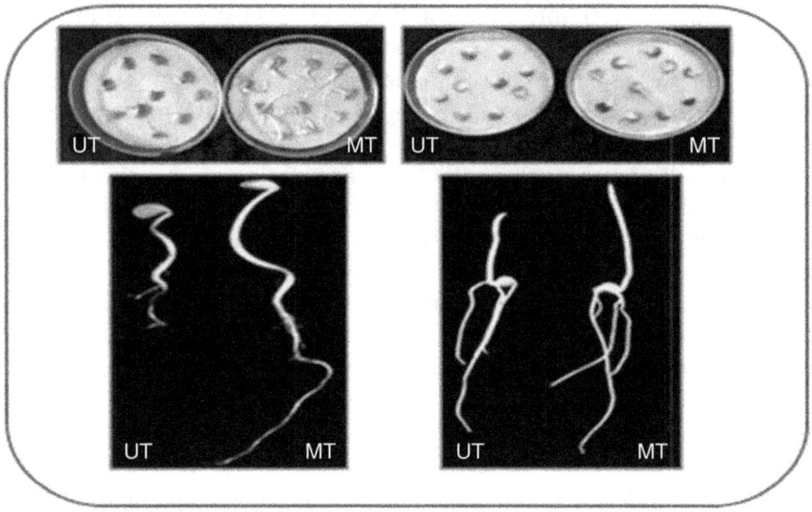

Figure 4.1 Effect of magnetopriming (static magnetic field treatment 200 mT for 1 h) on the percentage germination, speed of germination, and growth of root and shoot of soybean (left panels) and maize seedlings (right panels). UT, untreated; MT, magnetic field treated.

the ionic fluxes across cell membranes cause alterations in the mechanism of water uptake, as osmoregulation in embryo cells is controlled by the ionic transport across the membrane (Reina and Pascual, 2001).

Activity of Enzymes Related to Germinating Magnetoprimed Seeds

Starch breakdown during germination is the result of hydrolytic enzymes, of which α- and β-amylase are the most prominent. Through respiratory breakdown of substrates the activity of these two enzymes provides energy and carbon skeletons to the growing embryo (Nandi *et al.*, 1995). Bhatnagar and Deb (1978) and Vashisth and Nagarajan (2010) observed that wheat and sunflower seeds, treated with SMF, had significantly higher α-amylase activity than their controls. The rate of germination and early growth characteristics of soybean and maize seeds were increased under SMF treatment; this could be due to an increase in water uptake rate and higher activities of α-amylase and protease (Shine *et al.*, 2011; Shine and Guruprasad, 2012). During germination of cotton seeds an increased level of reducing sugars was observed in *Euphorbia* (Suda and Giorgini, 2000). Pittman and Ormrod (1970) also found more reducing sugars in germinating magnetically treated wheat seeds. Changes in amylase and nitrate reductase activities in germinating seeds treated by electromagnets of different field strengths were also observed (Levedev *et al.*, 1977). Wheat seeds exposed to MFs of 50 to 300 mT showed increased seedling vigor, respiratory quotient, and α-amylase activity as compared to control seeds (Bhatnagar and Deb, 1978). Also in rice, *Ramy1A* mRNA coding for amylase accumulated after 12 h of imbibition (Kaneko *et al.*, 2002). The beneficial effect of both static and oscillating MF of very low to high field intensity has been reported in different plant species. Studies have shown that germination characteristics of maize seeds exposed to MF are enhanced (Pittman, 1965; Flórez *et al.*, 2004). Stationary MF strengths of 50 mT and 200 mT for 2 h exposure increased the activities of hydrolyzing enzymes in *Helianthus annus*, which was responsible for the rapid seed germination, improved seedling vigor, and better root characteristics of treated seeds in this plant (Vashisth and Nagarajan, 2010). The activities of enzymes involved in carbohydrate metabolism were altered when soybean seeds were pre-treated with pulsed MF of 1500 nT at 10 Hz for 20 days and daily for 5 h (Radhakrishnan and Kumari, 2012).

The increased rate of germination in SMF-treated chickpea, soybean and maize seeds under salinity was associated with higher activity of α-amylase and protease (Thomas *et al.*, 2013; Kataria *et al.*, 2017b). It was observed that α-amylase activity increased in magnetoprimed seeds of a salt-sensitive wheat variety whereas no change was observed in β-amylase activity (Rathod, 2013). This suggests that α-amylase was responsible for the faster metabolism in magnetoprimed seeds. The fact that β-amylase did not make a significant contribution to starch hydrolysis indicates it is unnecessary, at least during the initial stages in germinating wheat seeds (Rathod, 2013). Salinity resulted in increased α-amylase activity as low seed osmotic potential led to faster seed water uptake, consequently hydrating the seed and helping it to germinate under salinity (Thomas *et al.*, 2013).

Proteases are involved in the hydrolysis of proteins in the germinating seeds; the process is initiated by endoproteases, which convert the water-insoluble storage proteins into soluble peptides that can be further hydrolyzed to amino acids by exopeptidases (Callis, 1995). The increase in hydrolyzed protein per gram fresh weight of seeds represents an increase in protease activity. Magnetoprimed seeds maintained higher protease activity than untreated seeds of soybean, chickpea, and maize (Shine *et al.*, 2012; Thomas *et al.*, 2013; Kataria *et al.*, 2015).

Reactive Oxygen Species (ROS) and Seed Germination After Magnetopriming

Reactive oxygen species (ROS) play a key signaling role in the major events of seed life, such as germination or dormancy release. ROS, such as superoxide radicals ($O_2^{\cdot -}$), hydrogen peroxide (H_2O_2), and hydroxyl radicals (·OH), are generated as a result of aerobic metabolism in mitochondria, peroxisomes, and the apoplastic space in the germinating seeds. For ROS to act as cellular messengers, seeds have evolved specific ROS-removing mechanisms to protect against the over-production of ROS.

However, in dry seed, and in spite of low metabolism, there is production of free radicals as ROS; concentrations of these increase following resumption of seed metabolism and their levels, concentration, and longevity are modified by MF (Hunt *et al.*, 2009). At low levels, ROS might be involved in MF perception (Okano, 2008), particularly the superoxide ion $O_2^{\cdot -}$ (Solovyov and Schulten, 2009) in the proposed radical-pair model (Solovyov *et al.*, 2007). Among the proposed MF perception models, the radical-pair one is thought to be the most relevant as it induces responses measurable for strengths weaker than the geomagnetic field (Timmel *et al.*, 1998). According to this model, responses to MF are the result of photo-induced radical-pair reactions in cryptochromes (Harris *et al.*, 2009) – blue light photoreceptors that are ubiquitous among living taxa and are involved in plant development and photomorphogenesis. Using *Arabidopsis thaliana* mutants continuously subjected to SMF (500 μT), Ahmed *et al.* (2007) showed the involvement of cryptochromes in hypocotyl reduction and anthocyanin accumulation. Also, MF activates the antioxidative defense mechanism, improving plant tolerance to biotic and abiotic stresses (Rochalska and Grabowska, 2007). This might be the result of stimulation of enzyme activities implicated in the detoxification of $O_2^{\cdot -}$ (superoxide dismutases, SODs) and H_2O_2 (catalases, peroxidases) (Atak *et al.*, 2007; Celik *et al.*, 2009).

In radish, the beneficial SMF treatment (100 mT/3 h) for both germination and seedling growth allowed stimulation of catalase activity counteracting the enhanced H_2O_2 production, resulting probably in increased SOD activity in plantlet organs (Touati *et al.*, 2013). In *Glycine max*, an increase in both SOD and catalase activities is observed, allowing efficient detoxification. This is thought to be the result of the influence of MF on the SOD metal cofactors, and in particular Zn^{2+} and Cu^{2+} (Celik *et al.*, 2009). The enhanced ROS detoxification leads, in beneficial SMF treatments, to oxidative damage reduction, as observed in sunflower, for which membrane integrity was ameliorated (Vashisth and Nagarajan, 2010) probably by lowering iron absorption (Hajnorouzi *et al.*, 2011), which prevents ·OH formation by the Fenton reaction.

Shine *et al.* (2012) indicated a significant enhancement in the production of free radicals after treating the seeds with MF. The electron paramagnetic resonance (EPR) spectra of adducts produced with $O_2^{\cdot-}$ and ·OH radicals at an early stage (just germinated and 8-day-old seedlings) provide evidence for the important role of ROS in enhancing the performance of seeds at this stage. These data are in agreement with the earlier findings on faba bean seeds and pea seeds, where increased levels of free radicals by pre-sowing treatment of seeds with laser light and MF was followed by faster water uptake and greater biomass (Podleśny *et al.*, 2001, 2005). Production of free radicals in the germinating seeds has been associated with transition from a quiescent stage to a metabolically active state, which would account for faster germination (Puntarulo *et al.*, 1991; Gidrol *et al.*, 1994). Free radicals are also involved in the elongation growth of plant organs (Schopfer, 2001).

Gupta *et al.* (2015) found that contents of superoxide ion and hydrogen peroxide increased two-fold in static and pulsed magnetoprimed seeds. The increased production of ROS in germinating seeds of magnetically treated maize and soybeans is facilitated by the oxidative reaction of peroxidase, whereas its antioxidative capacity helps in scavenging the excess free radicals produced (Shine *et al.*, 2012). In addition to $O_2^{\cdot-}$ and H_2O_2 there was also an enhancement in the ·OH radical after treatment with MF (Shine *et al.*, 2012). In their study, increased ROS production was accompanied by enhanced seedling growth with MF treatment. Higher ROS levels were observed in pre-exposure with an MF of 200 mT, which was accompanied by greater seedling length, compared to the 150 mT treatment. Seed germination was enhanced with enhanced ROS production, and their scavenging reduced germination (Chaudhuri and Kar, 2008). ·OH radicals are associated with cell wall loosening and elongation growth (Liszkay *et al.*, 2004). Thus, the homeostasis maintained between ROS production and scavenging has a promoting effect on seed germination in magnetically treated seeds. Several earlier studies have shown that ROS generation and seed germination are linked. ROS generated in seed germination are involved in endosperm weakening, protection against pathogens, Ca^{2+} signaling, gene expression, redox regulation, hormone signaling, cell wall elongation, and so forth (Bailly *et al.*, 2008; El-Maarouf Bouteau and Bailly, 2008).

Antioxidant Enzyme Activities During Germination After Magnetopriming

Although the ROS content is enhanced during germination of magnetically treated seeds, the ROS scavenging enzymes such as catalase (CAT) and superoxide dismutase (SOD) are reduced (Shine and Guruprasad, 2012; Shine *et al.*, 2012) whereas the activity of peroxidase enzyme is enhanced by MF treatment, which is predicted to be the result of enhanced NADPH activity (Radhakrishnan and Kumari, 2013). SOD is an important enzyme for deactivating $O_2^{\cdot-}$ by converting the radical to H_2O_2 (Scandalios, 1993). Although magnetic treatment enhanced $O_2^{\cdot-}$, the activity of SOD was reduced in soybean (Shine *et al.*, 2012). SOD increased linearly with different times of imbibition (4 to 36 h) in cherry tomato seedlings although the increment was higher in pulsed magnetic field (PMF) magnetoprimed seeds than SMF treated seeds (Gupta *et al.*, 2015).

Many other experiments on several plant species, including pea (Polovinkina *et al.*, 2011), radish (*Raphanus sativus*), *Leymus chinensis* (Xia and Guo, 2000), cucumber (Bhardwaj *et al.*, 2012), broad bean (Jouni *et al.*, 2012), corn (Anand *et al.*, 2012), wheat (Payez *et al.*, 2013) and *Coffea arabica* (Aleman *et al.*, 2014), were also concurrent with these findings. This suggests that exposure to increased MF causes accumulation of ROS and alteration of the enzyme activities that maintain the homeostasis between ROS production and scavenging. Also the nitric oxide (NO) concentration and nitric oxide synthase (NOS) are also reported to be increased in response to MF pre-treatment in mung bean seedlings (Chen *et al.*, 2011). Mung bean seedlings treated with 600 mT MF followed by cadmium stress showed that MF compensates for the toxicological effects of cadmium exposure, in a way related to the NO signal (Chen *et al.*, 2011).

The effect of weak permanent MFs (flux density of 185–650 μT) on the antioxidant system was reported in 5-day-old radish seedlings by Serdyukov and Novitskii (2013). They found that by increasing MF strength to 650 μT, SOD activity increased 135% more than the control, while CAT activity increased 135 or 150% more than control seedlings grown in the dark or light respectively. A static MF (30 mT) increased CAT activity but reduced ascorbate peroxidase (APX) activity in parsley (*Petroselium crispum* L.) cells (Rajabbeigi *et al.*, 2013).

Increased antioxidant enzyme activities of SOD, CAT, and glutathione reductase (GR) were described in magnetoprimed cucumber (*Cucumis sativus* L.) seeds (Bhardwaj *et al.*, 2012). In agreement with these findings, Shine *et al.* (2011) highlighted reduced production of $O_2^{\cdot-}$ in magnetoprimed soybean seeds. Consequently, MF treatments have the additional potential to be used for minimizing the drought- or disease-induced adverse effects on crop productivity. MF treatment resulted in enhanced ROS production mediated by cell wall peroxidase, while ascorbic acid content, SOD, and APX activity decreased in the hypocotyl of germinating soybean seeds (Shine *et al.*, 2012). An increase in the cytosolic peroxidase activity indicated that this antioxidant enzyme had a vital role in scavenging the increased H_2O_2 produced in seedlings from the magnetically treated seeds (Shine *et al.*, 2012). The antioxidant enzymes showed increased SOD activity, and oxidative reaction of peroxidase resulted in production of hydrogen peroxide during germination of magnetoprimed seeds (Gupta *et al.*, 2015). Increased ascorbate peroxidase activity during the initial stages and catalase in the later stages of germination helped in the scavenging of hydrogen peroxide in germinating magnetoprimed seeds (Gupta *et al.*, 2015).

Conclusions and Future Prospects

Seed priming has emerged as a promising crop management technique that increases the speed of germination thus ensuring synchronized field emergence of the crop. Magnetopriming (exposure of seeds to a magnetic field) is a non-invasive physical stimulant used for improving seedling vigor that helps in establishment of a crop stand under normal as well as stress conditions. Unlike other conventional priming techniques it avoids a seed hydration and dehydration cycle, allows storage of the seed at room temperatures, and is eco-friendly. In previous experiments on maize, soybean, cucumber, and chickpea under normal as well as under abiotic stress conditions, like water deficit and salinity, magnetopriming improved seed water absorption characteristics, resulted in faster hydration of enzymes (amylases and protease) leading to early germination and enhanced vigor of seedlings. Increased levels of $O_2^{\cdot-}$ and H_2O_2 in faster-germinating magnetoprimed seeds, under both normal growing conditions as well as abiotic stress conditions, suggested the role of ROS in oxidative signaling during the seed germination process. Enhancement, by SMF treatment, of seed germination and seedling vigor under both normal growing conditions and abiotic stress conditions may be due to the combined effect of enhanced activity of amylase and protease and enhanced levels of free radicals in the seeds.

Plants resort to many adaptive strategies in response to abiotic environmental stresses such as high salinity, dehydration, cold, heat, metal, osmotic stress, and UV-B stress that affect plant germination, growth, and development (Singh *et al.*, 2011; Tripathi *et al.*, 2012a,b, 2017; Singh *et al.*, 2015, 2017; Singh *et al.*, 2016). These include morphological and developmental changes, adjustments in ion transport, and changes in carbon and nitrogen metabolism. The noticeable increase in root growth in terms of root length and root:shoot ratio promoted by magnetic exposure indicates the alteration in that growth pattern of seedlings that might lead to their adaptation to abiotic stress conditions. Since the scarcity of roots in the deeper soil layers restricts utilization of soil water by crops under various stresses, improvement of root traits by MF treatment may prove to be a promising approach in combating the adverse effects of stress. Alleviation of the inhibitory effects of heat, drought, salt, and UV-B stress in magnetically treated plants has been reported (Baghel *et al.*, 2016, 2017; Kataria *et al.*, 2017 a,b). Magnetopriming can be effectively used as a pre-sowing treatment for mitigating the adverse effects of abiotic stresses during seed germination and early seedling growth. The mechanism underlying this enhancement in germination and related parameters is not well understood, but many models have been proposed to provide a theoretical understanding of the mechanism of magnetoreception in plants. Studies have revealed that phytoferritin occurs in plant cells as crystalline magnetite (Fe_3O_4), ε-Fe_2O_3, and hematite (α-Fe_2O_3) (McClean *et al.*, 2001) and may interact strongly with the magnetic fields than diamagnetic or paramagnetic materials. These particles can also affect the superoxide-generated free radicals (Scaiano *et al.*, 1997).

Enzyme-catalyzed reactions that involve intermediates with radical pairs (Grissom, 1995) and some porphyrins that form free radicals along the mitochondrial respiratory chain may also be influenced by external MF as it affects singlet-to-triplet conversions. The ROS thus produced may result in oxidative modification of proteins (Job *et al.*, 2005), which would favor faster protein mobilization thereby increasing seedling vigor. An interaction between ROS and hormone signaling pathways can affect gene expression related to germination or may change the redox status of the seed and thus initiate events leading to germination. Further studies on the molecular analysis of genotypic differences may help in explaining the effect of magnetopriming on the activity of enzymes. Early and enhanced production of superoxide radicals and hydrogen peroxide in magnetoprimed seeds suggests its role in the accelerated rate of germination. It has been suggested that nitric oxide (NO) and ROS are crucial for dormancy breaking, while ROS formation starts up on imbibition and the NO production is

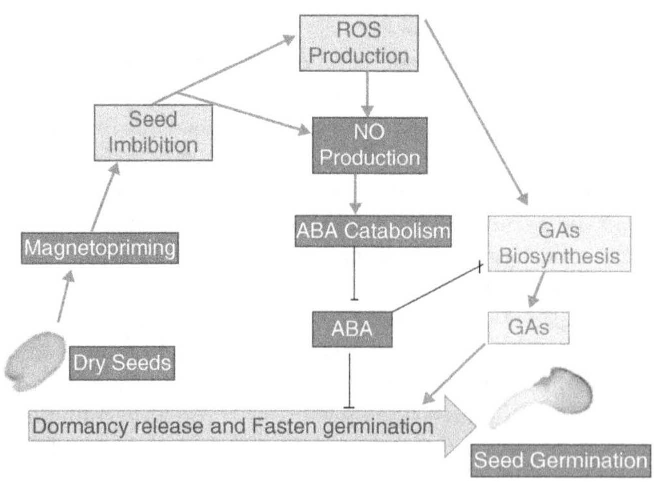

Figure 4.2 Proposed mechanism for seed germination by magnetopriming. Magnetoprimed seeds after imbibition accumulate reactive oxygen species (ROS) and nitric oxide (NO). ROS upregulate catabolism of abscisic acid (ABA) through NO, and also gibberellic acid (GA) biosynthesis. A high concentration of ABA also inhibits GA biosynthesis, but a balance of these two hormones jointly controls seed dormancy and hasten the germination of seeds through NO and ROS.

triggered by the progressing depletion of oxygen under seed coat upon germination. It may possible that MF trigger the NO production and NO may directly or indirectly interact with other signaling molecules such as ROS and cytosolic Ca^{2+}. Thus ROS and NO are important key players in the signaling of magnetopriming such as in dormancy alleviation and seed germination (Figure 4.2). These ROS and NO molecules may interact with seed hormones such as abscisic acid (ABA) and gibberellic acid (GA) to influence ABA/GA signaling pathways (see Figure 4.2) and also act as sensors that allow the seed to germinate faster. Further knowledge regarding regulation of No production by magnetopriming and signal transduction along with its beneficial potentials in seed germination would be of much interest. The characterization of NO-targeted proteins in various seed physiology context will undoubtedly reveal new area of research to explore for understanding the control of germination under non-stress as well as abiotic stress conditions also.

References

Aguilar CH, Dominguez P, Carballocarballo A, Alfredo CO, Ivanov J. (2009) Alternating magnetic field irradiation effects on three genotype maize seed field performance. *Acta Agrophysica* 14: 7–17.

Ahmad M, Galland P, Ritz T, Wiltschko R, Wiltschko W. (2007) Magnetic intensity affects cryptochrome-dependent responses in *Arabidopsis thaliana*. *Planta* 225: 615–624.

Ahamed MEM, Elzaawely AA, Bayoumi YA. (2013) Effect of magnetic field on seed germination, growth and yield of sweet pepper (*Capsicum annuum* L.). *Asian J Crop Science* 5: 286–294.

Aladjadjiyan A. (2002) Study of influence of magnetic field on some biological characteristics of *Zea mays*. *J Central Eur Agric* 3: 89–94.

Aladjadjiyan A. (2007) The use of physical methods for plant growing stimulation in Bulgaria. *J Cent Eur Agric* 3: 369–380.

Aleman EI, Moreira RO, Lima AA, Silva SC, Gonzalez-Olmedo J, Chalfun-Junior A. (2014) Effect of 60 Hz sinusoidal magnetic field on *in vitro* establishment multiplication and acclimatization phases of coffee Arabica seedling. *Bioelectromagnetics* 35: 414–425.

Alexander MP, Doijode SD. (1995) Electromagnetic field, a novel tool to increase germination and seedling vigor of conserved onion (*Allium cepa* L.) and rice (*Oryza sativa* L.) seeds with low viability. *Plant Genet Res Newslett* 104: 1–5.

Anand A, Nagarajan S, Verma APS, Joshi DK, Pathak PC, Bhardwaj J. (2012) Pre-treatment of seeds with static magnetic field ameliorates soil water stress in seedlings of maize (*Zea mays* L.). *Indian J Biochem Biophys* 49: 63–70.

AOSA (Association of Official Seed Analysts). (1983) *Seed Vigor Testing Handbook*. AOSA Handbook 32.

Atak C, Celik O, Olgun A, Alikamanoglu S, Rzakoulieva A. (2007) Effect of magnetic field on peroxidase activities of soybean tissue culture. *Biotechnol Biotechnol Eq* 21: 166–171.

Baghel L, Kataria S, Guruprasad KN. (2016) Static magnetic field treatment of seeds improves carbon and nitrogen metabolism under salinity stress in soybean. *Bioelectromagnetics* 37: 455–470.

Baghel L, Kataria S, Guruprasad KN. (2017) Effect of SMF pretreatment on growth, photosynthetic performance and yield of soybean under water stress. *Photosynthetica* 55: doi: 10.1007/s11099-017-0722-3.

Bailly C, El-Maarouf-Bouteau H, Corbineau F. (2008) From intracellular signaling networks to cell death: the dual role of reactive oxygen species in seed physiology. *C R Biol* 331: 806–814.

Belyavskaya NA. (2001) Ultrastructure and calcium balance in meristem cells of pea roots exposed to extremely low magnetic fields. *Adv Space Res* 28: 645–650.

Belyavskaya NA. (2004) Biological effects due to weak magnetic field on plants. *Adv Space Res* 34: 1566–1574.

Bhardwaj J, Anand A, Nagarajan S. (2012) Biochemical and biophysical changes associated with magnetopriming in germinating cucumber seeds. *Plant Physiol Biochem* 57: 67–73.

Bhatnagar D, Deb AR. (1977) Some aspects of pre-germination exposure of wheat seeds to magnetic fields: Germination and early growth. *Seed Res* 5: 129–137.

Bhatnagar D, Deb AR. (1978) Some aspects of pre-germination exposure of wheat seeds to magnetic field. II. Effect on some physiological processes. *Seed Res* 6: 14–22.

Cakmak T, Dumlupinar R, Erdal S. (2010) Acceleration of germination and early growth of wheat and bean seedlings grown under various magnetic field and osmotic conditions. *Bioelectromagnetics* 31: 120–129.

Callis J. (1995) Regulation of protein degradation. *Plant Cell* 7: 845–857.

Carbonell MV, Martínez E, Flórez M, Maqueda R, Pintor-López A, Amaya JM. (2008) Magnetic field treatments improve germination and seedling growth in *Festuca arundinaceae* Screb. and *Lolium perenne* L. *Seed Sci Technol* 36: 31–37.

Celestino C, Picazo ML, Toribio M. (2000) Influence of chronic exposure to an electromagnetic field on germination and early growth of *Quercus suber* seeds preliminary study. *Electro Magnetobiol* 19: 115–120.

Celik O, Bűyukuslu N, Atak C, Rzakoulieva A. (2009) Effects of magnetic field on activity of superoxide dismutase and catalase in *Glycine max* (L.). Merr. *Roots* 18: 175–182.

Chaudhuri A, Kar RK. (2008) Inhibition of seed germination by propyl gallate, a free radical scavenger and recovery of germination by hydrogen peroxide and ethylene in *Vigna radiata*. *World J Agri Sci* 4: 914–921.

Chen YP, Li R, He JM. (2011) Magnetic field can alleviate toxicological effect induced by cadmium in mungbean seedlings. *Ecotoxicology* 20: 760–769.

DeSouza A, Garcí D, Sueiro L, Gilart F, Porras E, Licea L. (2006) Pre-sowing magnetic treatments of tomato seeds increase the growth and yield of plants. *Bioelectromagnetics* 27: 247–257.

Dominguez PA, Hernandez AC, Cruz OA, *et al.* (2010) Influences of the electromagnetic field in maize seed vigor. *Rev Fitotec Mex* 33: 183–188.

Dorna H, Gorski R, Szopinska D, *et al.* (2010) Effect of a permanent magnetic field together with the shielding of an alternating electric field on carrot seed vigor and germination. *Ecol Chem Eng S* 17: 53–61.

EI-Maarouf Bouteau H, Bailly C. (2008) Oxidative signaling in seed germination and dormancy. *Plant Signal Behav* 3: 175–182.

Flórez M, Carbonell MV, Martínez E. (2004) Early sprouting and first stages of growth of rice seeds exposed to a magnetic field. *Electromagn Biol Med* 23: 167–176.

Florez M, Carbonell MV, Martinez E. (2007) Exposure of maize seeds to stationary magnetic fields: effects on germination and early growth. *Environ Exp Bot* 59: 68–75.

Garcia F, Arze LI. (2001) Influence of a stationary magnetic field on water relations in lettuce seeds. Part I: Theoretical considerations. *Bioelectromagnetics* 22: 589–595.

Gholami A, Sharafi S. (2010) Effect of magnetic field on seed germination of two wheat cultivars. *World Academy Sci Eng Tech* 62: 270–282.

Gidrol X, Lin WS, Degousee N, Yip SF, Kush A. (1994) Accumulation of reactive oxygen species and oxidation of cytokinin in germinating soybean seeds. *Eur J Biochem* 224: 21–28.

Grissom CB. (1995) Magnetic field effects in biology: a survey of possible mechanisms with emphasis on radical-pair recombination. *Chem Rev* 95: 3–24.

Gupta MK, Anand A, Paul V, Dahuja A, Singh AK. (2015) Reactive oxygen species mediated improvement in vigour of static and pulsed magneto-primed cherry tomato seeds. *Ind J Plant Physiol* 20: 205–212.

Hajnorouzi A, Vaezzadeh M, Ghanati F, Jamnezhad H, Nahidian B. (2011) Growth promotion and a decrease of oxidative stress in maize seedlings by a combination of geomagnetic and weak electromagnetic fields. *J Plant Physiol* 168: 1123–1128.

Harris SR, Henbest KB, Maeda K, *et al.* (2009) Effect of magnetic fields on cryptochrome-dependent responses in *Arabidopsis thaliana. J R Soc Interface* 6: 1193–1205.

Hunt RW, Zavalin A, Bhatnagar A, Chinnasamy S, Das CK. (2009) Electromagnetic biostimulation of living cultures for biotechnology, biofuel and bioenergy applications. *Int J Mol Sci* 10: 4515–4558.

Job C, Rajjou L, Lovigny Y, Belghazi M, Job D. (2005) Patterns of protein oxidation in *Arabidopsis* seeds and during germination. *Plant Physiol* 138: 790–802.

Jouni FJ, Abdolmaleki P, Ghanati F. (2012) Oxidative stress in broad bean (*Vicia faba* L.) induced by static magnetic field under natural radioactivity. *Mutat Res* 741: 116–121.

Kaneko M, Inukai Y, Ueguchi-Tanaka M, Itoh H, Izawa T, Kobayashi Y. (2002) Loss-of-function mutations of the rice *GAMYB* gene impair α-amylase expression in aleurone and flower development. *Plant Cell* 16: 33–44.

Kataria S, Baghel L, Guruprasad KN. (2015) Acceleration of germination and early growth characteristics of soybean and maize after pre-treatment of seeds with static magnetic field. *Int J Tropical Agri* 33: 985–992.

Kataria S, Baghel L, Guruprasad KN. (2017a) Alleviation of adverse effects of ambient UV stress on growth and some potential physiological attributes in soybean (*Glycine max*) by seed pre-treatment with static magnetic field. *J of Plant Growth Regulation* Doi: 10.1007/s00344-16-9657-3.

Kataria S, Baghel L, Guruprasad KN. (2017b) Pre-treatment of seeds with static magnetic field improves germination and early growth characteristics under salt stress in maize and soybean. *Biocatalysis and Agriculture Biotechnology* 10: 83–90.

Kato R, Kamada H, Asashma M. (1989). Effects of high and very low magnetic fields on the growth of hairy roots of *Daucus carotta* and *Atropa belladonna. Plant Cell Physiol* 30: 605–608.

Kavi PS. (1983) The effect of non-homogeneous, gradient magnetic field on the magnetic field susceptibility values *in situ* ragi seed material. *Mysore J Agri Sci* 17: 121–123.

Krawiec M, Krzysztof K, Salwina P, Magdalena K, Piotr B, Piotr K. (2013) Does the magnetic field improve the quality of radish seeds? *Acta Sci Pol Hortorum Cultus* 12: 93–102.

Lebedev IS, Litvinenko LG, Shiyan LT. (1977) After effect of a permanent magnetic field on photochemical activity of chloroplasts. *Soviet Plant Physiol* 24: 394–395.

Liszkay A, van der Zalm E, Schopfer P. (2004) Production of reactive oxygen intermediates $O_2^{\cdot-}$, H_2O_2 and ·OH by maize roots and their role in wall loosening and elongation growth. *Plant Physiol* 136: 3114–3123.

Maffei ME. (2014) Magnetic field effects on plant growth, development and evolution. *Front Plant Sci* 5: 445.

Mahajan TS, Pandey OP. (2014) Magnetic-time model at off-season germination. *Int Agrophys* 28: 57–62.

Martínez E, Carbonell MV, Amaya JMA. (2000) Static magnetic field of 125 mT stimulates the initial growth stages of barley (*Hordeum vulgare* L.). *Electro Magnetobiol* 19: 271–277.

Martínez E, Carbonell MV, Flórez M. (2002) Magnetic biostimulation of initial growth stages of wheat (*Triticum aestivum* L.). *Electro Magnetobiol Med* 21: 43–53.

Martínez E, Carbonell MV, Flórez M, Amaya JM, Maqueda R. (2009) Germination of tomato seeds (*Lycopersicon esculentum* L.) under magnetic field. *Int Agrophysi* 23: 45–49.

McClean RG, Schofield MA, Kean WF, *et al.* (2001) Botanical iron minerals: correlation between nanocrystal structure and modes of biological self-assembly. *Eur J Mineral* 13: 1235–1242.

McDonald MB. (2000) Seed priming. In: Black M, Bewley JD (eds), *Seed Technology and Its Biological Basis.* Sheffield Academic Press Ltd, Sheffield, UK, pp. 287–325.

Moon JD, Chung HS. (2000) Acceleration of germination of tomato seed by applying AC electric and magnetic fields. *J Electrostatics* 48: 103–114.
Mukherjee SP, Choudhari MA. (1983) Implications of water stressed induced changes in the levels of endogenous ascorbic acid and H_2O_2 in *Vigna* seedlings. *Physiol Plant* 58: 166–170.
Muraji M, Nishimura M, Tatebe W, Fujii T. (1992) Effect of alternating magnetic field on the growth of the primary root of corn. IEEE. *Trans Magn* 28: 1996–2000.
Nandi SK, Palni LMS, de Klerk JM. (1995) The influence of the embryonic axis and cytokinins on reserve mobilization in germinating lupin seeds. *J Exp Bot* 46: 329–336.
Okano H. (2008) Effects of static magnetic fields in biology: role of free radicals. *Front Biosci* 13: 6106–6125.
Payez A, Ghanati F, Behmanesh M, Abdolmaleki P, Hajnorouzi A, Rajabbeigi E. (2013) Increase of seed germination, growth and membrane integrity of wheat seedlings by exposure to static and a 10-kHz electromagnetic field. *Electromagn Biol Med* 32: 417–429.
Pietruszewski S. (1999) Magnetic biostimulation of spring wheat sowing material [in Polish]. Rozprawy Naukowe AR Lublin, Poland.
Pittman UJ. (1963) Magnetism and plant growth: I. Effect of germination and early growth of cereals seeds. *Can J Plant Sci* 43: 513–518.
Pittman UJ (1965) Magnetism and plant growth: III Effect of germination and early growth of corn and beans. *Can J Plant Sci* 45: 549–555.
Pittman UJ, Ormrod DP. (1970) Physiological and chemical features of magnetically treated winter wheat seeds and resultant seedlings. *Can J Plant Sci* 50: 211–217.
Podleśny J, Misiak L, Koper R. (2001) Concentration of free radicals in faba bean seeds after the pre-sowing treatment of the seeds with laser light. *Int Agrophys* 15: 185–189.
Podleśny J, Pietruszewski S, Podleśna A. (2004) Efficiency of the magnetic treatment of broad bean seeds cultivated under experimental plot conditions. *Int Agrophys* 18: 65–71.
Podleśny J, Pietruszewski S, Podleśna A. (2005) Influence of magnetic stimulation of seeds on the formation of morphological features and yielding of the pea. *Int Agrophys* 19: 1–8.
Polovinkina E, Kal'yasova E, Sinitsina YV, Veselov A. (2011) Effect of weak pulse magnetic fields on lipid peroxidation and activities of antioxidant complex components in pea chloroplasts. *Russ J Plant Physiol* 58: 1069–1073.
Pourakbar L, Hatami S. (2012) Exposure of *Satureia hortensis* L. seeds to magnetic fields: effect on germination growth characteristic, and activity of some enzymes. *J Stress Physiol Biochem* 21: 191–198.
Puntarulo S, Galleano M, Sanchez RA, Boveris A. (1991) Superoxide anion and hydrogen peroxide metabolism in soybean embryonic axes during germination. *Biochim Biophys Acta* 1074: 277–283.
Racuciu M, Creanga D, Horga I. (2008) Plant growth under static magnetic field influence. *Rom J Phys* 53: 353–359.
Radhakrishnan R, Kumari BR. (2012) Pulsed magnetic field: A contemporary approach offers to enhance plant growth and yield of soybean. *Plant Physiol Bio* 51: 139–144.
Radhakrishnan R, Kumari BR. (2013) Protective role of pulsed magnetic field against salt stress effects in soybean organ culture. *Plant Biosystems* 147: 135–140.
Rajabbeigi E, Ghanati F, Abdolmaleki P, Payez A. (2013) Antioxidant capacity of parsley cells (*Petroselium crispum* L.) in relation to iron-induced ferritin levels and static magnetic field. *Electromagn Biol Med* 32: 430–441.
Rajendra P, Nayak HS, Sashidhar RB, Subramanyam C, Devendarnath D, Gunasekaran B. (2005) Effects of power frequency electromagnetic fields on growth of germinating *Vicia faba* L. the broad bean. *Electromagn Biol Med* 24: 39–54.
Rathod G. (2013) Effect of magnetopriming on carbohydrate metabolism under salinity in wheat (*Triticum aestivum* L.). MSc thesis, Division of Plant Physiology, IARI, PUSA, New Delhi, India.

Reina FG, Pascual LA. (2001) Influence of a stationary magnetic field on water relations in lettuce seeds. Part I Theoretical considerations. *Bioelectromagnetics* 22: 589–595.
Reina FG, Pascual LA, Fundora IA. (2001) Influence of a stationary magnetic field on water relations in lettuce seeds. Part II: Experimental results. *Bioelectromagnetics* 22: 596–602.
Rochalska M. (2002) Magnetic field as a method of seed vigour estimation. *Acta Agrophysica* 62: 103–111.
Rochalska M, Grabowska K. (2007) Influence of magnetic fields on the activity of enzymes α- and β-amylase and glutathione S-transferase (GST) in wheat plants. *Int Agrophys* 21: 185–188.
Rochalska M, Orzeszko-Rywka A. (2005) Magnetic field treatment improves seed performance. *Seed Sci Technol* 33: 669–674.
Romana R, Dominik V, Igor J. (2000) Influence of aluminum in biologic effects of ELF magnetic field stimulation. *Electro Magneto Biol* 19: 57–68.
Ruzic R, Jerman I. (2002) Weak magnetic field decreases heat stress in cress seedlings. *Electromagn Biol Med* 21: 69–80.
Savostin PW. (1930) Magnetic growth relations in plants. *Planta* 12: 327.
Scaiano JC, Monahan S, Renaud J. (1997) Dramatic effect of magnetic particles on the dynamics of photogenerated free radicals. *Photochem Photobiol* 65: 759–762.
Scandalios JG. (1993) Oxygen stress and superoxide dismutases. *Plant Physiol* 105: 7–12.
Schopfer P, Plachy C, Frahry G. (2001) Release of reactive oxygen intermediates (superoxide radicals, hydrogen peroxide, and hydroxyl radicals) and peroxidase in germinating radish seeds controlled by light, gibberellin, and abscisic acid. *Plant Physiol* 125: 1591–1602.
Selim AH, El-Nady MF. (2011) Physio-anatomical responses of drought stressed tomato plants to magnetic fields. *Acta Astronaut* 69: 387–396.
Serdyukov YA, Novitskii YI. (2013) Impact of weak permanent magnetic field on antioxidant enzyme activities in radish seedlings. *Russ J Plant Physiol* 60: 69–76.
Shabrangi A, Majd A. (2009) Effect of magnetic fields on growth and antioxidant systems in agricultural plants. *PIERS Proceedings*, Beijing, China, March 23–27: pp. 1142–1147.
Shine MB, Guruprasad KN. (2012) Impact of pre-sowing magnetic field exposure of seeds to stationary magnetic field on growth, reactive oxygen species and photosynthesis of maize under field conditions. *Acta Physiol Plant* 34: 255–265.
Shine MB, Guruprasad KN, Anand A. (2011) Enhancement of germination, growth, and photosynthesis in soybean by pretreatment of seeds with magnetic field. *Bioelectromagnetics* 32: 474–484.
Shine MB, Guruprasad KN, Anand A. (2012) Effect of stationary magnetic field strengths of 150 and 200 mT on reactive oxygen species production in soybean. *Bioelectromagnetics* 33: 428–437.
Singh R, Singh S, Parihar P, *et al.* (2016) Reactive oxygen species (ROS): beneficial companions of plants' developmental processes. *Front Plant Sci* 7: 1299.
Singh S, Srivastava PK, Kumar D, Tripathi DK, Chauhan DK, Prasad SM. (2015) Morpho-anatomical and biochemical adapting strategies of maize (*Zea mays* L.) seedlings against lead and chromium stresses. *Biocatal Agric Biotechnol* 4: 286–295.
Singh S, Tripathi DK, Singh S, *et al.* (2017) Toxicity of aluminium on various levels of plant cells and organism: a review. *Environ Exp Bot* 137: 177–193.
Singh VP, Tripathi DK, Kumar D, Chauhan DK. (2011) Influence of exogenous silicon addition on aluminium tolerance in rice seedlings. *Biol Trace Elem Res* 144: 1260–1274.
Solovyov I, Schulten K. (2009) Magneto-reception through cryptochrome may involve superoxide. *Biophys J* 96: 4804–4813.
Solovyov I, Chandler D, Schulten K. (2007) Magnetic field effects in *Arabidopsis thaliana* cryptochrome-1. *Biophys J* 92: 2711–2726.
Soltani F, Kashi A, Arghavani M. (2006) Effect of magnetic field on *Asparagus officinalis* L. seed germination and seedling growth. *Seed Sci Technol* 34: 371–375.

Suda CN, Giorgini JF. (2000) Seed reserve composition and mobilization during germination and initial seedling development of *Euphorbia heterophylla*. *Rev Bras Fisiol Veg* 12: 226–245.

Taia W, Al-Zahrani H, Kotbi A. (2007) The effect of static magnetic forces on water contents and photosynthetic pigments in sweet basil *Ocimum basilicum* L. (Lamiaceae). *Saudi J Biol Sci* 14: 103–107.

Teixeira da Silva JA, Dobránszki J. (2015) How do magnetic fields affect plants in vitro? *In Vitro Cell Dev Biol Plant* 51: 233–240.

Thomas S, Anand A, Chinnusamy V, Dahuja A, Basu S. (2013) Magneto-priming circumvents the effect of salinity stress on germination in chickpea seeds. *Acta Physiol Plant* 35: 3401–3411.

Timmel CR, Till U, Brocklehurst B, McLauchlan KA, Hore PJ. (1998) Effects of weak magnetic field on free radical recombination reactions. *Mol Phys* 95: 71–89.

Touati MA, Boughanmi NG, Salem MB, Haouala R. (2013) Effects of moderate static magnetic field presowing treatment on seedling growth and oxidative status in two *Raphanus sativus* L. varieties. *African J Biotech* 12: 275–283.

Tripathi DK, Singh VP, Kumar D, Chauhan DK. (2012a) Impact of exogenous silicon addition on chromium uptake, growth, mineral elements, oxidative stress, antioxidant capacity, and leaf and root structures in rice seedlings exposed to hexavalent chromium. *Acta Physiol Plant* 34:279–289.

Tripathi DK, Singh VP, Kumar D, Chauhan DK. (2012b) Rice seedlings under cadmium stress: effect of silicon on growth, cadmium uptake, oxidative stress, antioxidant capacity and root and leaf structures. *Chem Ecol* 28: 281–291.

Tripathi DK, Singh S, Singh VP, Prasad SM, Dubey NK, Chauhan DK. (2017) Silicon nanoparticles more effectively alleviated UV-B stress than silicon in wheat (Triticum aestivum) seedlings. *Plant Physiol Biochem* 110: 70–81.

Vakharia DN, Davariya RL, Parameswaran M. (1991) Influence of magnetic treatment on groundnut yield and yield attributes. *Ind J Plant Physiol* 34: 131–136.

Varier A, Vari A-K, Dadlani M. (2010) The subcellular basis of seed priming. *Curr Sci* 99: 450–456.

Vashisth A, Nagarajan S. (2007) Effect of pre-sowing exposure to static magnetic field of maize (*Zea mays* L.) seeds on germination and early growth characteristics. *Pusa Agrisci* 30: 48–55.

Vashisth A, Nagarajan S. (2008) Exposure of seeds to static magnetic field enhances germination and early growth characteristics in chickpea (*Cicer arietinum* L.). *Bioelectromagnetics* 29: 571–578.

Vashisth A, Nagarajan S. (2010) Effect on germination and early growth characteristics in sunflower (*Helianthus annus*) seeds exposed to static magnetic field. *J Plant Physiol* 167: 149–156.

Vasilevski G. (2003) Perspectives of the application of biophysical methods in sustainable agriculture. *Bulgarian J Plant Physiol* Special Issue 2003: 179–186.

Wolff SA, Coelho LH, Karoliussen I, Jost AK. (2014) Effects of the extra terrestrial environment on plants: recommendations for future space experiments for the Melissa higher plant compartment. *Life* 4: 189–204.

Xi G, Fu ZD, Ling J. (1994) Change of peroxidase activity in wheat seedlings induced by magnetic field and its response under dehydration condition. *Acta Bot Sinica* 36: 113–118.

Xia L, Guo J. (2000) Effect of magnetic field on peroxidase activation and isozyme in *Leymus chinensis*. *Ying Yong Sheng Tai Xue Bao* 11: 699–702.

Yaycili O, Alikamanoglu S. (2005) The effect of magnetic field on *Paulownia* tissue cultures. *Plant Cell Tiss Org Cult* 83: 109–114.

Yinan L, Yuan L, Yongquing Y, Chunyang L. (2005) Effect of seed pre-treatment by magnetic field on the sensitivity of cucumber (*Cucumis sativus*) seedlings to ultraviolet-B radiation. *Environ Exp Bot* 54: 286–294.

5

Reactive Oxygen Species: Generation, Damage, and Quenching in Plants During Stress

Krishna Kumar Choudhary[1,2], *Nivedita Chaudhary*[1], *S.B. Agrawal*[1], and *Madhoolika Agrawal*[1]

[1] *Banaras Hindu University, Varanasi, India*
[2] *Agricultural Research Organization, Bet Dagan, Israel*

Introduction

Oxygen in the Earth's atmosphere is vital for the existence of most life forms, although in its ground state O_2 is relatively unreactive. During normal metabolic activities it is changed into reactive oxygen species (ROS). ROS are generated by normal cell metabolism, but production is accelerated under stress conditions and therefore equilibrium between ROS generation and removal may be disturbed (Apel and Hirt, 2004). Numerous defense strategies are employed by organisms to cope with the various stresses to which they are subject, and plants have evolved a range of physiological and biochemical adaptations to enable them to endure those stresses (Hirayama and Shinozaki, 2010). ROS are utilized for various cellular metabolic activities such as photosynthesis and respiration, in chloroplasts, peroxisomes, and mitochondria. Stress may arise from natural as well as anthropogenic activities, and a common feature of different stresses is increased ROS production. In plants, besides normal cellular metabolism, large quantities of ROS are generated as a result of numerous biotic and abiotic stress factors such as ozone pollution, ultraviolet-B radiation, drought, extreme temperatures, salinity, heavy metals, and pathogens (Arora *et al.*, 2002; V.P. Singh *et al.*, 2011; Tripathi *et al.*, 2012a,b, 2016a,b, 2017a–c; S. Singh *et al.*, 2015, 2017; R. Singh *et al.*, 2016). ROS damage plant function by causing lipid peroxidation, protein oxidation, and damage to nucleic acids, enzyme suppression, and stimulation of programmed cell death. Under normal conditions, ROS generation in cell compartments remains low. However, during stress ROS generation can exceed a certain level, thereby disrupting cellular homeostasis and enhancing ROS production (Sharma *et al.*, 2012).

Plants are naturally equipped with antioxidative defense systems consisting of an array of enzymatic and non-enzymatic antioxidants capable of dissipating ROS. For the maintenance of plant vigor, equilibrium between ROS production and their detoxification is required. Imbalance between ROS production and detoxification results in oxidative damage particularly to DNA, proteins, and lipids (Tripathy and Oelmüller, 2012). However, ROS are not simply a harmful product generated during stress and makes the system imperfect

Reactive Oxygen Species in Plants: Boon Or Bane - Revisiting the Role of ROS, First Edition. Edited by Vijay Pratap Singh, Samiksha Singh, Durgesh Kumar Tripathi, Sheo Mohan Prasad, and Devendra Kumar Chauhan.

rather acts an important role in the signaling (Reczek and Chandel, 2015). ROS function as important signaling molecules in regulating various processes such as growth, development, reaction to biotic and abiotic stresses, and programmed cell death (Ahmad *et al.*, 2008). Sensing of ROS occurs in various cellular compartments, and the interaction of ROS with their target molecules stimulates the expression of genes involved in the signal pathways (Laloi *et al.*, 2004).

Abiotic stress involves environmental stress factors such as ozone pollution and increased ultraviolet-B radiation. Other stress factors like heavy metals, salinity, extreme temperature, drought, and pathogens are also potential generators of ROS. The chief sites of ROS production in plants are mitochondria, chloroplasts, and peroxisomes. Apart from reporting the destructive roles of ROS, this chapter elaborates the quenching and combating of ROS-induced oxidative stress. We provide a comprehensive overview of the plant's antioxidative defense system, involving antioxidants (enzymatic and non-enzymatic) capable of detoxifying ROS by acting as an integrative network, using a series of redox reactions. Enzymatic antioxidants comprise of catalase (CAT), superoxide dismutase (SOD), guaiacol peroxidase (GP), ascorbate peroxidase (APX), monodehydroascorbate reductase (MDHAR), dehydroascorbate reductase (DHAR), and glutathione reductase (GR). Non-enzymatic antioxidants include ascorbate (AsA), glutathione (GSH), carotenoids, tocopherols, and phenolics. Normal plant functioning is achieved by maintaining the equilibrium between production of ROS and the plant's antioxidative capabilities. The versatile character of ROS is also overviewed to depict their role as ubiquitous "signaling molecules." The signaling functions of ROS and ROS-induced redox reactions are also described along with the gene expression. ROS signaling is assimilated with many signaling networks in plants, such as calcium signaling, protein kinase networks, cellular metabolic networks, and redox responses (Mittler *et al.*, 2011).

Types of ROS

ROS are derivatives of O_2 and are mainly produced in mitochondria, peroxisomes, and chloroplasts of plant cells. ROS play both harmful and beneficial roles, depending upon their formation and accumulation rates in plant cells. Increased ROS generation is liable to cause damage to plant cells; however, low concentrations lead to the initiation of numerous beneficial reactions in plants. In plants, out of the total O_2 consumed, about 1% is destined for production of the various ROS (Asada and Takahashi, 1987). When O_2 absorbs adequate energy, its unpaired electron gets activated and starts reverse spin leading to formation of singlet oxygen, $^1O_2$, while monovalent reduction leads to formation of the superoxide radical, $^{\cdot}O_2^{\cdot -}$, hydrogen peroxide, H_2O_2, and the hydroxyl radical, $^{\cdot}OH$ (Apel and Hirt, 2004). Here, we provide details of the various types of ROS formed inside plant cells.

Superoxide Radical ($O_2^{\cdot -}$)

Superoxide radical is mainly produced in the thylakoid membrane of chloroplasts by photosystem system I (PSI). Reaction of O_2 with cytochrome *c* oxidase and the alternative oxidase leads to release of H_2O and transfer of four electrons. However, when O_2 reacts with other components of the electron transport chain (ETC) only one electron is transferred, leading to $O_2^{\cdot -}$ formation (Puntarulo *et al.*, 1988). This $^{\cdot}O_2^{\cdot -}$ is further responsible for $^{\cdot}OH$ and $^1O_2$ formation (Elstner, 1987; Halliwell, 2006). After protonation this $O_2^{\cdot -}$ forms $HO_2^{\cdot}$, which attacks polyunsaturated fatty acids (PUFA) causing lipid peroxidation (Bielski *et al.*, 1983). $O_2^{\cdot-}$ is also transformed into $^{\cdot}OH$ via the Haber–Weiss and Fenton reactions.

Hydrogen Peroxide (H_2O_2)

Hydrogen peroxide in plants is generated mainly during photosynthesis and photorespiration. H_2O_2 is relatively stable and plays very important roles inside plant cells. Slesak *et al.* (2007) speculated on the involvement of H_2O_2 in the evolution of photosystem II (PSII) on Earth. At low concentrations, H_2O_2 triggers tolerance mechanisms in plants against various stress factors, whereas programmed cell death (PCD) is the result of higher H_2O_2 accumulation (Quan *et al.*, 2008).

Hydroxyl Radical ($^{\cdot}OH$)

Hydroxyl radicals are highly reactive ROS and can easily react with lipids, proteins, and DNA. $\cdot OH$ can also be produced by the transformation $O_2^{\cdot-}$ and H_2O_2 in the manifestation of some transition metals like Fe, through Fenton's reaction (Gill and Tuteja, 2010). As plants lacks the enzymatic mechanisms for quenching $^{\cdot}OH$, it can ultimately lead to cell death (Gill and Tuteja, 2010).

Singlet Oxygen ($^1O_2$)

This is an excited state of oxygen not related to electron transport to O_2. The chlorophyll triplet state due to uneven energy dissipation during photosynthesis is liable to cause $^1O_2$ formation (Gill and Tuteja, 2010). Also low concentration of intercellular CO_2 inside chloroplast could also leads to $^1O_2$ formation (Gill & Tuteja, 2010). The entire photosynthetic machinery, along with PSI and PSII, experiences severely damaging effects due to $^1O_2$. It has the capacity to react with different biomolecules such as proteins, lipids, nucleic acids, pigments, and so forth, and can alter PSII activity, finally causing cell death (Wagner *et al.*, 2004; Krieger-Liszkay *et al.*, 2008). Tocopherol, plastoquinone, and β-carotenes are well recognized quenchers of $^1O_2$. When this ROS is not quenched inside plant cells, it upregulates various defense genes against photo-oxidative damage (Krieger-Liszkay *et al.*, 2008).

ROS Production in Plants

Photosynthesis in sunlight is the main source of energy for plants. Oxidative strain is a usual phenomenon for all living beings. Plants are at risk of oxidative damage mainly due to the presence of photoreceptors, oxygenic environments, and PUFA in the chloroplasts (Gill and Tuteja, 2010). During light conditions chloroplasts and peroxisomes are key sites of ROS production (Foyer and Noctor, 2003), while mitochondria are chief sources during darkness (Møller, 2001). In this section, we discuss the various sites of ROS production.

Chloroplasts

The chloroplast is well organized and equipped with thylakoid membranes and light-harvesting complexes (Pfannschmidt, 2003). ROS ($O_2^{\cdot-}$, $^1O_2$, and H_2O_2) are generated in the chloroplast at various sites, but PSI and PSII related to ETCs are target sites. Various stress conditions lead to CO_2 limiting conditions, mainly liable for ROS generation. During normal conditions, electrons flow from excited photosystems to $NADP^+$, reducing it to NADPH, which subsequently reduces the ultimate electron acceptor, CO_2, in the Calvin–Benson cycle. During stressed conditions electron transport flows through ferredoxin to O_2, and this diverted flow of electrons leads to $O_2^{\cdot-}$ formation via the Mehler reaction (Wise and Naylor, 1987; Elstner, 1991). Further PSII studies have confirmed the participation of quinone A (QA) and quinone B (QB) in $O_2^{\cdot-}$ production (Takahashi *et al.*, 1988). Under low light conditions $^1O_2$ is the regular by-product at PSII (Buchert and Forreiter, 2010). H_2O_2 formation at the stromal membrane is the result of $O_2^{\cdot-}$ dismutation by Cu/Zn-SOD (Takahashi *et al.*, 1988). Further, H_2O_2 may be transformed to $\cdot OH$ depending upon the availability of Fe^{2+} at Fe-S centers with the help of the Fenton reaction.

Peroxisomes

Peroxisomes are small, spherical organelles bounded by a single lipid bilayer. They are concerned with the oxidation of long-chain fatty acids, and are major sites for ROS generation in plants. There are two main sites of $O_2^{\cdot-}$ production in the peroxisome, within the organelle's matrix and on the peroxisome membrane (Del Río *et al.*, 2002). In the matrix $O_2^{\cdot-}$ is produced due to the action of xanthine oxidase (XOD) during the tranformation of xanthine as well as hypoxanthine to uric acid (Corpas *et al.*, 2001). Generation of $O_2^{\cdot-}$ on the peroxisome membrane involves cytochrome *b* and NADH (with flavoprotein) along with the participation of MDHAR (Del Río *et al.*, 2002). H_2O_2 generation in peroxisomes is mainly dependent on the glycolate oxidase response, the β-oxidation of fatty acids, and the reactions of flavin oxidases and imbalance of $O_2^{\cdot-}$ (Huang *et al.*, 1983; Del Rio *et al.*, 2002, 2006).

Mitochondria

Mitochondria are the cell's "powerhouses" and are also major sites of ROS production, mainly at various sites in the ETC. ROS generation inside mitochondria is a general phenomenon, but various stresses causes modification of electron carriers in the ETC leading to enhancement of ROS formation (Noctor *et al.*, 2007; Blokhina and Fagerstedt, 2010). Complex I and III are identified as a major site for $O_2^{\cdot-}$ production in mitochondrial ETC, and this $O_2^{\cdot-}$ is further reduced to H_2O_2 by SOD activity (Raha and Robinson, 2000; Sweetlove and Foyer, 2004; Quan *et al.*, 2008). According to Møller (2001) about 1–5% of O_2 consumption by mitochondria is related with H_2O_2 generation, which further generates highly toxic $^{\cdot}OH$ by reaction with Fe^{2+} and Cu^+; these $^{\cdot}OH$ radicals are capable of migration from mitochondria through membrane penetration (Sweetlove and Foyer, 2004; Rhoads *et al.*, 2006). Any stress can change the physiology of plants, and mitochondria can play a principal role in controling ROS generation by energy dissipating systems (Gill and Tuteja, 2010).

Other Sources

There are also various other known sites of ROS formation in plants, like endoplasmic reticulum, plasma membrane, cell wall, and apoplast. In endoplasmic reticulum, cyt P450 with flavoproteins leads to ROS formation (Dybing *et al.*, 1976). Oxidoreductases, which take part in electron transport, are commonly present on plasma membranes and are mainly responsible for ROS production. NADPH oxidase has a major role in ROS metabolism during stress conditions in plants (Kwak *et al.*, 2003; Apel and Hirt, 2004). Cell walls are also a potential site, where NADH and peroxidases lead to ROS formation (Gross, 1977). Apoplastic ROS generation depends on enzymes of the cell wall (Apel and Hirt, 2004; Heyno *et al.*, 2011), for instance, oxalate oxidase leads to H_2O_2 generation from oxalic acid in the apoplastic region (Wojtaszek, 1997). Further studies have confirmed that generation of $\cdot OH$ in the apoplast is mainly dependent on cell wall peroxidases (Heyno *et al.*, 2011).

Various Stresses Generate ROS in Plants

ROS formation is a normal phenomenon for all living beings throughout their life cycle. Plants subject to stress conditions undergo increased ROS formation (Sharma *et al.*, 2010). Numerous stresses causes oxidative stress in plants and consequent stimulation of the antioxidative defense system (enzymatic and non-enzymatic) to provide protection against ROS (Figure 5.1). ROS production under several stress factors is discussed in this section.

UV-B Stress

UV-B radiation coming from the Sun to the Earth's surface is detrimental to plant growth, physiology, and yield (Agrawal and Rathore, 2007; Choudhary and Agrawal, 2015). Decreases in leaf size, chlorophyll content, photosystem II activities, stomatal conductance, and chlorophyll fluorescence have been observed (Choudhary and Agrawal, 2014a; Kakani *et al.*, 2003).

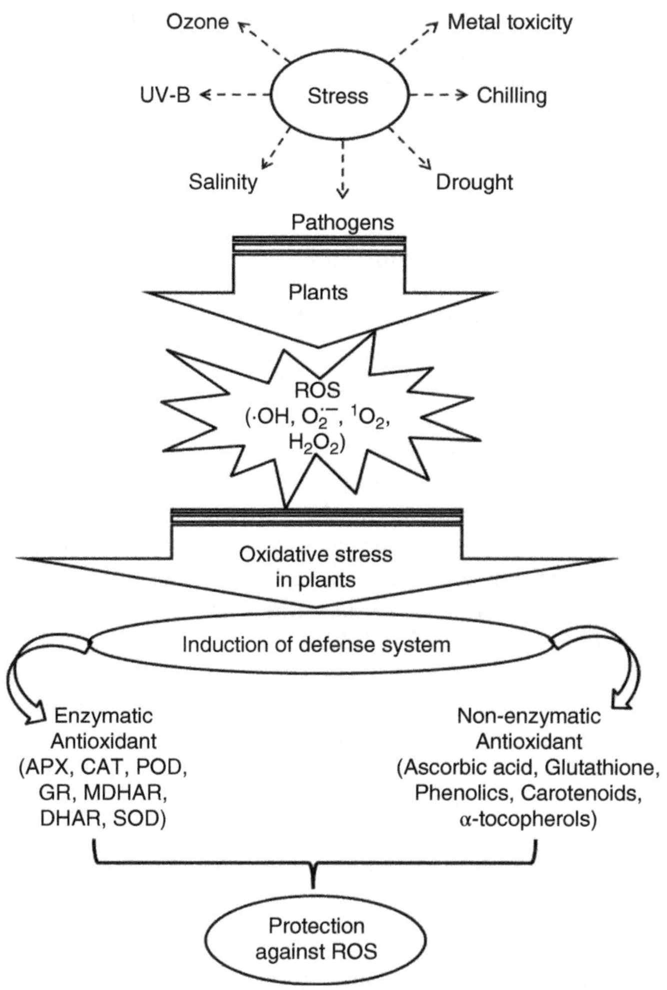

Figure 5.1 An overview of reactive oxygen species (ROS) production and its quenching in plants. APX, ascorbate peroxide; CAT, catalase; DHAR, dehydroascorbate reductase; GR, glutathione reductase; MDHAR, monodehydroascorbate reductase; POD, peroxidase; SOD, superoxide dismutase.

Elevated UV-B is known to cause increases in lipid peroxidation, solute leakage, and formation of ROS (e.g., $O_2^{\cdot-}$, $^1O_2$, H_2O_2, and $^{\cdot}OH$). Reduction in photosynthetic pigments and protein content, and increase in phenolics and non-photosynthetic pigments are well known UV-B effects on plants. Plants exposed to UV-B radiation often display low photosynthetic rates due to impaired functions of photosystems, especially PSII reactions in thylakoid membranes, a decrease in rubisco activity and the enzymatic processes of the Calvin–Benson cycle, ATPase activity, and stomatal limitations to CO_2 diffusion (Surabhi *et al.,* 2009). PSII is a complex of pigments and proteins; the core is made up of D1 and D2 proteins (Barber *et al.,* 1997). Degradation of D1 and D2 is driven even at UV-B fluence rates as low as 1 μmol m^{-2} s^{-1} (Jansen *et al.,* 1998; Bouchard *et al.,* 2006). A detailed analysis of different PSII properties revealed that donor and acceptor sites of the PSII reaction center were affected by UV-B, with the primary target for PSII damage by UV-B at acceptor sites (QA, QB, and plastoquinone pool) (Apostolova *et al.,* 2014). Photo-oxidative stress under UV-B could potentially disturb CO_2 fixation and reduce the content and activity of $NADP^+$ and ribulose bisphosphate carboxylase/oxygenase (rubisco). Consequently, the photosynthetic ETC is over-reduced, forming superoxide radical and singlet oxygen in the chloroplasts. To overcome these negative effects, plants induce antioxidative defense mechanisms (Agrawal *et al.,* 2009). Increased activities of different antioxidative enzymes (SOD, POX, APX, CAT, GR, etc.) enable ROS scavenging (Choudhary and Agrawal, 2014b; Takshak and Agrawal, 2014).

Ozone Stress

In the troposphere, O_3 (ozone) is an important pollutant and reported to be harmful to plants. Ozone penetrates leaves through their open stomata and follows the same path as carbon dioxide. Ozone is a powerful oxidizing agent able to react with molecules on cell walls and activate the production of the ROS, which causes numerous destructive effects inside cells, affecting lipids, proteins, and nucleic acids (Iriti and Faoro, 2008; Chaudhary and Agrawal, 2013; Singh *et al.,* 2014; Rai *et al.,* 2015). In a cell O_3 itself infrequently gets far-off and being a strong oxidant, can interact with components of the apoplast to produce ROS such as H_2O_2, $O_2^{\cdot-}$, $^{\cdot}OH$, and $HOO^{\cdot}$ radicals (Heath, 2008). Production of these ROS induced by exposure to O_3 seems to an example of an "oxidative burst" in the O_3 affected cells; therefore the appearance of symptoms and ROS accumulation at the injured sites in plants are detected along with localized cell death (Pasqualini *et al.,* 2003). Accumulation of H_2O_2 has been confirmed in tobacco leaves (Schraudner *et al.,* 1998), while in *Arabidopsis* the accumulation of both $O_2^{\cdot-}$ and H_2O_2 has been observed by Rao *et al.* (2000). ROS accumulation at definite sites and the visible lesions that develop suggested that ROS act as the key regulators of cell death (Overmyer *et al.,* 2000). Chronic dosage of O_3 is also capable of generating lesions in leaves, as observed by H_2O_2 localization in mung bean cultivars, although this is suppressed by a foliar spray of 10 M ascorbic acid (Chaudhary and Agrawal, 2014a,b). Maize cultivars exposed to ambient + 30 ppb of O_3 showed the localization of both H_2O_2 and $O_2^{\cdot-}$ (Singh *et al.,* 2014). Besides the localization of ROS, increments in the content of ROS (H_2O_2 and $O_2^{\cdot-}$) were observed in crop plants exposed to the elevated dose of O_3, such as mung bean, clover, and linseed under field conditions (Chaudhary and Agrawal, 2013; Tripathi *et al.,* 2011; Mishra and Agrawal, 2015).

Chilling Stress

Under low-temperature conditions, plants undergo various modifications such as the accumulation of soluble sugars, RNA chaperones, dehydrins, and ROS, and undergo numerous physiological disturbances. They may go on to develop chilling symptoms leading to the plant's death. Greater ROS production under chilling conditions is mainly related to the imbalance of absorption and use of light in the Calvin–Benson cycle (Logan *et al.,* 2006). Significant reductions in rubisco content and its activity have been observed in cucumber plants under chilling conditions (Zhou *et al.*, 2006). Lipid peroxidation along with oxidation of proteins by H_2O_2 and $O_2^{\cdot-}$ are common phenomena under chilling stress (Prasad, 1997; Zhang *et al.*, 2008a). To combat oxidative stress under chilling conditions, plants show enhanced activities of antioxidant enzymes (APX, GR, SOD, DHAR, MDHAR, etc.) (Fryer *et al.*, 1998; Zhang *et al.*, 2008b). Non-enzymatic antioxidants (GSH, AsA, carotenoids, and α-tocopherol) also have significant roles in overcoming chilling stress (Radyuk *et al.*, 2009). Under severe and long-duration cold stress plants are unable to scavenge ROS efficiently, which may lead to death of the plants (Zhang *et al.*, 2008a).

Drought Stress

Drought is also an important environmental constraint that increases oxidative load on plant species in many ways. During drought, CO_2 inhibition leads to disturbances in activities of the photosystems and ETC inside the chloroplast, resulting in increased ROS generation (Asada, 1999). Also, excess light energy dissipation toward PSII and antenna pigment leads to greater ROS formation (Foyer and Harbinson, 1994). Production of $^{\cdot}OH$ through H_2O_2 reduction with the help of Fe, by SOD and AsA in chloroplast, is also a potential threat under drought. Increased ROS under drought led to increased activities of SOD, APX, POD, DHAR, GR, AsA, and GSH (Wang *et al.*, 2012).

Salinity Stress

Salinity stress also leads to more ROS like $O_2^{\cdot-}$, $^1O_2$, H_2O_2, and $^{\cdot}OH$ by the impairment of ETCs in chloroplasts and mitochondria. Inhibition of carbon fixation due to reduced availability of CO_2 after stomatal closure is a common phenomenon under salinity stress. A low CO_2 to O_2 ratio in chloroplast leads to increased production of H_2O_2 (Hernandez *et al.*, 2000). However, under high CO_2 conditions low ROS generation resulted due to higher rates of assimilation and lowered photorespiration (Perez-Lopez *et al.*, 2009). Lipid peroxidation and oxidation of proteins and nucleic acids are common damaging effects of ROS under salinity stress (Hernandez *et al.*, 2000; Tanou *et al.*, 2009; Karray-Bouraoui *et al.*, 2011). SOD, CAT, GR, APX, and GP are common antioxidant enzymes that combat the negative effects of high salt conditions (Mishra *et al.*, 2013). Upregulation of many genes at proteomic and genomic levels has been observed under salinity stress (Wang *et al.*, 2008).

Heavy Metal Stress

Agricultural soils, on a global scale, are slightly to moderately polluted by heavy metals like Cd, Cu, Zn, Ni, Co, Cr, Pb, As, and so forth. The formation of ROS, like $O_2^{\cdot-}$, H_2O_2, and $^{\cdot}OH$, is the primary response of plants against heavy metal stress. Several metals take part in ROS generation through the Haber–Weiss reaction (Mithofer *et al.*, 2004), by the disruption of ETC (Qadir *et al.*, 2004), and CO_2 fixation (Moustakas *et al.*, 1994). Oxidative injury to nucleic acids, lipids, and proteins is a common phenomenon under metal stress. Higher activities of GP, SOD, GR, APX, DHAR, MDHAR, and non-enzymatic antioxidants have been observed in different plant species under metal stress (Gallego *et al.*, 1996; Yamamoto *et al.*, 1997; Sharma and Agrawal, 2005; Maheshwari and Dubey, 2009).

Pathogens

Pathogens also affect plant metabolism and productivity by the excess production of ROS. Plasma membrane-associated NADPH oxidase leads to an oxidative burst during incompatible plant–pathogen communication (Bhattacharjee, 2012). During pathogen attack, $O_2^{\cdot-}$ and

H_2O_2 generation in the apoplastic region have been observed in potato (Doke, 1983; Grant *et al.*, 2000). Although the primary ROS production site is the apoplast, ROS may be produced in chloroplasts and mitochondria after pathogen attack (Abdollahi and Ghahremani, 2011). ROS generation during pathogen interaction causes necrosis; also, SIPK/Ntf4/WIPK activation leads to ROS formation in chloroplast, which later participates in cell death signaling in plants (Liu *et al.*, 2007). Salicylic acid (SA) also plays a significant role in ROS accumulation by regulating CAT and APX activities during pathogen infection (Mittler *et al.*, 1999; Klessig *et al.*, 2000). Enhanced SOD, POD, CAT, and APX activities have been observed in pathogen infections in plants (Radwan *et al.*, 2010; Ashry and Mohamed, 2012).

Quenching of ROS in Plants

Plants exposed to different stress conditions like high or low temperature, heavy metals, drought, air pollutants, pathogens, salinity, and so forth generate increased concentrations of ROS such as $O_2^{\cdot-}$, $^1O_2$, H_2O_2, and $^{\cdot}OH$ (see Figure 5.1). As a protection mechanism against these toxic radicals, plants possess antioxidant defense systems, which comprise of enzymatic and non-enzymatic antioxidants. Enzymatic defense includes CAT, APX, SOD, GR, DHAR, and MDHAR, whereas non-enzymatic antioxidants are carotenoids, GSH, AsA, and tocopherols (Mittler *et al.*, 2004).

Enzymatic Defense System

Catalase (CAT; 1.11.1.6)

Catalase is universal enzyme present in all living organisms that utilize oxygen, whether animals, plants, or bacteria. An important enzyme, it plays a vital role in protecting plant cells against oxidative damage caused by ROS. Moreover, it was the first enzyme to be discovered and characterized (Sharma *et al.*, 2012). Its main role is in the decomposition of H_2O_2 to H_2O and O_2 with the help of its heme cofactor. One molecule of catalase has the capacity to convert about six million H_2O_2 molecules to H_2O and O_2 per minute. Three classes of catalase have been described to date. Class I is mainly present inside photosynthetic tissues and its regulation is light dependent. Class II is found in the vascular tissues, whereas class III occurs in young seedlings (Willekens *et al.*, 1995). Also, three catalase isoforms have been located in maize plants, namely CAT1 and CAT2 in peroxisomes and cytosol, and CAT3 in mitochondria (Scandalios, 1990). Scavenging of peroxisomal H_2O_2 is mainly done during photorespiratory oxidation, fatty acid β-oxidation, and purine catabolism. Increased H_2O_2 concentrations in stressful conditions is a common phenomenon in plants. To combat these increments of H_2O_2, alteration in catalase activity has been described under UV-B (Choudhary and Agrawal, 2014b; Takshak and Agrawal, 2014) and Cd (Agrawal and Mishra, 2007) stress.

Ascorbate Peroxidase (APX; 1.1.11.1)

Ascorbate peroxidase mainly detoxifies H_2O_2 in plant cells by utilizing ascorbate as a substrate and forming DHA and H_2O as products. APX is the central component of the ascorbic acid-glutathione (AsA-GSH) cycle, having an important role in controlling ROS production. Five different isoforms of APX are known in plants, located in cytosol, peroxisomes, mitochondria, thylakoid, and stroma (Gill and Tuteja, 2010). APX is known to have much better affinity to H_2O_2 as compared to CAT and POD, which makes it an efficient H_2O_2 scavenger (Wang *et al.*, 1999). Under pathogen attack, it has been shown that higher APX activity leads to increased

POD activity, which boosts tolerance to oxidative damage and pathogens (Sarowar *et al.*, 2005). Higher APX activities have been reported under various stress factors, including UV-B (Choudhary and Agrawal, 2016), ozone (Chaudhary and Agrawal, 2013), and metal toxicity (Agrawal and Mishra, 2007).

Superoxide Dismutase (SOD; 1.15.1.1)

SOD is the most important metalloenzyme present in all aerobic organisms, which provides defense against $O_2^{\cdot-}$ under various stresses (Scandalios, 1993). Cu/Zn-SOD, Mn-SOD, and Fe-SOD are the three known isoforms in plants, all leading to the dismutation of $O_2^{\cdot-}$ to O_2 and H_2O_2 (Fridovich, 1989). Fe-SOD and Mn-SOD are present in chloroplasts and mitochondria, respectively however, Cu/Zn-SOD is located in chloroplasts, mitochondria, peroxisomes, and cytosol (Sharma *et al.*, 2012). Various reports have suggested that Cu/Zn-SOD is present in dimer forms only and is cyanide sensitive, whereas Fe-SOD and Mn-SOD can be in dimer or tetramer form and are cyanide insensitive (Scandalios, 1993; Del Rio *et al.*, 1998). Increased SOD activity is a common defense strategy adopted by plants against different biotic and abiotic stresses. Increased SOD production by plants was observed under UV-B (Takshak and Agrawal, 2014; Choudhary and Agrawal, 2016) and ozone (Chaudhary and Agrawal, 2013).

Glutathione Reductase (GR; 1.6.4.2)

GR is the flavoprotein oxidoreductase, generally present in all eukaryotes and prokaryotes. It plays a vital role in ROS defense as a central component of the ascorbate-glutathione (AsA-GSH) cycle. In plant cells it is found chiefly in chloroplasts, but also occurs in mitochondria and cytosol in small amounts (Gill and Tuteja, 2010). GR catalyzes the formation of glutathione (GSH) through NADPH-dependent reduction of oxidized glutathione (GSSG) (Reddy and Raghavendra, 2006). GSSG consists of two GSH molecules joined together with disulfide bond, which can be transformed to GSH by GR activity. The balance between GR and GSH is an important factor in determining the tolerance of plants against any environmental stress (Chalapathi Rao and Reddy, 2008). The transformation of GSSG to GSH with GR involves two reactions (Ghisla and Massey, 1989). First, the flavin part of GSSG is reduced by NADPH, and after the flavin is oxidized the disulfide bridge gets reduced to produce thiolate and cysteine anion. The second step leads to GSSG reduction through interchange reactions of thiol disulfide. Induction of GR activity under environmental stress has been explained in various studies and well correlated with the defense mechanism of plants (Chaudhary and Agrawal, 2013; Choudhary and Agrawal, 2014a; Takshak and Agrawal, 2014)

Dehydroascorbate Reductase (DHAR; 1.8.5.1)

Dehydroascorbate reductase is a monomeric thiol found mainly in roots and green shoots as well as in dry seeds. DHAR is of utmost importance to maintain reduced AsA inside plants cells. It catalyzes the transformation of dehydroascorbate (DHA) to ascorbic acid (AsA) through utilization of GSH as substrate (Ushimaru *et al.*, 1997). DHAR overexpression leads to plants' tolerance against various stresses. During AsA regeneration from monodehydroascorbate (MDHA) via enzymatic or non-enzymatic pathways, there is always the possibility of DHA formation. This DHA has a very short life inside plant cells, and is rapidly transformed to AsA by the action of DHAR, or to 2,3-diketogulonic acid via irreversible hydrolysis (Sharma *et al.*, 2012). Also, DHAR overexpression is related to enhancement and maintenance of the AsA pool size (Chen *et al.*, 2003; Qin *et al.*, 2011). Plants with DHAR overexpression exhibited tolerance against metal toxicity (Yin *et al.*, 2010), ozone stress (Chen and Gallie, 2005), and salinity (Ushimaru *et al.*, 2006).

Monodehydroascorbate Reductase (MDHAR; 1.6.5.4)

This enzyme belongs to the oxidoreductase family, and is mainly present in chloroplasts, mitochondria, peroxisomes, and cytosol of plant cells. MDHA produced after the APX reaction has a very short lifetime; if not reduced rapidly, it will be transformed to AsA and DHA (Ushimaru *et al.*, 1997). MDHAR shows high affinity toward MDHA as electron acceptor; NADH is its first choice as electron donor, compared to NADPH. It is the only known enzyme that utilizes MDA (organic radical) as substrate and also reduces phenoxyl radicals produced after H_2O_2 and horseradish peroxidase reaction (Sakihama *et al.*, 2000). MDHAR in chloroplasts has two major roles; the first is AsA generation from MDHA, and the second is reduction of O_2 to $O_2^{\cdot -}$ in the absence of MDHA (Miyake *et al.*, 1998). Overexpression of MDHAR provides tolerance against chilling stress in tomato (Stevens *et al.*, 2008) and salinity stress in potato (Eltayeb *et al.*, 2007).

Guaiacol Peroxidase (GP; 1.11.1.7)

This enzyme contains heme protein and is present in plants, vertebrates, and fungi. It contains a prosthetic group (ferriprotoporphyrin IX) and oxidizes numerous aromatic substrates like guaiacol and pyragallol by utilizing H_2O_2 (Vianello *et al.*, 1997). Isozymes of GP are present mainly in vacuoles, cytosol, and cell wall (Asada, 1992). GP plays diverse roles inside plant cells, including cell wall lignification, ethylene biosynthesis, indole acetic acid degradation, wound healing, and tolerance against various stress factors (Kobayashi *et al.*, 1996). GP is well recognized for effective scavenging of reactive forms of oxygen and peroxy radicals under different stresses (Vangronsveld and Clijsters, 1994). Induced activity of GP is well correlated with its defensive role against ROS, under metal (Radotic *et al.*, 2000) and salinity (Tayefi-Nasrabadi *et al.*, 2011) stress.

Non-enzymatic Defense System

Ascorbic Acid (AsA)

Ascorbic acid, also called vitamin C, is an organic compound having antioxidant properties. Highest concentrations are found in mature leaves with completely developed chloroplasts (Smirnoff *et al.*, 2004). AsA is a low molecular weight and very efficient ROS quencher, because of its tendency to donate electrons easily in enzymatic and non-enzymatic reactions (Sharma *et al.*, 2012). AsA plays a vital role in removal of H_2O_2 through the AsA-GSH cycle, where AsA is transformed into MDHA and then to DHA (Pinto *et al.*, 2003). MDHA and DHA are highly unstable compounds; MDHA can be reduced to AsA with the help of MDHAR (Miyake and Asada, 1994) while DHA can be transformed to tartarate and oxalate at pH greater than 6.0 (Noctor and Foyer, 1998) or it can be transformed to AsA by DHAR (Asada, 1996). Most of the AsA is located in the cytoplasm, but some is present in the apoplast where it acts as first line of defense against external ROS (Barnes *et al.*, 2002). Membrane protection is also provided by AsA, as it reacts directly with H_2O_2 and $O_2^{\cdot -}$ to generate α-tocopherol (Noctor and Foyer, 1998). Modifications in the level of AsA in the face of various stresses have been described in several studies (Chaudhary and Agrawal, 2013; Maheshwari and Dubey, 2009).

Glutathione (GSH)

Glutathione, an important metabolite present in plants, acts as a strong antioxidant. GSH is capable of avoiding damage to cellular components due to $^1O_2$, H_2O_2 and $^{\cdot}OH$. It is a tripeptide with gamma peptide linkages (γGlu-Cys-Gly). Studies have demonstrated its presence in various cellular organelles and compartments, including chloroplasts, peroxisomes, mitochondria, cytosol, apoplast, vacuoles, and endoplasmic reticulum (Mittler and Zilinskas 1992; Jimenez *et al.*, 1998). Besides signal transduction, it also participates in gene expression under stress and sulfate transport (Mullineaux and Rausch, 2005; Rausch and Wachter, 2005); it also plays a vital role in cell differentiation, resistance against pathogens, cell death, senescence, and enzymatic regulation (Xiang *et al.*, 2001). GSH acts as substrate for various cellular reactions that produce the oxidized form of glutathione (GSSG). The balance between the utilization and production of GSH and GSSG is vital for the proper functioning of plant cells (Foyer and Noctor, 2005). Role of GSH in counteracting ROS production is reflected in several studies examining several stress factors (Xiang *et al.*, 2001; Agrawal, 2007).

Carotenoids

Carotenoids are the pigments mainly present in chloroplasts and chromoplasts of plants, and also in some photosynthetic microorganisms. Carotenoids have a photoprotective role through the dissipation of extra energy as heat, by scavenging ROS and reducing lipid peroxidation. Carotenoids also participate in light harvesting by absorbing light of 400 and 550 nm and transferring this energy to chlorophyll (Sieferman-Harms, 1987). They scavenge triplet sensitizer (3Chl*; an excited triplet configuration of chlorophyll), $^1O_2$, and other free radicals as antioxidants to prevent damage to the photosynthetic apparatus (Collins, 2001). Carotenoids also act as signaling molecules for the stability of PSI, light-harvesting complex, and thylakoid membrane stabilization (Niyogi *et al.*, 2001). The prevention of triplet chlorophyll formation is based on the chemical properties of carotenoids; however, the presence of an isoprene chain permits the easy uptake of energy and its dissipation as heat from the excited molecules (Mittler, 2002). An increased carotenoid content has been shown to be beneficial under different environmental stresses (Gomathi and Rakkiyapan, 2011; Choudhary and Agrawal, 2016).

Tocopherols

These are also major potential ROS scavengers in the plant cell, being mainly involved in the quenching of ROS such as $^1O_2$ and peroxy lipid radicals (Diplock *et al.*, 1989). Four isomers of tocopherol (α, β, γ, and δ) have been described in plants, of which α-tocopherol posseses maximum antioxidant capacity owing to the three methyl groups in its structure (Fukuzawa *et al.*, 1982). Only photosynthetic organisms can make tocopherols, which occur mainly in green parts of the plant. Tocopherols are very useful for the protection of PSII, as they interact chemically and physically with ROS to prevent lipid membranes (Ivanov and Khorobrykh, 2003). A single molecule of tocopherol has the capacity to quench about 220 molecules of $^1O_2$ (Munné-Bosch, 2005). The benzoquinone ring and phytyl chain of tocopherol play a significant role in the quenching of $^1O_2$ (Fryer, 1992). Studies have suggested that tocopherols accumulate under different stresses like chilling, drought, and salinity (Yamaguchi-Shinozaki and Shinozaki, 1994; Munné-Bosch *et al.*, 1999; Bafeel and Ibrahim, 2008).

Phenolics

Phenolics are widely dispersed and the most abundant secondary metabolites in the plant kingdom. Flavonoids, phenolic acids, and tannins are common phenolics present in plants. Plant phenolics are mainly derivatives of cinnamic acid and benzoic acid. The antioxidant properties of phenolics have two aspects, scavenging and suppressing ROS formation (Grace and Logan, 2000). These compounds are free radical acceptors and inhibit oxidation of lipids and other molecules by donation of hydrogen molecules. Under adverse environmental conditions, flavonoids are involved in neutralization of ROS before they prove harmful to plant cells (Løvdal *et al.*, 2010). Polyphenols can also decrease membrane fluidity by modifying the lipid structure (Arora *et al.*, 2000). Induction of phenolics under different stresses has been observed in several studies (Michalak, 2006; Choudhary *et al.*, 2013; Choudhary & Agrawal, 2014a).

ROS Induce Signaling in Plants

The production of ROS in plants is stimulated by various environmental stresses, which disturb the balance between production of ROS and their detoxification by enzymatic and non-enzymatic antioxidants that diminish the oxidative stress (Apel and Hirt, 2004). High levels of ROS formation lead to photo-oxidative damage to proteins, DNA, and lipids and eventually cell death occurs. But apart from these damaging consequences of ROS, they also act as signaling molecules required for normal cellular function, including developmental processes, the hypersensitive reaction, systemic acquired resistance against pathogens, stress hormone production, acclimation, and PCD (Bhattacharjee, 2005). ROS are a central component of the stress response, and the amounts of ROS decide their function – high concentrations lead to cell death whereas low concentrations initiate expression of the defense genes (Das and Roychoudhury, 2014) (Figure 5.2). ROS influence various processes in cells, and at different developmental stages of plants. The ROS-induced biological reactions depend on the ROS dose, intensity of the signal, production site, and the plant's development stage (Zaninotto *et al.*, 2006). In the process of signal transduction ROS play a major role and alter cellular components as they are extremely reactive molecules able to oxidize cellular components and cause damage. To combat any stress numerous enzymes required in the scavenging of ROS are influenced, either downregulated or overexpressed (Ahmad *et al.*, 2008).

When plants encounter different stress factors, cells generate the capacity to rapidly produce and scavenge diverse kinds of ROS in a synchronized way, facilitating fast and vigorous fluctuations in the amount of ROS (Mittler *et al.*, 2011).

Additional advantageous functions of ROS are to disseminate the signal automatically for long distances throughout the plant (Mittler *et al.*, 2011). The different molecular properties of ROS determine their mobility within cells. The superoxide radical is not able to cross membranes, but can change into H_2O_2, which transfers through membranes and acts as a

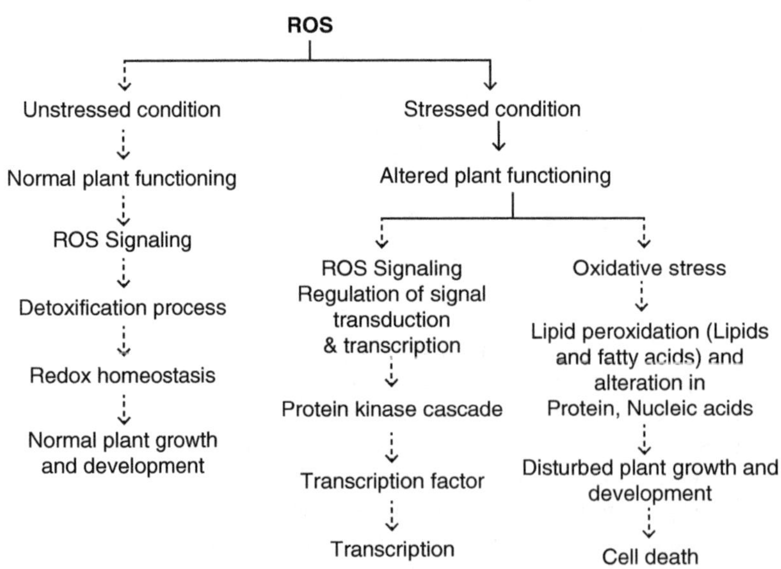

Figure 5.2 Role of reactive oxygen species (ROS) and the biological consequences with and without stress.

stable intermediate of ROS metabolism; it performs as a second messenger, since it can diffuse from the site of production (Varnova *et al.*, 2002). Furthermore, cells have evolved numerous mechanisms for prompt and controllable ROS production and exclusion (Grether-Beck *et al.*, 2000).

Signaling Functions of ROS

Hydrogen peroxide

Hydrogen peroxide (H_2O_2) is uncharged and relatively stable compared to other ROS. It is capable of traveling long distances and penetrating membranes, so can act as an efficient messenger within or between cells (Mullineaux and Lawson, 2009). In plant responses to stress, H_2O_2 activates genes encoding many antioxidant and signaling proteins leading to stimulation of APX, GR, CAT, mitogen-activated protein kinase (MAPK), and phosphatases. H_2O_2 production initiates the signalling from chloroplasts to nucleus through activation of MAP kinase cascade (Pfannschmidt *et al.*, 2009).

Singlet Oxygen

Singlet oxygen possesses a short lifespan, therefore its action requires the participation of other signaling components such as the proteins Executer 1 and 2. Singlet oxygen induces lipid peroxides, and their derivatives function as signal molecules releasing from chloroplasts (Kreslavski *et al.*, 2012). It also exits from chloroplasts into cytosol and even reaches the nucleus to stimulate expression of the nuclear *GPXH* gene encoding glutathione peroxidase. Therefore, singlet oxygen ($^1O_2$) not only causes oxidation of PUFA but also lipid oxidation of thylakoid membrane (Valdivieso and Mullineaux, 2010).

Superoxide Anion

Superoxide anion also has a short life and shows similar signaling characteristics as H_2O_2. Signal transduction by superoxide anion ($O_2^{\cdot-}$) has been shown when plants are deficient in Cu/Zn-SOD (Rizhsky *et al.*, 2003). Superoxide anion reacts rapidly with NO, producing peroxynitrite in chloroplasts, also fulfilling signaling functions (Kreslavski *et al.*, 2012).

Redox Responsive Proteins

Plants are able to sense, transduce, and translate the signals induced by ROS into suitable cellular reactions. This process involves redox-sensitive proteins (RSPs) that can undergo reversible oxidation-reduction depending on the redox state of the cell (Foyer and Noctor, 2003). Glutathione or thioredoxins are abundant redox-sensitive molecules, able to control the redox state in the cells of higher plants (Foyer and Noctor, 2005).

Plants naturally encounter various stresses and can deploy a repertoire of protective responses involving various cellular functions stimulated by ROS signaling. Plants commonly experience excess photochemical energy, therefore chloroplasts stimulate RSPs in the photosynthetic electron transport chain of cytochrome *bf* complex. Free radicals generated during light harvesting leads to oxidative stress, if not balanced through utilization and dissipation might leads to cell death. However, ROS can show an affirmative function in the response to excess photochemical energy through photoinhibition and by introducing proliferation in the degradation rate of D1 protein at photosystem II (PSII) reaction center (Karuse, 1994). Phosphorylation is another method for controlling the dissipation of surplus light energy from PSII. RSP initiates the signals in the cytochrome bf complex which stimulates kinase activation and activation of phosphorylation proceeded by the plastoquinone pool (Vener *et al.*, 1998). Therefore, ROS are significant contributors to cellular redox status, which participates in processes of controlling damage by acting as an “alarm” that initiates defense responses.

Depending on the degree of oxidative stress, linked signal transduction pathways are stimulated that can lead either to stress acclimation or to cell death (Shao *et al.*, 2008). Plant vigor may be affected by numerous biotic and abiotic stresses that precede signal transduction in plants. Initially, signaling comprises ion fluxes through the plasma membrane, increasing Ca^{2+} levels in the cytosol and activating MAPKs. The plasma membrane permits inward calcium flux as Ca^{2+} channels are transiently open. This Ca^{2+} signal might be related to regulatory mechanisms of ROS-producing enzymes (Mittler *et al.*, 2012).

ROS and Redox Signaling

Numerous signaling actions in cells are based on redox reactions, and ROS are directly associated with the cellular redox metabolism. ROS cause oxidation of lipids, DNA, and proteins besides other components, hence the need for redox homeostasis to be maintained by the cell’s antioxidants and antioxidative enzymes (Shao *et al.*, 2008). Increasing ROS concentration disturbs the redox equilibrium in the cell by conversion of antioxidants into their oxidized state, which are normally found in the reduced state and utilized for signaling resulted into the unbalanced redox metabolism of cell (Tripathy and Oelmüller, 2012). The mechanism is unclear, but an example in plastids shows the influence of ROS on the redox situation. Increased light intensity induces production of ROS and decline in plastoquinone pool, leading to stimulation of protein kinase and photoacclimation process along with activation of various antioxidative genes (Apel and Hert, 2004; Mittler *et al.*, 2004).

Mechanism of ROS Signaling

In the presence of stress, plants have developed an integral mechanism for ROS sensing, transduction, and translation of signals into suitable cellular responses. This requires the presence of redox-sensitive proteins for oxidation and reduction reactions (Shao *et al.*, 2006). ROS facilitate the molecular mechanisms of redox-sensitive regulation of proteins mediated by signals comprising MAP kinase regulated protein phosphorylation, heterotrimeric G-proteins, and protein Tyr phosphatases (Foyer and Noctor, 2005).

ROS Generation and Signaling Sites

ROS generation and scavenging in plants under stress (abiotic/biotic) depends on the nature of ROS signaling, which takes place inside cells between different organelles, and between cells over long distances. Plants have the capability to scavenge ROS for long-distance signaling, and this is possible only when uninterrupted ROS generation takes place. ROS are mainly used as a general signal to activate the cellular signaling networks, while other signals function together with ROS to convey specific signals to combat the stress (Mittler *et al.*, 2011).

ROS generated in chloroplasts and mitochondria induce alterations in the nuclear transcriptome, influencing gene expression by modifying transcription factors (Choudhury *et al.*, 2013). ROS induce expression of numerous genes, signifying that ROS act as biological signals to regulate stress (Laloi *et al.*, 2004). When plants encounter stress (abiotic/biotic) the ROS concentration is increased, which alters gene expression. Changes at the gene level occurs via oxidation of components stimulating the signaling pathways resulting in the activation of transcription factors or possibly those transcription factors that are redox sensitive (Choudhury *et al.*, 2013). Signaling components in the chloroplast are plastoquinone (PQ), ascorbate, glutathione, and ROS along with ferredoxin or thioredoxin (Pfannschmidt *et al.*, 1999).

In peroxisomes high amounts of H_2O_2 are generated naturally when plants encounter high light energy or any other stress, capable of stimulating the antioxidative enzymes such as catalase and ascorbate peroxidase. Photorespiration contributes to the scavenging causing a decline in catalase activity and leading to the accumulation of oxidized glutathione. Therefore the catalase deficiency is balanced by ascorbate and glutathione involving the signal transduction from chloroplast to peroxisomes (Foyer and Noctor, 2003).

ROS production in chloroplasts and peroxisomes is relatively high compared to mitochondria, although the quantity of oxidized protein is high in mitochondria due to its susceptibility to ROS (Møller & Sweetlove, 2010). Increased production of ROS is responsible for more oxidized proteins in mitochondrial electron transport complex which leads to programmed cell death (Tiwari *et al.*, 2002).

Gene Expression

In plants, stress generates ROS and appropriate amounts of ROS benefit plants in their responses to the stresses. Hydrogen peroxide is specifically involved in plant defense, stimulating both gene expression and via the activities of proteins such as MAP kinase, regulating transcription (Desikan *et al.*, 2000) (Figure 5.3). Gene expression in plant cells affected by ROS-stimulated signal transduction and altered redox state involves transcription factors, as promoter elements and DNA-binding factors acting as redox response elements to control the plants' gene expression.

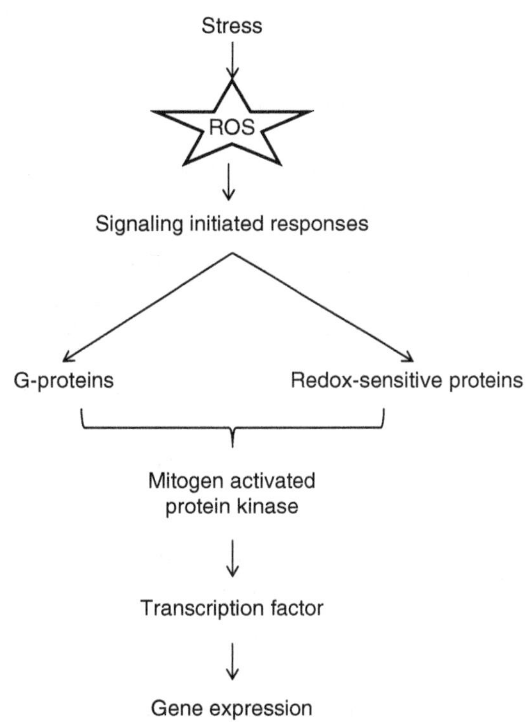

Figure 5.3 Reactive oxygen species (ROS) influence gene expression.

For example, glutathione S-transferase (GST) catalyzes the conjugation of GSH and stimulates the cellular detoxification pathway. GST gene expression is regulated by the pro-oxidant state of the cell, resulting in reduced GSH content (Daniel, 1993). Similarly, the oxidation of thiol group proteins and oxidation of Fe-S clusters are also integral to the function of redox-sensitive proteins. Another example of gene expression is observed during the photo-oxidative damage caused by the high light energy able to alter utilization, dissipation, and generation of the toxic radicals. This mechanism induces the signal transduction which leads to reduction in the plastoquinone pool along with kinase activation that changes the Fe-S protein linked with cytochrome *cf* complex (Vener *et al.*, 1998; Pfannschmidt *et al.*,1999).

Role of ROS Signaling

Signaling facilitated by ROS in plants induced by stress (biotic/abiotic) involves heterotrimeric G-proteins and protein phosphorylation controlled by definite MAP kinases and protein Tyr phosphatases (Foyer and Noctor, 2005; Pfannschmidt *et al.*, 2009). ROS, particularly H_2O_2, activates signaling protein kinases, especially MAPK; a protein kinase can phosphorylate a transcription factor and further regulate gene expression (Foyer and Noctor, 2003; Apel and Hirt, 2004).

ROS Induce Activation of MAPK Signaling Pathways

Mitogen-activated protein kinases are a class of plant serine/threonine protein kinases that perform an essential role in the transduction of numerous stress signals. They typically function as a cascade where MAPK is phosphorylated and activated by MAPK kinase (MAPKK), which is activated by MAPKK kinase (MAPKKK). These three kinases are interlinked and known as extracellular receptor kinases (Ahmad *et al.*, 2008) (Figure 5.4). Therefore, the signal transduction unit is conceived to stimulate active MAP kinase kinase kinase (MAPKKK), MAP kinase kinase (MAPKK), and MAP kinase (MAPK). A sequential phosphorylation-activation process initiates and transmits the signal from the MAPKKK to the target, in the form of a transcription factor (TF) whose activity and localization are influenced by phosphorylation. The proportions of phosphorylation activation and transmission designate that MAPKKKs can be stimulated by particular stimuli and the signaling pathways may assemble at the MAPKK level of the cascade. A single MAPKK could then phosphorylate several MAPKs. Signaling through MAPKKKs and MAPKKs might continue through other mechanisms as well besides phosphorylation of their direct downstream targets. MAPKs are controlled by protein phosphatases and able to target modification by ROS induced changes in protein phosphatase activity through MAPK signaling (Brosché *et al.*, 2010).

Therefore, MAPK signaling components are involved in eliciting responses to several signals including oxidative stresses. Activation of MAPK by phosphorylation normally leads to nuclear localization (Ahlfors *et al.*, 2004). Abiotic stress leads to the expression of genes affected by these kinase classes and stimulates the antioxidative defenses (Samuel *et al.*, 2000).

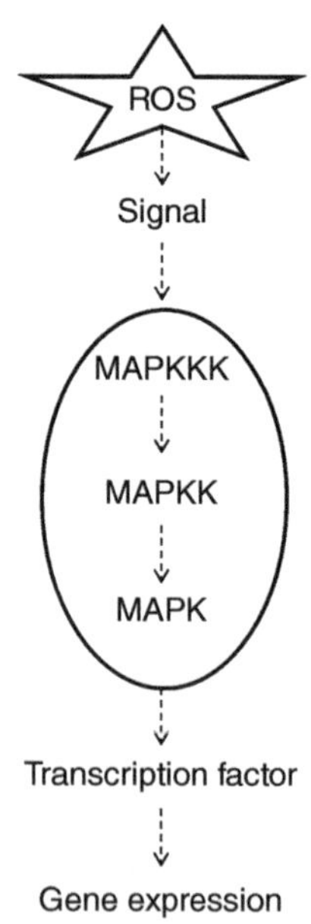

Figure 5.4 Signaling mechanisms through the mitogen-activated protein kinase (MAPK) pathway.

ROS Signaling Perspectives

Normally ROS are generated and induce metabolic activities by signaling the activation of plant metabolic pathways. Under stress condition ROS generation is accelerated in different cell organelles, such as chloroplasts, peroxisomes, and mitochondria. Oxidative stress is generated due to the accumulation of the ROS, with consequent impairment of cell membranes, nucleic acids, proteins, and lipids. Normally, plants are endowed with an array of enzymatic and non-enzymatic antioxidants such as CAT, SOD, MDHAR, DHAR, GR, glutathione peroxidase, α-tocopherols, ascorbic acid, glutathione, carotenoids, and flavonoids to manage stress. ROS perform a significant role in signaling between various metabolic pathways; however, their dual role, determined by their concentration in the cell, means they can have a good or bad effect on the plant. The consequent signaling affects several downstream processes, leading to the induction of stress-responsive genes. The role of ROS as second messengers has raised considerable interest, and understanding the integration of ROS signaling is a powerful tool for influencing plant growth and development in the presence of stress.

ROS Induce Programmed Cell Death (PCD) in Plants

PCD is a vital process for a plant's normal growth and function, for example, in trichome formation, floral organ abscission, embryo formation, leaf development, and so forth. The proportion of PCD and proliferation of cells influences the growth and development of the plant (Gadjev *et al.,* 2008). Programmed cell death in plants is significantly caused by high concentrations of ROS, while lower doses induce signals that mediate lowering of oxidative stress (Miller *et al.,* 2010).

Excess production of ROS alters the functioning of mitochondria and chloroplasts, which play a significant role in PCD during photosynthesis; these organelles transmit signals to the nucleus, which regulates expression of various genes according to the redox situation and activates PCD. Endoplasmic reticulum also takes part in PCD, mostly by provoking the mobilization of Ca^{2+} from ER to mitochondria (Suzuki *et al.,* 2012). Therefore, PCD is an active process controlled genetically, and its execution takes place in a well-organized, stepwise manner intended to destroy an individual cell without affecting other cells (Petrov *et al.,* 2015).

Conclusions

ROS production and elimination are common processes in plants, whether in normal or stressed situations. However, several environmental stresses (such as drought, chilling, UV-B, ozone, salinity, metal toxicity, and drought) lead to increased generation of ROS, and are generally responsible for the oxidative stress leading to cell death. Widespread overproduction of ROS is the result of disturbed metabolism in mitochondria, chloroplasts, plasma membrane, and peroxisomes, or various altered metabolic activities in different components of the plant cell. Besides ROS causing damage under abiotic/biotic stress conditions, they also function as signaling components to encourage prevention of damage. The role played by ROS depends upon the extent of production; at low concentrations they play the role of signaling molecules that induce various beneficial responses in plants. However, at higher concentrations ROS are contributors to oxidative damage to DNA, proteins, and lipids, finally leading to cell death.

In order to protect themselves, plants are equipped with enzymatic (e.g., SOD, GR, CAT, APX, GP, DHAR, and MDHAR) and non-enzymatic (e.g., ascorbic acid, carotenoids, glutathione, tocopherols, and phenolics) defense systems. Numerous studies on ROS production and scavenging show that overexpression of antioxidant enzymes leads to enhanced scavenging of ROS in plants against various stresses (biotic and abiotic). Detailed studies are required to elucidate the particular mechanisms of ROS production and scavenging, and the characterization of genes involved in cellular signaling, to develop plants with better tolerance to harsh environmental conditions.

Acknowledgments

The authors are grateful to the Head, Department of Botany, for facilities, and the funding agencies, namely the Council of Scientific and Industrial Research (CSIR), University Grants Commission (UGC), and the Department of Science and Technology (DST), Government of India, for providing financial help. The authors would also like to thank the researchers who helped us indirectly by providing their significant research work on ROS and signaling.

References

Abdollahi H, Ghahremani Z. (2011) The role of chloroplasts in the interaction between *Erwinia amylovora* and host plants. *Acta Hortic* 896: 215–221.

Agarwal S. (2007) Increased antioxidant activity in Cassia seedlings under UV-B radiation. *Biol Plantarum* 51: 157–160.

Agrawal SB, Mishra S. (2007) Effects of supplemental ultraviolet-B and cadmium on growth, antioxidants and yield of *Pisum sativum* L. *Ecotox Environ Safe* 72: 610–618.

Agrawal SB, Rathore D. (2007) Changes in oxidative stress defense in wheat (*Triticum aestivum* L.) and mung bean (*Vigna radiata* L.) cultivars grown with or without mineral nutrients and irradiated by supplemental ultraviolet-B. *Environ Exp Bot* 59: 21–23.

Agrawal SB, Singh S, Agrawal M. (2009) Ultraviolet-B induced changes in gene expression and antioxidants in plants. In: Jacquot J (ed.), *Advances in Botanical Research,* vol. 52. Academic Press, pp. 47–86.

Ahlfors R, Macioszek V, Rudd J, *et al.* (2004) Stress hormone-independent activation and nuclear translocation of mitogen activated protein kinases in *Arabidopsis thaliana* during ozone exposure. *Plant J* 40: 512–522.

Ahmad P, Sarwat M, Sharma S. (2008) Reactive oxygen species, antioxidants and signaling in plants. *J Plant Biol* 51: 167–173.

Apel K, Hirt H. (2004) Reactive oxygen species: metabolism, oxidative stress, and signal transduction. *Annu Rev Plant Biol* 55: 373–399.

Apostolova EL, Pouneva I, Rashkov G, Dankov K, Grigorova I, Misra AN. (2014) Effect of UV-B radiation on photosystem II functions in Antarctic and mesophilic strains of a green alga *Chlorella vulgaris* and a cyanobacterium *Synechocystis salina. Indian J Plant Physiol* 19: 111–118.

Arora A, Byrem TM, Nair MG, Strasburg GM. (2000) Modulation of liposomal membrane fluidity by flavonoids and isoflavonoids. *Arch Biochem Biophys* 373: 102–109.

Arora A, Sairam RK, Srivastava GC. (2002) Oxidative stress and antioxidative system in plants. *Curr Sci India* 82: 1227–1238.

Asada K. (1992) Ascorbate peroxidase: a hydrogen peroxide scavenging enzyme in plants. *Physiol Plantarum* 85: 235–241.

Asada K. (1996) Radical production and scavenging in the chloroplasts. In: Baker NR (ed.), *Photosynthesis and the Environment.* Kluwer Publishers, Dordrecht, pp. 123–150.

Asada K. (1999) The water-water cycle in chloroplasts: scavenging of active oxygens and dissipation of excess photons. *Annu Rev Plant Biol* 50: 601–639.

Asada K, Takahashi M. (1987) Production and scavenging of active oxygen in photosynthesis. In: Kyle J, Osmond CB, Arntzen CJ (eds), *Photoinhibition: Topics of Photosynthesis*, 9th edn. Elsevier, Amsterdam, pp. 227–287.

Ashry NA, Mohamed HI. (2012) Impact of secondary metabolites and related enzymes in flax resistance and/or susceptibility to powdery mildew. *Afr J Biotech* 11: 1073–1077.

Bafeel SO, Ibrahim MM. (2008) Antioxidants and accumulation of α-tocopherol induce chilling tolerance in *Medicago sativa. Int J Agric Biol* 10: 593–598.

Barber J, Nield J, Morris EP, Zheleva D, Hankamer B. (1997) The structure, function and dynamics of photosystem two. *Physiol Plantarum* 100: 817–827.

Barnes JD, Zheng Y, Lyons TM. (2002) Plant resistance to ozone: the role of ascorbate. In: Omasa K, Saji H, Youssefian S, Kondo N (eds), *Air Pollution and Plant Biotechnology*. Springer, Tokyo, pp. 235–254.

Bhattacharjee S. (2005) Reactive oxygen species and oxidative burst: Roles in stress, senescence and signal transduction in plants. *Curr Sci India* 89: 1–22.

Bhattacharjee S. (2012) The language of reactive oxygen species signaling in plants. *J Bot* 2012: Art. ID 985298.

Bielski BH, Arudi RL, Sutherland MW. (1983) A study of the reactivity of HO_2/O_2^- with unsaturated fatty acids. *J Biol Chem* 258: 4759–4761.

Blokhina O, Fagerstedt KV. (2010) Reactive oxygen species and nitric oxide in plant mitochondria: origin and redundant regulatory systems. *Physiol Plantarum* 138: 447–462.

Bouchard JN, Roy S, Campbell DA. (2006) UVB effects on the photosystem II-D1 protein of phytoplankton and natural phytoplankton communities. *Photochem Photobiol* 82: 936–951.

Brosché M, Overmyer K, Wrzaczek M, Kangasjärvi J. (2010) Stress signaling III: Reactive oxygen species (ROS). In: Pareek A, Sopory SK, Bohnert HJ, Govindjee (eds), *Abiotic Stress Adaptation in Plants: Physiological, Molecular and Genomic Foundation*. Springer, pp. 91–102.

Buchert F, Forreiter C. (2010) Singlet oxygen inhibits ATPase and proton translocation activity of the thylakoid ATP synthase CF1 CFo. *FEBS Lett* 584: 147–152.

Chalapathi Rao ASV, Reddy AR. (2008) Glutathione reductase: a putative redox regulatory system in plant cells. In: Khan NA, Singh S, Umar S (eds), *Sulfur Assimilation and Abiotic Stresses in Plants*. Springer, The Netherlands, pp. 111–147.

Chaudhary N, Agrawal SB. (2013) Intraspecific responses of six Indian clover cultivars under ambient and elevated levels of ozone. *Environ Sci Pollut R* 20, 5318–5329.

Chaudhary N, Agrawal SB. (2014) Cultivar specific variations in morphological and biochemical characteristics of mung bean due to foliar spray of ascorbic acid under elevated ozone, *Acta Physiol Plant* 36: 1793–1803.

Chen Z, Gallie DR. (2005) Increasing tolerance to ozone by elevating foliar ascorbic acid confers greater protection against ozone than increasing avoidance. *Plant Physiol* 138: 1673–1689.

Chen Z, Young TE, Ling J, Chang SC, Gallie DR. (2003) Increasing vitamin C content of plants through enhanced ascorbate recycling. *Proc Natl Acad Sci USA* 100: 3525–3530.

Choudhary KK, Agrawal SB. (2014a) Cultivar specificity of tropical mung bean (*Vigna radiata* L.) to elevated ultraviolet-B: changes in antioxidative defense system, nitrogen metabolism and accumulation of jasmonic and salicylic acids. *Environ Exp Bot* 99: 122–132.

Choudhary KK, Agrawal SB. (2014b) Ultraviolet-B induced changes in morphological, physiological and biochemical parameters of two cultivars of pea (*Pisum sativum* L.). *Ecotox Environ Safe* 100: 178–187.

Choudhary KK, Agrawal SB. (2015) Effect of elevated ultraviolet-B on four tropical soybean cultivars: quantitative and qualitative aspects with special emphasis on gas exchange, chlorophyll fluorescence, biomass and yield. *Acta Physiol Plant* 37: 31; doi: 10.1007/s11738-015-1780-4.

Choudhary KK, Agrawal SB. (2016) Assessment of fatty acid profile and seed mineral nutrients of two soybean (*Glycine max* L.) cultivars under elevated UV-B: Role of ROS, pigments and antioxidants. *Photochem Photobiol* 92: 134–143.
Choudhary KK, Pandey D, Agrawal SB. (2013) Deterioration of rhizospheric soil health due to elevated ultraviolet-B. *Arch Agron Soil Sci* 59: 1419–1437.
Choudhury S, Panda P, Sahoo L, Panda SK. (2013) Reactive oxygen species signaling in plants under abiotic stress. *Plant Signaling Behav* 8: 4.
Collins A. (2001) Carotenoids and genomic stability. *Mutat Res* 475: 1–28.
Corpas FJ, Barroso JB, Del Río LA. (2001) Peroxisomes as a source of reactive oxygen species and nitric oxide signal molecules in plant cells. *Trends Plant Sci* 6: 145–150.
Daniel V. (1993) Glutathione S-transferases: gene structure and regulation of expression. *Crit Rev Biochem Mol* 28: 173–207.
Das K, Roychoudhury A. (2014) Reactive oxygen species (ROS) and response of antioxidants as ROS-scavengers during environmental stress in plants. *Front Environ Sci* 2: 53.
Del Rio LA, Pastori GM, Palma JM, *et al.* (1998) The activated oxygen role of peroxisomes in senescence. *Plant Physiol* 116: 1195–1200.
Del Río LA, Corpas FJ, Sandalio LM, Palma JM, Gómez M, Barroso JB. (2002) Reactive oxygen species, antioxidant systems and nitric oxide in peroxisomes. *J Exp Bot* 53: 1255–1272.
Del Rio LA, Sandalio LM, Corpas FJ, Palma JM, Barroso JB. (2006) Reactive oxygen species and reactive nitrogen species in peroxisomes. Production, scavenging, and role in cell signalling. *Plant Physiol* 141: 330–335.
Desikan R, Neill SJ, Hancock JT. (2000) Hydrogen peroxide-induced gene expression in *Arabidopsis thaliana. Free Radical Bio Med* 28: 773–778.
Diplock T, Machlin LJ, Packer L, Pryor WA. (1989) Vitamin E: biochemistry and health implications. *Ann NY Acad Sci* 570: 372–378.
Doke N. (1983) Generation of superoxide anion by potato tuber protoplasts during the hypersensitive response to hyphal wall components of *Phytophthora infestans* and specific inhibition of the reaction by suppressors of hypersensitivity. *Physiol Plant Pathol* 23: 359–367.
Dybing E, Nelson JR, Mitchell JR, Sesame HA, Gillette JR. (1976) Oxidation of a methyldopa and other catechols by cytochromes R450-generated superoxide anion: possible mechanism of methyldopa hepatitis. *Mol Pharmacol* 12: 911–920.
Elstner EF. (1987) Metabolism of activated oxygen species. In: Davies DD (ed.), *Biochemistry of Plants.* Academic Press, London, pp. 253–315.
Elstner EF. (1991) Mechanism of oxygen activation in different compartments. In: Pell EJ, Steffen KL (eds), *Active Oxygen/Oxidative Stress and Plant Metabolism.* American Society of Plant Physiologists, Rockville, MD, pp. 13–25.
Eltayeb AE, Kawano N, Badawi GH, *et al.* (2007) Overexpression of monodehydroascorbate reductase in transgenic tobacco confers enhanced tolerance to ozone, salt and polyethylene glycol stresses. *Planta* 225: 1255–1264.
Foyer CH, Harbinson J. (1994) Oxygen metabolism and the regulation of photosynthetic electron transport. In: Foyer CH, Mullineaux P (eds), *Causes of Photooxidative Stresses and Amelioration of Defense Systems in Plants.* CRC Press, Boca Raton, FL, pp. 1–42.
Foyer CH, Noctor G. (2003) Redox sensing and signaling associated with reactive oxygen in chloroplasts, peroxisomes and mitochondria. *Physiol Plantarum* 119: 355–364.
Foyer CH, Noctor G. (2005) Redox homeostasis and antioxidant signaling: a metabolic interface between stress perception and physiological responses. *Plant Cell* 17: 1866–1875.
Fridovich I. (1989) Superoxide dismutases. An adaptation to a paramagnetic gas. J Biol Chem 264: 7761–7764.

Fryer MJ. (1992) The antioxidant effect of thylakoid vitamin-E (α-tocopherol). *Plant Cell Environ* 15: 381–392.

Fryer MJ, Andrews JR, Oxborough K, Blowers DA, Baker NR. (1998) Relationship between CO_2 assimilation, photosynthetic electron transport, and active O_2 metabolism in leaves of maize in the field during periods of low temperature. *Plant Physiol* 116: 571–580.

Fukuzawa K, Tokumura A, Ouchi S, Tsukatani H. (1982) Antioxidant activities of tocopherols on Fe^{2+}-ascorbate induced lipid peroxidation in lecithin liposomes. *Lipids* 17: 511–514.

Gadjev I, Stone JM, Gechev TS. (2008). Programmed cell death in plants: new insights into redox regulation and the role of hydrogen peroxide. *Int Rev Cell Mol Biol* 270: 87–144.

Gallego SM, Benavides MP, Tomaro ML. (1996) Effect of heavy metal ion excess on sunflower leaves: evidence for involvement of oxidative stress. *Plant Sci* 121: 151–159.

GhislaS, Massey V. (1989) Mechanisms of flavoprotein catalyzed reactions. *Eur J Biochem* 181: 1–17.

Gill SS, Tuteja N. (2010) Reactive oxygen species and antioxidant machinery in abiotic stress tolerance in crop plants. *Plant Physiol Biochem* 48: 909–930.

Gomathi R, Rakkiyapan P. (2011) Comparative lipid peroxidation, leaf membrane thermostability, and antioxidant system in four sugarcane genotypes differing in salt tolerance. *Int J Plant Physio Biochem* 3: 67–74.

Grace SG, Logan BA. (2000) Energy dissipation and radical scavenging by the plant phenylpropanoid pathway. *Philos Trans Roy Soc B* 355: 1499–1510.

Grant JJ, Yun BW, Loake GJ. (2000) Oxidative burst and cognate redox signalling reported by luciferase imaging: identification of a signal network that functions independently of ethylene, SA and Me-JA but is dependent on MAPKK activity. *Plant J* 24: 569–582.

Grether-Beck S, Bonizzi G, Schmitt-Brenden H, *et al.* (2000) Non-enzymatic triggering of the ceramide signaling cascade by solar UVA radiation. *EMBO J* 19: 5793–5800.

Gross GG. (1977) Cell wall-bound malate dehydrogenase from horseradish. *Phytochemistry* 16: 319–321.

Halliwell B. (2006) Reactive species and antioxidants. redox biology is a fundamental theme of aerobic life. *Plant Physiol* 141: 312–322.

Heath RL. (2008) Modification of the biochemical pathways of plants induced by ozone: What are the varied routes to change? *Environ Pollut* 155: 453–463.

Hernandez JA, Jimenez A, Mullineaux P, Sevilla F. (2000) Tolerance of pea (*Pisum sativum* L.) to long-term salt stress is associated with induction of antioxidant defences. *Plant Cell Environ* 23: 853–862.

Heyno E, Mary V, Schopfer P, Krieger-Liszkay A. (2011) Oxygen activation at the plasma membrane: relation between superoxide and hydroxyl radical production by isolated membranes. *Planta* 234: 35–45.

Hirayama T, Shinozaki K. (2010). Research on plant abiotic stress responses in the post-genome era: past, present and future. *Plant J* 61: 1041–1052.

Huang AHC, Trelease RN, Moore TS. (1983) *Plant Peroxisomes.* Academic Press, New York.

Iriti M, Faoro F. (2008) Oxidative stress, the paradigm of ozone toxicity in plants and animals. *Water Air Soil Pollut* 187: 285–301.

Ivanov BN, Khorobrykh S. (2003) Participation of photosynthetic electron transport in production and scavenging of reactive oxygen species. *Antioxidants Redox Signal* 5: 43–53.

Jansen MAK, Gaba V, Greenberg BM. (1998) Higher plants and UV-B radiation: balancing damage, repair and acclimation. *Trends Plant Sci* 3: 131–135.

Jimenez A, Hernandez JA, Pastori G, Del Rio LA, Sevilla F. (1998) Role of the ascorbate-glutathione cycle of mitochondria and peroxisomes in the senescence of pea leaves. *Plant Physiol* 118: 1327–1335.

Kakani VG, Reddy KR, Zhao D, Sailaja K. (2003) Field crop responses to ultraviolet-B radiation: a review. *Agr Forest Meteorol* 120: 191–218.
Karray-Bouraoui N, Harbaoui F, M Rabhi, *et al.* (2011) Different antioxidant responses to salt stress in two different provenances of *Carthamus tinctorius* L. *Acta Physiol Plant* 33: 1435–1444.
Karuse GH. (1994) The role of oxygen in photoinhibition and photosynthesis. In: Foyer CH, Mullineaux PM (eds), *Causes of Photooxidative Stress and Amelioration of Defense Systems in Plants*. CRC Press, Boca Raton, FL, pp. 43–76.
Klessig DF, Durner J, Noad R, *et al.* (2000) Nitric oxide and salicylic acid signaling in plant defense. *Proc Natl Acad Sci USA* 97: 8849–8855.
Kobayashi K, Kumazawa Y, Miwa K, Yamanaka S. (1996) ε-(γ-glutamyl) lysine cross-links of spore coat proteins and transglutaminase activity in *Bacillus subtilis*. *FEMS Microbiol Lett* 144: 157–160.
Kreslavski VD, Los DA, Allakhverdiev SI, Kuznetsov VlV. (2012) Signaling role of reactive oxygen species in plants under stress. *Russ J Plant Physiol* 59: 141–154.
Krieger-Liszkay A, Fufezan C, Trebst A. (2008) Singlet oxygen production in photosystem II and related protection mechanism. *Photosynth Res* 98: 551–564.
Kwak JM, Mori IC, Pei ZM, *et al.* (2003) NADPH oxidase AtrbohD and AtrbohF genes function in ROS-dependent ABA signaling in *Arabidopsis*. *EMBO J* 22: 2623–2633.
Laloi C, Apel K, Danon A. (2004) Reactive oxygen signalling: the latest news. *Curr Opin Plant Biol* 7: 323–328.
Liu Y, Ren D, Pike S, Pallardy S, Gassmann W, Zhang S. (2007) Chloroplast-generated reactive oxygen species are involved in hypersensitive response-like cell death mediated by a mitogen-activated protein kinase cascade. *Plant J* 51: 941–954.
Logan BA, Kornyeyev D, Hardison J, Holaday AS. (2006) The role of antioxidant enzymes in photoprotection. *Photosynth Res* 88: 119–132.
Løvdal T, Olsen KM, Slimestad R, Verheul M, Lillo C. (2010) Synergetic effects of nitrogen depletion, temperature, and light on the content of phenolic compounds and gene expression in leaves of tomato. *Phytochemistry* 71: 605–613.
Maheshwari R, Dubey RS. (2009) Nickel-induced oxidative stress and the role of antioxidant defence in rice seedlings. *Plant Growth Regul* 59: 37–49.
Michalak A. (2006) Phenolic compounds and their antioxidant activity in plants growing under heavy metal stress. *Pol J Environ Stud* 15: 523–530.
Miller EW, Dickinson BC, Chang EJ. (2010) Aquaporin-3 mediates hydrogen peroxide uptake to regulate downstream intracellular signaling. *Proc Natl Acad Sci USA* 107: 15681–15686.
Mishra AK, Agrawal SB. (2015) Biochemical and physiological characteristics of tropical mung bean (*Vigna radiata* L.) cultivars against chronic ozone stress: An insight to cultivar-specific response. *Protoplasma* 252: 797–811.
Mishra P, Kumari B, Dubey RS. (2013) Differential responses of antioxidative defense system to prolonged salinity stress in salt-tolerant and salt-sensitive Indica rice (*Oryza sativa* L.) seedling. *Protoplasma* 250: 3–19.
Mithofer A, Schulze B, Boland W. (2004) Biotic and heavy metal stress response in plants: evidence for common signals. *FEBS Letts* 566: 1–5.
Mittler R. (2002) Oxidative stress, antioxidants and stress tolerance. *Trends Plant Sci* 7: 405–410.
Mittler R, Zilinskas BA. (1992) Molecular cloning and characterization of a gene encoding pea cytosolic ascorbate peroxidase. *J Biol Chem* 267: 21802–21807.
Mittler R, Herr EH, Orvar BL, *et al.* (1999) Transgenic tobacco plants with reduced capability to detoxify reactive oxygen intermediates are hyper responsive to pathogen infection. *Proc Natl Acad Sci USA* 96: 14165–14170.

Mittler R, Vanderauwera S, Gollery M, Van Breusegem F. (2004) Reactive oxygen gene network of plants. *Trends Plant Sci* 9: 490–498.

Mittler R, Vanderauwera S, Suzuki N, *et al.* (2011) ROS signaling: the new wave? *Trends Plant Sci* 16: 6.

Mittler R, Finka A, Goloubinoff P. (2012) How do plants feel the heat? *Trends Biochem Sci* 37: 118–125.

Miyake C, Asada K. (1994) Ferredoxin-dependent photo reduction of the monodehydro ascorbate radical in spinach thylakoids. *Plant Cell Physiol* 35: 539–549.

Miyake C, Schreiber U, Hormann H, Sano S, Asada K. (1998) The FAD-enzyme mono dehydro ascorbate radical reductase mediates photo production of superoxide radicals in spinach thylakoid membrane. *Plant Cell Physiol* 39: 821–829.

Moller IM. (2001) Plant mitochondria and oxidative stress: electron transport, NADPH turnover, and metabolism of reactive oxygen species. *Annu Rev Plant Physiol Mol Biol* 52: 561–591.

Møller IM, Sweetlove LJ. (2010) ROS signaling – specificity is required. *Trends Plant Sci* 15: 370–374.

Moustakas M, Lanaras T, Symeonidis L, Karataglis S. (1994) Growth and some photosynthetic characteristics of field grown *Avena sativa* under copper and lead stress. *Photosynthetica* 30: 389–396.

Mullineaux PM, Lawson T. (2009) Measuring redox changes in vivo in leaves: prospects and technical challenges. *Methods Mol Biol* 476: 65–75.

Mullineaux PM, Rausch T. (2005) Glutathione, photosynthesis and the redox regulation of stress-responsive gene expression. *Photosynth Res* 86: 459–474.

Munné-Bosch S, Schwarz K, Alegre L. (1999) Enhanced formation of α-tocopherol and highly oxidized abietane diterpenes in water-stressed rosemary plants. *Plant Physiol* 121: 1047–1052.

Munné-Bosch S. (2005) The role of α-tocopherol in plant stress tolerance. *J Plant Physiol* 162: 743–748.

Niyogi KK, Shih C, Chow WS, Pogson BJ, DellaPenna D, Bjorkman O. (2001) Photoprotection in a zeaxanthin- and lutein-deficient double mutant of *Arabidopsis*. *Photosynth Res* 67: 139–145.

Noctor G, Foyer CH. (1998) Ascorbate and glutathione: keeping active oxygen under control. *Annu Rev Plant Biol* 49: 249–279.

Noctor G, De Paepe R, Foyer CH. (2007) Mitochondrial redox biology and homeostasis in plants. *Trends Plant Sci* 12: 125–134.

Overmyer K, Tuominen H, Kettunen R, *et al.* (2000) Ozone-sensitive *Arabidopsis* rcd1 mutant reveals opposite roles for ethylene and jasmonate signaling pathways in regulating superoxide-dependent cell death. *The Plant Cell* 12: 1849–1862.

Pasqualini S, Piccioni C, Reale L, Ederli L, Della Torre G, Ferranti F. (2003) Ozone-induced cell death in tobacco cultivar Bel-W3 plants, The role of programmed cell death in lesion formation. *Plant Physiol* 133: 122–134

Perez-Lopez U, Robredo A, Lacuesta M, *et al.* (2009) The oxidative stress caused by salinity in two barley cultivars is mitigated by elevated CO_2. *Physiol Plantarum* 135: 29–42.

Petrov V, Hille J, Mueller BR, Gechev TS. (2015) ROS-mediated abiotic stress-induced programmed cell death in plants. *Front Plant Sci* 6: 69.

Pfannschmidt T. (2003) Chloroplast redox signals: how photosynthesis controls its own genes. *Trends Plant Sci* 8: 33–41.

Pfannschmidt T, Nilsson A, Allen JF. (1999) Photosynthetic control of chloroplast gene expression. *Nature* 406: 731–734.

Pfannschmidt T, Bräutigam K, Wagner R, *et al.* (2009) Potential regulation of gene expression in photosynthetic cells by redox and energy state: approaches towards better understanding. *Ann Bot* 103: 599–607.

Pinto E, Sigaud-Kutner TCS, Leitão MAS, Okamoto OK, Morse D, Colepicolo P. (2003) Heavy metal induced oxidative stress in algae. *J Phycol* 39: 1008–1018.

Prasad TK. (1997) Role of catalase in inducing chilling tolerance in pre-emergent maize seedlings. *Plant Physiol* 114: 1369–1376.

Puntarulo S, Sánchez RA, Boveris A. (1988) Hydrogen peroxide metabolism in soybean embryonic axes at the onset of germination. *Plant Physiol* 86: 626–630.

Qadir S, Qureshi MI, Javed S, Abdin MZ. (2004) Genotypic variation in phytoremediation potential of Brassica juncea cultivars exposed to Cd stress. *Plant Sci* 167: 1171–1181.

Qin A, Shi Q, Yu X. (2011) Ascorbic acid contents in transgenic potato plants overexpressing two dehydroascorbate reductase genes. *Mol Biol Rep* 38: 1557–1566.

Quan L-J, Zhang B, Shi W-W, Li H-Y. (2008) Hydrogen peroxide in plants: a versatile molecule of the reactive oxygen species network. *J Integrat Plant Biol* 50: 2–18.

Radotic K, Ducic T, Mutavdzic D. (2000) Changes in peroxidase activity and isoenzymes in spruce needles after exposure to different concentrations of cadmium. *Environ Exp Bot* 44: 105–113.

Radwan DEM, Fayez KA, Mahmoud SY, Lu G. (2010) Modifications of antioxidant activity and protein composition of bean leaf due to Bean yellow mosaic virus infection and salicylic acid treatments. *Acta Physiol Plant* 32: 891–904.

Radyuk MS, Domanskaya IN, Shcherbakov RA, Shalygo NV. (2009) Effect of low above-zero temperature on the content of low-molecular antioxidants and activities of antioxidant enzymes in green barley leaves. *Russ J Plant Physiol* 56: 175–180.

Raha S, Robinson BH. (2000) Mitochondria, oxygen free radicals, disease and ageing. *Trends Biochem Sci* 25: 502–508.

Rai R, Agrawal M, Choudhary KK, Agrawal SB, Emberson L, Buker P. (2015) Application of ethylene diurea (EDU) in assessing the response of a tropical soybean cultivar to ambient O_3: Nitrogen metabolism, antioxidants, reproductive development and yield. *Ecotox Environ Safe* 112: 29–38.

Rao MV, Koch JR, Davis KR. (2000) Ozone: A tool for probing programmed cell death in plants, *Plant Mol Biol* 44: 345–358.

Rausch T, Wachter A. (2005) Sulfur metabolism: a versatile platform for launching defence operations. *Trends Plant Sci* 10: 503–509.

Reczek CR, Chandel NS. (2015) ROS-dependent signal transduction. *Curr Opin Cell Biol* 33: 8–13.

Reddy AR, Raghavendra AS. (2006) Photooxidative stress. In: Madhava Rao KV, Raghavendra AS, Reddy KJ (eds), *Physiology and Molecular Biology of Stress Tolerance in Plants*. Springer, pp. 157–186.

Rhoads DM, Umbach AL, Subbaiah CC, Siedow JN. (2006) Mitochondrial reactive oxygen species. Contribution to oxidative stress and inter organellar signalling. *Plant Physiol* 141: 357–366.

Rizhsky L, Liang H, Mittler R. (2003) The water-water cycle is essential for chloroplast protection in the absence of stress. *J Biol Chem* 278: 38921–38925.

Sakihama Y, Mano J, Sano S, Asada K, Yamasaki H. (2000) Reduction of phenoxyl radicals mediated by mono dehydroascorbate reductase. *Biochem Biophys Res Commun* 279: 949–954.

Samuel MA, Miles GP, Ellis BE. (2000) Ozone treatment rapidly activates MAP kinase signalling in plants. *Plant J* 22: 367–376.

Sarowar S, Kim EN, Kim YJ, *et al.* (2005) Overexpression of a pepper ascorbate peroxidase-like 1 gene in tobacco plants enhances tolerance to oxidative stress and pathogens. *Plant Sci* 169: 55–63.

Scandalios JG. (1990) Response of plant antioxidant defense genes to environmental stress. *Adv Genet* 28: 1–41.

Scandalios JG. (1993) Oxygen stress and superoxide dismutases. *Plant Physiol* 101: 7–12.

Schraudner M, Moeder W, Wiese C, *et al.* (1998) Ozone-induced oxidative burst in the ozone biomonitor plant, tobacco Bel-W3. *The Plant Journal* 16: 235–245.

Shao HB, Chu LY, Zhao CX, Guo QJ, Liu XA, Ribaut JM. (2006) Plant gene regulatory network system under abiotic stress. *Acta Biol Sezeged* 50: 1–9.

Shao Hb, Li CY, Ming-an S, Jaleel CA, Hong-mei M. (2008) Higher plant antioxidants and redox signaling under environmental stresses. C. R. Biologies 331: 433–441.

Sharma P, Jha AB, Dubey RS. (2010) Oxidative stress and antioxidative defense system in plants growing under abiotic stresses. In: Pessarakli M (ed.), *Handbook of Plant and Crop Stress*, 3rd edn. CRC Press, Taylor & Francis, pp. 89–138.

Sharma P, Jha AB, Dubey RS, Pessarakli M. (2012) Reactive oxygen species, oxidative damage, and antioxidative defense mechanism in plants under stressful conditions. *J Bot* 2012: Art. ID 217037.

Sharma RK, Agrawal M. (2005) Biological effects of heavy metals: An overview. *J Environ Biol* 26: 301–313.

Sieferman-Harms D. (1987) The light harvesting function of carotenoids in photosynthetic membrane. *Plant Physiol* 69: 561–568.

Singh AA, Agrawal SB, Shahi JP, Agrawal M. (2014) Assessment of growth and yield losses in two *Zea mays* L. cultivars (quality protein maize and non-quality protein maize) under projected levels of ozone. *Environ Sci Pollut R* 21: 2628–2641.

Singh R, Singh S, Parihar P, *et al.* (2016) Reactive oxygen species (ROS): beneficial companions of plants' developmental processes. *Front Plant Sci* 7: 1299.

Singh S, Srivastava PK, Kumar D, Tripathi DK, Chauhan DK, Prasad SM. (2015) Morpho-anatomical and biochemical adapting strategies of maize (*Zea mays* L.) seedlings against lead and chromium stresses. *Biocatal Agric Biotechnol* 4: 286–295.

Singh S, Tripathi DK, Singh S, *et al.* (2017) Toxicity of aluminium on various levels of plant cells and organism: a review. *Environ Exp Bot* 137: 177–193.

Singh VP, Tripathi DK, Kumar D, Chauhan DK. (2011) Influence of exogenous silicon addition on aluminium tolerance in rice seedlings. *Biol Trace Elem Res* 144: 1260–1274.

Slesak I, Libik M, Karpinska B, Karpinski S, Miszalski Z. (2007) The role of hydrogen peroxide in regulation of plant metabolism and cellular signalling in response to environmental stresses. *Acta Biochim Pol* 54: 39–50.

Smirnoff N, Running JA, Gatzek S. (2004) Ascorbate biosynthesis: a diversity of pathways. In: Asard H, May JM, Smirnoff N (eds), *Vitamin C: Its Functions and Biochemistry in Animals and Plants*. BIOS Scientific, New York, pp. 7–29.

Stevens R, Page D, Gouble B, Garchery C, Zamir D, Causse M. (2008) Tomato fruit ascorbic acid content is linked with monodehydroascorbate reductase activity and tolerance to chilling stress. *Plant Cell Environ* 31: 1086–1096.

Surabhi GK, Reddy KR, Singh SK. (2009) Photosynthesis, fluorescence, shoot biomass and seed weight response of three cowpea (*Vigna unguiculata* L. Walp.) cultivars with contrasting sensitivity to UV-B radiation. *Environ Exp Bot* 66: 160–171.

Suzuki N, Koussevitzky S, Mittler R, Miller G. (2012) ROS and redox signaling in the response of plants to abiotic stress. *Plant Cell Environ* 35, 259–270.

Sweetlove LJ, Foyer CH. (2004) Roles for reactive oxygen species and antioxidants in plant mitochondria. In: Day DA, Millar AH, Whelan J (eds), *Plant Mitochondria: From Genome to Function. Advances in Photosynthesis and Respiration*, vol. 1. Kluwer Academic Press, Dordrecht, pp. 307–320.

Takahashi M, Shiraishi T, Asada K. (1988) Superoxide production in aprotic interior of chloroplast thylakoids. *Arch Biochem Biophys* 267: 714–722.

Takshak S, Agrawal SB. (2014) Effect of ultraviolet-B radiation on biomass production, lipid peroxidation, reactive oxygen species, and antioxidants in *Withania somnifera. Biol Plantarum* 58: 328–334.

Tanou G, Molassiotis A, Diamantidis G. (2009) Induction of reactive oxygen species and necrotic death-like destruction in strawberry leaves by salinity. *Env Exp Bot* 65: 270–281.

Tayefi-Nasrabadi H, Dehghan G, Daeihassani B, Movafegi A, Samadi A. (2011) Some biochemical properties of guaiacol peroxidases as modified by salt stress in leaves of salt-tolerant and salt-sensitive safflower (*Carthamus tinctorius* L. cv.) cultivars. *Afr J Biotech* 10: 751–763.

Tiwari BS, Belenghi B, Levine A. (2002) Oxidative stress increased respiration and generation of reactive oxygen species, resulting in ATP depletion, opening of mitochondrial permeability transition, and programmed cell death. *Plant Physiol* 128: 1271–1281.

Tripathi DK, Singh VP, Kumar D, Chauhan DK. (2012a) Impact of exogenous silicon addition on chromium uptake, growth, mineral elements, oxidative stress, antioxidant capacity, and leaf and root structures in rice seedlings exposed to hexavalent chromium. *Acta Physiol Plant* 34: 279–289.

Tripathi DK, Singh VP, Kumar D, Chauhan DK. (2012b) Rice seedlings under cadmium stress: effect of silicon on growth, cadmium uptake, oxidative stress, antioxidant capacity and root and leaf structures. *Chem Ecol* 28: 281–291.

Tripathi DK, Singh S, Singh VP, Prasad SM, Chauhan DK, Dubey NK. (2016a) Silicon nanoparticles more efficiently alleviate arsenate toxicity than silicon in maize cultivar and hybrid differing in arsenate tolerance. *Front Environ Sci* 4: 46.

Tripathi DK, Singh VP, Prasad SM, Dubey NK, Chauhan DK, Rai AK. (2016b) LIB spectroscopic and biochemical analysis to characterize lead toxicity alleviative nature of silicon in wheat (Triticum aestivum L.) seedlings. *J Photochem Photobiol B: Biol* 154: 89–98.

Tripathi DK, Singh S, Singh VP, Prasad SM, Dubey NK, Chauhan DK. (2017a) Silicon nanoparticles more effectively alleviated UV-B stress than silicon in wheat (Triticum aestivum) seedlings. *Plant Physiol Biochem* 110: 70–81.

Tripathi DK, Singh S, Singh S, *et al.* (2017b) Nitric oxide alleviates silver nanoparticles (AgNps)-induced phytotoxicity in Pisum sativum seedlings. *Plant Physiol Biochem* 110: 167–177.

Tripathi DK, Mishra RK, Singh S, *et al.* (2017c) Nitric oxide ameliorates zinc oxide nanoparticles phytotoxicity in wheat seedlings: implication of the ascorbate–glutathione cycle. *Front Plant Sci* 8: 1.

Tripathi R, Sarkar A, Rai SP, Agrawal SB. (2011) Supplemental ultraviolet-B and ozone: Impact on antioxidants, proteome and genome of linseed (*Linum usitatissimum* L. cv. Padmini). *Plant Biol* 13: 93–104.

Tripathy BC, Oelmüller R. (2012) Reactive oxygen species generation and signaling in plants. *Plant Signaling Behav* 7: 1621–1633.

Ushimaru T, Maki Y, Sano S, Koshiba K, Asada K, Tsuji H. (1997) Induction of enzymes involved in the ascorbate-dependent antioxidative system, namely, ascorbate peroxidase, monodehydroascorbate reductase and dehydroascorbate reductase, after exposure to air of rice (*Oryza sativa*) seedlings germinated under water. *Plant Cell Physiol* 38: 541–549.

Ushimaru T, Nakagawa T, Fujioka Y, *et al.* (2006) Transgenic Arabidopsis plants expressing the rice dehydroascorbate reductase gene are resistant to salt stress. *J Plant Physiol* 163: 1179–1184.

Valdivieso G, Mullineaux PM. (2010) The role of reactive oxygen species in signalling from chloroplasts to the nucleus. *Physiol Plant* 138: 430–439.

Vangronsveld J, Clijsters H. (1994) Toxic effects of metals. In: Farago ME (ed.), *Plants and the Chemical Elements. Biochemistry, Uptake, Tolerance and Toxicity*. VCH Publishers, Weinheim, Germany, pp. 150–177.

Varnova E, Van Brcusegem F, Dat J, Belles-Bolx E, Inze D. (2002) The Role of Reactive Oxygen Species in Signal Transduction. In: Scheel D, Wasternack C (eds), *Plant Signal Transduction*. Oxford University Press, pp. 41–73.

Vener AV, Ohad I, Anderson B. (1998) Protein phosphorylation and redox sensing in chloroplast thylakoids. *Curr Opin Plant Biol* 1: 217–229.

Vianello A, Zancani M, Nagy G, Macri F. (1997) Guaiacol peroxidase associated to soybean root plasma membranes oxidizes ascorbate. *J Plant Physiol* 150: 573–577.

Wagner D, Przybyla D, Op den Camp R, *et al.* (2004) The genetic basis of singlet oxygen induced stress responses of *Arabidopsis thaliana*. Science 306: 1183–1185.

Wang J, Zhang H, Allen RD. (1999) Overexpression of an Arabidopsis peroxisomal ascorbate peroxidase gene in tobacco increases protection against oxidative stress. *Plant Cell Physiol* 40: 725–732.

Wang S, Liang D, Li C, Hao Y, Ma F, Shu H. (2012) Influence of drought stress on the cellular ultrastructure and antioxidant system in leaves of drought-tolerant and drought-sensitive apple root stocks. *Plant Physiol Biochem* 51: 81–89.

Wang X, Yang P, Gao Q, *et al.* (2008) Proteomic analysis of the response to high-salinity stress in *Physcomitrella patens*. *Planta* 228: 167–177.

Willekens H, Inze D, Van Montagu M, Van Camp W. (1995) Catalases in plants. *Mol Breeding* 1: 207–228.

Wise RR, Naylor AW. (1987) Chilling-enhanced photooxidation. Evidence for the role of singlet oxygen and superoxide in the breakdown of pigments and endogenous antioxidants. *Plant Physiol* 83: 278–282.

Wojtaszek P. (1997) Oxidative burst: an early plant response to pathogen infection. *Biochem J* 322: 681–692.

Xiang C, Werner BL, Christensen EM, Oliver DJ. (2001) The biological functions of glutathione revisited in *Arabidopsis* transgenic plants with altered glutathione levels. *Plant Physiol* 126: 564–574.

Yamaguchi-Shinozaki K, Shinozaki K. (1994) A novel cis element in an *Arabidopsis* gene is involved in responsiveness to drought, low-temperature, or high-salt stress. *Plant Cell* 6: 251–264.

Yamamoto Y, Hachiya A, Matsumoto H. (1997) Oxidative damage to membranes by a combination of aluminum and iron in suspension-cultured tobacco cells. *Plant Cell Physiol* 38: 1333–1339.

Yin L, Wang S, Eltayeb AE, *et al.* (2010) Overexpression of dehydroascorbate reductase, but not monodehydroascorbate reductase, confers tolerance to aluminium stress in transgenic tobacco. *Planta* 231: 609–621.

Zaninotto F, Camera SL, Polverari A, Delledonne M. (2006) Cross talk between reactive nitrogen and oxygen species during the hypersensitive disease resistance response. *Plant Physiol* 141: 379–383.

Zhang Y, Luo Y, Hou YX, Jiang H, Chen Q, Tang RH. (2008) Chilling acclimation induced changes in the distribution of H_2O_2 and antioxidant system of strawberry leaves. *Agr J* 3: 286–291.

Zhang Y, Tang HR, Luo Y. (2008) Variation in antioxidant enzyme activities of two strawberry cultivars with short term low temperature stress. *World J Agric Sci* 4: 458–462.

Zhou YH, Yu JQ, Mao WH, Huang LF, Song XS, Nogues S. (2006) Genotypic variation of rubisco expression, photosynthetic electron flow and antioxidant metabolism in the chloroplasts of chill-exposed cucumber plants. *Plant Cell Physiol* 47: 192–199.

6

Effects of Reactive Oxygen Species on Crop Productivity: an Overview

Marisha Sharma, Sunil K. Gupta, Farah Deeba, and Vivek Pandey

CSIR-National Botanical Research Institute, Rana Pratap Marg, Lucknow, India

Introduction

Global attention on crop productivity is essential, both in the present and the future, due to its contribution to nutrition and sustainable agriculture. About 62% of crop production is for food; 35% is used for animal feed, and 3% for fuel and other industrial uses. Cereal grains are major components of world production of edible dry matter and provide half of the world's protein (Figure 6.1). Wheat, rice, maize (corn), millet, and sorghum provide nearly 70% of food energy (calories) and about 90% of all protein consumed by the world's population. A crop production shift toward food use could potentially add about 50% more calories to the global food supply (Foley *et al.*, 2011). Over the last century, the global population has quadrupled from 1.8 billion in 1915 to the present-day 7.3 billion, and may reach 9.7 billion by 2050 (UN, 2015). This growth, along with rising incomes in developing countries, is a driving force behind rising global food demand. Food demand is expected to increase by between 59% and 98%, and to match the projected population growth, global agricultural food production needs to increase by between 50% and 70% by 2050. Presently, crop production is unable to meet the demand for food. Many factors, from climate change to urbanization, will also make it challenging to produce enough food. Among them climate change-driven water scarcity, rising global temperatures, and extreme weather will have severe long-term effects on crop productivity. Agricultural systems are currently degrading land, water, biodiversity, and climate on a global scale. The world's future food security and sustainability needs of billions of chronically malnourished people can be met by substantial increase in food production.

Crop plants are prevented from attaining their full genetic potential for growth and reproduction due to their suboptimal environments (Bray *et al.*, 2000; Rockstrom and Falkenmark, 2000). The difference between maximum and the average crop yields clearly depicts the yield shortfalls. Abiotic stress factors (drought, salinity, heat, cold, UV radiation, air pollutants, and nutrient stress) and biotic stress factors (pests and pathogens) reduce average yields for most major crops worldwide. Disasters caused by extreme weather worldwide can cause marked damage to crops and food system infrastructure thus destabilizing food systems and threatening local/global food security. The major risks of soil erosion, soil mining, and deforestation due to extensive farming and ranching systems also lead to declining yields and desertification. The reduction in harvested land and yields due to drought and heat are the major causes of the loss in crop production. At present, to a lesser or greater extent, more than 20–50% of

Reactive Oxygen Species in Plants: Boon Or Bane - Revisiting the Role of ROS, First Edition. Edited by Vijay Pratap Singh, Samiksha Singh, Durgesh Kumar Tripathi, Sheo Mohan Prasad, and Devendra Kumar Chauhan.

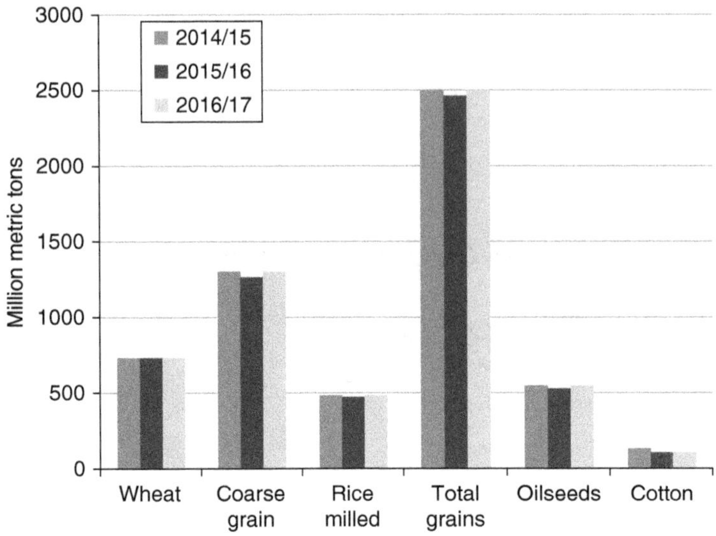

Figure 6.1 Production (million metric tons) of major crops worldwide from 2014/15 to 2016/17.

irrigated cropland is affected by salt and secondary salinity (Owens, 2001; Flowers, 2004). Low-temperature stress negatively influences crop performance and production (Xin and Browse, 2001). Worldwide losses in crop production due to cold stress amount to approximately US $2 billion each year (Sanghera *et al.*, 2011). Heavy metal contamination of soil and water is also one of the major abiotic stresses causing different deleterious effects on crops and lost productivity. These abiotic stresses lead to more than 50% of the yield reductions of major crops globally (Goel and Madan, 2014). Moreover, the biotic stresses (insects, diseases, and weeds) cause a 31–42% loss in crop yield, with an additional 6–20% post-harvest loss due to different pests, fungi, and bacterial rots (Tesfaw and Feyissa, 2014).

The common feature of the plants' responses to different stresses is the production of reactive oxygen species (ROS) (Jaspers and Kangasjärvi, 2010). The overproduction of ROS in plants causes impairment of redox homeostasis resulting in oxidative stress. The oxidative stress generated by ROS in plants has detrimental effects on crop production. The ROS comprise both free radicals – superoxide radical ($O_2^{\cdot-}$); hydroxyl radical ($\cdot OH$), perhydroxy radical ($HO_2\cdot$), and alkoxy radicals ($RO\cdot$) – and non-radical (molecular) forms – hydrogen peroxide (H_2O_2) and singlet oxygen ($^1O_2$). In chloroplasts, photosystems I and II (PSI and PSII), and complex I, ubiquinone, and complex III of the electron transport chain (ETC) in mitochondria, are the major sites for the production of $^1O_2$ and $O_2^{\cdot-}$. Plastids and specific regions like the apoplastic space and apex of polarized cells are also the sites of ROS production triggering different signaling cascades (Van Breusegem *et al.*, 2008). ROS pose a threat to cells by causing peroxidation of lipids, oxidation of proteins, damage to nucleic acids, enzyme inhibition, and activation of programmed cell death (PCD) (Mittler, 2002; Sharma & Dubey, 2005; Maheshwari & Dubey, 2009). Accumulation of ROS under various environmental stages is a major cause of crop productivity loss worldwide (Mittler, 2002; Gill *et al.*, 2011). Oxidative stress impairs directly or indirectly on various biochemical, physiological, and morphological functions in plants, and in turn interferes with crop productivity. The highly reactive ROS in oxidative stress are capable of causing damage to all cell structures (Gill & Tuteja, 2010). Impairment of cellular functions and damage to cellular structures negatively affects germination, plant growth, and yield. The oxidative stress directly affects photosynthesis, transpiration, respiration, energy metabolism, carbohydrate metabolism, lipid metabolism, protein and nucleic acid synthesis, stomatal conductance, cell wall integrity, and pigment concentration, together affecting plant development and growth (Flexas *et al.*, 2004; Garrett *et al.*, 2006; Chaves *et al.*, 2009; Compant *et al.*, 2010; Cramer *et al.*, 2011; V.P. Singh *et al.*, 2011; Walbot, 2011; Tripathi *et al.*, 2012a,b, 2016a,b, 2017a–c; Tullus *et al.*, 2012; S. Singh *et al.*, 2015, 2017; R. Singh *et al.*, 2016). Oxidative stress has been considered one of the major limiting factors in crop productivity. Now, the extreme challenge for the future is to increase crop productivity under the stressful conditions prevailing in the global climate change scenario. The increasing world population, industrialization, deterioration of cultivated land, and fresh water insufficiency are imposing additional pressures on future crop productivity demands. In this context, it is necessary to understand plant stress response mechanism(s) to augment productivity under stressful environments. Furthermore, it is imperative to develop stress-resistant plants having the ability to adapt and hence sustain growth and productivity.

Causes of ROS Overproduction and their Limitation on Productivity

The generation of ROS is an inevitable result of membrane-linked electron transport in chloroplasts, mitochondria, and plasma membranes by spilling of electrons on molecular oxygen. The production of ROS is aggravated by biotic and abiotic stresses causing oxidative stress in plants. The manifestation of oxidative stress ranges from impairment of metabolic and physiological processes to membrane damage and/or biomolecule damage (proteins and DNA) associated with aging and senescence in plants. The higher concentrations of ROS may lead to programmed cell death (PCD) (Gechev and Hille, 2005). Plant yield and productivity are adversely affected by the stress-induced PCD (Mittler and Blumwald, 2010). Oxidative stress is considered to be one of the major limitations to crop productivity worldwide (Vadez *et al.*, 2012). Hence, stress factors are the major threat to the yield potential of crops globally due to the deleterious effects of oxidative stress. The abiotic stresses include drought, salinity, light (UV radiation), heat, chilling, air pollutants, and heavy metals. Biotic stress in plants is caused by attack from pathogens like fungi, bacteria, viruses, and other pests. This section will give insight into different stresses causing the overproduction of ROS and the consequences of oxidative stress on crop productivity.

Abiotic Factors

Drought

The scarcity of available water in soil with continuous loss of water by evaporation or transpiration results in periods of drought. The semi-arid and arid areas of world contribute 40% of the total yield of all food grains (Thakurta, 2010). Rain-fed agriculture accounts about 80% of the world's agricultural land and feeds, and at least two-thirds of world production is under the threat of drought stress. Drought is a global issue, negatively influencing survival of agricultural crops and sustainable food production (Jaleel *et al.*, 2007a–d). Drought stress is characterized by a reduction of water content and leaf water potential with turgor loss, closure of stomata, decrease in cell enlargement and growth, with the arrest of photosynthesis, disturbance of metabolism, and finally death of the plant (Jaleel *et al.*, 2008). ROS play a crucial role in causing the deleterious effects on plants under water stress. The disequilibrium between light capture and its utilization results in a decrease of photosynthetic rate under drought conditions. The imbalance between the generation and utilization of electrons and excess light energy dissipated in photosynthetic systems results in the generation of ROS (Mattos, 2015). The well-known consequence of ROS overproduction is denaturation of functional and structural macromolecules impairing plant processes and ultimately growth. ROS under drought stress adversely affects membrane integrity, causing a broad range of biochemical, physiological, and molecular changes, impacting total protein quantity and quality, and reducing fitness and harvestable yield (Neumann, 2008; Harb *et al.*, 2010; Nawaz *et al.*, 2012, 2014). The carbon metabolism for energy production is negatively affected under drought leading to retardation of plant growth (Nawaz *et al.*, 2014). The decreased CO_2 fixation concomitant with increased electron leakage to triplet oxygen results in ROS accumulation under prolonged drought stress and may eventually lead to PCD (Gechev *et al.*, 2012). Drought has negative influence on the yield and quality of grains (Mittler, 2006; Balla *et al.*, 2011). Nouri-Ganbalani *et al.* (2009) estimated a 17% to 70% reduction in average grain yield due to drought stress. Globally, the areas sown with the major crops of barley, maize, rice, sorghum, soybean, and wheat are under increasing risk of drought, rising from 5–10% in 1965 to 15–25% of total sown area in 2006 (Li *et al.*, 2009). The drought-related yield reductions for the major crops are predicted to increase by more than 50% by 2050 (Li *et al.*, 2009).

Heat

Heat stress is often defined as a rise in temperature beyond a threshold level (usually 10–15 °C above ambient) for a period of time causing irreversible damage to plant growth and development. The increase of global mean surface temperature by the end of the twenty-first century (2081–2100) relative to 1986–2005 is likely to be 0.3–1.7 °C under representative concentration pathways (RCP)2.6, 1.1–2.6 °C under RCP4.5, 1.4–3.1 °C under RCP6.0, and 2.6–4.8 °C under RCP8.59. These global warming trends are likely to reduce yields by roughly 1.5% per decade worldwide (Lobell and Gourdji, 2012).

Heat stress causes various adverse alterations in plant growth, development, physiological processes, and yield due to excess generation of ROS (Hasanuzzaman *et al.*, 2012, 2013). ROS have potential to react with all biomolecules, such as pigments, proteins, lipids, and DNA, and with almost all constituents of cells (Moller *et al.*, 2007; Huang & Xu, 2008; Karuppanapandian *et al.*, 2011a,b). In many crop species, the adverse effect of high temperature stress is more prominent on reproductive development. The sudden decline in yield with temperature is mainly associated with pollen infertility (Young *et al.*, 2004; Zinn *et al.*, 2010). Impaired pollen development, male sterility, inadequate resources accumulation, and reduced grain weight and grain number are some major factors involved in reduced yield under heat stress conditions (Sakata and Higashitani, 2008; Wassmann *et al.*, 2009; Mahmood *et al.*, 2010; Zinn *et al.*, 2010).

Salinity

Salinity stress is defined as rise in the electrical conductivity (EC) of the saturation extract (EC_e) in the root zone beyond 4 dS m^{-1} (approximately 40 mM NaCl) at 25 °C and having <15% exchangeable sodium. Estimates show that 20% of total cultivated land worldwide and 33% of irrigated agricultural lands are salinity affected. Furthermore, the rate of increase in salinized areas is 10% annually due to various reasons, including low precipitation, high surface evaporation, weathering of native rocks, irrigation with saline water, and poor cultural practices. It has been estimated that by 2050 more than 50% of arable land will be afflicted by salinity. Salinity affects almost all aspects of plant development, from germination to vegetative and reproductive development along with morphological, physiological, and biochemical processes (Singh and Chatrath, 2001). Salinity imposes osmotic and oxidative stress along with ion toxicity, nutrient (N, Ca, K, P, Fe, Zn) deficiency, and reduction in water uptake from soil. Restricted gas exchange due to decreased stomatal conductance limits CO_2 uptake and availability for the

Calvin cycle (Geilfus *et al.*, 2015). As a consequence the pool of final electron acceptor at photosystem I (PSI), oxidized $NADP^+$, is depleted and electrons are transferred to O_2 to generate $O_2^{\bullet -}$. Photorespiration is enhanced under osmotic stress due to decrease in the CO_2/O_2 ratio in the mesophyll, accounting for over 70% of the H_2O_2 production (Noctor *et al.*, 2002; Ghannoum, 2009). Cellular homeostasis regulation under salinity includes control of ROS generation and readjustment of the cellular redox state. ROS under high salt conditions lead to metabolic disorders, cellular damage, and premature senescence or necrosis (Miller *et al.*, 2010; Habib *et al.*, 2016). The reasons for decreased growth, reduced fitness, and finally cell death under salinity are alteration in metabolism and excessive ROS accumulation.

Heavy Metals

Heavy metals (HMs) are non-biodegradable, persistent inorganic chemical constituents with an atomic mass over 20 and a density greater than $5\,g\,cm^{-3}$. HMs have cytotoxic, genotoxic, and mutagenic effects on plants when present in excess. Zinc (Zn), nickel (Ni), copper (Cu), vanadium (V), cobalt (Co), tungsten (W), and chromium (Cr) are trace elements in plant nutrition but toxic at high concentrations. Arsenic (As), mercury (Hg), silver (Ag), antimony (Sb), cadmium (Cd), lead (Pb), and uranium (U) have no known function as nutrients but are toxic to plants at high concentrations (Schützendübel and Polle, 2002; Rascio and Navari-Izzo, 2011; Tangahu *et al.*, 2011). HMs at toxic levels act as an impediment to metabolic processes in various ways, hampering normal plant functioning. HMs cause disturbance or displacement of the building blocks of proteins due to the formation of bonds between HMs and sulfhydryl groups (Hall, 2002). The functionality of essential metals in biomolecules such as pigments or enzymes gets disrupted, also adversely affecting the integrity of the cytoplasmic membranes (Hossain *et al.*, 2012; Ali *et al.*, 2013; Farid *et al.*, 2013). Vital functions in plants, such as photosynthesis, respiration, and enzymatic activities, are repressed under HMs. The increased generation of ROS is associated with the elevated levels of HMs disturbing the homeostasis within the plant cells (Hossain *et al.*, 2012; Sytar *et al.*, 2013). The oxidative stress causes multiple adverse effects such as redox imbalance, denaturation of cell structure and membranes, oxidation of proteins and lipids, oxidative DNA attack, and ion leakage, ultimately resulting in the activation of PCD pathways (Alvarez *et al.*, 2006; Flora *et al.*, 2008; Nagajyoti *et al.*, 2010; Sharma *et al.*, 2012). At the cellular level, heavy metal toxicity reduces crop productivity by causing oxidative stress.

Chilling/Cold Stress

Chilling/cold stress is experienced by plants at temperatures from 0 to 15 °C. Cold stress effects include poor germination, stunted seedlings, yellowing of leaves, withering, and reduced tillering. At the reproductive stage, cold stress delays heading and results in pollen sterility. These are key factors responsible for the reduction in grain yield of crops (Suzuki *et al.*, 2008). The imbalance between light absorption and light use under cold stress by inhibiting the activity of the Calvin–Benson cycle results in ROS accumulation. Additionally, the increased photosynthetic electron flux to and over-reduction of the respiratory ETC can also result in ROS accumulation causing oxidative stress (Hu *et al.*, 2008). Under severe chilling stress the defense system may not effectively remove overproduced ROS resulting in severe damage or even death (Zhang *et al.*, 2008). Cold stress imposition at the vegetative stage accounts for about 20% of yield loss (Cruz *et al.*, 2010). Chilling-sensitive plants showed up to 50% reduction in their productivity with no visible damage under a small drop in temperature.

Waterlogging Stress

Waterlogging and drought are the most severe global problems for agriculture. Waterlogging is when water is present in amounts exceeding the crop's optimum requirement. Yield losses are greatly exacerbated by waterlogging, including ethylene-induced abscission of flowers (Najeeb *et al.*, 2015). Waterlogging reduces nutrient availability, O_2 diffusion, and cellular respiration, which influences plant water relations and impairs biomass gain (Najeeb *et al.*, 2015). To cope with the waterlogging stress, plants sometimes develop tolerance. Waterlogging tolerance in plants is a function of tolerance to anaerobiosis and chemical toxicities (Setter *et al.*, 2009). The past decade has seen a wider recognition for quiescence as a strategy for survival during submergence, conserving energy (Voesenek and Bailey-Serres, 2013) during restricted O_2 supply. Plants tolerant to waterlogging stress exhibit certain adaptations, such as formation of aerenchyma and adventitious roots, the latter due to the interaction of the plant hormones auxin and ethylene (Ashraf, 2012).

Light and UV-B Radiation

High light input, and particularly its integral ultraviolet (UV) component, causes stress potentially leading to production of ROS, followed by DNA damage, impairment of pathogen resistance, and adverse affects on cellular processes (Jenkins, 2009; Li *et al.*, 2013). Moan (2001) reported that the depletion of the ozone layer due to the release of chlorofluorocarbons had increased the potential for UV-B damage in recent decades. The generation of ROS under UV-B radiation increases due to impairment of photosynthetic electron transport or enhanced activity of membrane-localized NADPH oxidases and peroxidases (Jenkins, 2009; Hideg *et al.*, 2013). Photosystem II (PSII) appears to be the most sensitive part of the system to increased UV-B radiation. Direct damage to DNA by high-energy UV-B photons or indirect damage by UV-B-induced overproduction of ROS are consequences of increased ground-level UV-B radiation. The deleterious effects of UVB, mainly by ROS, are largely partitioned between damage to the photosynthetic machinery and damage to the genome. UV-B exerts major deleterious effects on crop photosynthesis and productivity. The loss of crop productivity due to increased ground-level UV-B radiation worldwide is estimated to be 20–25%.

Ozone

Tropospheric ozone (O_3) is a damaging air pollutant that significantly impacts ecosystem health. The current annual average concentration of O_3 is roughly double that preceding the Industrial Revolution, and it ranges from 20 to 45 ppb across the globe (Vingarzan, 2004). Due to increases in precursor emissions from anthropogenic sources, background O_3 is predicted to increase by 0.5–2% per year over the 21st century (Solomon *et al.*, 2007). The damaging effects of O_3 on photosynthetic carbon assimilation, stomatal conductance, and plant growth result in reduced crop yields (Fiscus *et al.*, 2005; Ainsworth, 2008; Feng *et al.*, 2008). ROS are directly or indirectly derived from ozone exposure as a consequence of disturbance of the pro-oxidant/antioxidant balance leading to oxidative stress. Exposure to phytotoxic levels of O_3 may perturb the redox homeostasis within plant tissues giving rise to ROS, which react with biomolecules causing oxidative destruction of lipids, proteins, chlorophyll, DNA, and so forth, thereby reducing photosynthesis and other important physiological functions. This results in membrane disintegration and retarded growth affecting crop quality and yield (Fiscus *et al.*, 2005; Morgan *et al.*, 2006; Booker *et al.*, 2009; Fuhrer, 2009). ROS under ozone stress negatively influence vegetative growth as well as reproductive growth, limiting the productivity of plants worldwide. According to the upper boundary projection of surface O_3 concentrations, for the global year 2030 yield losses of wheat due to O_3 exposure range from 5.4 to 26%, from 15 to 19% for soybean, and from 4.4 to 8.7% for maize (Avnery *et al.*, 2011). Rising O_3 precursor emissions globally are projected to increase hemispheric-scale background O_3 concentrations, posing a threat to global food security.

Biotic Factors

Pathogens and Pests

Biotic stresses like insects, disease (viral, bacterial, and fungal), nematodes, mites, and so forth are major limitations to crop yield. The earliest cellular response to biotic stress is an oxidative burst involving production of ROS in apoplast and cellular organelles like chloroplasts and mitochondria (Grant *et al.*, 2000; Abdollahi and Ghahremani, 2011). Overproduction of ROS under pathogen attack leads to hypersensitive response cell death in plants (Liu *et al.*, 2007). The potential enzymatic sources of ROS production under biotic stress are apoplastic amine, diamine, and polyamine oxidase-type enzymes (Allan and Fluhr, 1997), a cell wall localized peroxidase that directly forms H_2O_2 (Bolwell and Wojtaszek, 1997), and a plasma membrane localized NADPH oxidase (Lamb and Dixon, 1997). Under biotic stress production of photo-activated perylenequinone toxins get induced utilizing the light to produce ROS which damage the host cells, allowing for successful pathogenesis (Daub and Chung, 2009). ROS cause damage to diverse macromolecules including lipids, proteins, nucleic acids and other cellular components, even leading to host cell death. Losses due to various diseases range from 20 to 30%, and potential yields loss is ~14% due to insect pests. Biotic stress is responsible for losses of 26–29% for soybean, wheat, and cotton, and 31, 37, and 40% for maize, rice, and potatoes, respectively, with the highest potential loss (34%) due to weeds (Oerke, 2006).

Targeted Approaches and Outcomes

Ensuring global food security in the face of climate change is one of the most important challenges in the twenty-first century – to supply sufficient food for an increasing population while sustaining the already stressed environment (Lal, 2005). In recent years, changes in temperature and rainfall due to climate change have occurred both globally and regionally, varying in amount and duration, and consequently exerting different impacts on agricultural production (Wolf, 2002). In India per capita demand for cereals has decreased while that of fruit and vegetables has increased. This reflects the growing diversification of high value cash crops having shorter growing period (may be due to unpredictable rainfall pattern). As a result there is more export of fruits and vegetables than cereals. It is projected that per capita demand for these commodity groups will increase by 63% by 2015 while those of cereals will decline by 3.6%. At the same time, the export of fruits and vegetables rose from US$101 million in 1980–81, to US$120 million in 1990–91, and US$513 million in 2003–04 (Government of India, 2005).

Yield loss is only one manifestation of plant stress, and one cannot talk about yield loss in isolation. Even in agricultural systems, the effect of stressors on yield is not the only issue of concern; quality and plant longevity are as important, if not more important, than yield for some crops. However, for many practical issues yield is taken as the final arbiter of stress, and it is the parameter most closely tied to economic impacts of stress. So focusing on yield loss as an index of plant stress is appropriate in an agricultural context.

Global Projections and Models

Model projections suggest that although increased temperature and decreased soil moisture will act to reduce global crop yields by 2050, the direct fertilization effect of a rising carbon dioxide concentration (CO_2) will offset these losses. The CO_2 fertilization factors used in models to project future yields were derived from enclosure studies conducted approximately 20 years ago.

Agronomic crop yield loss due to ambient O_3 in the USA is estimated to range from 5 to 15% (Heagle, 1989), worth US$3–5 billion annually (Fiscus *et al.*, 2005; US EPA, 2006). If anthropogenic

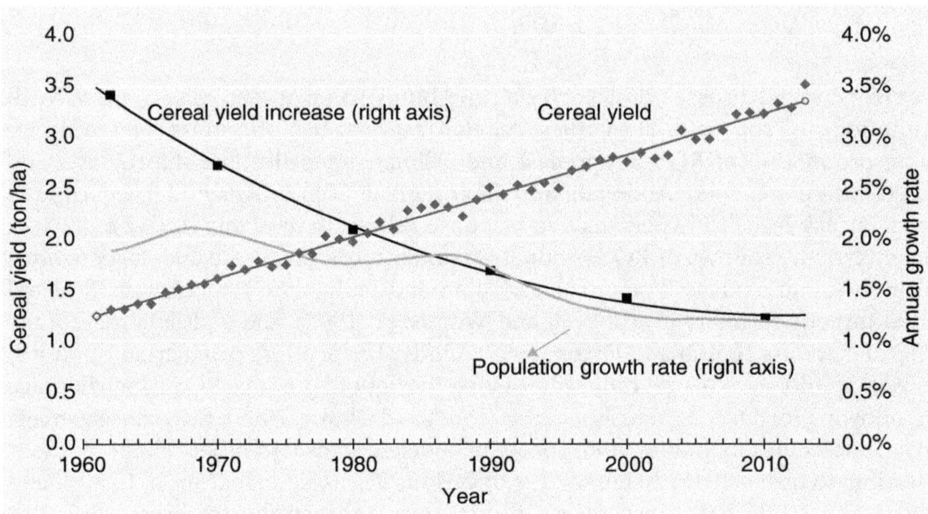

Figure 6.2 World cereal yield, annual yield increase, and population growth, 1961–2013. R square of the fitted trend line of cereal yield is 0.99. *Source:* Reproduced with permission of FAOSTAT.

O_3 was eliminated in the USA, the increased production value of eight major crops was estimated as US$2.8 to 5.8 billion in 1990 (Murphy *et al.*, 1999).

Weather and other factors cause yield to fluctuate greatly between years and geographical areas. However, aggregate grain yield growth is now down to about the level of global population growth, as Figure 6.2 shows. This means there is little room to accommodate income growth-induced additions to demand. These can force prices up, particularly when stocks are low and trade is curtailed, for example by export restrictions imposed by grain-surplus countries seeking to keep their domestic food prices low. The number of under-nourished people has increased globally. However, in five major regions – South Asia, East Asia, sub-Saharan Africa, the Near East and north Africa, and Latin America and the Caribbean – there has been a rapid decline in under-nourished people, especially for South and East Asia (Figure 6.3).

Photosynthetic Complexity

Crop productivity is generally regarded to be dependent on photosynthetic carbon assimilation and sink activity. Photosynthesis determines the strength of assimilate supply from the leaves, and the sink activity depends upon the utilization of this assimilate by the developing seeds. The impact on yield of hypothetical improvements in the molecular capacity of photosynthesis is calculated at each scale of organization from the molecular level to grain yield. The first question is: which parts of the photosynthetic process could be modified to enhance the amount of carbon gained? This is not an easy question to answer. The problem lies not just in the complexity of the processes that connect light absorption and capture through to carbon dioxide fixation and carbohydrate accumulation, but the interactions that occur between the component subsystems (Figure 8).

It has been known for some time that a large proportion of the limitation on carbon assimilation in plants using the C_3 cycle is due to the catalytic properties of the enzyme rubisco (Portis and Parry, 2007). Increasing leaf photosynthetic rates seems to be a straightforward way of increasing crop yields. Considerable physiological research has been carried out to select and breed for genotypes with superior photosynthetic rates, and was successful in identifying such cultivars in maize (Moss and Musgrave, 1971), wheat (Austin *et al.*, 1982), and soybean (Shibles *et al.*, 1987). In soybean, the trait is inherited quantitatively, that is, the broad-sense heritability

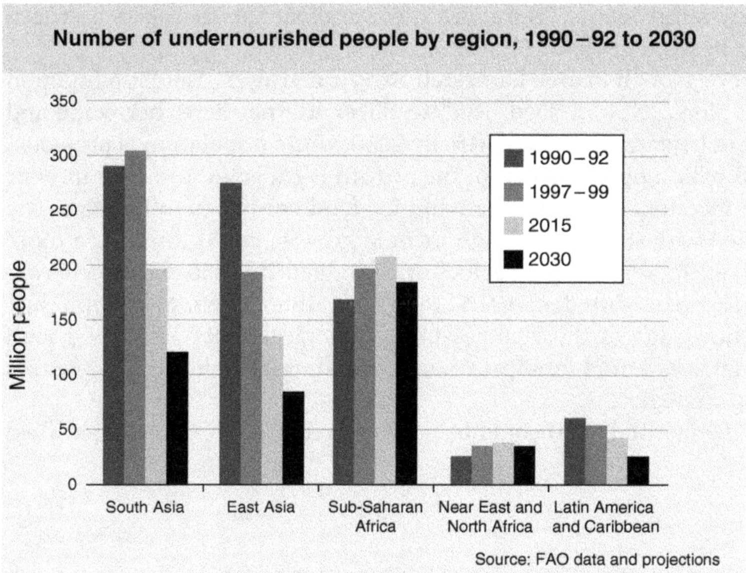

Figure 6.3 Projected and past undernourished people globally and by region from 1990–92 to 2030.
Source: Reproduced with permission of FAOSTAT.

for apparent photosynthesis ranged from 0.36 to 0.57 in soybean and from 0.29 to 0.42 in wheat (Wiebold *et al.*, 1981). Nevertheless, no increase in crop yield has been obtained as a result of breeding efforts based on genotypes with high leaf photosynthetic rates (Moss and Musgrave, 1971; Ford *et al.*, 1983).

ROS and Seed Germination

ROS play important roles at all stages of seed life, from germination (embryogenesis) to cell death (PCD). Many researches have shown that there is a great deal of compensation among seed mass, seed number, and seed growth rate (Donald and Hamblin, 1976). The process of seedling emergence from the seed embryonic axis is known as germination (Bewley and Black, 1994). Seed germination is a complex process regulated by extrinsic and intrinsic factors. Several reports have described ROS involvement in the early imbibition period, together with their signaling roles during seed germination. Knowledge of ROS functioning in seed longevity and seed aging is required because recent studies have shown that seed deterioration may be caused by pathways other than the accumulation of oxidative damage that leads to PCD.

Understanding the relative importance of beneficial and detrimental effects of ROS provides great scope for the improvement and maintenance of seed vigor and quality, factors that may ultimately increase crop yields. Improvements in seed growth traits that result in substantially increased harvest indices do not seem likely, as a negative relationship between grain number per unit land area and the average weight of these grains has frequently been found (Slafer *et al.*, 1996).

Approaches to Overcome Crop Yield Loss

ROS have beneficial as well as harmful effects on crop yields. Sometimes they increase the stress resistivity of a plant while at other times they are harmful to plants. To protect from ROS-induced damage, plants possess well-organized antioxidative defense systems comprising enzymatic and non-enzymatic antioxidants. Moreover, these antioxidants also play a crucial role in maintaining the redox balance in plants (Wu *et al.*, 2007; Zheng *et al.*, 2016).

Output growth is the result of growth in area harvested and yield growth. Yield is a function of soil quality, weather conditions, type of seed used, fertilizer use per acre, herbicide and pesticide use, the availability of irrigation, and so forth. By 2050, world population is projected to grow to between 9 and 10 billion people. Most of the growth is expected to occur in poor developing countries, where income elasticity of demand for food continues to be high. The population increase, combined with moderately high income growth, could result in a more than 70% increase in demand for food and other agricultural products by 2050. Crop yields are not increasing fast enough to keep up with demand. The effects of the "Green Revolution" half a century ago are fading. Knowledge of the changes in physiological traits associated with genetic gains in yield potential is essential to improve understanding of yield-limiting factors and to devise future breeding strategies.

Here we mention some of the methods that could be used to overcome ROS-mediated yield loss in crops.

Crop Transformation

The production and evaluation of genetically modified (GM) crops is an active area of research but the access of this technology to growers in many countries is currently restricted primarily because of political and bioethical issues. Gene transformation and genetic engineering are frequently offered as hopes for increasing crop yields worldwide, particularly in less developed areas plagued by low crop productivity and malnutrition (Lipton, 1999).

A crucial challenge for a transgenic approach to yield improvement is forecasting consequences in the expression of a transformed trait. As pointed out by Miflin (2000) "farmers cultivate phenotypes" so it is crucial to adequately document variations in phenotype that result from gene modification. Successful production of transgenic plants depends on the characteristics of the gene expression cassettes, the method chosen to deliver the DNA into recipient cells, and the tissue culture and selection techniques used to regenerate fertile plants from the transformed cells (Repellin *et al.*, 2001).

The genetic gain in yield potential in high-output wheat systems worldwide since the mid-1960s has been approximately 1% per year, for example, in Mexico (Waddington *et al.*, 1986), Argentina (Abbate *et al.*, 1998), Italy (Canevara *et al.*, 1994), France (Brancourt-Hulmel *et al.*, 2003), and the UK (Austin *et al.*, 1989; Shearman *et al.*, 2005).

On-farm field trials carried out with *Bacillus thuringiensis* (Bt) cotton in different states of India show that the technology substantially reduces pest damage and increases yields (Qaim and Zilberman, 2003). The yield gains are much higher than what has been reported for other countries where genetically modified crops were used mostly to replace and enhance chemical pest control.

Breeding Technology

The art of plant breeding was developed long before the birth of genetics and cytogenetics. To feed several billion people living on this planet, the production of high-quality food must increase with reduced inputs, but this accomplishment will be particularly challenging in the face of global environmental change. Breeding and agronomic improvements have, on average, achieved a linear increase in food production globally, at an average rate of 32 million metric tons per year (Alston *et al.*, 2009). New crop varieties will need to be more efficient in their use of reduced nitrogen than current varieties are (Peoples *et al.*, 1995). Yield stability is harder to select than improved yield because selection in breeding programs requires many years and many sites for evaluation. However, there is evidence for a genetic basis for yield stability and, hence, an opportunity for gain (Kraakman *et al.*, 2004).

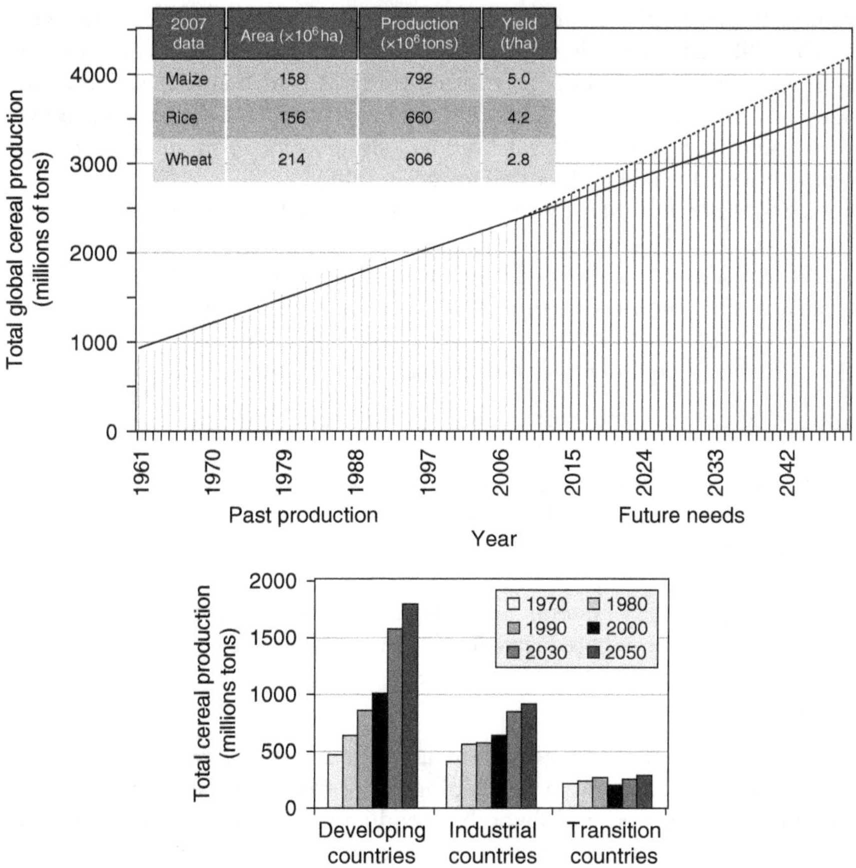

Figure 6.4 Cereal production targets. *Top:* Global cereal production has risen from 877 million metric tons in 1961 to 2351 million metric tons in 2007 (light gray). However, to meet predicted demands production will need to rise to over 4000 million metric tons by 2050 (dark gray). The rate of yield increase must move from the blue trend line (32 million metric tons per year) to the red dotted line (44 million metric tons per year) to meet this demand, an increase of 37%. The inset table shows the 2007 data for the three major cereals. *Bottom:* The greatest demand for yield increases will be from countries in the developing world. *Source*: Reproduced with permission of FAOSTAT (http://faostat.fao.org/).

Further technological developments are essential, and a major challenge will be to ensure that the technological advances already achieved are effectively deployed. Current and projected crop productions by 2050 have been given in Figure 6.4.

Ideotype breeding aimed at modifying the plant architecture is a time-tested strategy to achieve increases in yield potential. As an example, selection for short stature of cereals such as wheat, rice, and sorghum resulted in doubling of yield potential (Khush, 2005). A role of pre-breeding is to provide germplasm to breeders that produce significant increases in yield in stressful environments (e.g., Schroeder *et al.*, 2013).

Contribution of Alien Genes to Yield Potential

Genetic improvement of crops has traditionally been achieved through sexual hybridization between related species, which has resulted in numerous cultivars with high yields and superior agronomic performance. Conventional plant breeding sometimes combined with classical cytogenetic techniques, continues to be the main method of cereal crop improvement. In spring wheat, genes located on a fragment of rye chromosome have been reported to determine wide adaptation (Rabinovich, 1998) and high yield potential (Rajaram *et al.*, 1990; Villareal *et al.*, 1995, 1998). However, with the exception of the rooting traits, it is also anticipated that the identified traits will be relevant to enhancing yield potential in spring wheat. As outlined above, in order to raise yield potential in future, wheat breeders should aim to increase grains/m^2 while also increasing assimilate availability (source) for grain filling.

Marker-assisted recurrent selection (MARS) involves crossing in selected individuals at each cycle of crossing and selection. In this way, desirable alleles can be brought into the breeding scheme from many different sources. This technique has been applied to sunflower, soybean, and maize to introduce desirable alleles to several target loci into single elite lines (Eathington *et al.*, 2007).

The increased transformation frequencies for cereals have mainly been the result of:

- Systematic screenings of genotypes and explant tissues for suitability in transformation and regeneration systems.
- Reduced somaclonal variation by shortening the tissue culture period.
- Identification of useful scorable and selectable marker genes.
- Optimization of codon usage, transcriptional and translational signals to fit the monocot system.
- Improvement of direct DNA delivery systems, such as particle bombardment.
- Adaptation of the *Agrobacterium*-mediated transformation system to cereals.

Sources of Agricultural Growth

The past four decades have seen remarkable growth in agricultural productivity. This productivity increase has been particularly strong in developing countries and especially for cereals such as rice in Asia, wheat in irrigated and favorable production environments worldwide, and maize in Mesoamerica and selected parts of Africa and Asia (Pingali and Heisey, 2001). Other crops have also seen productivity increases. Furthermore, rising population pressure is squeezing agricultural land for cultivation and pastures. Finally, the agricultural sector is under significant adjustment pressure related to market liberalization and globalization.

Conclusion

Different biotic and abiotic factors, which cause oxidative stress, are mainly responsible for crop yield losses worldwide. Various crop models predict declines in crop productivity in coming decades. Therefore it is imperative to increase global crop production using various strategies including, but not limited to, improved breeding technologies, transgenic approaches, improved agricultural practices, use of microbial consortia, extending land under crop cultivation, and so forth. It is high time that plant physiologists, molecular biologists, and plant breeders work in tandem to tackle this global problem.

Acknowledgments

The authors are grateful to the Director of Council of Scientific and Industrial Research (CSIR) – National Botanical Research Institute (NBRI) for providing the necessary facilities. M.S. and S.K.G. are grateful to University Grant Commission (UGC) and CSIR, respectively, for senior research fellowships.

References

Abbate PE, Andrade FH, Lazaro L, *et al.* (1998) Grain yield increase in recent Argentine wheat cultivars. *Crop Science* 38: 1203–1209.

Abdollahi H, Ghahremani Z. (2011) The role of chloroplasts in the interaction between *Erwinia amylovora* and host plants. *Acta Hort* 896: 215–221.

Ainsworth EA. (2008) Rice production in a changing climate: a meta-analysis of responses to elevated carbon dioxide and elevated ozone concentration. *Glob Change Biol* 14: 1642–1650.

Ali H, Khan E, Sajad MA. (2013) Phytoremediation of heavy metals – concepts and applications. *Chemosphere* 91: 869–881.

Allan AC, Fluhr R. (1997) Two distinct sources of elicited reactive oxygen species in tobacco epidermal cells. *Plant Cell* 9: 1559–1572.

Alston JM, Beddow JM, Pardey PG. (2009) Agricultural research, productivity, and food prices in the long run. *Science* 325: 1209.

Alvarez R, Ortega-Villasante C, Alvarez-Fernandez A, Campo FFD, Hernandez LE. (2006) Stress responses of *Zea mays* to cadmium and mercury. *Plant Soil* 279: 41–50.

Ashraf, M.A. (2012) Waterlogging stress in plants: A review. *Afr J Agric Res* 7(13): 1976–1981.

Austin RB, Morgan CL, Ford MA, Bhagwat SG. (1982) Flag leaf photosynthesis of *Triticum aestivum* and related diploid and tetraploid species. *Ann Bot* 49: 177–189.

Austin RB, Ford MA, Morgan CL. (1989) Genetic improvement in the yield of winter wheat: a further evaluation. *J Agr Sci* 112: 295–301.

Avnery S, Mauzerall DL, Liu J, Horowitz LW. (2011) Global crop yield reductions due to surface ozone exposure: 2. Year 2030 potential crop production losses and economic damage under two scenarios of O_3 pollution. *Atmos Environ* 45: 2297–2309.

Balla K, Rakszegi M, Li ZG, Bekes F, Bencze S, Veisz O. (2011) Quality of winter wheat in relation to heat and drought shock after anthesis. *Czech J Food Sci* 29: 117–128.

Bewley JD, Black M. (1994) *Seeds. Physiology of Development and Germination*, 2nd edn. Plenum Press, New York.

Bolwell GP, Wojtaszek P. (1997) Mechanisms for the generation of reactive oxygen species in plant defence – a broad perspective. *Physiol Mol Plant Pathol* 51: 347–366.

Booker F, Muntifering R, McGrath M, *et al.* (2009) The ozone component of global change: potential effects on agricultural and horticultural plant yield, product quality and interactions with invasive species. *J Int Plant Biol* 51: 337–351.

Brancourt-Hulmel M, Doussinault G, Lecomte C, *et al.* (2003) Genetic improvement of agronomic traits of winter wheat cultivars released in France from 1946 to 1992. *Crop Sci* 43: 37–45.

Bray EA, Bailey-Serres J, Weretilnyk E. (2000) Responses to abiotic stresses. In: Gruissem W, Buchannan B, Jones R (eds), *Biochemistry and Molecular Biology of Plants.* American Society of Plant Physiologists, Rockville, MD, pp. 1158–1203.

Canevara MG, Romani M, Corbellini M, Perenzin M, Borghi B. (1994) Evolutionary trends in morphological, physiological, agronomical and qualitative traits of *Triticum aestivum* L. cultivars bred in Italy since 1900. *Eur J Agron* 3: 175–185.

Chaves MM, Flexas J, Pinheiro C. (2009) Photosynthesis under drought and salt stress: regulation mechanisms from whole plant to cell. *Ann Bot* 103: 551–560.

Compant S, Van Der Heijden MG, Sessitsch A. (2010) Climate change effects on beneficial plant–microorganism interactions. *FEMS Microbiol Ecol* 73: 197–214.

Cramer GR, Urano K, Delrot S, Pezzotti, M, Shinozaki K. (2011) Effects of abiotic stress on plants: a systems biology perspective. *BMC Plant Biol* 11: 163.

Cruz RPD, Duarte ITDL, Cabreira C. (2010) Inheritance of rice cold tolerance at the seedling stage. *Sci Agric* 67: 669–674.

Daub ME, Chung KR. (2009) Photoactivated perylenequinone toxins in plant pathogenesis. In: Deising H (ed.), *The Mycota V. Plant Relationships*, 2nd edn. Springer-Verlag, Heidelberg, Berlin, pp. 201–219.
Donald CM, Hamblin J. (1976) The biological yield and harvest index of cereals as agronomic and plant breeding criteria. *Adv Agronom* 38: 361–405.
Eathington SR, Crosbie TM, Edwards MD, Reiter RS, Bull JK. (2007) Molecular markers in a commercial breeding program. *Crop Sci* 1: 47.
Farid M, Shakoor MB, Ehsan S, Ali S, Zubair M, Hanif MA. (2013) Morphological, physiological and biochemical responses of different plant species to Cd stress. *Int J Chem Biochem Sci* 3: 53–60.
Feng ZZ, Kobayashi K, Ainsworth EA. (2008) Impact of elevated ozone concentration on growth, physiology and yield of wheat (*Triticum aestivum* L.): a meta-analysis. *Glob Change Biol* 14: 2696–2708.
Fiscus EL, Booker FL, Burkey KO. (2005) Crop responses to ozone: uptake, modes of action, carbon assimilation and partitioning. *Plant Cell Environ* 28: 997–1011.
Flexas J, Bota J, Loreto F, Cornic G, Sharkey TD. (2004) Diffusive and metabolic limitations to photosynthesis under drought and salinity in C3 plants. *Plant Biol* 6: 269–279.
Flora SJS, Mittal M, Mehta A. (2008) Heavy metal induced oxidative stress and its possible reversal by chelation therapy. *Indian J Med Res* 128: 501–523.
Flowers TJ. (2004) Improving crop salt tolerance. *J Exp Bot* 55: 307–319.
Foley JA, Ramankutty N, Brauman KA, *et al.* (2011) Solutions for a cultivated planet. *Nature* 478: 337–342.
Ford DM, Shibles R, Green DE. (1983) Growth and yield of soybean lines selected for divergent leaf photosynthetic ability. *Crop Sci* 23: 517–520.
Fuhrer J. (2009) Ozone risk for crops and pastures in present and future climates. *Naturwissenschaften* 96: 173–194.
Garrett KA, Dendy SP, Frank EE, Rouse MN, Travers SE. (2006) Climate change effects on plant disease: genomes to ecosystems. *Annu Rev Phytopathol* 44: 489–509.
Gechev TS, Hille J. (2005) Hydrogen peroxide as a signal controlling plant programmed cell death. *J Cell Biol* 168: 17–20.
Gechev TS, Dinakar C, Benina M, Toneva V, Bartels D. (2012) Molecular mechanisms of desiccation tolerance in resurrection plants. *Cell Mol Life Sci* 69: 3175–3186.
Geilfus CM, Mithöfer A, Ludwig-Müller J, Zörb C, Muehling KH. (2015) Chloride-inducible transient apoplastic alkalinizations induce stomata closure by controlling abscisic acid distribution between leaf apoplast and guard cells in salt-stressed *Vicia faba*. *New Phytol* 208: 803–816.
Ghannoum O. (2009) C4 photosynthesis and water stress. *Ann Bot* 103: 635–644.
Gill SA, Tuteja N. (2010) Reactive oxygen species and antioxidant machinery in abiotic stress tolerance in crop plants. *Plant Physiol Biochem* 48: 909–930.
Gill SS, Khan NA, Anjum NA, Tuteja N. (2011) Amelioration of cadmium stress in crop plants by nutrients management: Morphological, physiological and biochemical aspects. *Plant Stress* 5: 1–23.
Goel S, Madan B. (2014) Genetic engineering of crop plants for abiotic stress tolerance. In: Ahmed P, Rasool S (eds), *Emerging Technologies and Management of Crop Stress Tolerance*, Vol. 1: 99–123 Elsevier, Amsterdam.
Government of India. (2005) Economic Division, Ministry of Finance, *Economic Survey. Government of India*, New Delhi.
Grant JJ, Yun BW, Loake GJ. (2000) Oxidative burst and cognate redox signalling reported by luciferase imaging: identification of a signal network that functions independently of ethylene, SA and Me-JA but is dependent on MAPKK activity. *Plant J* 24: 569–582.
Habib SH, Kausar H, Saud HM. (2016) Plant growth-promoting rhizobacteria enhance salinity stress tolerance in okra through ROS-scavenging enzymes. *Biomed Res Int* 2016: 6284547.

Hall JL. (2002) Cellular mechanisms for heavy metal detoxification and tolerance. *J Exp Bot* 53: 1–11.

Harb A, Krishnan A, Ambavaram MMR, Pereira A. (2010) Molecular and physiological analysis of drought stress in *Arabidopsis* reveals early responses leading to acclimation in plant growth. *Plant Physiol* 154: 1254–1271.

Hasanuzzaman M, Hossain MA, da Silva JAT, Fujita M. (2012) Plant responses and tolerance to abiotic oxidative stress: antioxidant defenses is a key factor. In: Bandi V, Shanker AK, Shanker C, Mandapaka M (eds), *Crop Stress and Its Management: Perspectives and Strategies*. Springer, Berlin, pp. 261–316.

Hasanuzzaman M, Nahar K, Fujita M. (2013) Extreme temperatures, oxidative stress and antioxidant defense in plants. In: Vahdati K, Leslie C (eds), *Abiotic Stress—Plant Responses and Applications in Agriculture*. InTech, Rijeka, Croatia, pp. 169–205.

Heagle AS (1989) Ozone and crop yield. *Ann. Rev. Phytopath* 27: 397–423.

Hideg E, Jansen MA, Strid A. (2013) UV-B exposure, ROS, and stress: inseparable companions or loosely linked associates? *Trends Plant Sci* 18: 107–115.

Horton P. (2000) Prospects for crop improvement through the genetic manipulation of photosynthesis: morphological and biochemical aspects of light capture. *J Exp Bot* 51: 475–485.

Hossain MA, Piyatida P, da Silva JAT, Fujita M. (2012) Molecular mechanism of heavy metal toxicity and tolerance in plants: central role of glutathione in detoxification of reactive oxygen species and methylglyoxal and in heavy metal chelation. *J Bot* 2012: Article ID 872875.

Hu WH, Song XS, Shi K, *et al.* (2008) Changes in electron transport, superoxide mutase and ascorbate peroxidase isoenzymes in chloroplasts and mitochondria of cucumber leaves as influenced by chilling. *Photosynthetica* 46: 581–588.

Huang B, Xu C. (2008) Identification and characterization of proteins associated with plant tolerance to heat stress. *J Integr Plant Biol* 50: 1230–1237.

Jaleel CA, Manivannan P, Sankar B, *et al.* (2007a) Water deficit stress mitigation by calcium chloride in *Catharanthus roseus*; effects on oxidative stress, proline metabolism and indole alkaloid accumulation. *Colloids Surf B: Biointerfaces* 60: 110–116.

Jaleel CA, Manivannan P, Kishorekumar A, *et al.* (2007b) Alterations in osmoregulation, antioxidant enzymes and indole alkaloid levels in *Catharanthus roseus* exposed to water deficit. *Colloids Surf B: Biointerfaces* 59: 150–157.

Jaleel CA, Manivannan P, Sankar B, *et al.* (2007c) Induction of drought stress tolerance by ketoconazole in *Catharanthus roseus* is mediated by enhanced antioxidant potentials and secondary metabolite accumulation. *Colloids Surf B: Biointerfaces* 60: 201–206.

Jaleel CA, Manivannan P, Sankar B, *et al.* (2007d) Paclobutrazol enhances photosynthesis and ajmalicine production in *Catharanthus roseus*. *Process Biochem* 42: 1566–1570.

Jaleel CA, Sankar B, Murali PV, *et al.* (2008) Water deficit stress effects on reactive oxygen metabolism in *Catharanthus roseus*; impacts on ajmalicine accumulation. *Colloids Surf B: Biointerfaces* 62: 105–111.

Jaspers P, Kangasjärvi J. (2010) Reactive oxygen species in abiotic stress signaling. *Physiol Plant* 138: 405–413.

Jenkins GI. (2009) Signal transduction in responses to UV-B radiation. *Annu Rev Plant Biol* 6: 407–431.

Karuppanapandian T, Moon JC, Kim C, Manoharan K and Kim W. (2011a) Reactive oxygen species in plants: their generation, signal transduction, and scavenging mechanisms. *Aust J Crop Sci* 5(6): 709.

Karuppanapandian T, Wang HW, Prabakaran N, Jeyalakshmi K, Kwon M, Manoharan K and Kim W. (2011b) 2, 4-dichlorophenoxyacetic acid-induced leaf senescence in mung bean (Vigna radiata L. Wilczek) and senescence inhibition by co-treatment with silver nanoparticles. *Plant Physiol Biochem* 49(2): 168-177.

Khush GS. (2005) What it will take to feed 5.0 billion rice consumers in 2030. *Plant Molecular Biology* 59: 1–6.

Kraakman ATW, Niks RE, Van den Berg PM, Stam P, Van Eeuwijk FA. (2004) Linkage disequilibrium mapping of yield and yield stability in modern spring barley cultivars. *Genetics* 168: 435.

Lal R. (2005) Climate change, soil carbon dynamics, and global food security. In: Lal R, Stewart B, Uphoff N, *et al.* (eds), *Climate Change and Global Food Security*. CRC Press, Boca Raton, FL, pp. 113–143.

Lamb C, Dixon RA. (1997) The oxidative burst in plant disease resistance. *Annu Rev Plant Physiol Plant Mol Biol* 48: 251–275.

Li J, Yang L, Jin D, *et al.* (2013) UV-B-induced photomorphogenesis in *Arabidopsis*. *Protein Cell* 4: 485–492.

Li YP, Ye W, Wang M, Yan XD. (2009) Climate change and drought: a risk assessment of crop-yield impacts. *Clim Res* 39: 31–46.

Lipton M. (1999) Reviving global poverty reduction: what role for genetically modified plants? Sir John Crawford memorial lecture 1999. CGIAR Secretariat (available at: http://library.cgiar.org/bitstream/handle/10947/1250/lipton.pdf?sequence=1).

Liu Y, Ren D, Pike S, *et al.* (2007) Chloroplast-generated reactive oxygen species are involved in hypersensitive response-like cell death mediated by a mitogen-activated protein kinase cascade. *Plant J* 51: 941–954.

Lobell DB, Gourdji SM. (2012) The influence of climate change on global crop productivity. *Plant Physiol* 160: 1686–1697.

Maheshwari R, Dubey RS. (2009) Nickel-induced oxidative stress and the role of antioxidant defence in rice seedlings. *Plant Growth Regul* 59: 37–49.

Mahmood S, Wahid A, Javed F, Basra SMA. (2010) Heat stress effects on forage quality characteristics of maize (*Zea mays*) cultivars. *Int J Agric Biol* 12: 701–706.

Mattos LM. (2015) Oxidative stress in plants under drought conditions and the role of different enzymes. *Enz Eng* 5: 2.

Miflin B. (2000) Crop improvement in the 21st century. *J Exp Bot* 51: 1–8.

Miller G, Suzuki N, Ciftci-Yilmaz S, Mittler R. (2010) Reactive oxygen species homeostasis and signalling during drought and salinity stresses. *Plant Cell Environ* 33: 453–467.

Mittler R. (2002) Oxidative stress, antioxidants and stress tolerance. *Trends Plant Sci* 7: 405–410.

Mittler R. (2006) Abiotic stress, the field environment and stress combination. *Trends Plant Sci* 11: 15–19.

Mittler R, Blumwald E. (2010) Genetic engineering for modern agriculture: challenges and perspectives. *Annu Rev Plant Biol* 61: 443–462.

Moan J. (2001) Visible light and UV radiation. In: Brune D, Hellborg R, Persson BRR, Pääkkönen R (eds), *Radiation at Home, Outdoors and in the Workplace*. Scandinavian Science Publisher, Oslo, pp. 69–85.

Moller IM, Jensen PE, Hansson A. (2007) Oxidative modifications to cellular components in plants. *Ann Rev Plant Biol* 58: 459–481.

Morgan PB, Mies TA, Bollero GA, Nelson RL, Long SP. (2006) Season-long elevation of ozone concentration to projected 2050 levels under fully open-air conditions substantially decreases the growth and production of soybean. *New Phytol* 170: 333–343.

Moss DN, Musgrave RB. (1971) Photosynthesis and crop production. *Adv Agronom* 23: 317–336.

Murphy JJ, Delucchi MA, McCubbin DR, Kim HJ. (1999) The cost of crop damage caused by ozone air pollution from motor vehicles. *J Environ Man* 55: 365–376.

Nagajyoti PC, Lee KD, Sreekanth TVM. (2010) Heavy metals, occurrence and toxicity for plants: a review. *Environ Chem Lett* 8: 199–216.

Najeeb U, Bange MP, Tan DK, Atwell BJ. (2015) Consequences of waterlogging in cotton and opportunities for mitigation of yield losses. *AoB Plants* 1: 7.

Nawaz F, Ahmad R, Waraich EA, Naeem MS, Shabbir RN. (2012) Nutrient uptake, physiological responses, and yield attributes of wheat (*Triticum aestivum*) exposed to early and late drought stress. *J Plant Nutr* 35: 961–974.

Nawaz F, Ashraf MY, Ahmad R, Waraich EA, Shabbir RN. (2014) Selenium (Se) regulates seedling growth in wheat under drought stress. *Adv Chem* 2014: Art. no. 143567; doi: 10.1155/2014/143567.

Neumann PM. (2008) Coping mechanisms for crop plants in drought prone environments. *Ann Bot* 101: 901–907.

Noctor G, Veljovic-Jovanovic S, Driscoll S, Novitskaya L, Foyer C. (2002) Drought and oxidative load in the leaves of C3 plants: A predominant role for photorespiration? *Ann Bot* 89: 841–850.

Nouri-Ganbalani A, Nouri-Ganbalani G, Hassanpanah D. (2009) Effects of drought stress condition on the yield and yield components of advanced wheat genotypes in Ardabil, Iran. *J Food Agric Environ* 7: 228–234.

Oerke EC. (2006) Crop losses to pests. *J Agric Sci* 144: 31–43.

Owens S. (2001) Salt of the earth. Genetic engineering may help to reclaim agricultural land lost due to salinisation. *EMBO Rep* 2: 877–879.

Peoples MB, Mosier AR, Freney JR. (1995) Minimizing gaseous losses of nitrogen. In: Bacon PE (ed.), *Nitrogen Fertilization in the Environment.* Marcel Dekker, New York, pp. 505–602.

Pingali P, Heisey PW. (2001) Cereal-crop productivity in developing countries. In: Alston JM, Parday PG, Taylor MJ (eds), *Agricultural Science Policy. Changing Global Agendas* 56–84. The Johns Hopkins University Press, Baltimore and London.

Portis AR Jr, Parry MAJ. (2007) Discoveries in Rubisco (ribulose 1,5-bisphosphate carboxylase/ oxygenase): a historical perspective. *Photosynth Res* 94: 121–143.

Qaim M, Zilberman D. (2003) Yield effects of genetically modified crops in developing countries. *Science* 299: 900–902.

Rabinovich SV. (1998) Importance of wheat–rye translocations for breeding modern cultivars of *Triticum aestivum* L. *Euphytica* 100: 323–340.

Rajaram S, Villareal R, Mujeeb-Kazi A. (1990) The global impact of 1B/1R spring wheat. *Agronomy Abstracts.* ASA, Madison, WI, 105 pp.

Rascio N, Navari-Izzo F. (2011) Heavy metal hyperaccumulating plants: how and why do they do it? And what makes them so interesting? *Plant Sci* 180: 169–181.

Repellin A, Båga M, Jauhar PP, Chibbar RN. (2001) Genetic enrichment of cereal crops via alien gene transfer: new challenges. *Plant Cell Tiss Org* 64: 159–183.

Rockstrom J, Falkenmark M. (2000) Semiarid crop production from a hydrological perspective: gap between potential and actual yields. *Crit Rev Plant Sci* 19: 319–346.

Sakata T, Higashitani A. (2008) Male sterility accompanied with abnormal anther development in plants – genes and environmental stresses with special reference to high temperature injury. *Int J Plant Dev Biol* 2: 42–51.

Sanghera GS, Wani SH, Hussain W, Singh NB. (2011) Engineering cold stress tolerance in crop plants. *Curr Genomics* 12: 30–43.

Schroeder JI, Delhaize E, Frommer WB, *et al.* (2013) Using membrane transporters to improve crops for sustainable food production. *Nature* 497: 60–66.

Schützendübel A, Polle A. (2002) Plant responses to abiotic stresses: heavy metal-induced oxidative stress and protection by mycorrhization. *J Exp Bot* 53: 1351–1365.

Setter TL, Waters I, Sharma SK, *et al.* (2009) Review of wheat improvement for waterlogging tolerance in Australia and India: the importance of anaerobiosis and element toxicities associated with different soils. *Ann Bot* 103: 221–235.

Sharma P, Dubey RS. (2005) Drought induces oxidative stress and enhances the activities of antioxidant enzymes in growing rice seedlings. *Plant Growth Regul* 46: 209–221.

Sharma P, Jha AB, Dubey RS, Pessarakli. (2012) Reactive oxygen species, oxidative damage, and antioxidative defense mechanism in plants under stressful conditions. *J Bot* 2012, Art. ID 217037.
Shearman VJ, Sylvester-Bradley R, Scott RK, Foulkes MJ. (2005) Physiological processes associated with wheat yield progress in the UK. *Crop Sci* 45: 175–185.
Shibles R, Secor J, Ford DM. (1987) Carbon assimilation and metabolism. In: Wilcox JR (ed.), *Soybeans: Improvement, Production, and Uses*, 2nd edn. American Society of Agronomy, Madison, WI, pp. 535–588.
Singh KN, Chatrath R. (2001) Salinity tolerance. In: Reynolds MP, Monasterio JIO, McNab A (eds), *Application of Physiology in Wheat Breeding*. CIMMYT, Mexico, pp. 101–110.
Slafer GA, Calderini DF, Miralles DJ. (1996) Yield components and compensation in wheat: opportunities for further increasing yield potential. In: Reynolds MP *et al.* (eds), *Increasing Yield Potential in Wheat: Breaking the Barriers*. CIMMYT, Mexico, pp. 101–132.
Singh R, Singh S, Parihar P, *et al.* (2016) Reactive oxygen species (ROS): beneficial companions of plants' developmental processes. *Front Plant Sci* 7: 1299.
Singh S, Srivastava PK, Kumar D, Tripathi DK, Chauhan DK, Prasad SM. (2015) Morpho-anatomical and biochemical adapting strategies of maize (*Zea mays* L.) seedlings against lead and chromium stresses. *Biocatal Agric Biotechnol* 4: 286–295.
Singh S, Tripathi DK, Singh S, *et al.* (2017) Toxicity of aluminium on various levels of plant cells and organism: a review. *Environ Exp Bot* 137: 177–193.
Singh VP, Tripathi DK, Kumar D, Chauhan DK. (2011) Influence of exogenous silicon addition on aluminium tolerance in rice seedlings. *Biol Trace Elem Res* 144: 1260–1274.
Solomon S, Qin D, Manning M, *et al.* (2007) *The Physical Science Basis. Contribution of Working Group I to the Fourth Assessment Report of the Intergovernmental Panel on Climate Change.* Cambridge University Press, New York.
Suzuki K, Nagasuga K, Okada M. (2008) The chilling injury induced by high root temperature in the leaves of rice seedlings. *Plant Cell Physiol* 49: 433–442.
Sytar O, Kumar A, Latowski D, *et al.* (2013) Heavy metal-induced oxidative damage, defense reactions, and detoxification mechanisms in plants. *Acta Physiol Plant* 35: 985–999.
Tangahu BV, Abdullah SRS, Basri H, *et al.* (2011) A review on heavy metals (As, Pb, and Hg) uptake by plants through phytoremediation. *Int J Chem Eng* 2011: Art. ID 939161.
Tesfaw A, Feyissa T. (2014) Current trends in genetic manipulations to enhance abiotic and biotic stresses in tobacco. *Afr J Biotechnol* 13: 2095–2102.
Thakurta PG. (2010) The magic of millets in drought prone country. *Asian Age* 30: 1–2.
Tripathi DK, Singh VP, Kumar D, Chauhan DK. (2012a) Impact of exogenous silicon addition on chromium uptake, growth, mineral elements, oxidative stress, antioxidant capacity, and leaf and root structures in rice seedlings exposed to hexavalent chromium. *Acta Physiol Plant* 34: 279–289.
Tripathi DK, Singh VP, Kumar D, Chauhan DK. (2012b) Rice seedlings under cadmium stress: effect of silicon on growth, cadmium uptake, oxidative stress, antioxidant capacity and root and leaf structures. *Chem Ecol* 28: 281–291.
Tripathi DK, Singh S, Singh VP, Prasad SM, Chauhan DK, Dubey NK. (2016a) Silicon nanoparticles more efficiently alleviate arsenate toxicity than silicon in maize cultivar and hybrid differing in arsenate tolerance. *Front Environ Sci* 4: 46.
Tripathi DK, Singh VP, Prasad SM, Dubey NK, Chauhan DK, Rai AK. (2016b) LIB spectroscopic and biochemical analysis to characterize lead toxicity alleviative nature of silicon in wheat (Triticum aestivum L.) seedlings. *J Photochem Photobiol B: Biol* 154: 89–98.
Tripathi DK, Singh S, Singh VP, Prasad SM, Dubey NK, Chauhan DK. (2017a) Silicon nanoparticles more effectively alleviated UV-B stress than silicon in wheat (Triticum aestivum) seedlings. *Plant Physiol Biochem* 110: 70–81.

Tripathi DK, Singh S, Singh S, *et al.* (2017b) Nitric oxide alleviates silver nanoparticles (AgNps)-induced phytotoxicity in Pisum sativum seedlings. *Plant Physiol Biochem* 110: 167–177.

Tripathi DK, Mishra RK, Singh S, *et al.* (2017c) Nitric oxide ameliorates zinc oxide nanoparticles phytotoxicity in wheat seedlings: implication of the ascorbate–glutathione cycle. *Front Plant Sci* 8: 1.

Tullus A, Kupper P, Sellin A, *et al.* (2012) Climate change at northern latitudes: rising atmospheric humidity decreases transpiration, N-uptake and growth rate of hybrid aspen. *PLoS One* 7(8): e42648.

United Nations, Department of Economic and Social Affairs, Population Division. (2015) *World Population Prospects.* The 2015 Revision. New York: United Nations.

US EPA. (2006) Air Quality Criteria for Ozone and Related Photochemical Oxidants. EPA/600/R-05/004aF-cF.

Vadez V, Berger JD, Warkentin T, *et al.* (2012) Adaptation of grain legumes to climate change: a review. *Agron Sustain Dev* 32: 31–44.

Van Breusegem F, Bailey-Serres J, Mittler R. (2008) Unraveling the tapestry of networks involving reactive oxygen species in plants. *Plant Physiol* 147: 978–984.

Villareal RL, Del Toro E, Mujeeb-Kazi A, Rajaram S. (1995) The 1BL/1RS chromosome translocation effect on yield characteristics in a *Triticum aestivum* L. cross. *Plant Breeding* 114: 497–500.

Villareal RL, Banuelos O, Mujeeb-Kazi A, Rajaram S. (1998) Agronomic performance of chromosomes 1B and 1BL/1RS near-isolines in the spring bread wheat Seri M82. *Euphytica* 103: 195–202.

Vingarzan R. (2004) A review of surface ozone background levels and trends. *Atmos Environ* 38: 3431–3442.

Voesenek LA, Bailey-Serres J. (2013) Flooding tolerance: O_2 sensing and survival strategies. *Curr Opin Plant Biol* 16: 647–653.

Waddington SR, Ransom JK, Osmanzai M, Saunders DA. (1986) Improvement in the yield potential of bread wheat adapted to northwest Mexico. *Crop Sci* 26: 698–703.

Walbot V. (2011) How plants cope with temperature stress. *BMC Biol* 9: 79.

Wassmann R, Jagadish S, Sumfleth K, *et al.* (2009) Regional vulnerability of climate change impacts on Asian rice production and scope for adaptation. *Adv Agron* 102: 91–133.

Wiebold WJ, Shibles R, Green DE. (1981) Selection for apparent photosynthesis and related leaf traits in early generations of soybeans. *Crop Sci* 21: 969–973.

Wolf J. (2002) Comparison of two potato simulation models under climate change. I. Model calibration and sensitivity analyses. *Climate Res* 21: 173–186.

Wu F, Zhang G, Dominy P, Wu H, Bachir DML. (2007) Differences in yield components and kernel Cd accumulation in response to Cd toxicity in four barley genotypes. *Chemosphere* 70: 83–92.

Xin Z, Browse J. (2001) Cold comfort farm: the acclimation of plants to freezing temperatures. *Plant Cell Environ* 23: 893–902.

Young LW, Wilen RW, Bonham-Smith PC. (2004) High temperature stress of *Brassica napus* during flowering reduces micro- and mega-gametophyte fertility, induces fruit abortion, and disrupts seed production. *J Exp Bot* 55: 485–495.

Zhang Y, Tang HR, Luo Y. (2008) Variation in antioxidant enzyme activities of two strawberry cultivars with short-term low temperature stress. *World J Agric Sci* 4: 458–462.

Zheng M, Tao Y, Hussain S, *et al.* (2016) Seed priming in dry direct-seeded rice: consequences for emergence, seedling growth and associated metabolic events under drought stress. *Plant Growth Regul* 78: 167–178.

Zinn KE, Tunc-Ozdemir M, Harper JF. (2010) Temperature stress and plant sexual reproduction: uncovering the weakest links. *J Exp Bot* 61: 1959–1968.

Web References

FAO. World Food Situation. http://www.fao.org/worldfoodsituation/csdb/en/
National Academy Press. Transgenic Plants and World Agriculture. 2000. Available at: https://www.nap.edu/catalog/9889/transgenic-plants-and-world-agriculture
UN Department of Economic and Social Affairs. http://www.un.org/en/development/desa/news/population/2015-report.html

7

Reactive Oxygen Species and Photosynthetic Functioning: Past and Present

Cristina Sgherri[1]*, Calogero Pinzino*[2]*, and Mike Frank Quartacci*[1]

[1] *University of Pisa, Pisa, Italy*
[2] *Research National Council (CNR), Istituto di Chimica dei Composti Organo Metallici (ICCOM), Pisa, Italy*

Introduction

Oxygen, besides being fundamental for energy metabolism and respiration, is toxic to the same aerobic organisms since it may be converted into activated forms of oxygen, the reactive oxygen species (ROS), which are toxic for the cell. There are two main mechanisms involved in ROS formation. The reduction of oxygen to form superoxide ($O_2^{\cdot-}$), hydrogen peroxide (H_2O_2), and hydroxyl radicals ($^{\cdot}OH$) is the principal mechanism of oxygen activation in most biological systems. However, in photosynthetic plants, the formation of singlet oxygen ($^1O_2$) by the photosystems also has importance (Krieger-Liszkay, 2005; Sgherri and Pérez-López, 2013). Indeed, ROS are generated either by energy transfer or by electron transfer reactions. The former reaction leads to the formation of singlet oxygen, whereas the latter results in the sequential reduction from superoxide to hydroxyl radical.

Activated oxygen is formed by the dysfunctioning of enzymes or electron transport systems, as a result of perturbations in metabolism caused by chemical or environmental stresses (Sgherri *et al.*, 1993). Main sources of ROS can be identified as the electron transport chains of mitochondria and chloroplasts, due to leakage of electrons (Menconi *et al.*, 1995; Foyer, 1997). Different sites are recognized as responsible for ROS production, and for plant cells chloroplasts represent an important additional source for ROS increase compared to mammalian cells (Asada and Takahashi, 1987).

ROS cause oxidative stress, damaging the cell. The targets of oxidation are membrane lipids, proteins, and DNA (Verma and Dubey, 2003). Thus, free radicals ($R^{\cdot}$) – molecules containing an unpaired electron – are some of the most chemically reactive molecules known. Because of the need to pair its single electron, a free radical must abstract a second electron from a neighboring molecule. This causes the formation of another free radical giving rise to a self-propagating chain reaction (Sgherri and Pérez-López, 2013).

Under steady-state conditions these molecules are scavenged by different antioxidative defense components; these are often confined to particular compartments until the equilibrium between production and scavenging of ROS is perturbed, for instance, in the case of adverse environmental conditions. As a result of these disturbances, intracellular levels of ROS may rapidly rise (Sgherri *et al.*, 1996). In the plant kingdom every environmental stress determines an oxidative stress, and an oxidative burst is the response to each environmental

Reactive Oxygen Species in Plants: Boon Or Bane - Revisiting the Role of ROS, First Edition. Edited by Vijay Pratap Singh, Samiksha Singh, Durgesh Kumar Tripathi, Sheo Mohan Prasad, and Devendra Kumar Chauhan.

change affecting the plant cell (V.P. Singh *et al.*, 2011; Tripathi *et al.*, 2012a,b, 2016a,b, 2017a–c; S. Singh *et al.*, 2015, 2017; R. Singh *et al.*, 2016). It is generally accepted that the mechanism underlying the response of the plant cell to each environmental stressor is the same. Initially, plants respond with an increased production of ROS, which can determine a decrease or an increase of antioxidants and antioxidative enzymes (Mittler, 2002; V.P. Singh *et al.*, 2011; Tripathi *et al.*, 2012a,b, 2016a,b, 2017a–c; S. Singh *et al.*, 2015, 2017; R. Singh *et al.*, 2016), and the antioxidative response of the plant depends on stress intensity (Sharma and Dubey, 2005; V.P. Singh *et al.*, 2011; Tripathi *et al.*, 2012a,b, 2016a,b, 2017a–c; Sgherri and Pérez-López, 2013; S. Singh *et al.*, 2015, 2017; R. Singh *et al.*, 2016).

Activation of Oxygen and Free Radicals

A free radical is any species that contains one or more unpaired electrons (Halliwell and Gutteridge, 2006). An unpaired electron is one that occupies an atomic or molecular orbital by itself. Atmospheric oxygen in its ground state is a biradical because it contains two unpaired electrons with parallel spins, this feature making oxygen paramagnetic. Sometimes it can be "activated" and this occurs mostly by reactions involving transfer of only a single electron (monovalent reduction), according to the Pauli exclusion principle. Activation of oxygen by absorption of sufficient energy to reverse the spin of one of the unpaired electrons, or monovalent reduction, gives rise to the formation of ROS, harmful to living systems because they react with organic molecules of the cell (Verma and Dubey, 2003).

Generally, the term ROS includes not only the oxygen radicals but also some non-radical derivatives of O_2. Thus, ROS is a collective term that includes both oxygen radicals and certain non-radicals that are oxidizing agents and/or are easily converted into radicals. Hence, all oxygen radicals are ROS, but not all ROS are oxygen radicals (Halliwell and Gutteridge, 2006). Among non-radical ROS are singlet oxygen ($^1O_2$) and hydrogen peroxide (H_2O_2). The former can be generated by an input of energy that rearranges the electrons (Foote *et al.*, 1985). In this way, the spin restriction is removed and the oxidizing ability greatly increased, making $^1O_2$ more reactive toward organic molecules such as proteins, DNA, and lipids than its triplet counterpart (Foote *et al.*, 1985).

The second mechanism of activation is by the stepwise monovalent reduction of oxygen to form the superoxide radical anion ($O_2^{\cdot-}$), H_2O_2, hydroxyl radical ($^{\cdot}OH$), and, eventually, water according to the scheme shown in Figure 7.1. Thus, the two-electron reduction product of oxygen is H_2O_2, and the four-electron product is water. The first step in the reduction of oxygen forming superoxide is endothermic but subsequent reductions are exothermic.

Among ROS, reactivity toward biological molecules can be very different. In fact, H_2O_2 and $O_2^{\cdot-}$ react fast with few molecules, whereas $^{\cdot}OH$ reacts fast with almost everything (Halliwell and Gutteridge, 2006).

Although $O_2^{\cdot-}$ and H_2O_2 are highly selective in their reactions with biological molecules, they are particularly dangerous to the cell because in the presence of transition metals they can give rise to $^{\cdot}OH$, which attacks everything around it. The most studied metals involved in the reac-

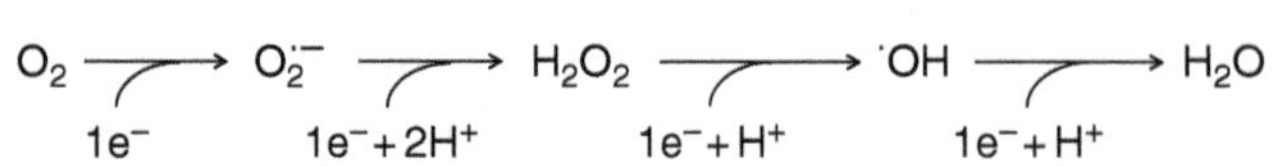

Figure 7.1 Univalent reductive chain reactions of oxygen. $O_2^{\cdot-}$, superoxide radical; H_2O_2, hydrogen peroxide; $^{\cdot}OH$, hydroxyl radical.

tion are iron, copper, cadmium, and mercury (Sgherri and Pérez-López, 2013). Fenton described in the late nineteenth century (Fenton, 1894, 1899) the oxidizing potential of hydrogen peroxide mixed with ferrous salts. Forty years later, Haber and Weiss (1934) identified the hydroxyl radical as the oxidizing species in these reactions. In the case of iron, the reactions involved are schematized below:

$$Fe^{3+} + O_2^{\cdot-} \rightarrow Fe^{2+} + O_2 \quad (7.1)$$

$$Fe^{2+} + H_2O_2 \rightarrow Fe^{3+} + OH^- + {}^{\cdot}OH \quad (7.2)$$

$$O_2^{\cdot-} + H_2O_2 \xrightarrow[Fe^{3+}]{} OH^- + {}^{\cdot}OH + O_2 \quad (7.3)$$

In biological systems the availability of ferrous ions limits the rate of reaction, but the recycling of iron from the ferric to the ferrous form by a reducing agent such as superoxide (Equation 7.1) can maintain an ongoing Fenton reaction leading to the generation of the hydroxyl radical, the agent of lipid peroxidation, which is dangerous for the cell (Verma and Dubey, 2003). Therefore, in the presence of trace amounts of iron, the reaction of superoxide and hydrogen peroxide will form the destructive hydroxyl radical and initiate the oxidation of organic molecules.

Oxidative Stress and Damage to Membranes

ROS-induced oxidations of biological molecules give rise to oxidative stress in the cell (Mittler, 2002). Many vital activities of the cell originate in the membrane, the structure and function of which are profoundly altered following oxidative stress. In particular, oxidative stress occurs in the plant cell when plants are subjected to biotic and abiotic stresses such as altered environmental conditions (Navari-Izzo *et al.*, 1992).

Oxidation of lipids mainly alters the physical state of the lipid bilayer, influencing the lipid-protein and the protein-protein associations, enzyme activities, and transport capacity of membranes. Moreover, lipid degradation alters the protein–lipid interaction and protein conformation, and initiates proteolysis (Navari-Izzo *et al.*, 1992).

It is generally accepted that ROS damage membrane lipids often exclusively by the mechanism of peroxidation. The lipid bilayer of the membranes is composed of phospholipids and glycolipids with fatty acid chains attached to carbons 1 and 2 of the glycerol backbone by an ester linkage. The peroxidation reactions differ among these fatty acids depending on the number and position of the double bonds on the acyl chain.

The peroxidation of lipids involves three distinct steps: initiation, propagation, and termination (Sgherri and Pérez-López, 2013). Non-enzymatic (chemical) lipid peroxidation is initiated by any primary free radical that has sufficient reactivity to extract a weakly bonded hydrogen atom from a reactive methylene group of a polyunsaturated fatty acid (Halliwell and Gutteridge, 1999). Strong oxidant species such as $^{\cdot}OH$, alkoxyl radicals ($RO^{\cdot}$), and peroxyl radicals ($ROO^{\cdot}$) may be involved. Chelated iron-oxygen complexes such as ferryl radicals may participate as well. The resulting fatty acid alkyl radical ($R^{\cdot}$), reacting with molecular oxygen, produces $ROO^{\cdot}$, which abstracts hydrogen from another polyunsaturated fatty acid and gives rise to hydroperoxide and a new alkyl radical. This cycle can be repeated many times resulting in the conversion of fatty acid side-chains into lipid monohydroperoxides. Thus, an initial triggering event can be greatly amplified if sufficient O_2 and unoxidized lipids are available.

The result of these chain reactions is the production of aldehydes, such as malondialdehyde, and hydrocarbons, such as ethane and ethylene, which are commonly measured end products of lipid peroxidation.

The peroxidation reactions in membrane lipids are terminated when the carbon or peroxy radicals cross-link to form conjugated products that are not radicals (Equation 7.4).

$$R^{\cdot} + R^{\cdot} \rightarrow R\text{-}R \tag{7.4}$$

An alternative to the classical mechanism of lipid peroxidation is the reaction of singlet oxygen and superoxide radical with the lipid bilayer. In fact, singlet oxygen can also react readily with unsaturated fatty acids producing a complex mixture of hydroperoxides (Bradley and Minn, 1992). However, oxidation of unsaturated fatty acids by singlet oxygen produces distinctly different products than oxidation by the hydroxyl radical (Bradley and Minn, 1992). Moreover, superoxide attacks the lipid bilayer cleaving the C-O bond, thus producing free fatty acids by de-esterification reactions (Senaratna *et al.*, 1985). Kinetic analysis indicates that superoxide attack on esters occurs by a nucleophilic addition mechanism (Afanas'ev, 1985).

Oxidative damage to proteins occurs when ROS attack amino acids. The consequent amino acid modifications induce fragmentation of the peptide chain, aggregation of cross-linked reaction products, and altered electrical charge, thus increasing susceptibility to proteolysis. Amino acids in a peptide differ in their susceptibility to attack, but sulfur-containing amino acids and thiol groups are considered the most susceptible sites (Navari-Izzo *et al.*, 1997). In general, activated oxygen can abstract an H atom from cysteine residues to form a thiyl radical that will cross-link to a second thiyl radical to form disulfide bridges. Alternatively, oxygen can add to a methionine residue to form methionine sulfoxide derivatives.

However, many amino acids can be subjected to specific irreversible modifications when a protein is oxidized. For example, tryptophan is readily cross-linked to form bityrosine products (Davies, 1987). Histidine, lysine, proline, arginine, and serine form carbonyl groups on oxidation (Stadtman, 1986).

Oxidative degradation of proteins can also be enhanced in the presence of transition metals, such as Fe, through a Fenton type reaction. In this case, the metal binds to a divalent cation binding site on the protein. The metal then reacts with hydrogen peroxide giving rise to hydroxyl radical, which rapidly oxidizes an amino acid residue at or near the cation binding site of the protein (Stadtman, 1986).

In plants, a common example of oxidative attack on proteins is represented by the various peptide components of photosystem II, among which is D1 protein, which has a high frequency of turnover. It is assumed that this turnover is a consequence of oxidative attack at specific sites on the protein (Barber and Andersson, 1992).

Electron Paramagnetic Resonance (EPR) as a Tool to Study ROS from Photosynthesis

Electron paramagnetic resonance (EPR) spectroscopy, also called electron spin resonance (ESR), is a technique used to study chemical species with unpaired electrons. It is the only direct method to detect the presence of free radicals and to identify the paramagnetic species, and provides information on the molecular structure near the unpaired electron; it plays an important role in the understanding of organic and inorganic radicals. Moreover, the EPR spectrum line shape gives insight to dynamic processes such as molecular motions or fluidity, besides probing the structure of "active sites" in metalloproteins.

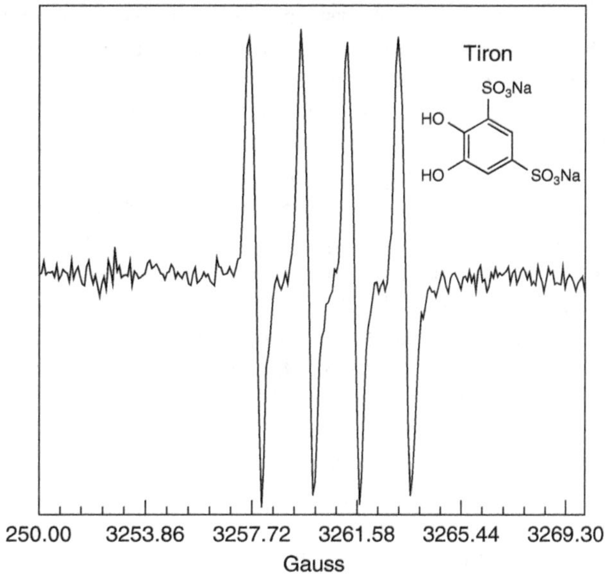

Figure 7.2 Electron paramagnetic resonance (EPR) spectrum of Tiron semiquinone.

Like most spectroscopic techniques, EPR spectrometers measure the absorption of electromagnetic radiation. An EPR spectrum is obtained by holding the frequency of radiation constant and varying the magnetic field. Absorption occurs when the magnetic field "tunes" the two spin states so that their energy difference is equal to the radiation. This is known as the field for resonance. A phase-sensitive detector is used in EPR spectrometers, which converts the normal absorption signal to its first derivative. Thus, the absorption signal is presented as its first derivative in the spectrum (Figure 7.2).

Also very important in EPR are hyperfine interactions. Besides the applied magnetic field, the free radical species is sensitive to its local "micro" environment. Additional information can be obtained from the so-called hyperfine interaction. The nuclei of the atoms in a molecule or complex usually have their own fine magnetic moments. Such magnetic moments can produce a local magnetic field intense enough to affect the electron. This interaction between the electron and the nuclei produces a local magnetic field, which is called the hyperfine interaction. Hyperfine interactions can be used to provide a great deal of information about the sample, such as the number and identity of atoms in a molecule or compound, as well as their distance from the unpaired electron. The EPR spectra have very different line shapes depending on many factors, such as the physical phase of samples and dynamic properties of molecules. Spectral simulation provides information on structure and dynamics from experimental data, that is, in studying the dependencies of spectral features on the magnetic parameters, or extracting accurate parameters from experimental spectra.

ROS are generated by the photosynthetic apparatus (Sgherri and Pérez-López, 2013) and, among the various methods used to study their generation, EPR spectroscopy is the most direct and sensitive technique. ROS such as superoxide and hydroxyl radical are short-lived free radicals with a half-life at 37 °C of 1 μs and 1 ns, respectively. Thus, they cannot be detected directly by EPR but require spin traps. The latter are compounds that react covalently with the radical products and form more stable adducts that also give rise to paramagnetic resonance spectra detectable by EPR spectroscopy. Typically, the detection of superoxide radical has used nitrones or nitroso spin traps (Finkelstein *et al.*, 1979). Tiron (1,2-dihydroxybenzene-3,5-disulfonate) semiquinone may also be used as an indicator free radical for the formation of superoxide radical in biological systems. In fact, Tiron is oxidized to an EPR-visible semiquinone (Figure 7.2) by superoxide radicals produced by xanthine oxidase (Ledenev *et al.*, 1986). The rapid rate of oxidation of the catechol by superoxide, compared to the rate of superoxide formation in most enzymatic and biological systems, makes it a suitable diagnostic probe (Greenstock and Miller, 1975; Sgherri *et al.*, 1993; Menconi *et al.*, 1995).

Otherwise, only spin trapping can specifically identify and quantify superoxide and hydroxyl radicals. One of the most commonly used spin traps, the 5,5-dimethyl-1-pyrroline-*N*-oxide (DMPO), reacts with superoxide, hydroxyl, and alkyl radicals, forming distinctive spin trap adducts. Recently, the new β-phosphorylated nitrone, 5-(diethoxyphosphoryl)-5-methyl-1-pyrroline-*N*-oxide (DEPMPO), has been applied and reported to have improved properties for trapping of superoxide, forming a more stable superoxide adduct with a half-life of 15 min (Sgherri *et al.*, 1999, 2015). DEPMPO is the most efficient spin trap for the *in vitro* and *in vivo* detection of O-, N-, S-, and C-centered free radicals (Sgherri *et al.*, 2015) and can be used to distinguish between superoxide-dependent and -independent mechanisms that lead to the hydroxyl radical (Sgherri *et al.*, 1999; Zvezdanović *et al.*, 2013). Figure 7.3 shows the appearance of superoxide and hydroxyl radicals after continuous illumination of thylakoids from wheat. In fact, in the experimental spectrum both the DEPMPO-OOH adduct and the DEPMPO-OH adduct can be distinguished. From the simulation of the EPR spectrum superoxide accounted for 53.8% whereas hydroxyl radical accounted for 46.2%. Those results were obtained assuming the presence of two conformers, A and B, of DEPMPO-OOH contributing 7.8% and 46.2%, respectively, with the following EPR hyperfine splitting constants: A: $a^P = 50.55$ G, $a^N = 12.63$ G, $a^H = 12.33$ G; B: $a^P = 49.65$ G, $a^N = 13.22$ G, $a^H = 10.65$ G. For the DEPMPO-OH adduct the following EPR hyperfine splitting constants were assumed: $a^P = 46.93$ G, $a^N = 14.00$ G, $a^H = 13.13$ G.

Besides spin trapping, another technique that makes use of EPR to study membranes is spin labeling. A spin label is an organic molecule that possesses an unpaired electron, usually on a nitrogen atom, and has the ability to bind to another molecule (Figure 7.4). Spin labels are

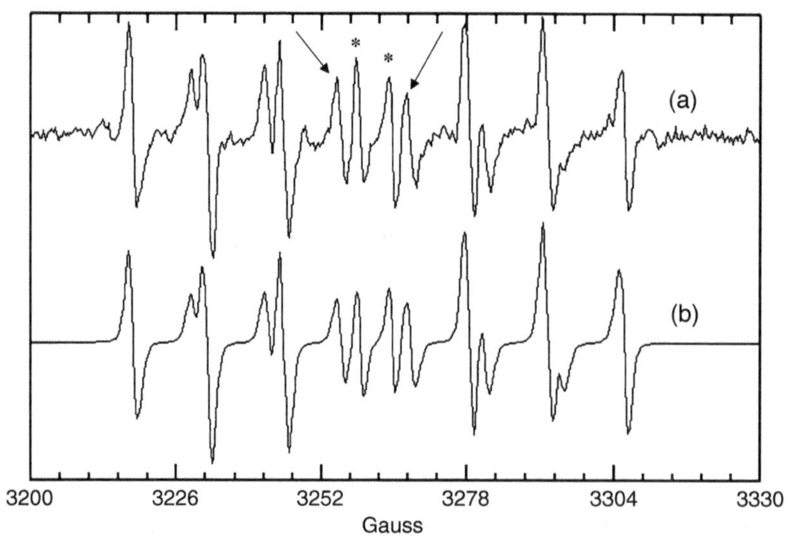

Figure 7.3 Experimental (a) and simulated (b) electron paramagnetic resonance (EPR) spectra of 5-(diethoxyphosphoryl)-5-methyl-1-pyrroline-*N*-oxide (DEPMPO) obtained after 5 min of continuous illumination of thylakoid membranes from wheat. Arrows show the pure lines corresponding to the superoxide adduct (DEPMPO-OOH); asterisks show the pure lines corresponding to the hydroxyl radical adduct (DEPMPO-OH).

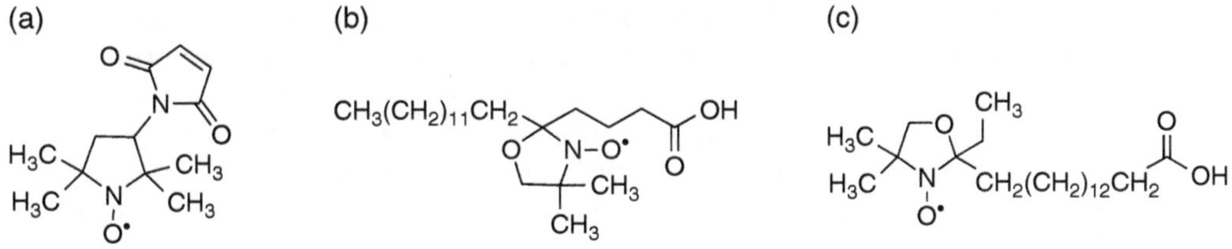

Figure 7.4 Examples of spin labels: (a) 3-maleimido proxyl; (b) 5-doxyl-stearic acid; (c) 16-doxyl-stearic acid.

normally used as tools for probing proteins or to study biological membrane-local dynamics using EPR spectroscopy. 3-Maleimido proxyl (3-maleimido-2,2,5,5-tetramethyl-1-pyrrolidinyloxy) free radical and spin-labeled fatty acids have been extensively used to understand the dynamic organization of thylakoid membranes (Navari-Izzo *et al.*, 2000; Calucci *et al.*, 2001). For example, stearic acid labeled with a nitroxyl spin label moiety at various carbons (5, 7, 9, 12, 13, 14, and 16th) with respect to the first carbon of the carbonyl group have been used to study the flexibility gradient of membrane lipids to understand membrane fluidity conditions at different depths of their lipid bilayer organization.

Oxidative Stress and Photosynthetic Functioning

Oxidative Stress and Thylakoid Membranes

Approximatively 60% of the dry weight of photosynthetic membranes is formed by proteins, largely made up of chlorophyll-protein complexes represented by the chlorophyll *a* protein complexes containing the reaction centers of photosystems I (PSI) and II (PSII), and the chlorophyll *a*/*b* light-harvesting complexes (Pribil *et al.*, 2014). As regards lipids, in contrast to the energy-transducing inner mitochondrial membrane, the thylakoid bilayer consists of four glycerolipids (Boudière *et al.*, 2014), which account for 70% of the total lipids. These include: two uncharged galactolipids, monogalactosyldiacylglycerol (MGDG) and digalactosyldiacylglycerol (DGDG), which together form more than two-thirds of the polar lipids; one anionic sulfolipid, sulfoquinovosyldiacylglycerol (SQDG); and two phospholipids, phosphatidylglycerol (PG) and phosphatidylcholine (PC) (Shimojima *et al.*, 2009). At the curved regions of the thylakoid membrane globular lipid droplets (plastoglobules) are present. They consist of a membrane lipid monolayer surface dotted with proteins surrounding a core of neutral lipids such as prenylquinones, carotenoids, fatty acid phytyl esters, and triacylglycerols (Eugeni Piller *et al.*, 2012). Membrane lipids can be divided into bound (boundary) lipids associated with membrane proteins, and free (bulk) lipids diffusing freely within the membranes (Yoshioka-Nishimura, 2016). Lipids are important for photosynthetic activity by being directly assembled into the photosynthetic machinery or forming the thylakoid membrane structure where photosynthesis occurs (Jones, 2007).

The fatty acid composition of the photosynthetic membranes is relatively simple (Boudière *et al.*, 2014). Both the MGDG and DGDG fractions usually contain a high proportion of polyunsaturated fatty acids, mostly 18-carbon trienoic acids (α-linolenic acid). SQDG, and to a lesser extent the two phospholipids, have considerable amounts of more saturated fatty acids. A notable feature of PG is the presence of a *trans*-hexadecenoic residue located exclusively at the *sn*-2 position of the glycerol backbone (Somerville *et al.*, 2000).

Photosynthetic cells are particularly susceptible to oxidative damage since they are sites of high oxygen concentration as a result of water oxidation in PSII, contain large amounts of polyunsaturated lipids, and can, via their photosynthetic pigments, absorb light and use this energy to generate harmful ROS. Following environmental stresses, two major events occur in chloroplasts: (i) molecular oxygen is converted to $^1O_2$ by excited chlorophyll in the reaction center of PSII by a phenomenon known as photoinhibition (Krieger-Liszkay 2005); and (ii) molecular oxygen is converted to $O_2^{\cdot-}$ by phylloquinone, ferredoxin, and reduced Fe-S clusters of the PSI reaction-center complex (Asada, 2006). Among ROS, $^{\cdot}OH$ readily reacts with all types of thylakoid cellular components, whereas the others appear to have a preferred cellular target with which they react (Schmidt and Schippers, 2015). Singlet oxygen is responsible for most chloroplast photo-oxidative damage (Triantaphylidès *et al.*, 2008), reacting directly with conjugated double bonds of polyunsaturated fatty acids to form lipid hydroperoxides (Foyer and Shigeoka, 2011). Superoxide radicals produced as a consequence of oxidative conditions attack primarily protein Fe-S centers of PSI (Braconi *et al.*, 2011), besides reacting with unsaturated fatty acids as well.

In isolation or combination, environmental factors (drought, salinity, UV radiation, high temperatures, excess light and metals) may cause direct or indirect damage to chloroplasts by triggering increased production of ROS, which ultimately react with membrane lipids (Gill and Tuteja, 2010; Anjum *et al.*, 2012; Sharma *et al.*, 2012). Changes in lipid composition may affect membrane fluidity and also the intrinsic protein activities as a result of alterations in the lipid environment in which they are embedded (Quartacci *et al.*, 2001). Even though it is unlikely that the fluidity of the bulk lipid phase has any remarkable effect on the function of membrane proteins, the altered composition and/or unsaturation of specific membrane domains might contribute to changing the lipid environment surrounding integral proteins (Mansour *et al.*, 1994).

Plants subjected to water stress of increasing intensity (from mild to very severe) showed adaptive capacity at the cellular level as increases in the DGDG to MGDG ratio and fatty acid unsaturation (Quartacci *et al.*, 1994; Gigon *et al.*, 2004). In pure lipid-water mixtures, MGDG (non-bilayer-forming lipid) molecules tend to form hexagonal phase structures due to their conical shape, whereas cylindrical DGDG (bilayer-forming lipid) forms lamellar phases. Their proportion is very important for the stability and ionic permeability of the chloroplast membranes. Under stress conditions, an increase in the DGDG to MGDG ratio could help to maintain the membrane in a bilayer conformation necessary for the biological functions of photosystems (Navari-Izzo *et al.*, 1995; Quartacci *et al.*, 1995; Torres-Franklin *et al.*, 2007). Also an enhanced fatty acid unsaturation degree is important for maintaining photosynthetic activity as the number of double bonds is strictly linked to membrane fluidity and efficient electron transport (Navari-Izzo *et al.*, 1995). It was suggested that under stress conditions the reduced microviscosity of the bilayer due to high unsaturation leads to an increased rate of diffusion of plastoquinone, considered to be the main long-range electron carrier between the two photosystems (Murphy, 1986). A higher fluidity of thylakoid membranes also stabilizes photosynthesis, influencing the tolerance of the photosynthetic oxygen evolution complex to altered conditions (Gombos *et al.*, 1994). In stress adaptation both the occurrence of bilayer/non-bilayer transformations and the lipid to protein ratio have a primary influence on the packaging of proteins (Gounaris *et al.*, 1983). In the presence of alterations of these two parameters, EPR measurements of 3-maleimido proxyl spin-labeled wheat thylakoid showed a higher protein spin label rotational correlation time accompanied by a decrease in the number of reduced sulfhydryl groups and more mobile proteins (Quartacci *et al.*, 1995; Navari-Izzo *et al.*, 2000). The enhanced protein rotational correlation time suggested a reduced membrane fluidity into which proteins were embedded, and such an effect differed depending on the sensitivity of the cultivars to water stress. In this situation, the functionality of the membrane itself was likely altered, as all the proteins that undergo conformational changes to act as enzymes and carriers or are components of ionic channels could not perform their activities.

When exposed to NaCl for 7 days, the total fatty acid content of the thylakoid membranes was reduced compared to that of the control plants, with an increase in saturated fatty acid levels (Shu *et al.*, 2015). The structure of the salt-stressed thylakoid membranes revealed that the grana thylakoid lamellae were disordered and that the stroma thylakoid lamellae were fractured owing to alterations of the membrane lipids.

The irradiation of rice plants with UV light resulted in a decline of xanthophyll, carotene, and chlorophyll levels (Lidon and Ramalho, 2011). Such changes were accompanied by an increase in the peroxidation level in the chloroplast membranes, reflecting a loss of protection against oxidative stress. Also in this case, UV radiation determined a reduction of the MGDG level as a consequence of higher ROS production.

Chloroplast lipids and, as a consequence, all membrane-related processes, are particularly sensitive to metal excess. Alteration of lipid composition and organization is reflected at the ultrastructural level as modifications of the chloroplast structure (Quartacci *et al.*, 2000; Djebali *et al.*, 2005). Several studies observed that metal-induced changes in the contents of chloroplast lipids and their unsaturation degree depend on the type and concentration of the metal (Ben Ammar *et al.*, 2007; Pál *et al.*, 2007). A decrease in the MGDG level was detected following Cd, Cu, Mn, and Pb excess (Quartacci *et al.*, 2000; Belkadhi *et al.*, 2015). As the MGDG to DGDG ratio influences membrane curvature, a change in this ratio may explain the alterations in grana stacking and/or grana-to-stroma thylakoid ratios observed under metal excess (Solymosi, 2012). Metal-induced changes in polar lipid composition of thylakoids may affect membrane fluidity. Spin-labeled membranes showed that following Cu supply only the fluidity of the surface region of the bilayer close to the polar head group was reduced. In contrast, the fluidity of the inner membrane region of the bilayer did not show any change, in agreement with the maintained unsaturation degree (Quartacci *et al.*, 2000). In its turn, an altered fluidity may influence the ionic permeability of the thylakoid membranes, thus disturbing solute transport and exchange (Devi and Prasad, 2006).

Increased lipid peroxidation might be related to a relatively high membrane fluidity. Indeed, fluidization of membranes allows for better penetration of oxygen inside the membranes, which can lead to an increase in the production of ROS (Subczynski *et al.*, 2009).

Besides altering thylakoid lipids, lipid peroxidation caused by environmental constraints can also lead to the release of pigment-protein complexes and plastocyanin from membranes (Siedlecka and Krupa, 1999).

Altered Functioning of Electron Transport Systems: Generation of ROS

Possible sources of ROS production in photosynthetic tissues may be the leakage of electrons from electron transport systems to oxygen, in particular from the photosynthetic electron transport system, which in turn may cause free organic radical production through lipid peroxidation (Sgherri and Pérez-López, 2013).

During environmental stresses, such as water deficit, a CO_2-limited carbon fixation occurs and there is decreased availability of oxidized nicotinamide adenine dinucleotide phosphate ($NADP^+$) as an electron acceptor of PSI. Under these conditions oxygen, in the presence of thylakoids, can compete with $NADP^+$ as a Hill reductant and can produce superoxide (Sgherri *et al.*, 1993).

EPR measurements have demonstrated that illuminated thylakoids can produce superoxide radicals (Sgherri *et al.*, 1993, 1999; Quartacci *et al.*, 1994; Zvezdanović *et al.*, 2013) (see Figure 7.3) and that the rise in $O_2^{\cdot-}$ production may reflect adverse molecular rearrangements in the thylakoid membranes that expose chlorophyll molecules normally inaccessible to oxygen (Sgherri *et al.*, 1996). EPR studies have also shown that the formation of superoxide by illuminated thylakoids from bean leaves increases about four times during the early stages of senescence and then declines again as senescence proceeds (Thompson *et al.*, 1987). Similarly, superoxide production from damaged thylakoids by water depletion was higher when water stress intensity was mild (Sgherri *et al.*, 1996). This increased production from stressed thylakoids is sensitive to superoxide dismutase and appears to be mediated by a photochemical reaction involving chlorophyll (McRae and Thompson, 1983; Sgherri *et al.*, 1999).

The linkage between the electron transport system of thylakoids and superoxide production is demonstrated by the presence of photosynthetic radiation. When the electron transport system is altered there is a different production of ROS (Sgherri *et al.*, 1996). Figure 7.5 shows that formation of Tiron semiquinone by thylakoid membranes was induced when the lamp was switched on, whereas it was inhibited when the lamp was switched off.

Thus, a lot of information can be gained monitoring EPR kinetics of Tiron semiquinone induction such as that related to electron transport system of thylakoids. Since a decay of superoxide formation can be evidenced even upon continuous illumination (Figure 7.6), the kinetics of Tiron semiquinone formation by photosynthetically active membranes can be classically distinguished in three stages, which reflect the physiological conditions of the membranes as well as oxygen consumption in the test tube (Menconi *et al.*, 1995). A rapid signal increase, a steady-state phase,

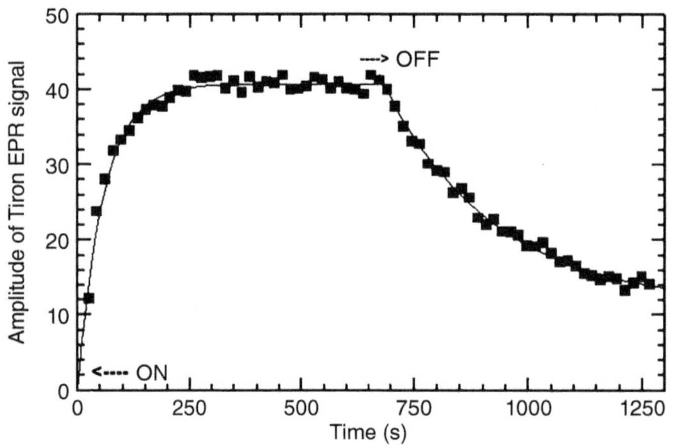

Figure 7.5 Kinetics of superoxide production by thylakoids from wheat upon illumination. Arrows indicate the switch on (ON) and the switch off (OFF) of the lamp.

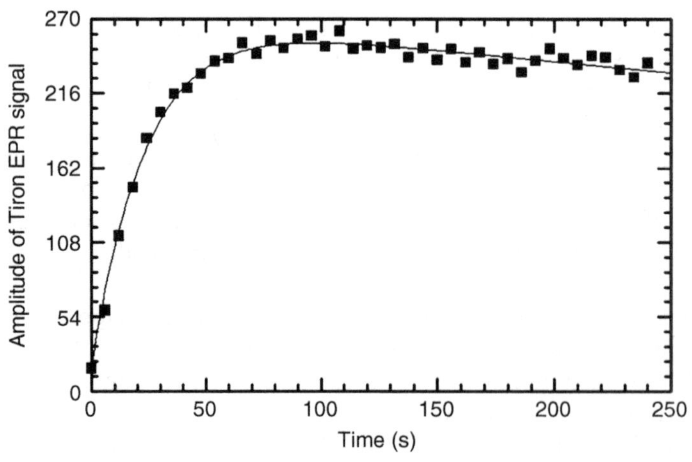

Figure 7.6 Kinetics of superoxide radical production upon continuous illumination of thylakoids from wheat.

and a signal decrease (Figure 7.6) were originated by the simultaneous action of first-order kinetics of formation and decay of Tiron semiquinone. The steady-state signal was reached over a period of about 3 min. This amplitude represents a condition where the decay rate balances the rate of electron transfer from Tiron to superoxide anion. The decay rate could be the result of the rates of disproportionation of the radical and the oxidation of the semiquinone by oxygen as well as of photoinhibition of the membranes and oxygen depletion in the reaction mixture (Sgherri *et al.*, 1996).

Thus, to evaluate the actual capacity of thylakoids to leak electrons toward oxygen, that is, the electron transport system, it was necessary to calculate, by simulation of the experimental points, the rate constant of first-order kinetics of superoxide formation (k_f). The maximum amplitude (A_0) of the EPR signal in the absence of decay rate can be considered as indicative of the total superoxide production in the detection system. A first-order process suggests an involvement of a reaction between Tiron semiquinone (limiting substrate) and other components of the reaction medium that were in excess (Miller and Rapp, 1973). From a non-linear best fitting of the experimental points shown in Figure 7.6, an A_0 of 258.2, a formation rate constant (k_f) of 4.2×10^{-2} s^{-1}, and a decay rate constant (k_d) of 6.9×10^{-4} s^{-1} can be drawn. The superoxide formation rate is defined as $k_f = \ln 2\ (t_{1/2})^{-1}$, where $t_{1/2}$ is the time required to obtain the $A_0/2$ value (Menconi *et al.*, 1995).

Sites of ROS Production: Past and Present

ROS are formed by the dysfunctioning of enzymes or electron transport systems, as a result of perturbations in metabolism caused by chemical or environmental stresses. The reduction of oxygen to form superoxide, hydrogen peroxide, and hydroxyl radicals is the principal mechanism of oxygen activation in most biological systems. However, in photosynthetic organisms, the formation of singlet oxygen by the photosystems is important (Krieger-Liszkay, 2005).

Photoreduction of oxygen to H_2O_2 in PSI was discovered over 60 years ago by Mehler (1951). Subsequently, the primary reduced product was identified to be superoxide anion, whose disproportionation produces H_2O_2 and O_2 (Asada *et al.*, 1974). Direct evidence of photoreduction of oxygen to superoxide came from EPR experiments with spin probes (Greenstock and Miller, 1975; Sgherri *et al.*, 1993) or spin traps (Sgherri *et al.*, 1999), and the suppression of the EPR signal after the addition of exogenous superoxide dismutase constituted a proof of its identity (Sgherri *et al.*, 1999). It was also found that an increase in light intensities results in an increase in both superoxide radical and hydrogen peroxide production by the photosynthetic electron transport chain (Ivanov *et al.*, 2012). In EPR experiments with the spin trap DEPMPO, the two adducts DEPMPO-OH and DEPMPO-OOH were distinguishable (Figures 7.3 and 7.7) demonstrating that the formation of both superoxide and hydroxyl radicals was involved during illumination of thylakoid membranes.

The reaction centers of PSI and PSII in chloroplast thylakoids have always been seen as the major generation sites of ROS. In contrast, PSII has long been considered only as the production site of singlet oxygen. Here, oxygen of the ground (triplet) state ($^3O_2$) is excited to the singlet state ($^1O_2$) by the reaction center chlorophyll of triplet excited state ($^3P680^*$) (Telfer *et al.*, 1994; Hideg *et al.*, 1998). However, Elstner (1991) postulated that PSII also can give rise to the generation of superoxide radical. Under physiological pH conditions the redox potential of the O_2/O_2^{-} couple is −0.33 V. Because for thermodynamic reasons electron donors to oxygen require redox potentials below this value, the thermodynamically feasible sites of superoxide production have been attributed to the donor side of PSI and the acceptor and donor sides of PSII (Ananyev *et al.*, 1994).

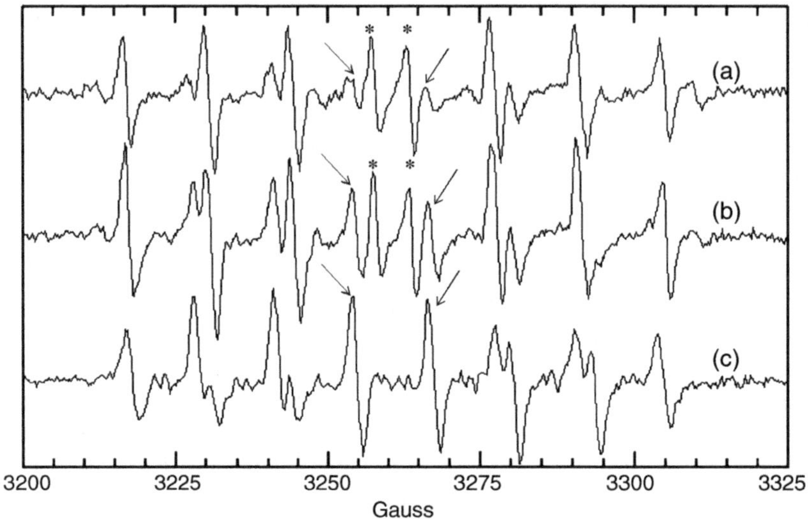

Figure 7.7 Experimental electron paramagnetic resonance (EPR) spectra of 5-(diethoxyphosphoryl)-5-methyl-1-pyrroline-*N*-oxide (DEPMPO) obtained after 5 min of continuous illumination of (a) PSII membrane fragments and (b) thylakoid membranes from wheat. (c) EPR spectrum of the superoxide adduct DEPMPO-OOH obtained in the presence of xanthine/xanthine oxidase. Arrows show the pure lines corresponding to the superoxide adduct (DEPMPO-OOH); asterisks show the pure lines corresponding to the hydroxyl radical adduct (DEPMPO-OH).

Thus, there are at least three sites within the chloroplast that can activate oxygen:

1) The reducing side of PSI, which is thought to contribute significantly to the monovalent reduction of oxygen under conditions where $NADP^+$ is limiting. This would occur, for example, if the Calvin–Benson cycle did not oxidize NADPH as rapidly as PSI supplied electrons (Sgherri *et al.*, 1993). This is the so-called "Mehler reaction," the most important mechanism of oxygen activation in the chloroplast.
2) Photoactivation of chlorophyll, which normally transfers its excitation energy to the photosystem reaction centers, but under conditions that prevent the captured light energy from being utilized in the electron transport systems, this energy can excite oxygen from the triplet to singlet form. These conditions include stomatal closure caused by drought, damage to the membrane transport systems, lack of specific nutrients, or the presence of xenobiotic chemicals such as pollutants or herbicides.
3) The oxidizing side of PSII, which facilitates four single electron transfers from water to the PSII reaction center, releasing triplet or ground-state oxygen. Leaks of electrons from this site to molecular oxygen, or release of partially reduced oxygen products are thought to make a relatively minor contribution to activated oxygen production, but nonetheless it has been demonstrated that certain alcohols can be reduced by PSII.

Evidence for the formation of superoxide, monitored with cytochrome *c* reduction, by PSII membrane fragments has been reported (Ananyev *et al.*, 1994; Chen *et al.*, 1995). Navari-Izzo *et al.* (1998), monitoring the generation of Tiron semiquinone, also observed that PSII particles, with and without DCMU, and thylakoids with added DCMU all showed the same formation rate constant for superoxide production.

Pospíšil *et al.* (2004) demonstrated the photogeneration of hydroxyl radicals in PSII membranes using EPR spin-trapping spectroscopy. The same authors' results indicated that hydroxyl radical is produced on the electron acceptor side of PSII by two different routes: (i) superoxide, which is generated by oxygen reduction on the acceptor side of PSII, interacts with a PSII metal center, probably the non-heme iron, to form an iron-peroxide species that is further reduced to hydroxyl radical by an electron from PSII, presumably via $QA^{\cdot-}$; (ii) superoxide dismutates to form free H_2O_2, which is then reduced to hydroxyl radical via the Fenton reaction in the presence of metal ions, the most likely being Mn^{2+} and Fe^{2+} released from photodamaged PSII.

Navari-Izzo *et al.* (1999) also showed that illumination of PSII membrane fragments of wheat gave rise to both superoxide and hydroxyl radical adducts of the spin trap DEPMPO (Figure 7.7). Moreover, the above authors demonstrated that, with time, the hydroxyl radical became predominant, demonstrating the conversion of superoxide into hydroxyl radical. An intrinsic activity of superoxide dismutase was also found in PSII, testifying that H_2O_2 is involved in the reaction.

Also recent studies have revealed that superoxide, in addition to $^1O_2$, is produced by PSII, causing photoinhibition (Bondarava *et al.*, 2010; Zulfugarov *et al.*, 2014). Roach and Krieger-Liszkay (2014) observed that besides being generated at the acceptor side of PSI, superoxide can also be produced *in vitro* at the level of the cytochrome *b6f* complex (cyt*b6f*) and at the acceptor side of PSII. In addition, cytochrome *b*559, an intrinsic protein subunit of PSII, can act, depending on its redox potential, as an oxygen reductase, as a superoxide reductase, or as a superoxide oxidase. However, the capacities of these pathways of O_2 reduction seem to play only very minor roles in an intact, functional electron transport chain in thylakoid membranes.

Maintaining Function: The Ascorbate/Glutathione Cycle and its Interrelation with Lipoic Acid in Chloroplast

By degrading lipids, proteins, and nucleic acids, ROS affect the functionality of photosynthetic membranes, which can be lethal to plant cells. Under physiological steady-state conditions these activated oxygen molecules are scavenged by different antioxidative defense components that are confined within the chloroplast. Thus, damage occurs when the equilibrium between production and scavenging of ROS is perturbed by a number of adverse environmental factors. The most studied antioxidant defense mechanism in chloroplast is the ascorbate/glutathione cycle (Asada–Halliwell–Foyer cycle), which allows chloroplasts to defend themselves from ROS generation by ascorbate (ASA) and glutathione (GSH) (Foyer and Halliwell, 1976). In fact, these molecules are involved in the detoxification of H_2O_2. Scavenging of H_2O_2 by ascorbate peroxidase (APX) is the first step of the GSH/ASA cycle, and APX can be found as either a soluble form or a thylakoid-bound form (Sgherri and Pérez-López, 2013). This activity yields the monodehydroascorbate radical (MDA). Being a radical, MDA can be converted (reduced) to ASA by monodehydroascorbate reductase (MDHAR) or disproportionated to ASA and dehydroascorbate (DHA) non-enzymatically. The reduction of DHA to ASA by the enzyme dehydroascorbate reductase (DHAR) at the expense of GSH results in the production of oxidized glutathione (GSSG). The final step of this cycle is the reduction of GSSG by glutathione reductase (GR) using NADPH as electron donor. At the end of the process, ASA and GSH are not consumed but regenerated and the net electron flow is from NADPH to H_2O_2 (Sgherri and Pérez-López, 2013). ASA is a major primary antioxidant reacting directly with $^{\cdot}OH$, $O_2^{\cdot-}$, and $^1O_2$ (Buettner and Jurkiewicz, 1996). Besides regenerating ASA via reduction of DHA, GSH can also react chemically with $^1O_2$ as well as with $O_2^{\cdot-}$ and tocopheroxyl radicals, functioning directly as a free radical scavenger (Noctor and Foyer, 1998). Both ASA and GSH are abundant, stable antioxidants, with appropriate redox potentials that make them able to interact with numerous cell components, and to exist generally in the reduced state.

Recently, it has been reported that the GSH/ASA cycle can be improved by lipoic acid (Sgherri and Pérez-López, 2013). Indeed, it has been hypothesized that the GSH/ASA cycle could be connected with the reduced form of lipoic acid (dihydrolipoic acid, DHLA), which is involved in ASA and GSH regeneration, reinforcing this cycle and strengthening the antioxidant network (Pérez-López et al., 2010). Its existence has been known since the 1930s, when it was discovered that a so-called "potato growth factor" was necessary for the growth of certain bacteria. Lipoic acid (LA) is a sulfur-containing cofactor, and in its reduced form, DHLA, two thiol groups per molecule are present. LA plays a pivotal role in energy metabolism, being involved in different multienzyme complexes such as pyruvate dehydrogenase, α-ketoglutarate dehydrogenase, branched-chain α-keto acid dehydrogenase, and glycine decarboxylase (Gueguen *et al.*, 2000). Overall, LA and DHLA act as a redox couple carrying electrons from the substrate of the dehydrogenase to NAD^+, and for this property the capacity to act also as antioxidants has been highlighted (Navari-Izzo *et al.*, 2002). In fact, LA has been found to be a moderate antioxidant and DHLA, a good antioxidant; in particular, the ability of DHLA to inhibit lipid peroxidation has been reported (Navari-Izzo *et al.*, 2002). The presence of lipoic acid in the stroma of wheat, both as free and protein-bound forms, has been demonstrated (Sgherri *et al.*, 2002). Its presence in chloroplasts certainly represents a significant contribution to the defense system of these organelles, which are the most exposed to oxidative stress and, therefore, strongly need the regeneration of soluble antioxidants. Moreover, lipoic acid, due to its solubility in both water and lipid phases, has the ability to connect the activity of antioxidants in the cell membrane (α-tocopherol) with antioxidants in the stroma (ASA and GSH), strengthening the antioxidant network. As powerful antioxidants, the functions of LA and DHLA involve: (i) quenching of ROS such as $O_2^{\cdot-}$, hydroperoxyl, and hydroxyl radicals; (ii) regeneration of endogenous and exogenous antioxidants involving vitamins C (ASA) and E (α-tocopherol) as well as GSH; (iii) chelation of redox metals including Cu(II) and Fe(II); (iv) repair of oxidized proteins (Packer *et al.*, 1995; Navari-Izzo *et al.*, 2002), thus contributing to the maintenance of photosynthetic functioning.

References

Afanas'ev IB. (1985) *Superoxide Ion: Chemistry and Biological Implications*, Vol. 1. CRC Press, Boca Raton, FL.

Ananyev G, Renger G, Wacker U, Klimov V. (1994) The photoproduction of superoxide radicals and the superoxide dismutase activity of photosystem II. The possible involvement of cytochrome b559. *Photosynth Res* 41: 327–338.

Anjum NA, Umar S, Ahmad A. (2012) *Oxidative Stress in Plants: Causes, Consequences and Tolerance*. IK International Publishing House, New Delhi.

Asada K. (2006) Production and scavenging of reactive oxygen species in chloroplasts and their functions. *Plant Physiol* 141: 391–396.

Asada K, Takahashi M. (1987) Production and scavenging of active oxygen in photosynthesis. In: Kyle DJ, Osmond CB, Arntzen CJ (eds), *Photoinhibition: Topics of Photosynthesis*, 9th edn. Elsevier, Amsterdam, The Netherlands, pp. 227–287.

Asada K, Kiso K, Yoshikawa K. (1974) Univalent reduction of molecular oxygen by spinach chloroplasts on illumination. *J Biol Chem* 249: 2175–2181.

Barber J, Andersson B. (1992) Too much of a good thing: light can be bad for photosynthesis. *Trends Biochem Sci* 17: 61–66.

Belkadhi A, De Haro A, Obregon S, Chaïbi W, Djebali W. (2015) Positive effects of salicylic acid pretreatment on the composition of flax plastidial membrane lipids under cadmium stress. *Environ Sci Pollut Res* 22: 1457–1467.

Ben Ammar W, Nouairi I, Zarrouk M, Jemal F. (2007) Cadmium stress induces changes in the lipid composition and biosynthesis in tomato (*Lycopersicon esculentum* Mill.) leaves. *Plant Growth Regul* 53: 75–85.

Bondarava N, Gross CM, Mubarakshina M, Golecki JR, Johnson GN, Krieger-Liszkay A. (2010) Putative function of cytochrome b559 as a plastoquinol oxidase. *Physiol Plant* 138: 463–473.

Boudière L, Michaud M, Petroutsos D, *et al.* (2014) Glycerolipids in photosynthesis: composition, synthesis and trafficking. *Biochim Biophys Acta* 1837: 470–480.

Braconi D, Bernardini G, Santucci A. (2011) Linking protein oxidation to environmental pollutants: redox proteomic approaches. *J Proteomics* 74: 2324–2337.

Bradley DE, Minn DB. (1992) Singlet oxygen oxidation of foods. *Cat Rev Food Sci Nutri* 31: 211–236.

Buettner GR, Jurkiewicz BA. (1996) *Handbook of Antioxidants*. Dekker, New York.

Calucci L, Navari-Izzo F, Pinzino C, Sgherri CLM. (2001) Fluidity changes in thylakoid membranes of durum wheat induced by oxidative stress: a spin probe EPR study. *J Phys Chem B* 105: 3127–3134.

Chen GH, Blubaugh DJ, Homann PH, Golbeck JH, Cheniae GM. (1995) Superoxide contributes to the rapid inactivation of specific secondary donors of the photosystem II reaction center during photodamage of manganese-depleted photosystem II membranes. *Biochem* 34: 2317–2332.

Davies KJA. (1987) Protein damage and degradation by oxygen radicals. I General aspects. *J Biol Chem* 162: 9895–9901.

Devi SR, Prasad MNV. (2006) Membrane lipid alterations in heavy metal exposed plants. In: Prasad MNV (ed.), *Heavy Metal Stress in Plants: From Biomolecules to Ecosystems*, 2nd edn. Springer, Berlin, pp 127–145.

Djebali W, Zarrouk M, Brouquisse R, *et al.* (2005) Ultrastructure and lipid alterations induced by cadmium in tomato (*Lycopersicon esculentum*) chloroplast membranes. *Plant Biol* 7: 358–368.

Elstner EF. (1991) Mechanisms of oxygen activation in different compartments of plant cells. In: Pell EJ, Steffen KL (eds), *Active Oxygen/Oxidative Stress and Plant Metabolism*. American Society of Plant Physiologists, Rockville, MD, pp. 13–25.

Eugeni Piller L, Abraham M, Dörmann P, Kessler F, Besagni C. (2012) Plastid lipid droplets at the crossroads of prenylquinone metabolism. *J Exp Bot* 63: 1609–1618.

Fenton HJH. (1894) Oxidation of tartaric acid in the presence of iron. *J Chem Soc* 65: 899.

Fenton HJH. (1899) Oxidation of certain organic acids in the presence of ferrous salts. *Proc Chem Soc* 25: 224.

Finkelstein E, Rosen GM, Rauckman EJ, Paxton J. (1979) Spin trapping of superoxide. *Molec Pharmacol* 16: 676–685.

Foote CS, Valentine JS, Greenberg A, Liebman JF (eds) (1985) *Active Oxygen in Chemistry*. Chapman & Hall, New York.

Foyer CH. (1997) Oxygen metabolism and electron transport in photosynthesis. In: Scandalios J (ed.), *Molecular Biology of Free Radical Scavenging Systems*. Cold Spring Harbor Laboratory Press, New York, NY, pp. 587–621.

Foyer CH, Halliwell B. (1976) The presence of glutathione and glutathione reductase in chloroplasts: a proposed role in ascorbic acid metabolism. *Planta* 133: 21–25.

Foyer CH, Shigeoka S. (2011) Understanding oxidative stress and antioxidant functions to enhance photosynthesis. *Plant Physiol* 155: 93–100.

Gigon A, Matos AR, Laffray D, Zuily-Fodil Y, Pham-Thi AT. (2004) Effect of drought stress on lipid metabolism in the leaves of *Arabidopsis thaliana* (ecotype Columbia). *Ann Bot* 94: 345–351.

Gill SS, Tuteja N. (2010) Reactive oxygen species and antioxidant machinery in abiotic stress tolerance in crop plants. *Plant Physiol Biochem* 48: 909–930.

Gombos Z, Wada H, Hideg E, Murata N. (1994) The unsaturation of membrane lipids stabilizes photosynthesis against heat stress. *Plant Physiol* 104: 563–567.

Gounaris K, Mannock DA, Sen A, Brain APR, Williams WP, Quinn PJ. (1983) Polyunsaturated fatty acyl residues of galactolipids are involved in the control of bilayer/non-bilayer lipid transition in higher plant chloroplasts. *Biochim Biophys Acta* 732: 229–242.

Greenstock CL, Miller RW. (1975) The oxidation of Tiron by superoxide anion. Kinetics of the reaction in aqueous solution and in chloroplasts. *Biochem Biophys Acta* 396: 11–16.

Gueguen V, Macherel D, Jaquinod M, Douce R, Bourguignon J. (2000) Fatty acid and lipoic acid biosynthesis in higher plant mitochondria. *J Biol Chem* 275: 5016–5025.

Haber F, Weiss J. (1934) The catalytic decomposition of hydrogen peroxide by iron salts. *Proc Royal Soc A* 147: 332.

Halliwell B, Gutteridge JMC. (1999) *Free Radicals in Biology and Medicine*, 3rd edn. Oxford Science Publications, NY.

Halliwell B, Gutteridge JMC. (2006) *Free Radicals in Biology and Medicine*, 4th edn. Clarendon Press, Oxford.

Hideg E, Kalai T, Hideg K, Vass I. (1998) Photoinhibition of photosynthesis in vivo results in singlet oxygen production detection via nitroxide-induced fluorescence quenching in broad bean leaves. *Biochemistry* 237: 11405–11411.

Ivanov B, Kozuleva M, Mubarakshina M. (2012) Oxygen metabolism in chloroplast. In: Bubulya P (ed.), *Cell Metabolism – Cell Homeostasis and Stress Response.* InTech (www.intechopen.com), pp. 39–72.

Jones MR. (2007) Lipids in photosynthetic reaction centres: structural roles and functional holes. *Prog Lipid Res* 46: 56–87.

Krieger-Liszkay A. (2005) Singlet oxygen production in photosynthesis. *J Exp Bot* 56: 337–346.

Ledenev AN, Konstantinov AA, Popova E, Ruuge EK. (1986) A simple assay of the superoxide generation rate with Tiron as an EPR-visible radical scavenger. *Biochem Int* 13: 391–396.

Lidon FC, Ramalho JC. (2011) Impact of UV-B irradiation on photosynthetic performance and chloroplast membrane components in *Oryza sativa* L. *J Photochem Photobiol B-Biol* 104: 457–466.

Mansour MMF, van Hasselt PR, Kuiper PJC. (1994) Plasma membrane lipid alterations induced by NaCl in winter wheat roots. *Physiol Plant* 92: 473–478.

McRae DG, Thompson JE. (1983) Senescence-dependent changes in superoxide anion production by illuminated chloroplast from bean leaves. *Planta* 158: 185–193.

Mehler AH. (1951) Studies on reactivities of illuminated chloroplasts. I. Mechanism of the reduction of oxygen and other Hill reagents. *Arch Biochem Biophys* 33: 65–77.

Menconi M, Sgherri CLM, Pinzino C, Navari-Izzo F. (1995) Activated oxygen production and detoxification in wheat plants subjected to a water deficit programme. *J Exp Bot* 46: 1123–1130.

Miller RW, Rapp U. (1973) The oxidation of catechols by reduced flavins and dehydrogenases. *J Biol Chem* 248: 6084–6090.

Mittler R. (2002) Oxidative stress, antioxidants and stress tolerance. *Trends Plant Sci* 7: 405–410.

Murphy DJ. (1986) The molecular organization of the photosynthetic membranes of higher plants. *Biochim Biophys Acta* 864: 33–95.

Navari-Izzo F, Sgherri CLM, Galleschi L. (1992) Plasma-membrane proteins in water-stressed sunflower seedlings. *Agr Med* 122: 306–312.

Navari-Izzo F, Ricci F, Vazzana C, Quartacci MF. (1995) Unusual composition of thylakoid membranes of the resurrection plant *Boea hygroscopica*: Changes in lipids upon dehydration and rehydration. *Physiol Plant* 94: 135–142.

Navari-Izzo F, Meneguzzo S, Loggini B, Vazzana C, Sgherri CLM. (1997) The role of glutathione system during dehydration of *Boea hygroscopica*. *Physiol Plant* 99: 23–30.

Navari-Izzo F, Quartacci MF, Pinzino C, Dalla Vecchia F, Sgherri CLM. (1998) Thylakoid-bound and stromal antioxidative enzymes in wheat treated with excess copper. *Physiol Plant* 104: 630–638.

Navari-Izzo F, Pinzino C, Quartacci MF, Sgherri CLM. (1999) Superoxide and hydroxyl radical generation, and superoxide dismutase in PSII membrane fragments from wheat. *Free Rad Res* 31: S3–9.

Navari-Izzo F, Quartacci MF, Pinzino C, Rascio N, Vazzana C, Sgherri CLM. (2000) Protein dynamics in thylakoids of the desiccation-tolerant plant *Boea hygroscopica* during dehydration and rehydration. *Plant Physiol* 124: 1427–1436.

Navari-Izzo F, Quartacci MF, Sgherri C. (2002) Lipoic acid: a unique antioxidant in the detoxification of activated oxygen species *Plant Physiol Biochem* 40: 463–470.

Noctor G, Foyer CH. (1998) Ascorbate and glutathione: keeping active oxygen under control. *Annu Rev Plant Physiol Plant Mol Biol* 49: 249–279.

Packer L, Witt EH, Tritschler HJ. (1995) Alpha-lipoic acid as a biological antioxidant. *Free Rad Biol Med* 19: 227–250.

Pál M, Leskó K, Janda T, Páldi E, Szalai G. (2007) Cadmium induced changes in the membrane lipid composition of maize plants. *Cereal Res Commun* 35: 1631–1642.

Pérez-López U, Robredo A, Lacuesta M, *et al.* (2010) Lipoic acid and redox status in barley plants subjected to salinity and elevated CO_2. *Physiol Plant* 139: 256–268.

Pospíšil P, Arató A, Krieger-Liszkay A, Rutherford AW. (2004) Hydroxyl radical generation by photosystem II. *Biochemistry* 43: 6783–6792.

Pribil M, Labs M, Leister D. (2014) Structure and dynamics of thylakoids in land plants. *J Exp Bot* 65: 1955–1972.

Quartacci MF, Sgherri CLM, Pinzino C, Navari-Izzo F. (1994) Superoxide radical production in wheat plants differently sensitive to drought. *Proc R Soc Edinburgh* 102B: 287–290.

Quartacci MF, Pinzino C, Sgherri CLM, Navari-Izzo F. (1995) Lipid composition and protein dynamics in thylakoids of two wheat cultivars differently sensitive to drought. *Plant Physiol* 108: 191–197.

Quartacci MF, Pinzino C, Sgherri CLM, Dalla Vecchia F, Navari-Izzo F. (2000) Growth in excess copper induces changes in the lipid composition and fluidity of PSII-enriched membranes in wheat. *Physiol Plant* 108: 87–93.

Quartacci MF, Cosi E, Navari-Izzo F. (2001) Lipids and NADPH-dependent superoxide production in plasma membrane vesicles from roots of wheat grown under copper deficiency or excess. *J Exp Bot* 52: 77–84.

Raison JK, Roberts JKM, Berry JA. (1982) Correlations between the thermal stability of chloroplast (thylakoid) membranes and the composition and fluidity of their polar lipids upon acclimation of the higher plant *Nerium oleander* to growth temperature. *Biochim Biophys Acta* 688: 218–228.

Roach T, Krieger-Liszkay A. (2014) Regulation of photosynthetic electron transport and photoinhibition. *Curr Protein Pept Sci* 15: 351–362.

Schmidt R, Schippers JHM. (2015) ROS-mediated redox signaling during cell differentiation in plants. *Biochem Biophys Acta* 1850: 1497–1508.

Senaratna T, McKersie BD, Stinson RH. (1985) Simulation of dehydration injury to membranes from soybean axes by free radicals. *Plant Physiol* 77: 472–474.

Sgherri C, Pérez-López U. (2013) Glutathione/ascorbate cycle and its interrelation with lipoic acid: could it explain why plants do not get cancer? In: Labrou N, Flemetakis E (eds), *Glutathione: Biochemistry, Mechanisms of Action and Biotechnological Implications.* Nova Science Publishers, Inc., New York, pp. 233–261.

Sgherri CLM, Pinzino C, Navari-Izzo F. (1993) Chemical changes and $O_2^{\cdot-}$ production in thylakoid membranes under water stress. *Physiol Plant* 87: 211–216.

Sgherri CLM, Pinzino C, Navari-Izzo F. (1996) Sunflower seedlings subjected to increasing stress by water deficit: changes in $O_2^{\cdot-}$ production related to the composition of thylakoid membranes. *Physiol Plant* 96: 446–452.

Sgherri CLM, Pinzino C, Samaritani E, Navari-Izzo F. (1999) Activated oxygen generation from thylakoids: a novel spin trap. *Free Rad Res* 31: S199–204.

Sgherri C, Quartacci MF, Izzo R, Navari-Izzo F. (2002) Relation between lipoic acid and cell redox status in wheat grown in excess copper. *Plant Physiol Biochem* 40: 591–597.

Sgherri C, Scattino C, Pinzino C, Tonutti P, Ranieri AM. (2015) Ultraviolet-B radiation applied to detached peach fruit: a study of free radical generation by EPR spin trapping. *Plant Physiol Biochem* 96: 124–131.

Sharma P, Dubey RS. (2005) Drought induces oxidative stress and enhances the activities of antioxidant enzymes in growing rice seedlings. *Plant Growth Regul* 46: 209–221.

Sharma P, Jha AB, Dubey RS, Pessarakli M. (2012) Reactive oxygen species, oxidative damage, and antioxidative defense mechanism in plants under stressful conditions. *J Bot* 2012, Art. ID 217037.

Shimojima M, Ohta H, Nakamura Y. (2009) Biosynthesis and function of chloroplast lipids. In: Wada H, Murata N (eds), *Lipids in Photosynthesis. Essential and Regulatory Functions.* Springer Science, Dordrecht, pp 35–55.

Shu S, Yuan Y, Chen J, *et al.* (2015) The role of putrescine in the regulation of proteins and fatty acids of thylakoid membranes under salt stress. *Sci Rep* 5, Art. no. 14390.

Siedlecka A, Krupa Z. (1999) Cd/Fe interaction in higher plants—its consequences for the photosynthetic apparatus. *Photosynthetica* 36: 321–331.

Singh R, Singh S, Parihar P, *et al.* (2016) Reactive oxygen species (ROS): beneficial companions of plants' developmental processes. *Front Plant Sci* 7: 1299.

Singh S, Srivastava PK, Kumar D, Tripathi DK, Chauhan DK, Prasad SM. (2015) Morpho-anatomical and biochemical adapting strategies of maize (*Zea mays* L.) seedlings against lead and chromium stresses. *Biocatal Agric Biotechnol* 4: 286–295.

Singh S, Tripathi DK, Singh S, *et al.* (2017) Toxicity of aluminium on various levels of plant cells and organism: a review. *Environ Exp Bot* 137: 177–193.

Singh VP, Tripathi DK, Kumar D, Chauhan DK. (2011) Influence of exogenous silicon addition on aluminium tolerance in rice seedlings. *Biol Trace Elem Res* 144: 1260–1274.

Solymosi B. (2012) Soil metals, chloroplasts, and secure crop production: a review. *Agron Sustain Dev* 32: 245–272.

Somerville C, Browse J, Jaworski JG, Ohlrogge JB. (2000) Lipids. In: Buchanan BB, Gruissem W, Jones RL (eds), *Biochemistry and Molecular Biology of Plants.* American Society of Plant Physiologists, Rockville, MD, pp. 456–527.

Stadtman ER. (1986) Oxidation of proteins by mixed-function oxidation systems: implication in protein turnover, aging and neutrophil function. *Trends Biochem Sci* 11: 11–12.

Subczynski WK, Widomska J, Feix JB. (2009) Physical properties of lipid bilayers from EPR spin labeling and their influence on chemical reactions in a membrane environment. *Free Radic Biol Med* 46: 707–718.

Telfer A, Bishop SM, Phillips D, Barber J. (1994) Isolated photosynthetic reaction center of photosystem II as a sensitizer for the formation of singlet oxygen. *J Biol Chem* 269: 13244–13253.

Thompson JE, Legge RL, Barber RF. (1987) The role of free radicals in senescence and wounding. *New Phytol* 105: 317–344.

Torres-Franklin ML, Gigon A, Fernandes de Melo D, Zuily-Fodil Y, Pham-Thi AT. (2007) Drought stress and rehydration affect the balance between MGDG and DGDG synthesis in cowpea leaves. *Physiol Plant* 131: 201–210.

Triantaphylidès C, Krischke M, Hoeberichts FA, *et al.* (2008) Singlet oxygen is the major reactive oxygen species involved in photooxidative damage to plants. *Plant Physiol* 148: 960–968.

Tripathi DK, Singh VP, Kumar D, Chauhan DK. (2012a) Impact of exogenous silicon addition on chromium uptake, growth, mineral elements, oxidative stress, antioxidant capacity, and leaf and root structures in rice seedlings exposed to hexavalent chromium. *Acta Physiol Plant* 34: 279–289.

Tripathi DK, Singh VP, Kumar D, Chauhan DK. (2012b) Rice seedlings under cadmium stress: effect of silicon on growth, cadmium uptake, oxidative stress, antioxidant capacity and root and leaf structures. *Chem Ecol* 28: 281–291.

Tripathi DK, Singh S, Singh VP, Prasad SM, Chauhan DK, Dubey NK. (2016a) Silicon nanoparticles more efficiently alleviate arsenate toxicity than silicon in maize cultivar and hybrid differing in arsenate tolerance. *Front Environ Sci* 4: 46.

Tripathi DK, Singh VP, Prasad SM, Dubey NK, Chauhan DK, Rai AK. (2016b) LIB spectroscopic and biochemical analysis to characterize lead toxicity alleviative nature of silicon in wheat (Triticum aestivum L.) seedlings. *J Photochem Photobiol B: Biol* 154: 89–98.

Tripathi DK, Singh S, Singh VP, Prasad SM, Dubey NK, Chauhan DK. (2017a) Silicon nanoparticles more effectively alleviated UV-B stress than silicon in wheat (Triticum aestivum) seedlings. *Plant Physiol Biochem* 110: 70–81.

Tripathi DK, Singh S, Singh S, *et al.* (2017b) Nitric oxide alleviates silver nanoparticles (AgNps)-induced phytotoxicity in Pisum sativum seedlings. *Plant Physiol Biochem* 110: 167–177.

Tripathi DK, Mishra RK, Singh S, *et al.* (2017c) Nitric oxide ameliorates zinc oxide nanoparticles phytotoxicity in wheat seedlings: implication of the ascorbate–glutathione cycle. *Front Plant Sci* 8: 1.

Verma S, Dubey RS. (2003) Lead toxicity induces lipid peroxidation and alters the activities of antioxidant enzymes in growing rice plants. *Plant Sci* 164: 645–655.

Yoshioka-Nishimura M. (2016) Close relationships between the PSII repair cycle and thylakoid membrane dynamics. *Plant Cell Physiol* 57: 1115–1122.

Zulfugarov IS, Tovuu A, Eu YJ, *et al.* (2014) Production of superoxide from photosystem II in a rice (*Oryza sativa* L.) mutant lacking PsbS. *BMC Plant Biol* 14: 242.

Zvezdanović JB, Marković DZ, Veljović Jovanović S, Radovanović B. (2013) UV-induced oxygen free radicals production in spinach thylakoids. *Advanced Technol* 2: 45–50.

8

Reactive Oxygen Species and Response of the Calvin–Benson Cycle: An Overview

Shivam Yadav and Neelam Atri

Banaras Hindu University, Varanasi, India

Introduction

Reactive oxygen species (ROS) have been an inevitable by-product of aerobic metabolism from around 2.7 billion years ago, following the introduction of O_2 by photosynthetic organisms (Halliwell and Gutteridge, 1989). ROS include not only numerous free radicals such as hydroxyl radical (·OH) and superoxide anion ($O_2^{\cdot-}$) but also non-radical molecules like singlet oxygen ($^1O_2$) and hydrogen peroxide (H_2O_2). Unlike O_2, these species are highly toxic and reactive, causing oxidative damage of cells (Asada and Takahashi, 1987).

Although ROS are known for their destructive nature, other roles have also been identified in recent years such as regulation of growth, cell cycles, abiotic and biotic stress responses, hormone signaling, and apoptosis (Dat *et al.*, 2000; Kovtun *et al.*, 2000; Pei *et al.*, 2000; Mittler, 2002; Mullineaux and Karpinski, 2002; Neill *et al.*, 2002; Torres *et al.*, 2002; Foreman *et al.*, 2003; Jiang *et al.*, 2003; Kwak *et al.*, 2003; Overmyer *et al.*, 2003; V.P. Singh *et al.*, 2011; Tripathi *et al.*, 2012ab, 2016a,b, 2017a–c; S. Singh *et al.*, 2015, 2017; R. Singh *et al.*, 2016). Their participation in either destructive or signaling pathways greatly depends on the subtle balance between ROS generation and scavenging. Reports suggest that about 1% of O_2 used by plants is diverted to produce ROS in chloroplasts, mitochondria, and peroxisomes (Asada and Takahashi, 1987).

Although ROS are generated in all organisms, they pose a major challenge for plants as compared to other organisms due to (i) their sedentary, stationary lifestyle under a constantly changing environment; and (ii) the unavoidable leakage of electrons onto O_2 from the electron transport activities of chloroplasts, the electron transport chain of mitochondria, and the various metabolic processes of different cellular compartments that are also responsible for formation of ROS (Foyer and Harbinson, 1994; Foyer, 1997; del Río *et al.*, 2006; Blokhina and Fagerstedt, 2010; Heyno *et al.*, 2011). The present chapter is an attempt to describe the responses of the Calvin–Benson cycle and each and every aspect of the interrelation between ROS and the Calvin–Benson cycle.

Reactive Oxygen Species in Plants: Boon Or Bane - Revisiting the Role of ROS, First Edition. Edited by Vijay Pratap Singh, Samiksha Singh, Durgesh Kumar Tripathi, Sheo Mohan Prasad, and Devendra Kumar Chauhan.

ROS Generation in Plants: A Consequence of Limited CO_2 Fixation/Calvin–Benson Cycle

The major generators of ROS in plants, located in different compartments, are (i) the electron transport chain in mitochondria and chloroplasts; (ii) a few oxidases and peroxidases (NADPH and NADH oxidase, lipoxygenase, xanthine oxidase, etc.); and (iii) photosensitizers, for example, chlorophyll molecules (Blokhina *et al.*, 2003).

Light-dependent photosynthetic processes contribute maximally to ROS generation in chloroplasts. CO_2 fixation in the Calvin–Benson cycle is highly sensitive to environmental stresses (Long *et al.*, 1994; Murata *et al.*, 2007). Various studies worldwide have reported high-temperature inhibition of rubisco-activase, thus suppressing activation of rubisco. Singh and co-workers reported Cd-induced inhibition of CO_2 fixation as one of the early signals of CO_2 toxicity (Singh *et al.*, 2015). Similar results were found with other abiotic stresses such as salinity, arsenic, and UV-B (Pandey *et al.*, 2012; Rai *et al.*, 2014; Srivastava *et al.*, 2015). Thus under such environmental stresses photosynthetic fixation of CO_2 is severely limited, which results in decreased utilization of NADPH. Accumulation of NADPH causes a decline in the level of $NADP^+$. Since $NADP^+$ is the only acceptor of electron in photosystem I (PSI), a reduction in $NADP^+$ causes transfer of electrons from ferredoxin (PSI) to molecular O_2, thus initiating the generation of ROS. This process is called the Mehler reaction:

$$2O_2 + 2Fd_{red.} \rightarrow 2O_2^{\bullet -} + 2Fd_{ox.}$$

A similar kind of electron leakage may also occur from 2Fe-4S clusters in the electron transport chain of PSI. Thus disruption of photosynthetic CO_2 fixation is directly linked to production of ROS (Radmer and Kok, 1976; Asada and Badger, 1984; Allahverdiyeva *et al.*, 2005)

ROS-Mediated Inhibition of the Calvin–Benson Cycle

ROS are known to strongly inhibit CO_2 fixation, whether formed endogenously by a pseudocyclic electron transport chain or added exogenously to chloroplast suspension (Forti and Gerola, 1977; Kaiser, 1979). This inhibition of the Calvin–Benson cycle is caused by an interaction of ROS with either intermediates or enzymes of the Calvin–Benson cycle (Kaiser, 1979). This section focuses on ROS-induced inhibition of the Calvin–Benson cycle and its underlying mechanism.

Reactive Oxygen Species (ROS)-Induced Modification of Rubisco

ROS are known to modify the structural integrity of rubisco, as witnessed by various studies. Mehta and co-workers reported ROS-induced degradation of rubisco (Mehta *et al.*, 1992). Similarly Garcia-Ferris and Moreno (1994) reported modification and breakdown of rubisco in *Euglena gracilis*.

Desimone *et al.* (1996) reported that ROS can induce partial degradation of the large subunit of rubisco in etiolated chloroplasts of barley. However, the role of ROS in the destruction of Rubisco remains open. A few workers suggest that certain modifications of rubisco enhance its susceptibility to proteolysis by chloroplastic enzymes (Moreno *et al.*, 1995). Similar results were also found for other biological systems (Wolff *et al.*, 1986; Goldberg, 1992; Rivett, 1993).

The above findings are consistent with the fact that rubisco degradation is correlated with protein denaturation and with cross-linking of -SH groups (Garcia-Ferris and Moreno, 1994).

Table 8.1 Calvin–Benson cycle enzymes inhibited by reactive oxygen species (ROS).

Enzyme	Reference
Ribulose 5-phosphokinase	Kaiser, 1979
Glyceraldehyde-3-phosphate dehydrogenase	Forti and Gerola, 1977
Sedoheptulose bisphosphatase	Heldt *et al.*, 1978
Fructose 1,6-bisphosphatase	Heldt *et al.*, 1978

Moreover, studies with isolated rubisco showed that ROS produce structural changes in protein (Eckardt and Pell, 1995) and that those modifications cause proteolysis by externally added proteases, for example, trypsin (Peñarrubia and Moreno, 1990).

Either addition of 0.1% SDS in chloroplasts or lowering pH below 5 stimulates rubisco degradation in chloroplasts. These observations clearly suggest that both factors are responsible for ROS-induced rubisco degradation.

ROS-Induced Inhibition of other Enzymes of the Calvin–Benson Cycle

Several studies have demonstrated a ROS-induced increment in proteolysis by activation of peptidases and by oxidatively "tagging" proteins for attack by latent proteases (Casano *et al.*, 1990, 1994).

Various studies have demonstrated inhibition of Calvin–Benson cycle enzymes by H_2O_2 (Table 8.1). ROS cause damage to these enzymes not only at the protein level but also at the gene level. ROS generation induced by the phytotoxin coronatine is able to decrease expression of various Calvin–Benson cycle genes (*RBC2A*, *RCA1*, *GAPB*, *SFBA*, and *SBPase*), *FBPase* (fructose 1,6-bisphosphatase) and *PRK* (phosphoribulokinase).

Mechanism of Enzyme Inhibition by ROS: Protein Oxidation

ROS are known to cause covalent modification of proteins, defined as protein oxidation, which can be either reversible or irreversible. Particularly the residues Arg, His, Lys, Pro, Thr, and Trp undergo oxidation, which generates free carbonyl groups and increases susceptibility to proteolysis (Moller *et al.*, 2007). Protein carbonylation is a result of direct oxidation of amino acid side chains, for example, lysine to aminoadipic semialdehyde (Shringarpure and Davies, 2002). Proteins comprising S-containing amino acids and thiol groups are more likely to be attacked by ROS (e.g., Cys and Met are reactive towards O_2 and $\cdot OH$). Active oxygen can remove an H atom from cysteine residues and forms a thiyl radical, which further cross-links another thiyl radical and forms disulfide bridges (Hancock *et al.*, 2006). On the other hand, oxygen can be added to methionine forming methionine sulfoxide derivatives (Sadanandom *et al.*, 2000).

Effect of ROS on Calvin–Benson Cycle Intermediates

The impact of ROS on Calvin–Benson cycle enzymes directly affects various intermediates. Increased levels of fructose-1,6-bisphosphate (FBP) and sedoheptulose-1,7-bisphosphate (SBP) were reported during chilling of tomato leaves in the light (Ishiga *et al.*, 2009). These increases were associated with direct inhibition of FBPase and SBPase through ROS.

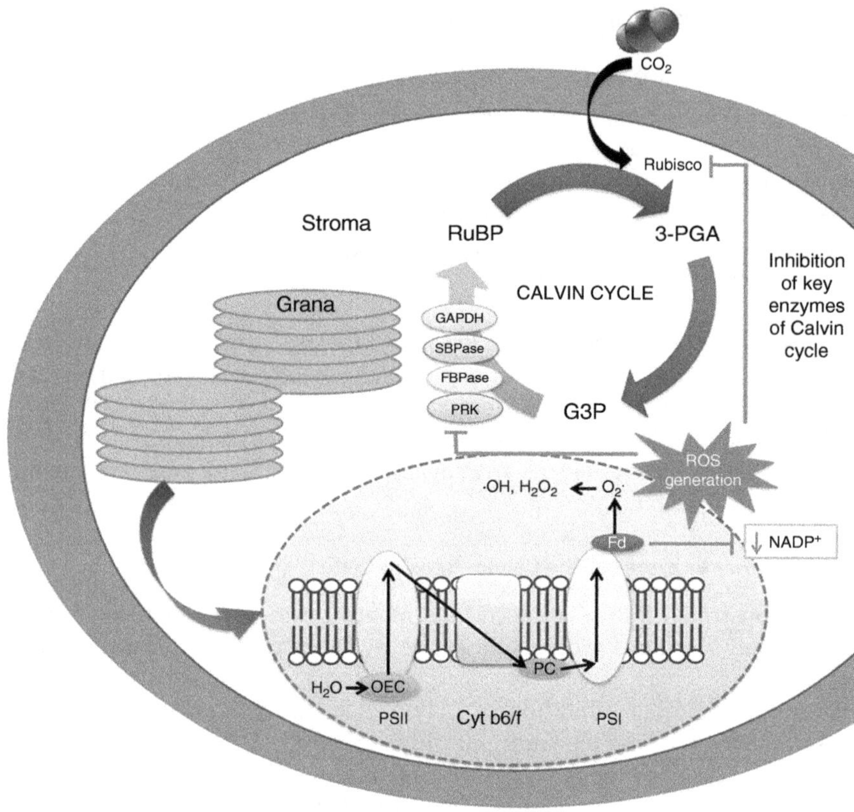

Figure 8.1 Generation of reactive oxygen species (ROS) and overall impact of ROS on the Calvin–Benson cycle as represented in the chloroplast. Cyt, cytochrome; GAPDH, glyceraldehyde-3-phosphate dehydrogenase; FBPase, fructose 1,6-bisphosphatase; Fd, ferredoxin; G3P, glyceraldehyde 3-phosphate; OEC, oxygen-evolving complex; PC, Plastocyanin; PSI, photosystem I; PSII, photosystem II; PRK, phosphoribulokinase; RuBP, ribulose 1,5-bisphosphate; SBPase, sedoheptulose bisphosphatase.

In summary, ROS generation during the light reaction and the overall impact of ROS on the Calvin–Benson cycle is presented in Figure 8.1.

Conclusion

ROS are inevitably produced as a by-product of normal cell metabolism. Light-driven photosynthetic processes are the main providers of ROS in chloroplasts. Chloroplasts with numerous ROS-producing cores are a major source of ROS generation in plants, controlled by a differentiated ROS scavenging system. Uncontrolled production or inefficient quenching of ROS in chloroplasts arising under stress conditions, result in a detrimental effect on diverse vital molecular targets in the photosynthetic system. Nevertheless, at low concentrations ROS act as signaling molecules responsible for various responses in plant cells, including stress responses. Rapid progress had been made in recent years on understanding responses of the light reaction of photosynthesis, but responses of the Calvin–Benson cycle/dark reaction toward ROS are still little known. Future progress in genomics, proteomics, and metabolomics under conditions of oxidative stress will help to gain a clear understanding of ROS-mediated responses, specifically with regard to the Calvin–Benson cycle.

References

Allahverdiyeva Y, Mamedov F, Mäenpää P, Vass I, Aro EM. (2005) Modulation of photosynthetic electron transport in the absence of terminal electron acceptors: characterization of the rbcL deletion mutant of tobacco. *Biochim Biophys Acta* 1709: 69–83.

Asada K, Badger MR. (1984) Photoreduction of $^{18}O_2$ and $H_2{}^{18}O$ with concomitant evolution of $^{16}O_2$ in intact spinach chloroplasts: evidence for scavenging of hydrogen peroxide by peroxidase. *Plant Cell Physiol* 25: 1169–1179.

Asada K, Takahashi M. (1987) Production and scavenging of active oxygen in photosynthesis. In: Kyle DJ, Osmond CB (eds), *Photoinhibition (Topics in Photosynthesis)*, Vol. 9. Elsevier, pp. 227–287.

Blokhina O, Fagerstedt KV. (2010) Reactive oxygen species and nitric oxide in plant mitochondria: origin and redundant regulatory systems. *Plant Physiol* 138: 447–462.

Blokhina O, Virolainen E, Fagerstedt KV. (2003) Antioxidants, oxidative damage and oxygen deprivation stress: a review. *Ann Bot* 91: 179–194.

Casano LM, GÓmez LD, Trippi VS. (1990) Oxygen- and light-induced proteolysis in isolated oat chloroplasts. *Plant Cell Physiol* 31: 377–382.

Casano LM, Lascano RH, Trippi VS. (1994) Hydroxyl radicals and a thylakoid-bound endopeptidase are involved in light- and oxygen-induced proteolysis in oat chloroplasts. *Plant Cell Physiol* 35: 14–152.

Dat J, Vandenabeele S, Vranová E, Van Montagu M, Inzé D, Van Breusegem F. (2000) Dual action of the active oxygen species during plant stress responses. *Cell Mol Life Sci* 57: 779–795.

Del Río LA, Sandalio LM, Corpas FJ, Palma JM, Barroso JB. (2006) Reactive oxygen species and reactive nitrogen species in peroxisomes: Production, scavenging, and role in cell signaling. *Plant Physiol* 141: 330–335.

Desimone M, Henke A, Wagner E. (1996) Oxidative stress induces partial degradation of the large subunit of ribulose-l,5-bisphosphate carboxylase/oxygenase in isolated chloroplasts of barley. *Plant Physiol* 111: 789–796.

Eckardt NA, Pell EJ. (1995) Oxidative modification of Rubisco from potato foliage in response to ozone. *Plant Physiol Biochem* 33: 273–282.

Foreman J, Demidchik V, Bothwell JH, *et al.* (2003) Reactive oxygen species produced by NADPH oxidase regulate plant cell growth. *Nature* 422: 442–446.

Forti G, Gerola P. (1977) Inhibition of photosynthesis by azide and cyanide and the role of oxygen in photosynthesis. *Plant Physiol* 59: 859–862.

Foyer CH. (1997) Oxygen metabolism and electron transport in photosynthesis. In: Scandalios J (ed.), *Molecular Biology of Free Radical Scavenging Systems.* Cold Spring Harbor Laboratory Press, New York, pp. 587–621.

Foyer CH, Harbinson J. (1994) Oxygen metabolism and the regulation of photosynthetic electron transport. In: Foyer CH, Mullineaux P (eds), *Causes of Photooxidative Stresses and Amelioration of Defense Systems in Plants.* CRC Press, Boca Raton, FL, pp. 1–42.

García-Ferris C, Moreno J. (1994) Oxidative modification and breakdown of ribulose-1,5-bisphosphate carboxylase/oxygenase induced in *Euglena gracilis* nitrogen starvation. *Planta* 193: 208–215.

Goldberg AL. (1992) The mechanism and functions of ATP-dependent proteases in bacterial and animal cells. *Eur J Biochem* 203: 9–23.

Halliwell B, Gutteridge JMC. (1989) *Free Radicals in Biology and Medicine.* Clarendon Press, Oxford.

Hancock J, Desikan R, Harrison J, Bright J, Hooley R, Neill S. (2006) Doing the unexpected: proteins involved in hydrogen peroxide perception. *J Exp Bot* 57: 1711–1718.

Heldt HW, Cben CJ, Liley RM, Portis A. (1978) The role of fructose and sedoheptulose bisphosphatase in the control of CO_2 fixation. Evidence from the effects of Mg^{2+} concentration, pH, and H_2O_2. In: Hall DO, Coombs J, Goodwin TW (eds), *Proceedings of the 4th International Congress on Photosynthesis*. The Biochemical Society, London, pp. 469–478.

Heyno E, Mary V, Schopfer P, Krieger-Liszkay A. (2011) Oxygen activation at the plasma membrane: relation between superoxide and hydroxyl radical production by isolated membranes. *Planta* 234: 35–45.

Ishiga Y, Uppalapati SR, Ishiga T, Elavarthi S, Martin B, Bender CL. (2009) The phytotoxin coronatine induces light-dependent reactive oxygen species in tomato seedlings. *New Phytol* 181: 147–160.

Jiang K, Meng YL, Feldman LJ. (2003) Quiescent center formation in maize roots is associated with an auxin-regulated oxidizing environment. *Development* 130: 1429–1438.

Kaiser WM. (1979) Reversible inhibition of the Calvin cycle and activation of oxidative pentose phosphate cycle in isolated intact chloroplasts by hydrogen peroxide. *Planta* 145: 377–382.

Kovtun Y, Chiu WL, Tena G, Sheen J. (2000) Functional analysis of oxidative stress activated mitogen-activated protein kinase cascade in plants. *Proc Natl Acad Sci USA* 97: 2940–2945.

Kwak JM, Mori IC, Pei ZM, *et al.* (2003) NADPH oxidase AtrbohD and AtrbohF genes function in ROS-dependent ABA signaling in *Arabidopsis*. *EMBO J* 22: 2623–2633.

Long SP, Humphries S, Falkowski PG. (1994) Photoinhibition of photosynthesis in nature. *Annu Rev Plant Physiol Plant Mol Biol* 45: 633–662.

Mehta RA, Fawcett TW, Porath D, Mattoo A. (1992) Oxidative stress causes rapid membrane translocation and *in vivo* degradation of ribulose-1,5-bisphosphate carboxylase/oxygenase. *J Biol Chem* 267: 2810–2816.

Mittler R. (2002) Oxidative stress, antioxidants and stress tolerance. *Trends Plant Sci* 7: 405–410.

Moller IM, Jensen PE, Hansson A. (2007) Oxidative modifications to cellular components in plants. *Annu Rev Plant Biol* 58: 459–481.

Moreno J, Peñarrubia L, Garcia-Ferris C. (1995) The mechanism of redox regulation of ribulose-1,5-bisphosphate carboxylase/oxygenase turnover. A hypothesis. *Plant Physiol Biochem* 33: 121–127.

Mullineaux P, Karpinski S. (2002) Signal transduction in response to excess light: getting out of the chloroplast. *Curr Opin Plant Biol* 5: 43–48.

Murata N, Takahashib S, Nishiyamac Y, Allakhverdiev SI. (2007) Photoinhibition of photosystem II under environmental stress. *Biochim Biophys Acta* 1767: 414–421.

Neill S, Desikan R, Hancock J. (2002) Hydrogen peroxide signalling. *Curr Opin Plant Biol* 5: 388–395.

Overmyer K, Brosché M, Kangasjärvi J. (2003) Reactive oxygen species and hormonal control of cell death. *Trends Plant Sci* 8: 335–342.

Pandey S, Rai R, Rai LC. (2012) Proteomics combines morphological, physiological and biochemical attributes to unravel the survival strategy of *Anabaena* sp. PCC 7120 under arsenic stress. *J Proteomics* 75: 921–937.

Pei ZM, Murata Y, Benning G, *et al.* (2000) Calcium channels activated by hydrogen peroxide mediate abscisic acid signalling in guard cells. *Nature* 406: 731–734.

Peñarrubia L, Moreno J. (1990) Increased susceptibility of ribulose-1,5-bisphosphate carboxylase/oxygenase to proteolytic degradation caused by oxidative treatments. *Arch Biochem Biophys* 281: 319–323.

Radmer RJ, Kok B. (1976) Photoreduction of O_2 primes and replaces CO_2 assimilation. *Plant Physiol* 58: 336–340.

Rai S, Agrawal C, Shrivastava AK, Singh PK, Rai LC. (2014) Comparative proteomics unveils cross species variations in *Anabaena* under salt stress. *J Proteomics* 98: 254–270.

Rivett AJ. (1993) Proteasomes: multicatalytic proteinase complexes. *Biochem J* 291: 1–10.

Sadanandom A, Poghosyan Z, Fairbairn DJ, Murphy DJ. (2000) Differential regulation of plastidial and cytosolic isoforms of peptides methionine sulfoxide reductase in *Arabidopsis*. *Plant Physiol* 123: 1299–1309.

Shringarpure R, Davies KJ. (2002) Protein turnover by the proteasome in aging and disease. *Free Radic Biol Med* 32: 1084–1089.

Shrivastava AK, Chatterjee A, Yadava S, Singh PK, Singh S, Rai LC. (2015) UV-B stress induced metabolic rearrangements explored with comparative proteomics in three *Anabaena* species. *J Proteomics* 127(PtA): 122–133.

Singh PK, Shrivastava AK, Chatterjee A, *et al.* (2015) Cadmium toxicity in diazotrophic *Anabaena* spp. adjudged by hasty up-accumulation of transporter and signaling and severe down-accumulation of nitrogen metabolism protein. *J Proteomics* 127(Pt A): 134–146.

Singh R, Singh S, Parihar P, *et al.* (2016) Reactive oxygen species (ROS): beneficial companions of plants' developmental processes. *Front Plant Sci* 7: 1299.

Singh S, Srivastava PK, Kumar D, Tripathi DK, Chauhan DK, Prasad SM. (2015) Morpho-anatomical and biochemical adapting strategies of maize (*Zea mays* L.) seedlings against lead and chromium stresses. *Biocatal Agric Biotechnol* 4: 286–295.

Singh S, Tripathi DK, Singh S, *et al.* (2017) Toxicity of aluminium on various levels of plant cells and organism: a review. *Environ Exp Bot* 137: 177–193.

Singh VP, Tripathi DK, Kumar D, Chauhan DK. (2011) Influence of exogenous silicon addition on aluminium tolerance in rice seedlings. *Biol Trace Elem Res* 144: 1260–1274.

Torres MA, Dangl JL, Jones JD. (2002) *Arabidopsis* gp91phox homologues AtrbohD and AtrbohF are required for accumulation of reactive oxygen intermediates in the plant defense response. *Proc Natl Acad Sci USA* 99: 517–522.

Tripathi DK, Singh VP, Kumar D, Chauhan DK. (2012a) Impact of exogenous silicon addition on chromium uptake, growth, mineral elements, oxidative stress, antioxidant capacity, and leaf and root structures in rice seedlings exposed to hexavalent chromium. *Acta Physiol Plant* 34: 279–289.

Tripathi DK, Singh VP, Kumar D, Chauhan DK. (2012b) Rice seedlings under cadmium stress: effect of silicon on growth, cadmium uptake, oxidative stress, antioxidant capacity and root and leaf structures. *Chem Ecol* 28: 281–291.

Tripathi DK, Singh S, Singh VP, Prasad SM, Chauhan DK, Dubey NK. (2016a) Silicon nanoparticles more efficiently alleviate arsenate toxicity than silicon in maize cultivar and hybrid differing in arsenate tolerance. *Front Environ Sci* 4: 46.

Tripathi DK, Singh VP, Prasad SM, Dubey NK, Chauhan DK, Rai AK. (2016b) LIB spectroscopic and biochemical analysis to characterize lead toxicity alleviative nature of silicon in wheat (Triticum aestivum L.) seedlings. *J Photochem Photobiol B: Biol* 154: 89–98.

Tripathi DK, Singh S, Singh VP, Prasad SM, Dubey NK, Chauhan DK. (2017a) Silicon nanoparticles more effectively alleviated UV-B stress than silicon in wheat (Triticum aestivum) seedlings. *Plant Physiol Biochem* 110: 70–81.

Tripathi DK, Singh S, Singh S, *et al.* (2017b) Nitric oxide alleviates silver nanoparticles (AgNps)-induced phytotoxicity in Pisum sativum seedlings. *Plant Physiol Biochem* 110: 167–177.

Tripathi DK, Mishra RK, Singh S, *et al.* (2017c) Nitric oxide ameliorates zinc oxide nanoparticles phytotoxicity in wheat seedlings: implication of the ascorbate–glutathione cycle. *Front Plant Sci* 8: 1.

Wolff SP, Garner A, Dean RT. (1986) Free radicals, lipids and protein degradation. *Trends Biochem Sci* 11: 27.

9

Role of Reactive Oxygen Species in Photophosphorylation and Damage to D1 Protein: Past and Present

Sunil K. Gupta, Marisha Sharma, Farah Deeba, and Vivek Pandey

CSIR-National Botanical Research Institute, Rana Pratap Marg, Lucknow, India

Introduction

The oxygen-containing compounds that are chemically reactive are called reactive oxygen species (ROS). Examples include peroxides, superoxide radical, hydroxyl radical, and singlet oxygen. Cell metabolism generates ROS at low rates as normal by-products. ROS are produced in response to abiotic and biotic stresses during the oxidative burst (Perez and Brown, 2014). Plant responses to different stresses are highly complex and involve changes at the molecular, cellular, and physiological levels. Recent evidence shows that plants respond to multiple stresses differently from how they respond to individual stresses, activating a specific programme of gene expression relating to the exact environmental stress encountered. Various abiotic stresses lead to the overproduction of ROS in plants; these are highly reactive and toxic and cause damage to proteins, lipids, carbohydrates, and DNA, which ultimately results in oxidative stress (V.P. Singh *et al.*, 2011; Tripathi *et al.*, 2012a,b, 2016a,b, 2017a–c; S. Singh *et al.*, 2015, 2017; R. Singh *et al.*, 2016). Salinity stress often enhances the generation of ROS leading to metabolic disorders, cellular damage, and premature senescence or necrosis (Møller *et al.*, 2007; Jaleel *et al.*, 2009; Miller *et al.*, 2010; Habib *et al.*, 2016). ROS also influence the expression of a number of genes and therefore control many processes like growth (Foreman *et al.*, 2003), cell cycle (Chatterjee *et al.*, 2009), programmed cell death (PCD) (Jacobson, 1996), abiotic stress responses (Gill and Tuteja, 2010), pathogen defense (Torres *et al.*, 2006), systemic signaling (Mittler *et al.*, 2011), and development (Cona *et al.*, 2006). The formation of ROS has been detected in almost all cellular compartments, including chloroplasts (Edreva, 2005), mitochondria (Mittler *et al.*, 2004), peroxisomes (Schrader and Fahimi, 2006), and cytoplasm (Suzuki *et al.*, 2011). ROS can also be produced in the apoplast, through the activity of NADPH oxidases (Miller *et al.*, 2009; Suzuki *et al.*, 2011). ROS are scavenged by an antioxidant system including antioxidant compounds (ascorbate, glutathione, tocopherols, etc.) and antioxidant enzymes like superoxide dismutase (SOD), ascorbate peroxidase (APX), and catalase (CAT) (Foyer and Noctor, 2003; Tripathi *et al.*, 2016a,b, 2017a–c; R. Singh *et al.*, 2016; S. Singh *et al.*, 2017). It is not clear exactly how plant cells sense ROS, as specific ROS receptors have not yet been identified. One possibility is that ROS directly modify redox-sensitive proteins, including phosphatases (Apel and Hirt, 2004), and heat-shock transcription factors (Miller and Mittler, 2006).

The evolution of photosynthetic organisms resulted in a gradual, but massive increase in the level of atmospheric molecular oxygen (O_2) that enabled aerobic life on Earth. All life on Earth

Reactive Oxygen Species in Plants: Boon Or Bane - Revisiting the Role of ROS, First Edition. Edited by Vijay Pratap Singh, Samiksha Singh, Durgesh Kumar Tripathi, Sheo Mohan Prasad, and Devendra Kumar Chauhan.

is based on redox reactions (reduction – the gain of an electron; and oxidation – the loss of an electron) using reductive processes to store energy and oxidative processes to release it. Photosynthesis is considered one of the most important metabolic processes in plants and its performance is greatly affected under stress conditions (Baker, 1991). The two primary processes involved in formation of ROS during photosynthesis are the direct photoreduction of O_2 to the superoxide radical by reduced electron transport components associated with photosystem I (PSI), and reactions linked to the photorespiratory cycle, including rubisco in the chloroplast and glycolate-oxidase and CAT-peroxidase reactions in peroxisomes. When plants are exposed to light intensities that exceed the capacity of CO_2 assimilation, over-reduction of the electron transport chain leads to inactivation of photosystem II (PSI)I and the inhibition of photosynthesis. PSII is a multi-subunit thylakoid membrane protein complex that is responsible for the electron transfer from water to plastoquinone (PQ) with the help of light. PSII is located mostly in the stacked areas of the thylakoid membrane, called grana membranes, and more than 40 proteins have been found to be associated with PSII, either stably or transiently (Nelson and Yocum, 2006; Shi *et al.*, 2012).The minimal reaction center complex of PSII, capable of charge separation, contains five proteins: D1, D2, α and β subunits of Cyt b559, and PsbI subunit (Nanba and Satoh, 1987). The reaction center proteins D1 and D2 bind all the redox-active cofactors that are required for PSII electron transport. Excitation energy, captured by the light-harvesting complex (LHC) and the PSII internal antenna proteins CP43 and CP47, induces charge separation in PSII, which in turn enables the oxygen-evolving complex (OEC) to oxidize water molecules and provide electrons to the electron transfer chain. The reaction center protein, D1, is a major target for photo-oxidative damage in the PSII complex (Edelman and Mattoo, 2008). Obviously, all of three ROS components, including superoxide anion, hydrogen peroxide, and hydroxyl radical, are responsible for the degradation of oxidized D1 protein, and the protection of the D1 protein degradation by phosphorylation is accompanied by the inhibition of superoxide anion production. In the PSII repair cycle, D1 is therefore degraded rapidly from the PSII complex by proteolysis and subsequently replaced by newly synthesized D1 (Baena-Gonzalez and Aro, 2002; Aro *et al.*, 2005; Kato and Sakamoto, 2009). Reversible phosphorylation of PSII core proteins has long been recognized in higher plants, and is regarded as a regulatory mechanism in PSII repair (Aro *et al.*, 2005; Edelman and Mattoo, 2008).

Plants may use two strategies to protect the photosynthetic apparatus against photoinhibition: first, the thermal dissipation of excess excitation energy in the PSII antennae (non-photochemical quenching), and second, the ability of PSII to transfer electrons to various acceptors within the chloroplast (photochemical quenching) (Ort and Baker, 2002). However, in a changing environment, plants are able to adapt their metabolism in order to cope with stress and maintain a steady-state balance between energy production and consumption (Suzuki *et al.*, 2011). An increased rate of ROS generation is a common plant response to various environmental stresses. ROS are by-products of the plant aerobic metabolism and contribute to biological processes involved in growth and development (del Río, 2015). In stress conditions, inhibition of both PSII and PSI electron transport activities occurs and also a decrease in the number of active PSII reaction centers (Chang *et al.*, 2012). In this chapter, we describe the biochemistry of ROS and their production sites, and ROS-scavenging antioxidant defense machinery. The role of ROS in photophosphorylation and the effects on D1 protein, and its damage and protective mechanism are also discussed.

Generation of ROS

Ever since the introduction of molecular oxygen (O_2) into our atmosphere by O_2-evolving photosynthetic organisms around 2.7 billion years ago, ROS have been the unwelcome companions of aerobic life (Halliwell and Gutteridge, 1989). Plants frequently experience hypoxia when oxygen (O_2) availability limits aerobic metabolism. This usually occurs in plant organs like vascular bundles, stomata, mesophyll tissue, and so forth, and in cell organelles (i.e., mitochondria, chloroplasts) with a very high metabolic activity, or in bulky organs where a steep O_2 gradient is observed because of a low gas diffusion rate (Van Dongen and Licausi, 2015). Plants constantly produce ROS in chloroplasts, mitochondria, peroxisomes, and other sites of the cell because of various metabolic activities such as photosynthesis and respiration. Along with these organelles, peroxidases and amine oxidases present in cell walls and NADPH oxidase located in the plasma membrane produce ROS, often in response to stress signals. Oxygen is continuously produced inside the chloroplast due to photosynthetic electron transport and simultaneously removed by reduction and assimilation.

When the availability of CO_2 is restricted inside the leaves due to various stresses, ribulose-1,5-bisphosphate carboxylase/oxygenase catalyzes a competitive reaction in which oxygen is favored over CO_2 as a substrate. This oxygenation reaction leads to the formation of glycolate, which is transported to peroxisomes and produces H_2O_2 after subsequent oxidation by glycolate oxidase.

Plasma membrane NADPH-dependent oxidases contain a multimeric flavocytochrome that forms an electron transport chain capable of reducing O_2 to oxygen radical ($O_2^{\cdot-}$). Inhibitors of this oxidase impair ROS production. In addition to NADPH, pH-dependent cell wall peroxidase, oxalate oxidase, and amine oxidases have also been reported to produce ROS in the apoplast (Hernández *et al.*, 2001; Díaz-Vivancos *et al.*, 2006; Shapiguzov *et al.*, 2012).

Mitochondrial contribution to ROS production in green plants is not significant (Purvis, 1997). The mitochondrial electron transport chain can produce ROS, by transferring a single electron to oxygen and producing $O_2^{\cdot-}$. However, mitochondrial alternative oxidase (AOX) catalyzes the O_2-dependent oxidation of ROS. Lack of AOX induction caused increased ROS production, and tobacco plants overexpressing AOX had less hypersensitive lesions than wild type plants in response to virus infection (Robson and Vanlerberghe, 2002)

ROS Scavenging in the Plant Cell

Plants are equipped with antioxidants and antioxidative enzymes and other small molecules to scavenge harmful ROS. Imbalance between ROS production and their detoxification by enzymatic and non-enzymatic reactions causes oxidative stress. Enhanced net ROS formation results in damage to DNA, proteins, and lipids and ultimately cell death.

Major ROS-scavenging enzymes of plants include superoxide dismutase (SOD) (Kliebenstein *et al.*, 1999), ascorbate peroxidase (APX) (Yan *et al.*, 2003; Wang *et al.*, 2005, 2006; Xu *et al.*, 2008), catalase (CAT) (Garg and Manchanda, 2009), glutathione reductase (GR) (Melchiorre *et al.*, 2009), glutathione peroxidase (GPX) (Miller *et al.*, 2003), lipid peroxidase (LPO), guaiacol peroxidase (GPOX) (Asada, 1999), monodehydroascorbate reductase (MDHAR) (Schutzendübel *et al.*, 2001), dehydroascorbate reductase (DHAR) (Chen and Gallie, 2005), glutathione *S*-transferases (GST) (Noctor *et al.*, 2002), and peroxiredoxin (PrxR). Together with the antioxidants ascorbic acid and glutathione (Noctor and Foyer, 1998), these enzymes provide cells with highly efficient machinery for detoxifying $O_2^{\cdot-}$ and H_2O_2 (Bhattacharya *et al.*, 2000).

Stress

Plants have evolved to live in environments where they are often exposed to different stress factors, either alone or in combination. These environmental stress conditions can have a deleterious impact on plant growth and yield. Plant responses to different stresses are highly complex and involve changes at cellular, physiological, and molecular levels.

Abiotic Stress

Abiotic stress is defined as the negative impact of non-living factors on the living organisms in a specific environment. The non-living variable must influence the environment beyond its normal range of variation to adversely affect the population performance or individual physiology of the organism in a significant way. Among the abiotic stresses, water deficit and excess levels of various metals in the soil have received considerable attention (Emamverdian *et al.*, 2015; Pandey *et al.*, 2016). The situation gets compounded when plants face multiple stresses simultaneously in various combinations. The stressful factors either alone or in combination adversely affect germination of seeds, plant growth, or productivity (Farooq *et al.*, 2009; Mahmood *et al.*, 2012). Abiotic stresses include drought, salinity, temperature, flooding, metals, pollution, and so forth.

Biotic Stress

There are various living organisms, such as bacteria, viruses, fungi, parasites, insects, weeds, and cultivated or native plants, that directly or indirectly affect a plant's physiology and its functions. These biotic stresses cause damage to cells by generating ROS in plants.

ROS Production Site

Chloroplast

In higher plants and algae, photosynthesis takes place in chloroplasts, which contain a highly organized thylakoid membrane system that harbors all components of the light-capturing photosynthetic apparatus and provides all structural properties for optimal light harvesting (Pfannschmidt, 2003). In thylakoids, the reaction centers of photosystems I and II (PSI and PSII) are the major generation sites of ROS (Figure 9.1). Various abiotic stresses (heat, salinity, ozone, metals, etc.) lead to the overproduction of ROS in plants; these are highly reactive and toxic and cause damage to proteins (e.g., D1-protein), lipids (membrane structure), carbohydrates, and DNA. The ROS comprise both free radicals – $O_2^{\cdot-}$ (superoxide radical), ·OH (hydroxyl radical), $HO_2\cdot$ (perhydroxy radical), and RO· (alkoxy radicals) – and non-radical (molecular) forms – H_2O_2 (hydrogen peroxide) and $^1O_2$ (singlet oxygen). In chloroplasts, PSI and PSII are the major sites for the production of $^1O_2$ and $O_2^{\cdot-}$. The main generation site of $O_2^{\cdot-}$ is PSI at the level of ferredoxin-NADP$^+$ reductase, but the reduced form of MDHAR also produces $O_2^{\cdot-}$ (del Río, 2015). Furthermore, the generation of $O_2^{\cdot-}$ and ·OH radicals by PSII has also been reported, and P680, pheophytin, and protein QA have been proposed as being responsible for superoxide production (Miller *et al.*, 2009).

Mechanism of ROS Production

The photolysis of water in the chloroplast produces free oxygen molecules in PSII (Equation 9.1), whereas Equation 9.2 indicates that light is necessary for superoxide free radical ($2O_2^{\cdot-}$) production at PSI. With the oxidation of ferredoxin, assisted by iron-sulfur protein under conditions of restricted NADP supply, autoxidation of oxygen occurs, producing $O_2^{\cdot-}$ (Equation 9.3):

$$2H_2O \rightarrow 4e^- + O_2 + 4H^+ \quad (9.1)$$

$$2O_2 + 2e \rightarrow 2O_2^{\cdot-} \quad (9.2)$$

$$Fd_{(red)} + O_2 \rightarrow Fd_{(oxd)} + O_2^{\cdot-} \quad (9.3)$$

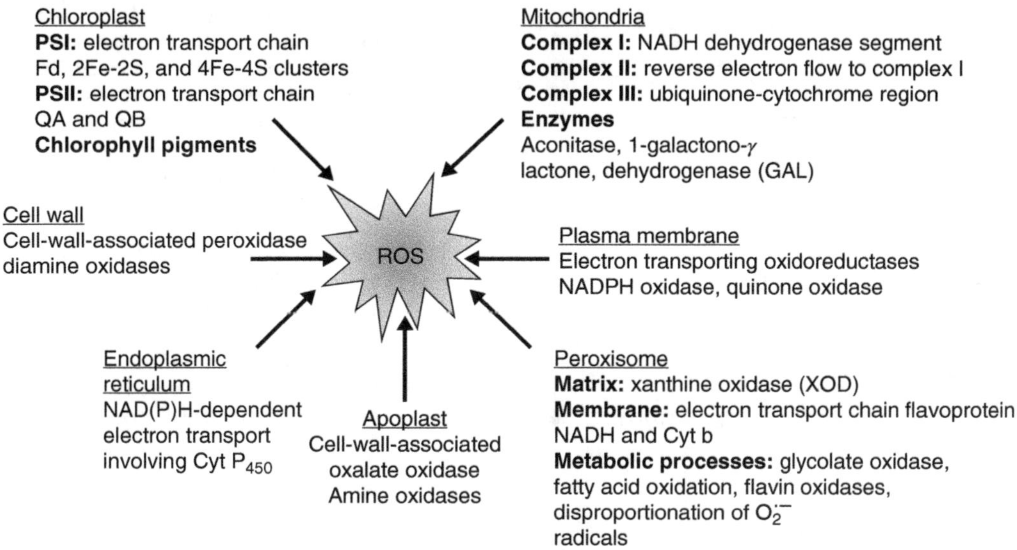

Figure 9.1 Sites of production of reactive oxygen species (ROS) in plants. ROS are produced at several locations in the cell, including chloroplasts, mitochondria, plasma membrane, peroxisomes, apoplast, endoplasmic reticulum, and cell wall. *Source:* Sharma *et al.*, 2012. Reproduced with permission of the *Journal of Botany.*

Reduced ferredoxin reacts with superoxide free radicals to form hydrogen peroxide. This is what Mehler (1951) found in the reaction now named after him (Equation 9.4):

$$Fd_{(red)} + O_2^{\cdot-} + 2H^+ \rightarrow Fd_{(oxd)} + H_2O_2 \quad (9.4)$$

Singlet oxygen is produced in PSII (P680) by the excitation of the ground triplet state ($^3O_2$) into the singlet state ($^1O_2$), as shown in Equation 9.5:

$$^3O_2 + ^3P680^* \rightarrow ^1O_2 + ^3P680 \quad (9.5)$$

In conditions of high illumination or low CO_2, electrons become excited and produce ROS under different stresses (Asada, 2006). According to the Fenton reaction, oxidative stress can lead to the release of intracellular bound iron and copper and generate highly toxic $\cdot OH$ (Halliwell and Gutteridge, 1992).

Mitochondria

ROS are produced as side products of the plant's aerobic metabolism and contribute to biological processes involved in growth and development (del Río, 2015). The formation of ROS has been detected in almost all cellular compartments, including mitochondria (Mittler *et al.*, 2004; Suzuki *et al.*, 2011). Plant mitochondria as "energy factories" are believed to be major sites of ROS production, such as H_2O_2, as well as targets for ROS (Rasmusson *et al.*, 2004).In mitochondria, complex I, ubiquinone, and complex III of the electron transport chain (ETC) are the major sites for the generation of $O_2^{\cdot-}$. Several enzymes present in the mitochondrial matrix can produce ROS. Some of them produce ROS directly, for example, aconitase, whereas some others like 1-galactono-γ-lactone dehydrogenase (GAL), are able to feed electrons to the ETC (Andreyev *et al.*, 2005; Rasmusson *et al.*, 2008).

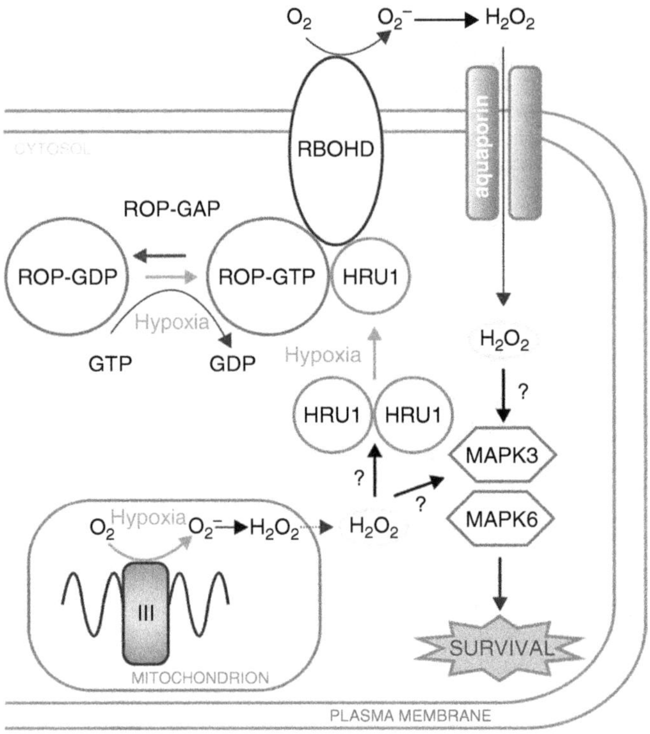

Figure 9.2 Proposed model of reactive oxygen species (ROS) production and amplification under conditions of low O_2 in plant cells (orange arrows indicate hypoxia-dependent mechanisms). GDP, guanosine diphosphate; GTP, guanosine triphosphate; HRU1, hypoxia responsive universal stress protein 1; MAPK, mitogen-activated protein kinase; RBOHD, respiratory burst oxidase homolog D; Rho-like small G-protein of plants (ROP) – GTPase activating protein (GAP); *Source:* Gonzali *et al.*, 2015. Reproduced with permission of *Nature Plants.*

In mitochondria, hypoxia leads to the production of superoxide ($O_2^{\cdot-}$) at mETC complex III; in the intermembrane space $O_2^{\cdot-}$ is dismutated to hydrogen peroxide (H_2O_2), which has a greater chance of reaching the cytosol (Chang *et al.*, 2011). In the cytosol, the hypoxia responsive universal stress protein 1 (HRU1)-HRU1 dimer dissociates under hypoxia via an unknown mechanism. HRU1 migrates to the plasma membrane. HRU1 interacts with the respiratory burst oxidase homolog D (RBOHD) and the hypoxia-activated Rho-like small G protein of plants (GTP-ROP2) (Gonzali *et al.*, 2015). RBOHD produces $O_2^{\cdot-}$, which is rapidly dismutated to H_2O_2. H_2O_2 can migrate from the apoplast to the cytosol, likely through aquaporin channels (Bienert and Chaumont, 2014). H_2O_2 rapidly and transiently activates mitogen activated protein kinases (MPKs), MPK3 and MPK6, which are associated with a higher plant survival under stress (Chang *et al.*, 2011) (Figure 9.2).

Peroxisomes

ROS are also produced inside the peroxisomes, which are single membrane-bounded subcellular organelles with an essentially oxidative type of metabolism; they are probably the major site of intracellular ROS production. Plant peroxisomes play important roles in a variety of metabolic reactions, such as photorespiration, fatty acid β-oxidation, the glyoxylate cycle, and generation–degradation of hydrogen peroxide (Hayashi and Nishimura, 2003; del Río *et al.*, 2002, 2006; Hu *et al.*, 2007). Peroxidases are regarded as stress enzymes, and increased peroxidase activity in plants is attributed to oxidative reactions occurring in cells due to an increase in the level of peroxides and free radicals (Sharma *et al.*, 2010).

Endoplasmic Reticulum

In endoplasmic reticulum, NAD(P)H-dependent electron transport involving cyt P450 produces $O_2^{\cdot-}$ (Mittler, 2002). Organic substrate, RH, reacts first with cyt P450 and then is

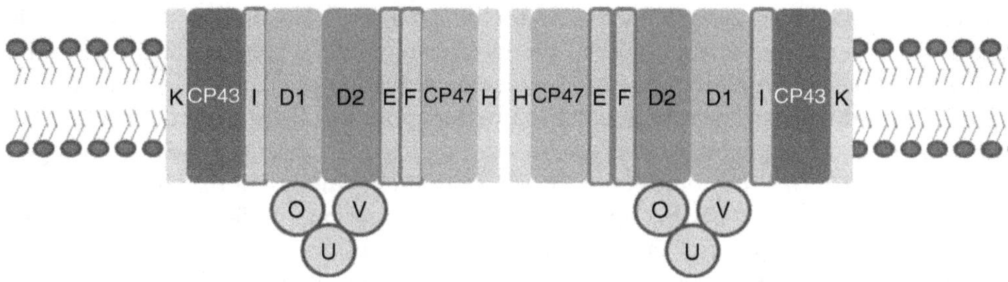

Figure 9.3 Assembly of the dimeric photosystem II (PSII) complex in *Synechocystis* sp. PCC 6803. The low molecular mass (LMM) PsbE, PsbF, PsbH, PsbI, and PsbK subunits and the extrinsic PsbO, PsbU, and PsbV subunits are designated by the appropriate upper case letters. For clarity, assembly factors and many of the LMM subunits are not included. Cytochrome *b*559 is composed of a heterodimer of the PsbE and PsbF subunits. *Source:* Adapted from Nixon *et al.*, 2010.

reduced by a flavoprotein to form a radical intermediate (Cyt P450R$\cdot^-$). Triplet oxygen can readily react with this radical intermediate as each has one unpaired electron. This oxygenated complex (Cyt P450-ROO$^-$) may be reduced by cytochrome *b* or occasionally the complexes may decompose releasing $O_2^{\cdot-}$ (Abouzari and Fakheri, 2015).

Apoplast

The cell wall-associated oxalate oxidase, also known as germin, releases H_2O_2 and CO_2 from oxalic acid (Wojtaszek, 1997). This enzyme is reported to be involved in apoplastic hydrogen peroxide accumulation during interactions between different plant species and fungi (Lane, 2002), that is, biotic stress. Amine oxidase-like enzymes may contribute to defense responses occurring in the apoplast following biotic stress, mainly through H_2O_2 production (Cona *et al.*, 2006).

Cell Walls

The generation of ROS by cell-wall located peroxidases has been shown during the hypersensitive response (HR) triggered in cotton by the bacterium *Xanthomonas campestris* pv. *malvacearum* (Martinez *et al.*, 1998) and in *Arabidopsis* by potassium (K) deficiency stress (Kim *et al.*, 2010).

Plasma Membrane

NAD(P)H-mediated production of $O_2^{\cdot-}$ in plasma membranes is attributed to the action of at least two enzymes, an NADPH oxidase and, in the presence of menadione, a quinone reductase (Heyno *et al.*, 2011).

Effects of ROS

Harmful Effects

One of the major toxic effects of excessive ROS is damage to cellular membranes by the process of lipid peroxidation. Species such as $\cdot$OH, $HO_2^{\cdot-}$, and $OONO^-$, but not $O_2^{\cdot-}$, can extract an H from methylene (-CH_2^-), which creates the carbon radical -$\cdot CH^-$. This carbon radical then attacks other -CH_2^- groups in lipid molecules and creates a chain reaction that alters the fluidity and shape of the membrane. Lipid peroxides can also damage DNA and proteins.

ROS can cause oxidative damage to nuclear, mitochondrial, and chloroplastic DNA. DNA is the cell's genetic material and any damage to the DNA can result in changes in the encoded proteins, which may lead to malfunctions or complete inactivation of the encoded proteins.

Protein oxidation disrupts receptors, enzyme function, and signal transduction pathways. The amino acid tyrosine is particularly prone to attack by ROS, especially reactive nitrogen species such as $OONO^-$ (Ischiropoulos *et al.*, 1992). Potentially important functional sites for oxidation of proteins are the -SH groups, because the formation of -S-S- bonds between different protein strands or parts of the same strand can result in conformational changes in the protein that alter its function. Direct modification involves modulation of a protein's activity through nitrosylation, carbonylation, disulfide bond formation, and glutathionylation. Proteins can be modified indirectly by conjugation with breakdown products of fatty acid peroxidation (Yamauchi *et al.*, 2008).

Beneficial Effects

ROS signaling is shown to be an integral part of the acclimation response of plants in stress conditions. It is used to sense stress due to enhanced ROS production caused by metabolic imbalances, as well as to actively send different signals via enhanced production of ROS at the apoplast by different RBOH proteins. The use of ROS as signaling molecules by plant cells suggests that, during the course of evolution, plants were able to achieve a high degree of control over ROS toxicity and are now using ROS as signaling molecules. Controlling ROS toxicity while enabling ROS such as H_2O_2 or $O_2^{\cdot-}$ to act as signaling molecules appears to require a large gene network composed of at least 152 genes in *Arabidopsis* (Mittler *et al.*, 2004).

In recent years, a new role for ROS has been identified: the control and regulation of biological processes, such as growth, cell cycle, programmed cell death, hormone signaling, biotic and abiotic stress responses, and development (Kovtun *et al.*, 2000; Neill *et al.*, 2002; Kwak *et al.*, 2003).

Photosystem II (PSII): Structure, Function, Damage, and Repair

Photosystem II (PSII) is a highly structured pigment-protein complex in the thylakoid membranes of chloroplasts, consisting of more than 40 polypeptides. PSII initiates photosynthesis and catalyzes the primary photochemistry leading to oxygen evolution and electron flow in oxygenic phototrophs (Mattoo *et al.*, 1989; Barber *et al.*, 1997). The core of the PSII reaction center (P680) complex constitutes membrane-spanning polypeptides D1, D2, cytochrome *b*559, and two low-molecular-weight polypeptides (Vermaas and Ikeuchi, 1991). The heterodimer core of D1/D2 proteins, spanning the thylakoid membrane five times, dominates the reaction center of PSII, coordinating the electron donors (primary and secondary) and acceptors involved in photochemically induced electron transfer from water to plastoquinone. CP43 and CP47, spanning the thylakoid membrane six times, are the proteins carrying antenna chlorophyll molecules involved in the energy transfer from the distal antenna to the reaction center (Bricker, 1990). In addition to the protein subunits, there are chlorophylls, pheophytins, β-carotenes, lipids, plastoquinones, heme and non-heme iron, manganese atoms, calcium atoms (Mn_4Ca cluster), Cl^- ions (in the vicinity of the Mn_4Ca cluster), bicarbonate ions, and detergents in a monomer (Umena *et al.*, 2011).

The reaction center, composed of D1/D2 heterodimer and cyt *b*559, is the primary center for photochemical and subsequent electron transfer reactions. The light energy captured by antenna pigments excites the primary donor P680, followed by the primary charge separation between P680 and the intermediate acceptor, Pheo a (pheophytin a), generating the radical pair (P680+/Pheo). A tightly bound plastoquinone QA (primary acceptor) after transfer of electrons from reduced Pheo generates reduced QA. QA subsequently reduces the secondary acceptor, QB (a loosely bound plastoquinone) through intermediate Fe^{2+}. The second photochemical reaction generates doubly reduced QB, QBH2 (a plastoquinol) through two coupled protonations. The plastoquinol is replaced with a plastoquinone in the lipid bilayer of the thylakoid membranes after getting released from the binding pockets. $P680^+$ on donor side is reduced by an electron from a cluster of four manganese atoms (Mn-cluster) resulting in oxygen evolution. The splitting of two water molecules at the Mn-cluster by accumulating four oxidizing equivalents generates one oxygen molecule and four protons.

Photosystem II (PS II) is prone to various environmental stresses. Strong visible light is the most prominent cause of photoinhibition. Although light is essential for photosynthesis, higher intensity or excessive light decreases the efficiency of energy production. Under low and moderate light levels, the activity of PSII is relatively constant due to the quick repair of light-induced damage preventing deleterious effects. However, under strong light conditions the photodamage to PSII exceeds the capacity for repair. In the photoinhibition mechanism, over-reduction of the acceptor side of PSII takes place inducing charge recombination between the oxidized primary electron donor $P680^+$ and the reduced primary electron acceptor $Pheo^-$, leading to the formation of triplet state P680 (Yoshioka *et al.*, 2006). Subsequently, triplet state P680 reacts with molecular oxygen forming highly reactive singlet oxygen, which is very destructive to the D1 protein. The illumination of PSII generates $P680^+$ and $chlorophyll^+$, endogenous cationic radicals, impairing the oxygen-evolving system by damaging D1 protein. The photodamaged D1 protein of the reaction center must be proteolyzed and then replaced with a functional D1 molecule to repair the damaged PSII due to photoinhibition. Some cellular components are also involved in protection of PSII against photoinhibition. DegP2 and FtsH, two proteases, are involved in degradation of photodamaged D1 protein. Degp2 is a GTP-dependent protease that is associated with the stromal side of thylakoid membrane, and FtsH is an ATP-dependent zinc metalloprotease involved in degradation of photodamaged D1. Several other abiotic and biotic stresses, like heat, cold, drought, and pathogen attack, can generate oxidative stress producing ROS, which causes irreversible damage to the D1 protein of PSII. The deleterious effect of these ROS reduces the maximum efficiency of PSII.

D1 Protein: Biogenesis, Structure, and Assembly

D1 and D2 proteins play a central role in the photochemistry of PSII, acting as a reaction center, carrying the primary and secondary electron donors and acceptors necessary for charge separation. These proteins are composed of five α-helices spanning the thylakoid membrane (named from A to E) with their N- and C-termini exposed to the stromal and luminal sides, respectively (Trebst, 1986; Ferreira *et al.*, 2004; Loll *et al.*, 2005). The chloroplast multigene psbA family encodes D1 protein. A family of three psbA gene (*psb*AI, AII, and AIII) encode for two distinct forms of the PSII D1 protein. The transcripts of *psb*AI predominate and encode for D1 form 1 protein, which under excess excitation is rapidly replaced by transcripts from *psb*AII and *psb*AIII, encoding the D1 form 2 protein (Golden *et al.*, 1986). The transient exchange of D1 forms causes fractional closure of PSII centers thereby reversing the photoinhibition; as the cell acclimates to new growth conditions, D1:2 is again replaced by D1:1 (Clarke *et al.*, 1993, 1995; Campbell *et al.*, 1995). D1 translation takes place on polyribosomes associated with the thylakoid membrane. D1 is synthesized with a 16-amino-acid-long carboxyl-terminal extension, which is C-terminally processed in stromal lamellae by peptidase CtpA (Anbudurai *et al.*, 1994) producing a mature D1 protein of 343 amino acid residues in cyanobacteria (Ivleva *et al.*, 2000). The C-terminal extension consists of 8–9 amino acid residues in eukaryotes (Takahashi *et al.*, 1988). The mature D1 protein is of 32 kDa gets translocated to grana in chloroplasts and provide terminal carboxyl moiety as one of the ligands to (Mn_4-Ca) water oxidation assembly. The D1 binding pocket binds QB, the secondary plastoquinone, through hydrogen bonds with one keto oxygen to the Nδ atom of His215, and with the other keto oxygen to the hydroxyl group of Ser264 and the backbone NH group of Phe265. The ring stacking of the quinone head group with the phenyl side chain of Phe255, and hydrophobic interactions with Phe211, Met214, Phe265, Leu271, and Phe274, further stabilizes the binding (Rea *et al.*, 2009). D1 associates with the low molecular mass (LMM) subunit PsbI and D2 protein, which is associated with cytochrome *b*559, a heterodimer of the PsbE and PsbF subunits. On either side of the above assembly are CP43 and CP47, each containing six transmembrane α-helices, which bind chlorophyll *a* (16 molecules in CP47 N and 13 molecules in CP43) and β-carotene. CP43 further associates with the LMM PsbK subunit, and CP47 associates with the LMM PsbH subunit. CP43 participates with D1 in the ligation of the $CaMn_4$ cluster. Three extrinsic subunits (PsbO, PsbU, and PsbV) are bound to the luminal side of the complex at the D1-D2 dimer contributing to the stability of $CaMn_4$ cluster. This complete assembly associates in dimer form at the thylakoid membrane to form the PSII complex. (Nixon *et al.*, 2010).

Turnover: Reversible Phosphorylation and Dephosphorylation of D1 Proteins

Phosphorylation and dephosphorylation of PSII core proteins, in particular the D1 protein, along with their subsequent degradation are essential components of the quality control and turnover of PSII as the PSII centers are prone to damage at all light intensities (Tikkanen *et al.*, 2008; Fristedt *et al.*, 2009).

There are five highly organized steps of PSII turnover:

1) Reversible phosphorylation and dephosphorylation of PSII core proteins by the STN8 kinase (Bonarddi *et al.*, 2005; Vainonen *et al.*, 2005) and PBCP phosphatase (Samol *et al.*, 2012), respectively.
2) Monomerization and disassembly of PSII complexes.
3) Proteolytic degradation of damaged D1 protein.
4) De novo synthesis of D1 protein and subsequent insertion into the partially disassembled PSII complex.
5) Reassembly of the PSII complex.

The phosphorylation of D1 protein prevents accumulation of photodamaged PSII complexes in the thylakoid membranes. Simultaneously, the conformation of D1 protein is altered resulting in reduced ability of quinones and herbicides to bind the QB site (Giardi *et al.*, 1992). ROS production is modulated due to the conformational change in the QB-binding pocket caused by the phosphorylation of D1 protein; the QB site plays an important role in the production of $O_2^{\cdot-}$ (Zhang *et al.*, 2003).

ROS-Mediated Damage to D1 Protein and Photophosphorylation

ROS are normal products of cellular metabolism. They comprise a group of free radicals, reactive molecules, and ions that are derived from O_2; about 1% of O_2 consumed by plants is diverted to produce ROS (Asada and Takahashi, 1987). Several locations in chloroplasts generate various forms of ROS; the main source is the electron transport chain (ETC) in PSI and PSII. In the light, charge recombination at the reaction center of PSII generates highly reactive $^1O_2$ via triplet chlorophyll reacting with the D1 protein of photosystem II (Hatz *et al.*, 2007; Vass, 2011). In PSII, leakage of electrons from QA and QB, on the acceptor side of the ETC, to O_2 contributes to the production of $O_2^{\cdot-}$ (Cleland and Grace, 1999). Earlier studies indicated that $^1O_2$ can be directly involved in damage and cleavage of the D1 reaction center subunits (Mishra and Ghanotakis, 1994; Okada *et al.*, 1996), which is further supported by singlet oxygen involvement as an intermediate in the triggering of D1 protein degradation (Trebst *et al.*, 2002). Inactivation of PSII can also take place under illumination by two mechanisms: (i) the Mn_4Ca cluster of water oxidation is unable to keep up with the rate of withdrawal of electrons by $P680^+\cdot$; and (ii) light-dependent impairment of the Mn_4Ca cluster leads to accumulation of $P680^+\cdot$ and Tyr-$Z^{+\cdot}$ ($YZ^{+\cdot}$) of D1 protein (strong oxidizing radicals), damaging their protein environment (Jegerschold *et al.*, 1990; Debus, 2001; Diner, 2001; Hakala-Yatkin *et al.*, 2010). These radicals ($O_2^{\cdot-}$, $P680^{+\cdot}$, and Tyr-$Z^{+\cdot}$) cause damage preferentially to D1 then D2, CP43, and PsbH subunits of PSII (Jansen *et al.*, 1996; Bergantino *et al.*, 2003; Nelson *et al.*, 2014). Tikkanen *et al.* (2008) suggested that phosphorylation of PSII reaction center protein D1, generally termed photophosphorylation, is a prerequisite for repair of damaged PSII. The efficient migration of damaged PSII complexes from grana to stroma lamellae for repair requires photophosphorylation of damaged D1 protein. The damaged D1 protein is dephosphorylated and successively degraded by a D1-specific protease in stroma lamellae. The phosphorylation of D1 protein prevents oxidative damage to photosynthetic proteins in the thylakoid membranes by photodamaged PSII. The altered conformation of D1 protein after phosphorylation also results in reduced ability of quinones to bind to the QB site, thereby modulating $O_2^{\cdot-}$ production and thus preventing further ROS production (Chen *et al.*, 2012).

Mechanism of PSII Repair

The unstacking of highly stacked grana under high light levels, accompanied by outward bending of thylakoids, harboring PSII, causes damage to the PSII core and initiates the PSII repair cycle (Yoshioka-Nishimura *et al.*, 2014). The damaged PSII complex under high light undergoes phosphorylation of D1 protein, catalyzed mainly by the STN8 kinase. Phosphorylation triggers the disassembly of the PSII super-complex in the grana core (Vainonen *et al.*, 2005; Tikkanen *et al.*, 2008; Wunder *et al.*, 2013). The phosphorylated and disassembled PSII migrates from the stacked grana to the stroma thylakoids through the grana margins. After migration the PSII complexes are dephosphorylated by the PSII core phosphatase (PBCP) (Samol *et al.*, 2012), which is necessary for degradation of the D1 protein (Puthiyaveetil *et al.*, 2014). FtsH proteases at grana margins and Deg proteases inside the lumen recognize dephosphorylated PSII (D1 protein) and degrade damaged D1 protein immediately (Yoshioka *et al.*, 2010; Kato *et al.*, 2012; Kirchhoff, 2013; Yoshioka-Nishimura *et al.*, 2014). In the final stage of the repair cycle, newly synthesized D1 is inserted into the PSII and reassembly of the PSII super-complex is accomplished in unstacked thylakoids (Aro *et al.*, 2005; Danielsson and Albertsson, 2009).

Irreversible Damage to D1 Protein

Under various environmental stresses, ROS production increases due to imbalance of cellular homeostasis, causing oxidative damage. Environmental stresses like drought, salinity, and extreme temperatures, individually as well as in combination with high light stress, limit CO_2 fixation causing increased ROS production. The limitation to CO_2 fixation leads to reduced regeneration of $NADP^+$ through the Calvin–Benson cycle. Due to lack of the electron acceptor $NADP^+$, over-reduction of the photosynthetic ETC occurs, which leads to enhanced leakage of electrons to O_2. The inhibition of photosynthetic activity under stress is due to an imbalance between light capture and its utilization, resulting in dissipation of excess light energy in the

PSII core leading to generation of ROS (Foyer and Noctor, 2000). Irreversible damage to D1 protein can be caused due to excessive production of ROS under various stresses or high light intensity. The damage exceeds the repair capability as photo-inhibition prevails over protection and repair (Barber and Andersson, 1992; Aro *et al.*, 1993). Irreversible aggregation or cross-linking of the photodamaged reaction center binding D1 protein is the main cause of irreversible damage (Yamamoto *et al.*, 2008; Khatoon *et al.*, 2009; Chan *et al.*, 2012). The irreversible aggregation of the D1 protein is the result of covalent cross-linking of the protein with nearby polypeptides after photo-oxidative damage to the D1 protein. The three cross-linked products of the D1 protein are: (i) the cross-linked products between D1 and D2 (Yamamoto *et al.*, 2008); (ii) D1 and the α-subunit of cytochrome *b*559 (Barbato *et al.*, 1995); and (iii) D1 and the core antenna chlorophyll binding protein CP43 (Yamamoto and Akasaka, 1995; Yamamoto *et al.*, 2008). The cross-linked products formed in the thylakoids are difficult to remove by proteases (Ishikawa *et al.*, 1999; Ferjani *et al.*, 2001; Lupinkova and Komenda, 2004). ROS inhibit de novo synthesis of D1 protein required to maintain the turnover of D1 protein and PSII repair (Nishiyama *et al.*, 2001, 2004). The inhibition of de novo D1 protein synthesis takes place at the level of psbA gene translation (Trebitsh, 2001; Zhang and Aro, 2002). ROS-induced inhibition of repair is reported to be mainly by interruption in redox signaling for the reduction of disulfide bonds in elongation factors (EF-G or EF-2) required for activation of elongation process in translation (Yamazaki *et al.*, 2004; Kojima *et al.*, 2007). The rate of these ROS-mediated damages exceeds the rate of repair, causing photoinhibition and resulting in the reduction of the maximum quantum efficiency of photosynthesis, thus directly affecting growth and yield.

Plant Protection Against ROS-Mediated Damage

Plants have many adaptive mechanisms for protecting PSII, specifically D1 protein, from ROS-mediated photodamage. Excess energy is dissipated as heat through non-photochemical quenching (NPQ), serving as safety valve; this requires a chlorophyll-binding protein of PSII, PsbS, zeaxanthin, and a carotenoid involved in the xanthophyll cycle for preventing oxidative damage (Bugos and Yamamoto, 1996; Li *et al.*, 2000; Demmig-Adams and Adams, 2002; Niyogi *et al.*, 2005). NPQ appears to play a role in preventing the generation of ROS by reducing PSII-mediated electron transport. Mullineaux (2005) suggested that partial unstacking or swelling of thylakoids under stress or high illumination may prevent possible energy transfer from LHCII in one membrane layer to the adjacent membrane layer, and the over-excitation of PSII. Partial unstacking and swelling of the thylakoids increases the area of the grana margins for facilitating proteolytic degradation of photodamaged D1 proteins (Yamamoto *et al.*, 2013). Proline, β-carotene, tocopherol, and plastoquinone play the roles of $^1O_2$ scavengers or quenchers (Alia *et al.*, 1991; Loll *et al.*, 2005). The inhibition of rubisco activase under stress and high light can be prevented by glycine-betaine, thus ameliorating the photo-inhibition due to oxidative stress (Yang *et al.*, 2005, 2007). Glycine-betaine protects the oxygen-evolving complex of PSII and dissociation of the extrinsic proteins under stress by stabilizing the conformation of proteins (Papageorgiou and Murata, 1995). Gill and Tuteja (2010) and Foyer and Shigeoka (2011) suggested the role of antioxidants like ascorbate, glutathione, and so forth, and antioxidative enzymes like Cu/Zn superoxide dismutase, catalase, ascorbate peroxidase, glutathione peroxidase, and so forth in scavenging of ROS. A chloroplast targeted heat-shock protein 70 (HSP70) contributes to the photoprotection and repair of PSII, participating in transmembrane protein transport, and they prevent aggregation of denatured proteins and allow the disassembly of multimolecular complexes (Schroda *et al.*, 1999). The redistribution of chloroplasts in leaves changes the rotation of the long axis toward light, thereby protecting PSII from high-intensity photons. Changes in fatty acid unsaturation in lipids of thylakoid membranes during the recovery of PSII after photodamage have been postulated by many studies (Tasaka *et al.*, 1996; Nishiyama *et al.*, 1999). Mehta *et al.* (2011) reported another mechanism of photoprotection involving changes in the heterogeneity of the antenna and acceptor side of PSII for the photosynthetic machinery under stress.

Conclusion

Components of the photosynthetic apparatus receive excess light energy that can ultimately engender photo-inhibition. Therefore chloroplasts are equipped with molecular systems to minimize accumulation of photodamaged proteins. Protein quality control plays an important role in the photosynthetic apparatus because its components receive excess light energy and are susceptible to photo-oxidative damage.

The coordinated PSII repair cycle (PSII disassembly, D1 degradation and synthesis, and PSII reassembly) is necessary to mitigate photo-inhibition. D1 modification, fragmentation, and cross-linking were suppressed by ROS scavengers, supporting the direct role of ROS in these phenomena. Obviously, all of three ROS components, namely superoxide anion, hydrogen peroxide, and hydroxyl radical, are responsible for the degradation of oxidized D1 protein, and the protection of the D1 protein against degradation by phosphorylation is accompanied by the inhibition of superoxide anion production. The phosphorylation of D1 is apparently valuable in these two ways: in addition to the migration of damaged PSII from grana to stroma thylakoids, protection against undesirable proteolysis might well be required.

Acknowledgments

The authors are grateful to the Director, CSIR-NBRI, for providing the necessary facilities. S.K.G. and M.S. are grateful to CSIR and UGC, respectively, for senior research fellowships.

References

Abouzari A, Fakheri BA. (2015) Reactive oxygen species: generation, oxidative damage, and signal transduction. *Intl J Life Sci* 9: 3–17.

Alia, Saradhi PP, Mohanty P. (1991) Proline enhances primary photochemical activities in isolated thylakoid membranes of Brassica juncea by arresting photoinhibitory damage. *Biochem Biophys Res Commun* 181: 1238–1244.

Anbudurai PR, Mor TS, Ohad I, Shestakov SV, Pakrasi HB. (1994) The ctpA gene encodes the C-terminal processing protease for the D1 protein of the photosystem II reaction center complex. *Proc Natl Acad Sci USA* 91: 8082–8086.

Andreyev AY, Kushnareva YE, Starkov AA. (2005) Mitochondrial metabolism of reactive oxygen species. *Biochemistry (Moscow)* 70: 200–214.

Apel K, Hirt H. (2004) Reactive oxygen species: metabolism, oxidative stress, and signal transduction. *Annu Rev Plant Biol* 55: 373–399.

Aro EM, Virgin I, Andersson B. (1993) Photoinhibition of photosystem II. Inactivation, protein damage and turnover. *Biochim Biophys Acta* 1143: 113–134.

Aro EM, Suorsa M, Rokka A, *et al.* (2005) Dynamics of photosystem II: a proteomic approach to thylakoid protein complexes. *J Exp Bot* 56: 347–356.

Asada K. (1999) The water-water cycle in chloroplasts: scavenging of active oxygens and dissipation of excess photons. *Annu Rev Plant Physiol Plant Mol Biol* 50: 601–639.

Asada K. (2006) Production and scavenging of reactive oxygen species in chloroplasts and their functions. *Plant Physiol,* 141(2): 391–396.

Asada K, Takahashi M. (1987) Production and scavenging of active oxygen in photosynthesis. In: Kyle DJ, Osmond CB, Arntzen CJ (eds), *Photoinhibition: Topics of Photosynthesis,* 9th edn. Elsevier, Amsterdam, The Netherlands, pp. 227–287.

Baena-Gonzalez E, Aro EM. (2002) Biogenesis, assembly and turnover of photosystem II units. *Philos Trans R Soc Lond B Biol Sci* 357: 1451–1459; discussion 1459–1460.

Baker NR. (1991) A possible role for photosystem II in environmental perturbations of photosynthesis. *Physiol Plant* 81: 563–570.

Barbato R, Friso G, Ponticos M, Barber J. (1995) Characterization of the light-induced cross-linking of the alpha-subunit of cytochrome b559 and the D1 protein in isolated photosystem II reaction centers. *J Biol Chem* 270: 24032–24037.

Barber J, Andersson B. (1992) Too much of a good thing: light can be bad for photosynthesis. *Trends Biochem Sci* 17: 61–66.

Barber J, Nield J, Morris EP, Zheleva D, Hankamer B. (1997) The structure, function and dynamics of photosystem two. *Physiol Plant* 100: 817–827.

Bergantino E, Brunetta A, Touloupakis E, Segalla A, Szabo I, Giacometti GM. (2003) Role of the PSII-H subunit in photoprotection: novel aspects of D1 turnover in *Synechocystis* 6803. *J Biol Chem* 278: 41820–41829.

Bhattacharya A, Ghosal S, Bhattacharya SK. (2000) Antioxidant activity of tannoid principles of *Emblica officinalis* (amla) in chronic stress induced changes in rat brain. *Ind J Exp Biol* 38(09): 877–880.

Bienert GP, Chaumont F. (2014) Aquaporin-facilitated transmembrane diffusion of hydrogen peroxide. *Biochem Biophys Acta* 1840: 1596–1604.

Bonarddi V, Pesaresi P, Becker T, *et al.* (2005) Photosystem II core phosphorylation and photosynthetic acclimation require two different protein kinases. *Nature* 43: 1179–1182.

Bricker TM. (1990) The structure and function of CPa-1 and CPa-2 in photosystem II. *Photosynth Res* 24: 1–13.

Bugos RC, Yamamoto HY. (1996) Molecular cloning of violaxanthin de-epoxidase from romaine lettuce and expression in *Escherichia coli. Proc Natl Acad Sci USA* 93: 6320–6325.

Campbell D, Zhou G, Gustafsson P, Oquist G, Clarke AK. (1995) Electron transport regulates exchange of two forms of photosystem II D1 protein in the cyanobacterium *Synechococcus. EMBO J* 14: 5457–5466.

Chan T, Shimizu Y, Pospisil P, *et al.* (2012) Quality control of photosystem II: lipid peroxidation accelerates photoinhibition under excessive illumination. *PLoS ONE* 7: 52100.

Chang IH, Cheng KT, Huang PC, Lin YY, Cheng LJ, Cheng TS. (2012) Oxidative stress in greater duckweed (*Spirodelapolyrhiza*) caused by long-term NaCl exposure. *Acta Physiol Plant* 34: 1165–1176.

Chang R, Jang CJH, Branco-Price C, Nghiem P, Bailey-Serres J. (2011) Transient MPK6 activation in response to oxygen deprivation and reoxygenation is mediated by mitochondria and aids seedling survival in *Arabidopsis. Plant Mol Bio* 78: 109–122.

Chatterjee S, Kundu S, Sengupta S, Bhattacharyya A. (2009) Divergence to apoptosis from ROS induced cell cycle arrest: effect of cadmium. *Mutation Res* 663: 22–31.

Chen L, Jia H, Tian Q, *et al.* (2012) Protecting effect of phosphorylation on oxidative damage of D1 protein by down-regulating the production of superoxide anion in photosystem II membranes under high light. *Photosynth Res* 112: 141–148.

Chen Z, Gallie DR. (2005) Increasing tolerance to ozone by elevating foliar ascorbic acid confers greater protection against ozone than increasing avoidance. *Plant Physiol* 138: 1673–1689.

Clarke AK, Soitamo A, Gustafsson P, Oquist G. (1993) Rapid interchange between two distinct forms of cyanobacterial photosystem II reaction-center protein D1 in response to photoinhibition. *Proc Natl Acad Sci USA* 90: 9973–9977.

Clarke AK, Campbell D, Gustafsson P, Oquist G. (1995) Dynamic responses of photosystem II and phycobilisomes to changing light in the cyanobacterium *Synechococcus* sp. PCC 7942. *Planta* 197: 553–562.

Cleland RE, Grace SC (1999) Voltammetric detection of superoxide production by photosystem II. *FEBS Lett* 457: 348–352.

Cona A, Rea G, Angelini R, Federico R, Tavladoraki P. (2006) Functions of amine oxidases in plant development and defence. *Trends Plant Sci* 11: 80–88.

Danielsson R, Albertsson PA. (2009) Fragmentation and separation analysis of the photosynthetic membrane from spinach. *Biochim Biophys Acta* 1787: 25–36.

Debus RJ. (2001) Amino acid residues that modulate the properties of tyrosine YZ and the manganese cluster in the water oxidizing complex of photosystem II. *Biochim Biophys Acta* 1503: 164–186.

del Río LA. (2015) ROS and RNS in plant physiology: an overview. *J Exp Bot* 66: 2827–2837.

del Río LA, Corpas FJ, Sandalio LM, Palma JM, Gómez M, Barroso JB. (2002) Reactive oxygen species, antioxidant systems and nitric oxide in peroxisomes. *J Exp Bot* 53(372): 1255–1272.

del Río LA, Sandalio LM, Corpas FJ, Palma JM, Barroso JB. (2006) Reactive oxygen species and reactive nitrogen species in peroxisomes. Production, scavenging, and role in cell signalling. *Plant Physiol* 141: 330–335.

Demmig-Adams B, Adams III WW. (2002) Antioxidants in photosynthesis and human nutrition. *Science* 298: 2149–2153.

Díaz-Vivancos P, Rubio M, Mesonero V, *et al.* (2006) The apoplastic antioxidant system in *Prunus*: response to long-term plum pox virus infection. *J Exp Bot* 57: 3813–3824.

Diner BA. (2001) Amino acid residues involved in the coordination and assembly of the manganese cluster of photosystem II. Proton-coupled electron transport of the redox-active tyrosines and its relationship to water oxidation. *Biochim Biophys Acta* 1503: 147–163.

Edelman M, Mattoo AK. (2008) D1-protein dynamics in photosystem II: the lingering enigma. *Photosynth Res* 98: 609–620.

Edreva A. (2005) Generation and scavenging of reactive oxygen species in chloroplasts: a submolecular approach. *Agricul Eco Env* 106: 119–133.

Emamverdian A, Ding Y, Mokhberdoran F, Xie Y. (2015) Heavy metal stress and some mechanisms of plant defense response. *Scientific World Journal* 26: 2015.

Farooq MA, Wahid N, Kobayashi D, Fujita BSMA. (2009) Plant drought stress: effects, mechanisms and management. *Agron Sustain Dev* 29: 185–212.

Ferjani A, Abe S, Ishikawa Y, *et al.* (2001) Characterization of the stromal protease(s) degrading the cross-linked products of the D1 protein generated by photoinhibition of photosystem II. *Biochim Biophys Acta* 1503: 385–395.

Ferreira KN, Iverson TM, Maghlaoui K, Barber J, Iwata S. (2004) Architecture of the photosynthetic oxygen-evolving center. *Science* 303: 1831–1838.

Foreman J, Demidchik V, Bothwell JH, *et al.* (2003) Reactive oxygen species produced by NADPH oxidase regulate plant cell growth. *Nature* 422: 442–446.

Foyer CH, Noctor G. (2000) Oxygen processing in photosynthesis: regulation and signalling. *New Phytol* 146: 359–388.

Foyer CH, Noctor G. (2003) Redox sensing and signalling associated with reactive oxygen in chloroplasts, peroxisomes and mitochondria. *Physiol Plant* 119: 355–364.

Foyer CH, Shigeoka S. (2011) Understanding oxidative stress and antioxidant functions to enhance photosynthesis. *Plant Physiol* 155: 93–100.

Fristedt R, Willig A, Granath P, Crèvecoeur M, Rochaix JD, Vener AV. (2009) Phosphorylation of photosystem II controls functional macroscopic folding of photosynthetic membranes in *Arabidopsis*. *Plant Cell* 21: 3950–3964.

Garg N, Manchanda G. (2009) ROS generation in plants: boon or bane? *Plant Biosys* 143: 8–96.

Giardi MT, Rigoni F, Barbato R. (1992) Photosystem II core phosphorylation heterogeneity, differential herbicide binding, and regulation of electron transfer in photosystem II preparations from spinach. *Plant Physiol* 100: 1948–1954.

Gill SS, Tuteja N. (2010) Reactive oxygen species and antioxidant machinery in abiotic stress tolerance in crop plants *Plant Physiol Biochem* 48: 909–930.

Golden SS, Brusslan J, Haselkorn R. (1986) Expression of a family of psbA genes encoding a photosystem II polypeptide in the cyanobacterium Anacystis nidulans R2. *EMBO J* 5: 2789–2798.

Gonzali E, Loreti E, Cardarelli F, Novi G, Parlanti S, Pucciariello C, Perata P. (2015) The Universal Stress Protein HRU1 mediates ROS homeostasis under anoxia. *Nature Plants* 1: 15151.

Habib SH, Kausar H, Saud HM. (2016) Plant growth-promoting rhizobacteria enhance salinity stress tolerance in okra through ROS-scavenging enzymes. *Biomed Res Int* 2016: 6284547.

Hakala-Yatkin M, Mäntysaari M, Mattila H, Tyystjärvi E. (2010) Contributions of visible and ultraviolet parts of sunlight to photoinhibition. *Plant Cell Physiol* 51: 1745–1753.

Halliwell B, Gutteridge JMC. (1989) *Free Radicals in Biology and Medicine*. Clarendon Press, Oxford.

Halliwell B, Gutteridge JMC. (1992) *Free Radicals in Biology and Medicine*, 3rd edn. Oxford University Press, Oxford.

Hatz S, Lambert JDC, Ogilby PR. (2007) Measuring the lifetime of singlet oxygen in a single cell: addressing the issue of cell viability. *Photochem Photobiol Sci* 6: 1106–1116.

Hayashi M, Nishimura M. (2003) Entering a new era of research on plant peroxisomes. *Curr Opin Plant Biol* 6: 577–582.

Hernández JA, Ferrer MA, Jiménez A, Barceló AR, Sevilla F. (2001) Antioxidant systems and $O_2^{\cdot-}/H_2O_2$ production in the apoplast of pea leaves. Its relation with salt-induced necrotic lesions in minor veins. *Plant Physiol* 127: 817–831.

Heyno E, Mary V, Schopfer P, Krieger-Liszkay A. (2011) Oxygen activation at the plasma membrane: relation between superoxide and hydroxyl radical production by isolated membranes. *Planta* 234: 35–45.

Hu JP. (2007) Toward understanding plant peroxisome proliferation. *Plant Sig Behav* 2: 308–310.

Ischiropoulos H, Zhu L, Chen J, *et al.* (1992) Peroxynitrite-mediated tyrosine nitration catalysed by superoxide dismutase. *Arch Biochem Biophys* 298: 431–437.

Ishikawa Y, Nakatani E, Henmi T, *et al.* (1999) Turnover of the aggregates and cross-linked products of the D1 protein generated by acceptor-side photoinhibition of photosystem II. *Biochim Biophys Acta* 1413: 147–158.

Ivleva NB, Shestakov SV, Pakrasi HB. (2000) The carboxyl-terminal extension of the precursor D1 protein of photosystem II is required for optimal photosynthetic performance of the cyanobacterium *Synechocystis* sp. PCC 6803. *Plant Physiol* 124: 1403–1412.

Jacobson MD. (1996) Reactive oxygen species and programmed cell death. *Trends Biochem Sci* 21:83–86.

Jaleel CA, Riadh K, Gopi R, *et al.* (2009) Antioxidant defense responses: physiological plasticity in higher plants under abiotic constraints. *Acta Physiol Planta* 31: 427–436.

Jansen MA, Gaba V, Greenberg BM, Mattoo AK, Edelman M. (1996) Low threshold levels of ultraviolet-B in a background of photosynthetically active radiation trigger rapid degradation of the D2 protein of photosystem-II. *Plant J* 9: 693–699.

Jegerschold C, Virgin I, Styring S. (1990) Light-dependent degradation of the D1 protein in photosystem II is accelerated after inhibition of the water splitting reaction. *Biochemistry* 29: 6179–6186.
Kato Y, Sakamoto W. (2009) Protein quality control in chloroplasts: a current model of D1 protein degradation in the photosystem II repaircycle. *J Biochem* 146: 463–469.
Kato Y, Sun X, Zhang L, Sakamoto W. (2012) Cooperative D1 degradation in the photosystem II repair mediated by chloroplastic proteases in *Arabidopsis. Plant Physiol* 159: 1428–1439.
Khatoon M, Inagawa K, Pospisil P, *et al.* (2009) Quality control of photosystem II: thylakoid unstacking is necessary to avoid further damage to the D1 protein and to facilitate D1 degradation under light stress in spinach thylakoids. *J Biol Chem* 284: 25343–25352.
Kim MJ, Ciani S, Schachtman DP. (2010) A peroxidase contributes to ROS production during *Arabidopsis* root response to potassium deficiency. *Mol Plant* 3: 420–427.
Kirchhoff H. (2013) Structural constraints for protein repair in plant photosynthetic membranes. *Plant Signal Behav* 8: 23634.
Kliebenstein DJ, Dietrich RA, Martin AC, Last RL, Dangl JL. (1999) LSD1 regulates salicylic acid induction of copper zinc superoxide dismutase in Arabidopsis thaliana. *Mol Plant-Microbe Interac* 12: 1022–1026.
Kojima K, Oshita M, Nanjo Y, *et al.* (2007) Oxidation of elongation factor G inhibits the synthesis of the D1 protein of photosystem II. *Mol Microbiol* 65: 936–947.
Kovtun Y, Chiu WL, Tena G, Sheen J. (2000) Functional analysis of oxidative stress activated mitogen-activated protein kinase cascade in plants. *Proc Natl Acad Sci USA* 97: 2940–2945.
Kwak JM, Mori IC, Pei ZM, *et al.* (2003) NADPH oxidase AtrbohD and AtrbohF genes function in ROS-dependent ABA signaling in *Arabidopsis. EMBO J* 22: 2623–2633.
Lane BG. (2002) Oxalate, germins, and higher-plant pathogens. *IUBMB Life* 53: 67–75.
Li X-P, Björkman O, Shih C, *et al.* (2000) A pigment-binding protein essential for regulation of photosynthetic light harvesting. *Nature* 403: 391–395.
Loll B, Kern J, Zouni A, Saenger W, Biesiadka J, Irrgang KD. (2005) The antenna system of photosystem II from *Thermosynechococcus elongatus* at 3.2 A resolution. *Photosynth Res* 86: 175–184.
Lupinkova L, Komenda J. (2004) Oxidative modifications of the photosystem II D1 protein by reactive oxygen species: from isolated protein to cyanobacterial cells. *Photochem Photobiol* 79: 152–162.
Mahmood S, Wahid A, Rasheed R, Hussain I, Basra SMA. (2012) Possible antioxidative role of endogenous vitamins biosynthesis in heat stressed maize (*Zea mays*). *Int J Agric Biol* 14: 705–712.
Martinez C, Montillet JL, Bresson E, *et al.* (1998) Apoplastic peroxidase generates superoxide anions in cells of cotton cotyledons undergoing the hypersensitive reaction to *Xanthomonas campestris* pv. *malvacearum* 18. *Mol Plant-Microbe Interact* 11: 1038–1047.
Mattoo AK, Marder J, Edelman M. (1989) Dynamics of the photosystem II reaction center. *Cell* 56: 241–246.
Mehler AH. (1951) Studies on reactions of illuminated chloroplasts: I. Mechanism of the reduction of oxygen and other hill reagents. *Arch Biochem Biophys* 33: 65–77.
Mehta P, Kreslavski V, Bharti S, Allakhverdiev SI, Jajoo A. (2011) Analysis of salt stress induced changes in photosystem II heterogeneity by prompt fluorescence and delayed fluorescence in wheat (*Triticum aestivum*) leaves. *J Photochem Photobiol B Biol* 104: 308–313.
Melchiorre M, Robert G, Trippi V, Racca R, Lascano HR. (2009) Superoxide dismutase and glutathione reductase overexpression in wheat protoplast: photooxidative stress tolerance and changes in cellular redox state. *Plant Growth Regul* 57: 57–68.

Miller G, Mittler R. (2006) Could heat shock transcription factors function as hydrogen peroxide sensors in plants? *Ann Bot* 98: 279–288.

Miller G Schlauch K, Tam R, *et al.* (2009) The plant NADPH oxidase RBOHD mediates rapid systemic signaling in response to diverse stimuli. *Sci Signal* 2: 45.

Miller G, Suzuki N, Ciftci-Yilmaz S, Mittler R. (2010) Reactive oxygen species homeostasis and signalling during drought and salinity stresses. *Plant Cell Environ* 33: 453–467.

Miller PR, Gan Y, McConkey BG, McDonald CL. (2003) Pulse crops for the northern Great Plains. *Agron J* 95(4): 980–986.

Miller S, Walker SW, Arthur JR, *et al.* (2002) Selenoprotein expression in endothelial cells from different human vasculature and species. *Biochim Biophys Acta* 1588: 85–93.

Mishra NP, Ghanotakis DF. (1994) Exposure of a photosystem II complex to chemically generated singlet oxygen results in D1 fragments similar to ones observed during aerobic photoinhibition. *Biochim Biophys Acta* 1187: 296–300.

Mittler R. (2002) Oxidative stress, antioxidants and stress tolerance. *Trends Plant Sci* 7: 405–410.

Mittler R, Vanderauwera S, Gollery M, Van Breusegem F. (2004) Reactive oxygen gene network of plants. *Trends Plant Sci* 9: 490–498.

Mittler R, Vanderauwera S, Suzuki N, *et al.* (2011) ROS signaling: the new wave? *Trends Plant Sci* 16: 300–309.

Møller IM, Jensen PE, Hansson A. (2007) Oxidative modifications to cellular components in plants. *Annu Rev Plant Biol* 58: 459–481.

Mullineaux CW. (2005) Function and evolution of grana. *Trends Plant Sci* 10: 521–525.

Nanba O, Satoh K. (1987) Isolation of a photosystem II reaction center consisting of D-1 and D-2 polypeptides and cytochrome b-559. *Proc Natl Acad Sci USA* 84: 109–112.

Neill S, Desikan R, Hancock J. (2002) Hydrogen peroxide signalling. *Curr Opin Plant Biol* 5: 388–395.

Nelson CJ, Alexova R, Jacoby RP, Millar AH. (2014) Proteins with high turnover rate in barley leaves estimated by proteome analysis combined with in planta isotope labeling. *Plant Physiol* 166: 91–108.

Nelson N, Yocum CF. (2006) Structure and function of photosystems I and II. *Annu Rev Plant Biol* 57: 521–565.

Nishiyama Y, Los' DA, Murata N. (1999) PsbU, a protein associated with photosystem II, is required for the acquisition of cellular thermotolerance in *Synechococcus* species PCC 7002. *Plant Physiol* 120: 301–308.

Nishiyama Y, Yamamoto H, Allakhverdiev SI, Inaba M, Yokota A, Murata N. (2001) Oxidative stress inhibits the repair of photodamage to the photosynthetic machinery. *EMBO J* 20: 5587–5594.

Nishiyama Y, Allakhverdiev SI, Yamamoto H, Hayashi H, Murata N. (2004) Singlet oxygen inhibits the repair of photosystem II by suppressing translation elongation of the D1 protein in *Synechocystis* sp. PCC 6803. *Biochemistry* 43: 11321–11330.

Nixon PJ, Michoux F, Yu J, Boehm M Komenda J. (2010) Recent advances in understanding the assembly and repair of photosystem II. *Ann Bot* 106: 1–16.

Niyogi X-PLi, Rosenberg V, Jung H-S. (2005) Is PsbS the site of non-photochemical quenching in photosynthesis? *J Exp Bot* 56: 375–382.

Noctor G, Foyer CH. (1998) Ascorbate and glutathione: keeping active oxygen under control. *Annu Rev Plant Physiol Plant Mol Biol* 49: 249–279.

Noctor G, Gomez L, Vanacker H, Foyer CH. (2002) Interactions between biosynthesis, compartmentation, and transport in the control of glutathione homeostasis and signalling. *J Exp Bot* 53: 1283–1304.

Okada K, Ikeuchi M, Yamamoto N, Ono TA, Miyao M. (1996) Selective and specific cleavage of the D1 and D2 proteins of photosystem II by exposure to singlet oxygen: factors responsible for the susceptibility to cleavage of the proteins *Biochim Biophys Acta* 1274: 73–79.

Ort RD, Baker NR. (2002) A photoprotective role for O_2 as an alternative electronic sink in photosynthesis? *Curr Opin Plant Biol* 5:193–98.

Pandey P, Srivastav, RK, Rajpoot R, *et al.* (2016) Water deficit and aluminum interactive effects on generation of reactive oxygen species and responses of antioxidative enzymes in the seedlings of two rice cultivars differing in stress tolerance. *Environ Sci Pollut Res* 23: 1516.

Papageorgiou GC, Murata N. (1995) The unusually strong stabilizing effects of glycine betaine on the structure and function of the oxygen-evolving photosystem II complex. *Photosynth Res* 44: 243–252.

Perez IB, Brown PJ. (2014) The role of ROS signaling in cross-tolerance: from model to crop. *Front Plant Sci* 23: 754.

Pfannschmidt T. (2003) Chloroplast redox signals: how photosynthesis controls its own genes. *Trends Plant Sci* 8: 33–41.

Purvis AC. (1997) Role of the alternative oxidase in limiting superoxide production by plant mitochondria. *Physiol Plant* 100: 165–170.

Puthiyaveetil S, Woodiwiss T, Knoerdel R, *et al.* (2014) Significance of the photosystem II core phosphatase PBCP for plant viability and protein repair in thylakoid membranes. *Plant Cell Physiol* 55: 1245–1254.

Rasmusson AG, Soole KL, Elthon TE. (2004) Alternative NAD(P)H dehydrogenases of plant mitochondria. *Annu Rev Plant Biol* 55: 23–39.

Rasmusson AG, Geisler DA, Møller IM. (2008) The multiplicity of dehydrogenases in the electron transport chain of plant mitochondria. *Mitochondrion* 8: 47–60.

Rea G, Polticelli F, Antonacci A, *et al.* (2009) Structure-based design of novel *Chlamydomonas reinhardtii* D1-D2 photosynthetic proteins for herbicide monitoring. *Protein Science* 18: 2139–2151.

Robson CA, Vanlerberghe GC. (2002) Transgenic plant cells lacking mitochondrial alternative oxidase have increased susceptibility to mitochondria-dependent and -independent pathways of programmed cell death. *Plant Physiol* 129: 1908–1920.

Samol I, Shapiguzov A, Ingelsson B, *et al.* (2012) Identification of a photosystem II phosphatase involved in light acclimation in *Arabidopsis*. *Plant Cell* 24: 2596–2609.

Schrader M, Fahimi HD. (2006) Peroxisomes and oxidative stress. *BBA-Mol Cell Res* 1763: 1755–1766.

Schroda M, Vallon O, Wollman FA, Beck CF. (1999) A chloroplast-targeted heat shock protein 70 (HSP70) contributes to the photoprotection and repair of photosystem II during and after photoinhibition. *The Plant Cell* 11: 1165–1178.

Schutzendübel A, Schwanz P, Teichmann T, *et al.* (2001) Cadmium-induced changes in antioxidative systems, H_2O_2 content and differentiation in pine (*Pinus sylvestris*) roots. *Plant Physiol* 127: 887–898.

Shapiguzov A, Vainonen J, Wrzaczek M, Kangasjärvi J. (2012) ROS-talk–how the apoplast, the chloroplast, and the nucleus get the message through. *Front Plant Sci* 3: 292.

Sharma P, Jha AB, Dubey RS, Pessarakli M. (2010) Oxidative stress and antioxidative defense system in plants growing under abiotic stresses. In: Pessarakli M (ed.), *Handbook of Plant and Crop Stress*, 3rd edn. Taylor & Francis, pp. 89–138.

Sharma P, Jha AB, Dubey RS, Pessarakli M. (2012) Reactive oxygen species, oxidative damage, and antioxidative defense mechanism in plants under stressful conditions. *J Bot* 2012: Art. ID 217037.

Shi LX, Hall M, Funk C, Schroder WP. (2012) Photosystem II, a growing complex: updates on newly discovered components and low molecular mass proteins. *Biochim Biophys Acta* 1817: 13–25.

Singh R, Singh S, Parihar P, *et al.* (2016) Reactive oxygen species (ROS): beneficial companions of plants' developmental processes. *Front Plant Sci* 7: 1299.

Singh S, Srivastava PK, Kumar D, Tripathi DK, Chauhan DK, Prasad SM. (2015) Morpho-anatomical and biochemical adapting strategies of maize (*Zea mays* L.) seedlings against lead and chromium stresses. *Biocatal Agric Biotechnol* 4: 286–295.

Singh S, Tripathi DK, Singh S, *et al.* (2017) Toxicity of aluminium on various levels of plant cells and organism: a review. *Environ Exp Bot* 137: 177–193.

Singh VP, Tripathi DK, Kumar D, Chauhan DK. (2011) Influence of exogenous silicon addition on aluminium tolerance in rice seedlings. *Biol Trace Elem Res* 144: 1260–1274.

Suzuki N, Miller G, Morales J, Shulaev V, Torres MA, Mittler R. (2011) Respiratory burst oxidases: the engines of ROS signalling. *Curr Opin Plant Biol* 14: 1–9.

Takahashi M, Shiraishi T, Asada K. (1988) COOH-terminal residues of D1 and the 44 kDa CPa-2 at spinach photosystem II core complex. *FEBS Lett* 240: 6–8.

Tasaka Y, Gombos Z, Nishiyama Y, *et al.* (1996) Targeted mutagenesis of acyl-lipid desaturases in *Synechocystis*: evidence for the important roles of polyunsaturated membrane lipids in growth, respiration and photosynthesis. *EMBO J* 15: 6416–6425.

Tikkanen M, Nurmi M, Kangasjärvi S, Aro EM. (2008) Core protein phosphorylation facilitates the repair of photodamaged photosystem II at high light. *Biochim Biophys Acta* 1777: 1432–1437.

Torres MA, Jones JD, Dangl JL. (2006) Reactive oxygen species signaling in response to pathogens. *Plant Physiol* 141: 373–378.

Trebitsh AD. (2001) Translation of chloroplast psbA mRNA is regulated by signals initiated by both photosystems II and I. *Proc Natl Acad Sci USA* 21: 12289–12294.

Trebst A. (1986) The topology of the plastoquinone and herbicide binding peptides of photosystem II in the thylakoid membrane. *Z Naturforsch* 41c: 240–245.

Trebst A, Depka B, Holländer-Czytko H. (2002) A specific role for tocopherol and of chemical singlet oxygen quenchers in the maintenance of photosystem II structure and function in *Chlamydomonas reinhardtii*. *FEBS Lett* 516: 156–160.

Tripathi DK, Singh VP, Kumar D, Chauhan DK. (2012a) Impact of exogenous silicon addition on chromium uptake, growth, mineral elements, oxidative stress, antioxidant capacity, and leaf and root structures in rice seedlings exposed to hexavalent chromium. *Acta Physiol Plant* 34: 279–289.

Tripathi DK, Singh VP, Kumar D, Chauhan DK. (2012b) Rice seedlings under cadmium stress: effect of silicon on growth, cadmium uptake, oxidative stress, antioxidant capacity and root and leaf structures. *Chem Ecol* 28: 281–291.

Tripathi DK, Singh S, Singh VP, Prasad SM, Chauhan DK, Dubey NK. (2016a) Silicon nanoparticles more efficiently alleviate arsenate toxicity than silicon in maize cultivar and hybrid differing in arsenate tolerance. *Front Environ Sci* 4: 46.

Tripathi DK, Singh VP, Prasad SM, Dubey NK, Chauhan DK, Rai AK. (2016b) LIB spectroscopic and biochemical analysis to characterize lead toxicity alleviative nature of silicon in wheat (Triticum aestivum L.) seedlings. *J Photochem Photobiol B: Biol* 154: 89–98.

Tripathi DK, Singh S, Singh VP, Prasad SM, Dubey NK, Chauhan DK. (2017a) Silicon nanoparticles more effectively alleviated UV-B stress than silicon in wheat (Triticum aestivum) seedlings. *Plant Physiol Biochem* 110: 70–81.

Tripathi DK, Singh S, Singh S, *et al.* (2017b) Nitric oxide alleviates silver nanoparticles (AgNps)-induced phytotoxicity in Pisum sativum seedlings. *Plant Physiol Biochem* 110: 167–177.

Tripathi DK, Mishra RK, Singh S, *et al.* (2017c) Nitric oxide ameliorates zinc oxide nanoparticles phytotoxicity in wheat seedlings: implication of the ascorbate–glutathione cycle. *Front Plant Sci* 8: 1.

Umena Y, Kawakami K, Shen JR, Kamiya N. (2011) Crystal structure of oxygen-evolving photosystem II at a resolution of 1.9 Å. *Nature* 473: 55–60.

Vainonen JP, Hansson M, Alexander AV. (2005) STN8 protein kinase in *Arabidopsis thaliana* is specific in phosphorylation of photosystem II core proteins. *J Biol Chem* 280: 33679–33686.

Van Dongen JT, Licausi F. (2015) Oxygen sensing and signaling. *Annu Rev Plant Biol* 66: 345–367.

Vass I. (2011) Role of charge recombination processes in photodamage and photoprotection of the photosystem II complex. *Physiol Plant* 142: 1–11.

Vermaas WF, Ikeuchi M. (1991) Photosystem II. In: Bogorad L, Vasil IK (eds) *The Photosynthetic Apparatus: Molecular Biology and Operation*. Academic Press, San Diego, 25–111.

Wang P, Duan W, Takabayashi A, Endo T, Shikanai T, Ye JY, Mi H. (2006) Chloroplastic NAD (P) H dehydrogenase in tobacco leaves functions in alleviation of oxidative damage caused by temperature stress. *Plant Physiol* 141(2): 465–474.

Wang Y, Wisniewski M, Meilan R, Cui M, Webb R, Fuchigami L. (2005) Overexpression of cytosolic ascorbate peroxidase in tomato confers tolerance to chilling and salt stress. *J Am Soc Hortic Sci* 130: 167–173.

Wojtaszek P. (1997) Oxidative burst: an early plant response to pathogen infection. *Biocheml J* 322: 681–692.

Wunder T, Xu WT, Liu QP, Wanner G, Leister D, Pribil M. (2013) The major thylakoid protein kinases STN7 and STN8 revisited: effects of altered STN8 levels and regulatory specificities of the STN kinases. *Front Plant Sci* 4: 417.

Xu WF, Shi WM, Ueda A, Takabe T. (2008) Mechanisms of salt tolerance in transgenic *Arabidopsis thaliana* carrying a peroxisomal ascorbate peroxidase gene from barley. *Pedosphere* 18: 486–495.

Yamamoto Y, Akasaka T. (1995) Degradation of antenna chlorophyll-binding protein CP43 during photoinhibition of photosystem II. *Biochemistry* 34: 9038–9045.

Yamamoto Y, Aminaka R, Yoshioka M, Khatoon M, *et al.* (2008) Quality control of photosystem II: impact of light and heat stresses. *hotosynth Res* 98: 589–608.

Yamamoto Y, Hori H, Kai S, *et al.* (2013) Quality control of photosystem II: reversible and irreversible protein aggregation decides the fate of photosystem II under excessive illumination. *Front Plant Sci* 4: 433.

Yamauchi Y, Furutera A, Seki K, Toyoda Y, Tanaka K, Sugimoto Y. (2008) Malondialdehyde generated from peroxidized linolenic acid causes protein modification in heat-stressed plants. *Plant Physiol Biochem* 46: 786–793.

Yamazaki D, Motohashi K, Kasama T, Hara Y, Hisabori T. (2004) Target proteins of the cytosolic thioredoxins in *Arabidopsis thaliana*. *Plant Cell Physiol* 45: 18–27.

Yan J, Wang J, Tissue D, Holaday AS, Allen R, Zhang H. (2003) Photosynthesis and seed production under water-deficit conditions in transgenic tobacco plants that overexpress an *Arabidopsis* ascorbate peroxidase gene. *Crop Sci* 43: 1477–1483.

Yang X, Liang Z, Lu C. (2005) Genetic engineering of the biosynthesis of glycine betaine enhances photosynthesis against high temperature stress in transgenic tobacco plants. *Plant Physiol* 138: 2299–2309.

Yang X, Wen X, Gong H, *et al.* (2007) Genetic engineering of the biosynthesis of glycine betaine enhances thermotolerance of photosystem II in tobacco plants. *Planta* 225: 719–733.

Yoshioka M, Uchida S, Mori H, *et al.* (2006) Quality control of photosystem II cleavage of reaction center D1 protein in spinach thylakoids by FtsH protease under moderate heat stress. *J Biol Chem* 281: 21660–21669.

Yoshioka M, Nakayama Y, Yoshida M, *et al.* (2010) Quality control of photosystem II: FtsH hexamers are localized near photosystem II at grana for the swift repair of damage. *J Biol Chem* 285: 41972–41981.

Yoshioka-Nishimura M, Nanba D, Takaki T, *et al.* (2014) Quality control of photosystem II: direct imaging of the changes in the thylakoid structure and distribution of FtsH proteases in spinach chloroplasts under light stress. *Plant Cell Physiol* 55: 1255–1265.

Zhang L, Aro E-M. (2002) Synthesis, membrane insertion and assembly of the chloroplast-encoded D1 protein into photosystem II. *FEBS Lett* 512: 13–18.

Zhang S, Weng J, Pan J, Tu T, Yao S, Xu C. (2003) Study on the photo-generation of superoxide radicals in photosystem II with EPR spin trapping techniques. *Photosynth Res* 75: 41–48.

10

Reactive Oxygen Species and Antioxidants: A Continuous Scuffle within the Cell

Supriya Tiwari

Banaras Hindu University, Varanasi, India

Introduction

With the introduction of molecular oxygen (O_2) into the atmosphere, aerobic respiration came into existence, which was accompanied by the inevitable processes of generation of reactive oxygen species (ROS). ROS include free radicals such as superoxide anions ($O_2^{\cdot-}$) and hydroxyl radicals (·OH) as well as non-radical molecules such as hydrogen peroxide (H_2O_2) and singlet oxygen ($^1O_2$). In plants, ROS are continuously being produced by the inevitable leakage of electrons via electron transport activities occurring in different subcellular compartments like chloroplasts, mitochondria, and plasma membrane, or as a by-product of various metabolic processes (Blokhina and Fagerstedt, 2010; Heyno *et al.*, 2011). Environmental stresses such as drought, salinity, pollution, metal toxicity, chilling, pathogen attack, and so forth cause enhanced generation of ROS in plants (Mittler, 2002; Hu *et al.*, 2008; V.P. Singh *et al.*, 2011; Tripathi *et al.*, 2012a,b, 2016a,b, 2017a–c; S. Singh *et al.*, 2015, 2017; R. Singh *et al.*, 2016). It has been estimated that 1–2% of O_2 consumption is used up in the formation of ROS in tissues (Bhattacharjee, 2005). The reactive nature of ROS may pose a threat to normal cellular functions by damaging the nucleic acids, oxidizing proteins, inhibiting enzymes, and causing lipid peroxidation, and if uncontrolled leading to the cell death (Han *et al.*, 2009; Mishra *et al.*, 2011; V.P. Singh *et al.*, 2011; Tripathi *et al.*, 2012a,b, 2016a,b, 2017a–c; S. Singh *et al.*, 2015, 2017; R. Singh *et al.*, 2016). To deal with such a situation, the cells have evolved an efficient antioxidant defense mechanism comprising of several enzymatic and non-enzymatic antioxidants that coordinate to scavenge/detoxify the generated ROS (Foyer and Noctor, 2005; Navrot *et al.*, 2007). Therefore, under normal conditions, there exists a tight equilibrium between ROS generation and detoxification in the cell (Apel and Hirt, 2004). Excess ROS generation under any kind of biotic/abiotic stress creates an imbalance between ROS generation and detoxification, resulting in a physiological condition called "oxidative stress." In biological systems, oxidative stress is characterized by the presence of elevated levels of oxidizing agents that are able to extract electrons from essential organic molecules and disturb cellular function (Tripathy and Oelmüller, 2012). Once the cellular antioxidants are exhausted, accumulation of ROS may have lethal consequences, responsible for huge losses in crop productivity around the world (Mittler, 2002; Apel and Hirt, 2004; Tuteja, 2010).

Besides acting as "harmful oxidizing agents," ROS play a significant role in fundamental plant processes such as stomatal closure, response to environmental stresses, and modulation of

Reactive Oxygen Species in Plants: Boon Or Bane - Revisiting the Role of ROS, First Edition. Edited by Vijay Pratap Singh, Samiksha Singh, Durgesh Kumar Tripathi, Sheo Mohan Prasad, and Devendra Kumar Chauhan.

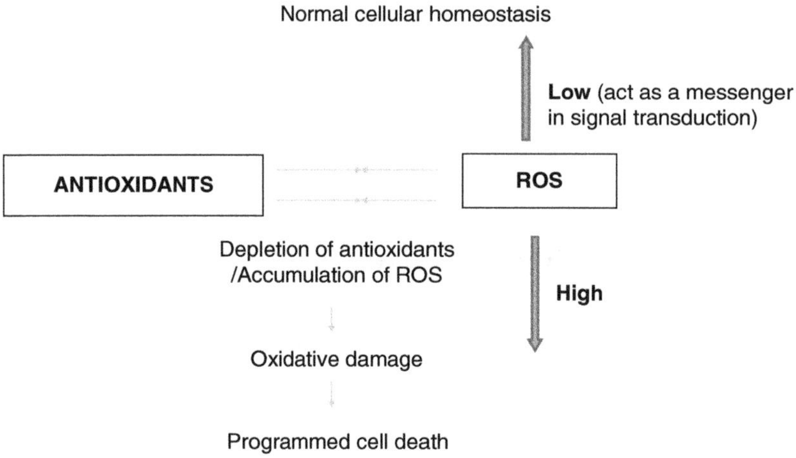

Figure 10.1 Effects of cellular concentration of reactive oxygen species (ROS) in plants.

protein activity or gene expression (Apel and Hirt, 2004; Moller *et al.*, 2007). At low concentrations ROS act as important signaling molecules; they are not only generated spontaneously as a by-product of various metabolic processes of the cell, but are also produced "intentionally" by certain cellular enzyme complexes, such as NADPH oxidases, amine oxidases, xanthine oxidases, and so forth (Apel and Hirt, 2004; Nanda *et al.*, 2010). It is important to note that whether ROS act as harmful oxidizing factors or as signaling molecules depends upon their proper site and time of production. At low concentrations, ROS act as signals for defense pathways, eliciting specific cellular responses. It has been shown that ROS influence the expression of a number of genes and signal transduction pathways, which suggests that cells have evolved specific strategies to utilize ROS as biological stimuli and signals that activate and control various genetic stress response programs (Petrov *et al.*, 2013). However, when ROS generation exceeds the antioxidant capacity of the cell, it can cause cell death, either by ROS toxicity leading to necrosis or through specific ROS-activated cell death-inducing signaling events leading to programmed cell death (PCD) (Quan *et al.*, 2008; Karuppanapandian *et al.*, 2011) (Figure 10.1). In this review, we discuss the bipartite roles of ROS, which make them very interesting molecules in plant developmental processes. The review focuses on the antagonistic nature of the ROS generated in the subcellular compartments of the cell and the intrinsic cellular antioxidant defense machinery. We will also discuss in detail the various sites and

Table 10.1 Non enzymatic antioxidants of the cell.

	Non enzymatic antioxidant	Primary ROS scavenged	Organelles in which located
1.	Ascorbic acid (AA)	H_2O_2, $O_2^{\bullet-}$	Chlorophyll, Cytoplasm, Mitochondria, Peroxisomes, Apoplasts
2.	Glutathione (GSH)	H_2O_2	Chlorophyll, Cytoplasm, Mitochondria, Peroxisomes, Apoplasts
3.	Carotenoids (CARs)	$^1O_2$	Chloroplasts
4.	Tocopherols (TOCs)	$^1O_2$	Membranes
5.	Flavinoids	H_2O_2	Vacuoles

metabolic processes responsible for production of ROS, and the different antioxidants that serve to scavenge the ROS, thus maintaining the normal cellular homeostasis.

Antioxidant Defense Machinery

In response to the ROS generated under normal or stressed conditions, plant cells and their organelles, like chloroplasts, mitochondria, and peroxisomes, utilize the antioxidant defense machinery to protect themselves against "oxidative stress." Antioxidants significantly delay or prevent the oxidation of oxidizable substrates with low concentrations of ROS (Halliwell, 2006). A number of experiments have proved that the stimulation of cellular antioxidant defense systems is essential for protection against various stresses (Tuteja, 2007; Khan and Singh, 2008; S. Singh *et al.*, 2008, 2017; R. Singh *et al.*, 2016; Tripathi *et al.*, 2016a,b, 2017a–c). ROS scavenging pathways from different cellular compartments are coordinated (Pang and Wang, 2008). The antioxidant defense machinery operating in plants has both enzymatic and non-enzymatic components.

1) *Non-enzymatic components of antioxidant defense system:* these include ascorbate (AA) and glutathione (GSH) (water-soluble) as well as carotenoids and tocopherols (lipid-soluble) (Mittler, 2004; Singh *et al.*, 2016) (Table 10.1).
2) *Enzymatic components of antioxidant defense system:* these include enzymes like superoxide dismutase (SOD), catalase (CAT), glutathione peroxidase (GPX), guaiacol peroxidase (GuPOX), enzymes of the AsA-GSH cycle such as ascorbate peroxidase (APX), monodehydroascorbate reductase (MDHAR), dehydroascorbate reductase (DHAR), and glutathione reductase (GR) (Noctor and Foyer, 1998). These enzymes are found in different subcellular compartments and are stimulated by the presence of ROS. Table 10.2 summarizes the different antioxidant enzymes and the functions they play in scavenging of ROS.

The different components of the antioxidant defense machinery and their functions are discussed in the subsequent sections of the chapter.

Table 10.2 Enzymatic antioxidants of the cell.

	Enzymatic antioxidants	Reactions Catalyzed	Organelles in which located
1.	Superoxide dismutase (SOD)	$O_2^{\cdot -} + O_2^{\cdot -} + 2H+ \rightarrow 2H_2O_2 + O_2$	Cytoplasm, Chlroplast, Peroxisomes, Mitochondria
2.	Ascorbate peroxidase (APX)	$H_2O_2 + AA \rightarrow 2H_2O + DHA$	Cytoplasm, Chlroplast, Peroxisomes, Mitochondria
3.	Catalase (CAT)	$2H_2O_2 \rightarrow O_2 + 2H_2O$	Peroxisomes, Mitochondria, Chloroplasts
4.	Glutahione peroxidase (GPX)	$H_2O_2 + DHA \rightarrow 2H_2O + GSSG$	Chloroplsts, Cytoplasm, Mitochondria, Endoplasmic Reticulum
5.	Monodehydroascorbate reductase (MDHAR)	$2MDHA + NADH \rightarrow 2AA + NAD^+$	Chloroplasts, Mitochondria, Cytoplasm
6.	Dehydroascorbate reductase (DHAR)	$DHA + 2GSH \rightarrow AA + GSSG$	Chloroplast, Mitochondria, Cytoplasm
7.	Glutathione reductase (GR)	$GSSG + NADPH \rightarrow 2GSH + NADP^+$	Cytoplasm, Chloroplast, Mitochondria

Reactive Oxygen Species

The process of aerobic respiration and its efficient energy generation system utilizes O_2 as the final electron acceptor, which acts as the primary source of ROS in cells (Temple *et al.*, 2005). The O_2 molecule itself is not a reactive molecule as in its ground state it has two unpaired electrons with parallel spins, which makes it paramagnetic, and hence unlikely to participate in any biological reaction (Apel and Hirt, 2004). Activation of O_2 may occur by two processes: (i) absorption of sufficient energy to reverse the spin of one of the unpaired electrons, or (ii) by non-valent stepwise reduction of molecular oxygen. The former pathway leads to the formation of singlet oxygen ($^1O_2$) whereas in the latter, O_2 is sequentially reduced to superoxide anions ($O_2^{\cdot -}$), hydrogen peroxide (H_2O_2), hydroxyl ions ($^{\cdot}OH$), and so forth (Figure 10.2).

In its ground state O_2 absorbs sufficient energy to reverse the spin of one of its unpaired electrons, leading to the formation of the singlet state ($^1O_2$) in which two electrons have opposite spins. Once the spin restriction is removed, $^1O_2$ can participate in reactions involving transfer of electrons. Formation of ROS by molecular oxygen occurs through its stepwise reduction. Due to spin restrictions, O_2 in its ground state cannot accept four electrons simultaneously to form H_2O. It accepts one electron at a time, and hence during the reduction of O_2, stable intermediates are formed in a stepwise manner (Valko *et al.*, 2005). Reduction of O_2 results in the formation of superoxide anions ($O_2^{\cdot -}$), the primary ROS formed in the cell, which initiates a cascade of reactions to form secondary ROS (Halliwell, 2006). At low pH, dismutation of $O_2^{\cdot -}$ is unavoidable, with one $O_2^{\cdot -}$ donating its extra electron to another $O_2^{\cdot -}$ along with protonation to form H_2O_2. $O_2^{\cdot -}$ can also be alternatively protonated to form $HO_2^{\cdot}$. Through the Haber–Weiss mechanism or Fenton's reaction, H_2O_2 is again protonated to form $^{\cdot}OH$, which is the most reactive chemical species in the biological world (see Figure 10.2).

ROS Generation and Antioxidant Scavenging

As ROS have a very short lifespan and diffuse only a short distance before reacting with a cellular molecule, the site of ROS generation is very important. They are produced in both unstressed and stressed conditions, in different cell organelles with a highly oxidizing metabolic activity or with an intense rate of electron flow, such as mitochondria, chloroplasts, and peroxisomes. In addition, peroxidase and amine oxidase present in the cell walls and NADPH oxidase found in plasma membrane also produce ROS, often in response to stress signals.

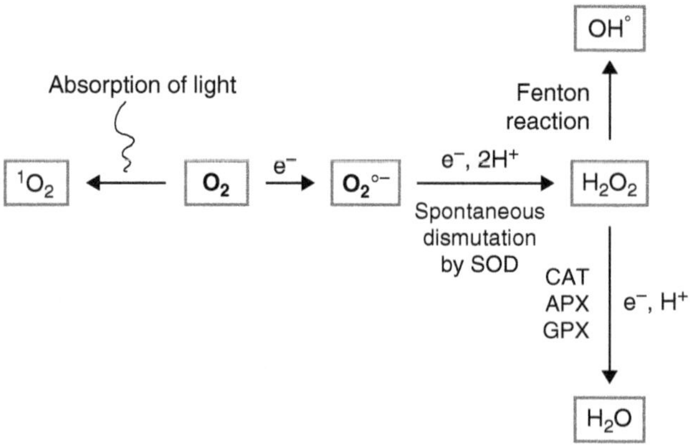

Figure 10.2 Metabolic pathways involved in the reduction of molecular oxygen (O_2) leading to generation of various intermediate reactive oxygen species (ROS) in plants. APX, ascorbate peroxidase; CAT, catalase; GPX, glutathione peroxidase; SOD, superoxide dismutase.

When the plant is illuminated, chloroplasts and peroxisomes are the main sources of ROS generation (Foyer and Noctor, 2005), whereas in dark ROS are mostly produced by mitochondria. It is established that 1–5% of the O_2 consumed by mitochondria is utilized in ROS generation (Moller, 2001). Important sites of ROS generation within the cell are considered below.

Mitochondria

Mitochondria are the most important sites of production of ROS, specifically H_2O_2 (Rasmusson *et al.*, 2008). The cellular environment of plant mitochondria is distinct from that of animal mitochondria due to the presence of photosynthesis, which creates an O_2-rich environment. In addition, photorespiration in plant mitochondria also creates favorable conditions for ROS formation (Noctor *et al.*, 2006). The mitochondrial electron transport chain (ETC), although important in aerobic respiration, is also an important source of ROS generation (Rhoads *et al.*, 2006). The first evidence of ROS generation by the ETC was given by Jensen (1966). ROS are produced under normal respiratory conditions and production can be enhanced in response to various biotic and abiotic stresses. The mitochondrial ETC harbors electrons with sufficient free energy to directly reduce O_2 (Rhoads *et al.*, 2006).

In mitochondria, complex I (NADH dehydrogenase segment) and complex III (ubiquinone-cytochrome region) of the respiratory chain are the main sites of ROS generation. In the flavo-protein region of complex I, direct reduction of O_2 to $O_2^{\cdot-}$ takes place (Arora *et al.*, 2002). Complex I is the entry point for electrons from NADH to the respiratory chain, from where they pass to the CoQ reduction site (Sazanov, 2007). Production of $O_2^{\cdot-}$ from complex I depends upon the NADH/NAD$^+$ ratio. Under normal conditions, the NADH/NAD$^+$ ratio is low and only a small amount of $O_2^{\cdot-}$ is produced (Votyakova and Reynolds, 2001). However, inhibition of the respiratory chain at complex I builds up NADH due to low ATP demand. As a result the NADH/NAD$^+$ ratio will increase, leading to more $O_2^{\cdot-}$ formation (Liu *et al.*, 2002). Complex III funnels the electrons from the CoQ pool to cytochrome *c* and produces $O_2^{\cdot-}$ from O_2. It is believed that fully reduced ubiquinone donates electrons to cytochrome c_1 and leaves an unstable, highly reducing ubisemiquinone radical, which is favorable for electron leakage to O_2, thereby leading to formation of $O_2^{\cdot-}$ (Murphy, 2009). Under normal conditions, the electron transport system (ETS) and ATP production in mitochondria are synchronized; however, under stress conditions the two processes are uncoupled, leading to over-reduction of electron carriers, and thus more ROS formation (Blokhina and Fagerstedt, 2010). Complex II of the ETC also produces its share of ROS by the phenomenon of reverse electron flow (REF). When NAD$^+$-linked substrates for complex I are limited, electron transport can occur from CoQ to complex I and reduce NAD$^+$ to NADH (Turrens, 2003). RET was first associated with H_2O_2 production, but later $O_2^{\cdot-}$ was found to be the most significant ROS produced by complex I through RET. Although the exact site and mechanism of $O_2^{\cdot-}$ production through RET is unknown, this process accounts for the maximum rate of $O_2^{\cdot-}$ production in mitochondria (Lambert and Brand, 2004).

In addition to ETC, the enzymes present in the mitochondrial matrix also contribute to the ROS pool. Some of them produce ROS directly, for example, aconitase, whereas some others like 1-galactono-γ-lactone dehydrogenase (GAL), are able to feed electrons to the ETC (Rasmusson *et al.*, 2008). The mitochondrial alternative oxidase (AOX) catalyzes the O_2-dependent oxidation of ROS. Studies have shown that reduced AOX in mitochondria leads to increased ROS production (Maxwell *et al.*, 1999). Sweetlove *et al.* (2008) have shown that H_2O_2 treatment of *Arabidopsis* cells caused increased levels of AOX in mitochondria. Similarly H_2O_2 accumulation in catalase-deficient tobacco also stimulated mitochondrial AOX levels (Sweetlove *et al.*, 2008).

Mitochondria were believed to be an important source of H_2O_2 production; however, it was later established that H_2O_2 was actually formed by the dismutation of superoxide anions (Quan *et al.*, 2008). In aqueous solution, $O_2^{\cdot-}$ is moderately reactive, but is reduced by the enzyme superoxide dismutase (SOD) to form more reactive H_2O_2 (Quan *et al.*, 2008). This H_2O_2 can react with reduced Fe^{2+} and Cu^{+} to produce highly toxic $\cdot OH$, which can penetrate membranes and leave the mitochondrion (Rhoads *et al.*, 2006). There are eight sites that are recognized as possessing the ability to produce $O_2^{\cdot-}$ (Brand, 2010). While seven of these sites release the $O_2^{\cdot-}$ produced directly into the mitochondrial matrix, one site on complex III releases $O_2^{\cdot-}$ into the intermembrane space. Intermembrane ROS have an easy approach to the cytosol as they need to cross only a single membrane, while the matrix ROS have to cross both outer and inner mitochondrial membranes (Muller *et al.*, 2004).

The $O_2^{\cdot-}$ generated in the mitochondrial matrix is converted to H_2O_2 by SOD (MnSOD) in the matrix (del Río *et al.*, 2003). $O_2^{\cdot-}$ can also be produced in the intermembrane space from complex III, owing to generation from ubisemiquinone at the outer ubiquinone binding site of complex III. The $O_2^{\cdot-}$ thus produced in the intermembrane space is acted upon by Cu/ZnSOD to form H_2O_2. The H_2O_2 is degraded in the matrix by GPX using reducing equivalents obtained from the oxidation of reduced glutathione, a reaction that utilizes DHAR (GSH) (Li *et al.*, 2000) (Figure 10.3). Oxidized glutathione (GSSG) is reduced by GR, which obtains its equivalents from NADPH oxidation. H_2O_2 generated in the matrix can oxidize proteins, lipids, or mitochondrial DNA. The proteins oxidized by H_2O_2 are repaired by thioredoxin (TRX) and glutaredoxin (GRX), which are subsequently reduced by glutathione. Lipid hydroperoxides, which are formed by the lipid peroxidation of the mitochondrial membrane lipids by the oxidative action action of H_2O_2, are reduced by GPX. It has been suggested that GSH plays a pivotal role in scavenging ROS in mitochondria, the level of which is maintained by the availability of NADPH (see Figure 10.3).

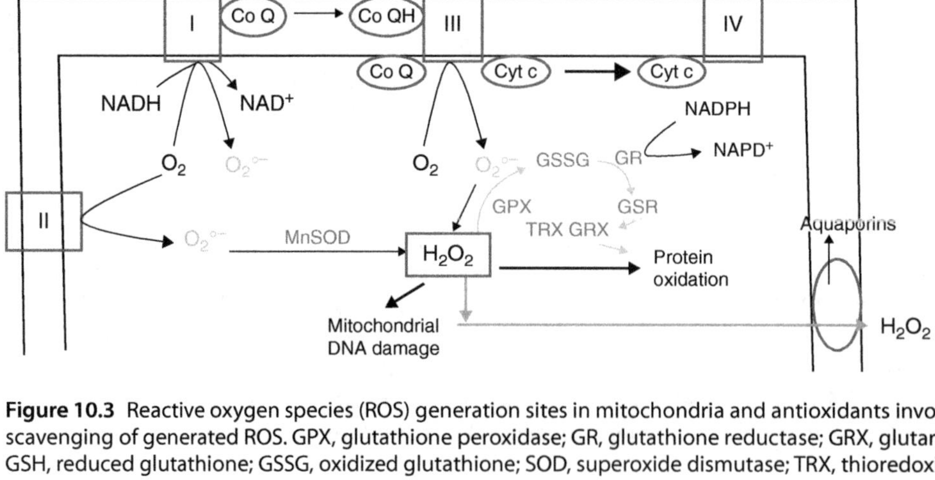

Figure 10.3 Reactive oxygen species (ROS) generation sites in mitochondria and antioxidants involved in scavenging of generated ROS. GPX, glutathione peroxidase; GR, glutathione reductase; GRX, glutaredoxin; GSH, reduced glutathione; GSSG, oxidized glutathione; SOD, superoxide dismutase; TRX, thioredoxin.

Chloroplasts

Chloroplasts contain a highly organized membrane system that harbors all the components of the light-capturing photosynthetic system, and are equipped with all the required structural features of light harvesting (Pfannschmidt, 2003). Oxygen generated in the chloroplasts during photosynthesis can accept electrons passing through the photosystems (PS), thus resulting in the formation of $O_2^{\cdot-}$. Therefore, the photosystems are important sites of ROS generation in the chloroplasts. Production of ROS by these sources is enhanced in plants by conditions limiting CO_2 fixation, such as drought, salt, and temperature stresses, as well as by the combination of these conditions with high-light stress. In the thylakoid membranes, ROS are formed when the absorption of light by chlorophyll exceeds the light utilization capacity of the photosynthetic apparatus (Asada, 2006). Normally, the electron flow from the excited photosystem centers is directed to $NADP^+$, which is reduced to NADPH. They then enter the Calvin–Benson cycle and reduce the final electron acceptor, CO_2. In situations of overloading of the ETC, a part of the electron flow is diverted from ferredoxin to O_2, through PSI, reducing it to $O_2^{\cdot-}$ via the Mehler reaction (Elstner, 1991). Leakage of electrons to O_2 may also occur from 2Fe-2S and 4Fe-4S clusters in the ETC of PSI. Electron leakage from Q_A and Q_B of PSII also contributes to the production of $O_2^{\cdot-}$ (Cleland and Grace, 1999). Once $O_2^{\cdot-}$ is formed, it generates more harmful ROS, either by protonation ($HO_2\cdot$) on the internal membrane of the lumen or through enzymatic dismutation by Cu/ZnSOD to H_2O_2 on the external membrane of stroma. H_2O_2 can further be converted to OH by Fenton's reaction at Fe-S centers depending upon the availability of Fe^{2+} (Takahashi *et al.*, 1988). Another important ROS, $^1O_2$, is generated in the chloroplasts when energy transfer to the PSII reaction system is limited or the ETC between the photosystems is inhibited. In this condition chlorophyll serves as an effective photosynthesizer that transfers the excitation energy to O_2 (Takahashi *et al.*, 1988). This transfer of energy is accompanied by the formation of singlet oxygen ($^1O_2$). $^1O_2$ is formed by the transfer of sufficient energy from the triplet excited state of chlorophyll ($^3Cl^*$) to the triplet ground state of molecular oxygen ($^3O_2$), so that $^1O_2$ is formed, whose outermost pair of electrons have antiparallel spins. Within the chloroplasts, $^1O_2$ is mainly produced by chlorophyll (Chl), the most abundant light-absorbing pigment in land plants, present in both light-harvesting complex (LHC) and photosynthetic reaction centers. Insufficient transfer of energy results in generation of the triplet state of chlorophyll (Chl*), which reacts with triplet O_2 to produce highly reactive $^1O_2$. In the presence of excess light energy, Q_A and Q_B of the ETC are over-reduced (Barber and Andersen, 1992), which inhibits charge separation between P680 and pheophytin. This favors the formation of Chl* at the P680 reaction center leading to the formation of $^1O_2$ (Aro *et al.*, 1993; Ohad *et al.*, 1994). The chlorophylls bound to the proteins in the PSII antenna complex are oriented in such a way that efficient energy transfer to PSII antenna complex is maintained. Under these circumstances, energy is transferred at a very fast rate to the neighboring chlorophyll molecules, which restricts the formation of triplet chlorophyll molecules. However, when the energy transfer to the neighboring molecules is restricted, the excess energy is utilized in changing the orientation of spin of excited electrons, that is, conversion of the singlet excited state with antiparallel spin to the triplet state with parallel spin (Pospíšil, 2012). $^1O_2$ is also formed when $O_2^{\cdot-}$ interacts with ·OH radicals (Skovsen *et al.*, 2005). In addition to these, chlorophyll biosynthesis intermediates like protochlorophyllides and protoporphyrin IX also produce $^1O_2$ in the thylakoids of the chloroplasts (Tripathy *et al.*, 2007).

Important ROS-scavenging antioxidants found in chloroplasts, which constitute the organelle's defense machinery, are SOD and enzymes of the AsA-GSH cycle. Several plants contain Fe-SOD in addition to Cu/ZnSOD, but the presence of MnSOD in chloroplasts is not reported. Chloroplastic APX is classified as either thylakoid bound (tAPX) or the stroma localized form

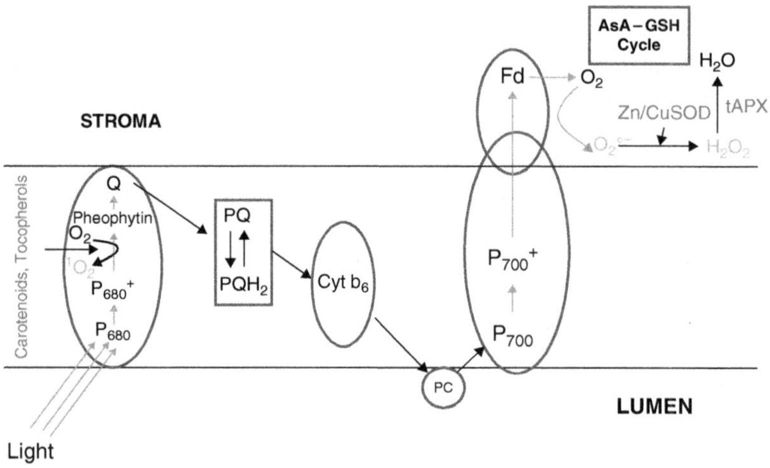

Figure 10.4 Reactive oxygen species (ROS) generation sites in chloroplasts and antioxidants involved in scavenging the generated ROS. AsA-GSH cycle, ascorbate-glutathione cycle; SOD, superoxide dismutase; tAPX, thylakoid bound ascorbate peroxidase.

(sAPX) (Asada, 2006). PSI, which is the main site of $O_2^{\cdot-}$ production, is associated with Cu/ZnSOD and tAPX as the main antioxidant scavengers of ROS (Ogawa *et al.*, 1995). $O_2^{\cdot-}$ produced at PSI is converted to H_2O_2, which is quickly reduced through the v (AsA-GSH) pathway using tAPX and MDHAR enzymes. Therefore PSI attached to Cu/ZnSOD, tAPX bound to the vicinity of PSI, and Fe-dependent reduction of MDHA form the first lines of defense of the thylakoid ROS scavenging system. Other enzymes associated with the stroma (stromal scavenging system) form the second lines of defense (Figure 10.4). Plants with reduced levels of tAPX are more sensitive to stress conditions (Tarantino *et al.*, 2005), while plants over-expressing tAPX are more tolerant to stress (Yabuta *et al.*, 2002; Murgia *et al.*, 2004). Similarly, plants with knockdown mutants of Cu/ZnSOD showed small chloroplasts and reduced photo-synthetic activity. $^1O_2$ formed at PSII and PSI is efficiently quenched by β-carotene and α-tocopherol (Krieger-Liszkay *et al.*, 2008). Excess formation of $^1O_2$ activates the alternative pathway, which serves to bring Chl* to its ground state before passing on the energy to O_2 (Bartoli *et al.*, 2004). The PSII reaction center and antenna subunit complex possess 11 molecules of β-carotene. In the reaction center, β-carotene participates in direct quenching of $^1O_2$ (Asada, 2006). However, in PSII, the distance between β-carotene and P680 does not allow direct quenching of $^1O_2$ (Loll *et al.*, 2005). Therefore, β-carotene participates in the cyclic electron flow with cyt b_{559} in PSII, which suppress the generation of $^1O_2$ (Telfer *et al.*, 2003). Tocopherols also quench $^1O_2$ and in turn are themselves oxidized (Trebst, 2003). The ascorbate pool present in chloroplasts is regenerated via the AsA-GSH pathway, which helps in regeneration of tocopherols from their oxidized radicals to their original state, which is again ready to quench the $^1O_2$ (Munne- Bosch and Alegre, 2002).

Peroxisomes

Peroxisomes, like mitochondria and chloroplasts, also show an oxidative type of metabolism and are therefore important sites of ROS generation. In peroxisomes from pea leaves and watermelon cotyledons, at least two sites of $O_2^{\cdot-}$ generation have been identified using biochemical and electron spin resonance spectroscopy (ESR) methods (del Río *et al.*, 2002). Firstly, the matrix contains an enzyme, xanthine oxidase (XOS), which catalyzes the oxidation

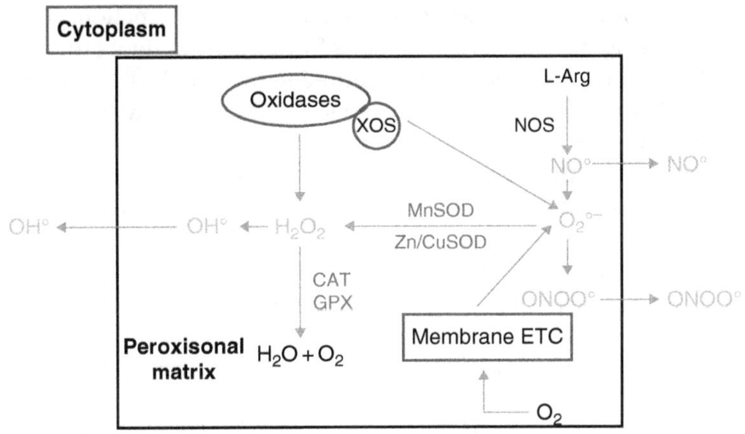

Figure 10.5 Reactive oxygen species (ROS) generation sites in peroxisomes and antioxidants involved in scavenging the generated ROS. CAT, catalase; ETC, electron transport chain; GPX, glutathione peroxidase; NOS, nitric oxide synthase; SOD, superoxide dismutase; XOS, Xanthine oxidase.

of xanthine and hypoxanthine to uric acid (Corpas *et al.*, 2001). The second site is in the peroxisome membranes, where a small NAD(P)H-dependent ETC is composed of a flavoprotein NADH and cytochrome *b*, and here $O_2^{\cdot-}$ is produced by the peroxisome ETC (Lopez-Huertas *et al.*, 1999). MDHAR also participates in $O_2^{\cdot-}$ production by the peroxisome membrane (Lopez-Huertas *et al.*, 1999). In addition to these, several metabolic processes taking place in the peroxisomes, like photorespiratory glycolate oxidase reaction, fatty acid β-oxidation, the enzymatic oxidation of flavin oxidases, the disproportionation of $O_2^{\cdot-}$ radicals, and so forth, also account for the production of a significant amount of ROS, especially H_2O_2 (del Río *et al.*, 2006). It has further been demonstrated that ·NO radicals are also produced in peroxisomes by the oxidation of L-arginine, catalyzed by nitric oxide synthase (NOS) (Stolz *et al.*, 2002). ·NO can react with $O_2^{\cdot-}$ to form peroxynitrite ($ONOO^-$), a powerful oxidant. ·NO and H_2O_2 both have a potential to penetrate the peroxisomal membrane (Schrader and Fahimi, 2006).

CAT, GPX, and SOD are the important antioxidant enzymes responsible for scavenging the peroxisomal ROS and maintaining the ROS-antioxidant balance in the peroxisomes (Figure 10.5). CAT is a classical enzyme of peroxisomes, metabolizing H_2O_2 (Oshino *et al.*, 1973). Beside CAT, GPX also functions in the catalytic destruction of H_2O_2 in peroxisomes, with a concomitant conversion of GSH to GSSG. H_2O_2 and ·OH radicals can also damage the peroxisomal membranes by lipid peroxidation of unsaturated fatty acids, forming hydroperoxides. CAT has an important protective function against the toxic effects of these peroxides and removes them efficiently (Siraki *et al.*, 2002). MnSOD and Cu/ZnSOD perform the function of dismutation of $O_2^{\cdot-}$ (del Río *et al.*, 2006).

Endoplasmic Reticulum (ER)

The endoplasmic reticulum is an organelle that is involved in protein folding. Any kind of misfolding in proteins results in protein aggregation in the ER lumen, triggering a condition called ER stress (Ushioda *et al.*, 2008). The lumen of the ER, in contrast to the cytosol, has a highly oxidizing environment with a high ratio of GSSG:GSH. Glutathione exists in a reduced state (GSH) due to the cytosolic NADPH-dependent reaction catalyzed by glutathione reductase (Moroder *et al.*, 1996). The oxidizing environment of the ER not only facilitates the formation of disulfide bonds, but also prevents the aggregation of unfolded proteins (van der Vlies *et al.*, 2003).

Two mechanisms have been proposed for ROS generation during disulfide bond formation. First, ROS are generated as a by-product during the transfer of electrons from a protein thiol group to molecular O_2, mediated by ERO1 (endoplasmic reticulum oxidoreductin 1) and PDI (protein disulfide isomerase). Alternatively, ROS are generated during protein misfolding due to depletion of GSH (Santos *et al.*, 2009). Accumulation of unfolded proteins in the ER elicits Ca^{2+} leakage into the cytosol, increasing ROS production in the mitochondria (Malhotra and Kaufman, 2007). In addition, $O_2^{\cdot-}$ is also produced in the ER by NAD(P)H-dependent electron transport involving cyt P450 (Mittler, 2002).

A major component of ER antioxidant machinery is GSH, which controls not only the activation state of ERO1 but also the PDI reduction rate using nascent protein thiols present in the ER (Kojer and Riemer, 2014). Lombardi *et al.* (2012) showed that GSH for these processes is obtained from the cytosolic influx. GSH in turn is rapidly oxidized by the activated ERO1-PDI machinery leading to a GSSH build-up, which in turn can be used to maintain a pool of oxidized PDI. It has been suggested that the efficient conversion of GSH to GSSH in ER could act as a driving force for GSH entry by maintaining a concentration gradient (Banhegyi *et al.*, 1999). H_2O_2 generated in the ER undergoes catalytic destruction during the interconversion of GSH and GSSG using GPX as the main antioxidant enzyme.

Plasma Membrane

The presence of certain electron transporters in the plasma membrane makes it a suitable site for ROS generation. Studies have shown that the plant plasma membrane exhibits enzyme activities for reducing O_2 to $O_2^{\cdot-}$ using NAD(P)H as a reductant (Bérczi and Møller, 2000). The ability of plasma membrane to generate ROS can be attributed to the presence of NAD(P)H oxidase, which can transport the electrons from the cytosolic side of the membrane to the apoplastic side, where the reduction of O_2 to $O_2^{\cdot-}$ occurs (Sagi and Fluhr, 2006). Through a series of reactions $O_2^{\cdot-}$ gives rise to other ROS like H_2O_2 and $\cdot OH$ (Halliwell and Gutteridge, 2007). Thus, the formation of $O_2^{\cdot-}$ by the plasma membrane can initiate the generation of an extremely toxic species, $\cdot OH$. Heyno *et al.* (2011) studied isolated plasma membranes from the growing and the non-growing zones of hypocotyls and roots of etiolated soybean seedlings as well as coleoptiles and roots of etiolated maize seedlings. Using EPR spin-trapping techniques and specific dyes they observed the production of reactive oxygen species ($O_2^{\cdot-}$, H_2O_2, and $\cdot OH$), mediated by NAD(P)H oxidase. NAD(P)H oxidase has been proposed to play an important role in generation and accumulation of ROS under stress conditions (Kwak *et al.*, 2003; Apel and Hirt, 2004).

In plants, NAD(P)H oxidases are also known as respiratory burst oxidase homologs (RBOHs) and constitute a multigene family (Sagi and Fluhr, 2006). RBOHs play an important role in apoplastic ROS generation in response to pathogenic stress, which is often biphasic in nature (Kadota *et al.*, 2004). The initial phase takes place within minutes and is due to rapid activation of RBOHs due to Ca^{2+} and phosphorylation (Torres *et al.*, 2002). It was observed that ROS generation was impaired in RBOH-D deficient mutants of *Arabidopsis*, suggesting the significance of RBOH in rapid production of apoplastic ROS upon pathogen infection (Nuhse *et al.*, 2007). The slower phase of ROS generation is attributed to the transcriptional activation of RBOH genes and accumulation of RBOH proteins. The regulation of RBOH genes is controlled by signal transduction processes occurring in the cell (Asai *et al.*, 2008).

Apoplast

In addition to the $O_2^{\cdot-}$ generated by NAD(P)H oxidase of plasma membrane, the cell wall-based enzymes have been proved to be responsible for apoplastic ROS production. The ROS generated

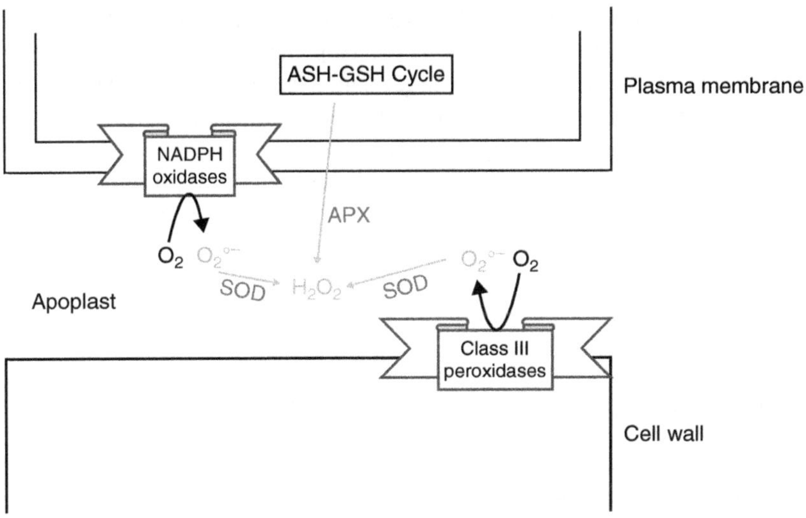

Figure 10.6 Reactive oxygen species (ROS) generation sites in apoplast and antioxidants involved in scavenging the generated ROS. APX, ascorbate peroxidase; ASH-GSH cycle, ascorbate-glutathione cycle; SOD, superoxide dismutase.

through cell wall-mediated reactions diffuse back as H_2O_2 into the cytoplasm via the aquaporin-mediated way (Dynowski *et al.*, 2008). Important cell wall-associated enzymes are class III peroxidases, amine oxidases, polyamine oxidases, oxalate oxidases, and so forth. Class III peroxidases are important cell wall localized enzymes that catalyze single-electron oxidation of several substrates (Passardi *et al.*, 2004). Peroxidases utilize NADH as an electron donor, which is oxidized to NAD radical, which then reduces O_2 to $O_2^{\cdot-}$. Heyno *et al.* (2011) suggested that apoplastic OH· generation depends mostly upon the activity of peroxidases localized in the cell wall. Certain other enzymes like amine oxidases and polyamine oxidases are abundant in the cell wall of leguminaceous and gramineous plants, respectively, and catalyze the oxidative deamination of di- or polyamines, with H_2O_2 as a common product (Angelini *et al.*, 2010). Polyamine oxidases are present not only in members of the Poaceae family but also in *Arabidopsis* (Fincato *et al.*, 2011) and in tobacco (Yoda *et al.*, 2006; Moschou *et al.*, 2008) and are involved in the terminal catabolism of polyamines, producing H_2O_2 (Planas-Portell *et al.*, 2013). Oxalate oxidases are cell wall-linked enzymes that belong to the germin protein family and are generally confined to the Poaceae; they catalyze the oxidation of oxalate to CO_2 and H_2O_2 (Bernier and Berna, 2001).

SOD, APX, and AA are responsible for scavenging the apoplasm-localized ROS. The source of AA in the apoplasm is the component of the symplastic AsA-GSH cycle, AA being imported out in the apoplasm where it forms the substrate for APX activity, resulting in H_2O_2 destruction (Figure 10.6).

Conclusion

The generation of ROS is an inevitable part of cellular physiology. There are specific sites for ROS generation within the cell, important ones being mitochondria, chloroplasts, peroxisomes, plasma membrane, and apoplast. During normal physiological activities, ROS generation is under control, but production increases several fold under stress conditions. ROS are highly reactive and create oxidative stress in plants, which can exert toxic effects in the cell.

These species may alter cell membrane properties and cause oxidative damage to nucleic acids, lipids, and proteins, which may render them non-functional. To protect themselves from the oxidative nature of ROS, cells have developed a well-organized system of antioxidants that serves to detoxify the harmful ROS, thus maintaining cellular homeostasis. The balance between the cellular ROS and cellular antioxidants determines if normal homeostasis will be maintained or if the cell will experience oxidative stress. The greater the potential of the antioxidative defense system to scavenge the ROS generated under stress, the more tolerant a plant is in enduring harsh environmental conditions.

References

Angelini R, Cona A, Federico R, *et al.* (2010) Plant amine oxidases "on the move": An update. *Plant Physiol Biochem* 48: 560–564.

Apel K, Hirt H. (2004) Reactive oxygen species: metabolism, oxidative stress, and signal transduction. *Ann Rev Plant Biol* 55: 373–399.

Aro EM, Virgin I, Andersson B. (1993) Photoinhibition of photosystem II. Inactivation, protein damage and turnover. *Biochim Biophys Acta* 1143: 113–134.

Arora A, Sairam RK, Srivastava GC. (2002) Oxidative stress and antioxidative system in plants. *Curr Sci* 82: 1227–1238.

Asada K. (2006) Production and scavenging of reactive oxygen species in chloroplasts and their functions. *Plant Physiol* 141: 391–396.

Asai S, Ohta K, Yoshioka H. (2008) MAPK signaling regulates nitric oxide and NADPH oxidase-dependent oxidative bursts in *Nicotiana benthamiana. Plant Cell* 20: 1390–1406.

Banhegyi G, Lusini L, Puskas F, *et al.* (1999) Preferential transport of glutathione versus glutathione disulfide in rat liver microsomal vesicles. *J Biol Chem* 274: 12213–12216.

Barber J, Andersson B. (1992) Too much of a good thing: light can be bad for photosynthesis *Trends Biochem Sci* 17: 61–66.

Bartoli CG, Gómez F, Martinez DE, Guiamet JJ. (2004) Mitochondria are the main target for oxidative damage in leaves of wheat (*Triticum aestivum* L.). *J Exp Bot* 55: 1663–1669.

Bérczi A, Møller IM. (2000) Redox enzymes in the plant plasma membrane and their possible roles *Plant Cell Environ* 23: 1287–1302.

Bernier F, Berna A. (2001) Germins and germin-like proteins: plant do-all proteins. But what do they do exactly? *Plant Physiol Biochem* 39: 545–554.

Bhattachrjee S. (2005) Reactive oxygen species and oxidative burst: roles in stress, senescence and signal transduction in plants. *Curr Sci* 89: 1113–1121.

Blokhina O Fagerstedt KV. (2010) Reactive oxygen species and nitric oxide in plant mitochondria: origin and redundant regulatory systems. *Physiol Plant* 138: 447–462.

Brand MD. (2010) The sites and topology of mitochondrial superoxide production *Exp Gerontol* 45: 466–472.

Cleland RE, Grace SC. (1999) Voltammetric detection of superoxide production by photosystem II. *FEBS Lett* 457: 348–352.

Corpas FJ, Barroso JB, del Río LA. (2001) Peroxisomes as a source of reactive oxygen species and nitric oxide signal molecules in plant cells. *Trends Plant Sci* 6: 145–150.

del Río LA, Corpas FJ, Sandalio LM, *et al.* (2002) Reactive oxygen species, antioxidant systems and nitric oxide in peroxisomes. *J Exp Bot* 53: 1255–1272.

del Río LA, Sandalio LM, Altomare DA, Zilinskas BA. (2003) Mitochondrial and peroxisomal manganese superoxide dismutase: differential expression during leaf senescence. *J Exp Bot* 54: 923–933.

del Río LA, Sandalio LM, Corpas FJ, Palma JM, Barroso JB. (2006) Reactive oxygen species and reactive nitrogen species in peroxisomes. Production, scavenging, and role in cell signaling. *Plant Physiol* 141: 330–335.

Dynowski M, Schaaf G, Loque D, *et al.* (2008) Plant plasma membrane water channels conduct the signalling molecule H_2O_2 *Biochem J* 414: 53–61.

Elstner EF. (1991) Mechanisms of oxygen activation in different compartments of plant cells. In: Pell EJ, Steffen KL (eds), *Active Oxygen/Oxidative Stress and Plant Metabolism*. American Society of Plant Physiologists, Rockville, MD, pp. 13–25.

Fincato P, Moschou PN, Spedaletti V, *et al.* (2011) Functional diversity inside the Arabidopsis polyamine oxidase gene family. *J Exp Bot* 62: 1155–1168.

Foyer CH, Noctor G. (2005) Redox homeostasis and antioxidant signaling: a metabolic interface between stress perception and physiological responses. *Plant Cell* 17: 1866–1875.

Halliwell B. (2006) Reactive species and antioxidants. redox biology is a fundamental theme of aerobic life. *Plant Physiol* 141: 312–322.

Halliwell B, Gutteridge JMC. (2007) *Free Radicals in Biology and Medicine*, 4th edn. Clarendon, Oxford.

Han C, Liu Q, Yang Y. (2009) Short-term effects of experimental warming and enhanced ultraviolet-B radiation on photosynthesis and antioxidant defense of *Picea asperata* seedlings. *Plant Growth Reg* 58: 153–162.

Heyno E, Mary V, Schopfer P, Krieger-Liszkay A. (2011) Oxygen activation at the plasma membrane: relation between superoxide and hydroxyl radical production by isolated membranes. *Planta* 234: 35–45.

Hu WH, Song XS, Shi K, Xia XJ, Zhou YH, Yu JQ. (2008) Changes in electron transport, superoxide dismutase and ascorbate peroxidase isoenzymes in chloroplasts and mitochondria of cucumber leaves as influenced by chilling. *Photosynthetica* 46: 581–588.

Jensen PK. (1966) Antimycin-insensitive oxidation of succinate and reduced nicotinamide-adenine dinucleotide in electron-transport particles. I. pH dependency and hydrogen peroxide formation. *Biochim Biophys Acta* 122: 157–166.

Kadota Y, Goh T, Tomatsu H, Tamauchi R, Higashi K, Muto S, Kuchitsu K. (2004) Cryptogein-induced initial events in tobacco BY-2 cells: pharmacological characterization of molecular relationship among cytosolic Ca^{2+} transients, anion efflux and production of reactive oxygen species. *Plant Cell Physiol* 45: 160–170.

Karuppanapandian T, Wang HW, Prabakaran N, *et al.* (2011) 2,4-dichlorophenoxyacetic acid-induced leaf senescence in mung bean (*Vigna radiata* L. Wilczek) and senescence inhibition by co-treatment with silver nanoparticles. *Plant Physiol Biochem* 49: 168–177.

Khan NA, Singh S (eds). (2008) *Abiotic Stress and Plant Responses*. IK International, New Delhi.

Kojer K, Riemer J. (2014) Balancing oxidative protein folding: the influences of reducing pathways on disulfide bond formation. *Biochim Biophys Acta* 1844: 1383–1390.

Krieger-Liszkay A, Fufezan C, Trebst A. (2008) Singlet oxygen production in photosystem II and related protection mechanism. *Photosynth Res* 98: 551–564.

Kwak JM, Mori IC, Pei ZM, *et al.* (2003) NADPH oxidase AtrbohD and AtrbohF genes function in ROS-dependent ABA signaling in *Arabidopsis*. *EMBO J* 22: 2623–2633.

Lambert AJ, Brand MD. (2004) Inhibitors of the quinone-binding site allow rapid superoxide production from mitochondrial NADH:ubiquinone oxidoreductase (complex I). *J Biol Chem* 279: 39414–39420.

Li WJ, Feng H, Fan JH, *et al.* (2000) Molecular cloning and expression of a phospholipid hydroperoxide glutathione peroxidase homolog in *Oryza sativa*. *Biochim Biophys Acta* 1493: 225–230.

Liu Y, Fiskum G, Schubert D. (2002) Generation of reactive oxygen species by the mitochondrial electron transport chain. *J Neurochem* 80: 780–787.

Loll B, Kern J, Saenger W, *et al.* (2005) Towards complete cofactor arrangement in the 3.0A° resolution structure of photosystem II. *Nature* 438: 1040–1044.

Lombardi A, Marshall RS, Castellazzi CL, *et al.* (2012) Redox regulation of glutenin subunit assembly in the plant endoplasmic reticulum. *Plant J* 72: 1015–1026.

Lopez-Huertas E, Corpas FJ, Sandalio LM, del Río LA. (1999) Characterization of membrane polypeptides from pea leaf peroxisomes involved in superoxide radical generation. *Biochem J* 337: 531–536.

Malhotra JD, Kaufman RJ. (2007) Endoplasmic reticulum stress and oxidative stress: A vicious cycle or a double-edged sword? *Antioxid Redox Signal* 9: 2277–2293.

Maxwell DP, Wang Y, McIntosh L. (1999) The alternative oxidase lowers mitochondrial reactive oxygen production in plant cells. *Proc Natl Acad Sci USA* 96: 8271–8276.

Mishra S, Jha AB, Dubey RS. (2011) Arsenite treatment induces oxidative stress, upregulates antioxidant system, and causes phytochelatin synthesis in rice seedlings. *Protoplasma* 248: 565–577.

Mittler R. (2002) Oxidative stress, antioxidants and stress tolerance *Trends Plant Sci* 7: 405–410.

Mittler R, Vanderauwera S, Gollery M, Brensegen FV. (2004) Reactive oxygen gene network of plants. *Trends in Plant Science* 9 (10): 490–498.

Moller IM. (2001) Plant mitochondria and oxidative stress: Electron transport, NADPH turnover, and metabolism of reactive oxygen specie.s *Annu Rev Plant Physiol* 52: 561–591.

Moller IM, Jensen PE, Hansson A. (2007) Oxidative modifications to cellular components in plants. *Annu Rev Plant Biol* 58: 459–481.

Moroder L, Besse D, Musiol HJ, *et al.* (1996) Oxidative folding of cystine-rich peptides *vs.* regioselective cysteine pairing strategies. *Biopolymers* 40: 207–234.

Moschou PN, Paschalidis KA, Delis ID, *et al.* (2008) Spermidine exodus and oxidation in the apoplast induced by abiotic stress is responsible for H_2O_2 signatures that direct tolerance responses in tobacco. *Plant Cell* 20: 1708–1724.

Muller FL, Liu Y, Van Remmen H. (2004) Complex III releases superoxide to both sides of the inner mitochondrial membrane. *J Biol Chem* 279: 49064–49073.

Munne-Bosch S, Alegre L. (2002) The function of tocopherols and tocotrienols in plants. *CRC Crit Rev Plant Sci* 21: 31–57.

Murgia I, Tarantino D, Vannini C, Bracale M, Carrabvieri S, Soave C. (2004) *Arabidopsis thaliana* plants over expressing thylakoidal ascorbate peroxidase show resistance to paraquat-induced photooxidative stress and to nitric oxide-induced cell death. *Plant J* 38: 940–995.

Murphy MP. (2009) How mitochondria produce reactive oxygen species. *Biochem J* 417: 1–13.

Nanda AK, Andrio E, Marino D, Pauly N, Dunand C. (2010) Reactive oxygen species homeostasis and signaling during plant-microorganism early interactions. *J Integr Plant Biol* 52: 195–204.

Navrot N, Rouhier N, Gelhaye E, Jaquot JP. (2007) Reactive oxygen species generation and antioxidant systems in plant mitochondria. *Physiol Plant* 129: 185–195.

Noctor G, Foyer CH. (1998) Ascorbate and glutathione: keeping active oxygen under control. *Ann Rev Plant Biol* 49: 249–279.

Noctor G, Paepe RD, Foyer CH. (2006) Mitochondrial redox biology and homeostasis in plants. *Trends Plant Sci* 12: 125–134.

Nuhse TS, Bottrill AR, Jones AM. (2007) Quantitative phosphoproteomic analysis of plasma membrane proteins reveals regulatory mechanisms of plant innate immune responses. *Plant J* 51: 931–940.

Ogawa K, Kenematsu K, Takabe K, Asada K. (1995) Attachment of CuZn superoxide dismutase to thylakoid membranes at the superoxide generation (PSI) in spinach chloroplasts: detection by immuno-gold labeling after rapid freezing and substitution method. *Plant Cell Physiol* 36: 565–573.

Ohad I, Keren N, Zer H, Gong H, Mor TS. (1994) Light induced degradation of the photosystem II reaction center D1 protein in vivo: an integrative approach. In: Baker NR, Bowyer JR (eds), *Photoinhibition of Photosynthesis: From Molecular Mechanisms to the Field*. BIOS Scientific Publishers, Oxford, pp. 161–177.

Oshino N, Chance B, Sies H, *et al.* (1973) The role of H_2O_2 generation in perfused rat liver and the reaction of catalase compound I and hydrogen donors. *Arch Biochem Biophys* 154: 117–131.

Pang CH, Wang BS. (2008) Oxidative stress and salt tolerance in plants. In: Luttge U, Beyschlag W, Murata J (eds), *Progress in Botany*. Springer, Berlin, pp. 231–245.

Passardi F, Longet D, Penel C, *et al.* (2004) The class III peroxidase multigenic family in rice and its evolution in land plants. *Phytochemistry* 65: 1879–1893.

Petrov V, Hille J, Meuller-Roeber B. (2013) ROS-mediated abiotic stress-induced programmed cell death in plants. *Front Plant Sci* 6: 1–16.

Pfannschmidt T. (2003) Chloroplast redox signals: how photosynthesis controls its own genes. *Trends Plant Sci* 8: 33–41.

Planas-Portell J, Gallart M, Tiburcio AF, *et al.* (2013) Copper containing amine oxidases contribute to terminal polyamine oxidation in peroxisomes and apoplast of *Arabidopsis thaliana*. *BMC Plant Biol* 13: 109.

Pospíšil P. (2012) Molecular mechanisms of production and scavenging of reactive oxygen species by photosystem II. *Biochim Biophys Acta* 1817: 218–231.

Quan L-J, Zhang B, Shi WW, Li HY. (2008) Hydrogen peroxide in plants: a versatile molecule of the reactive oxygen species network. *J Integrat Plant Biol* 50: 2–18.

Rasmusson AG, Geisler DA, Møller IM. (2008) The multiplicity of dehydrogenases in the electron transport chain of plant mitochondria. *Mitochondrion* 8: 47–60.

Rhoads DM, Umbach AL, Subbaiah CC, Siedow JN. (2006) Mitochondrial reactive oxygen species. Contribution to oxidative stress and interorganellar signaling. *Plant Physiol* 141: 357–366.

Sagi M, Fluhr R. (2006) Production of reactive oxygen species by plant NADPH oxidases. *Plant Physiol* 141: 336–340

Santos CX, Tanaka LY, Wosniak J, *et al.* (2009) Mechanisms and implications of reactive oxygen species generation during the unfolded protein response: roles of endoplasmic reticulum oxidoreductases, mitochondrial electron transport, and NADPH oxidase. *Antioxid Redox Signal* 11: 2409–2427.

Sazanov LA. (2007) Respiratory complex I: mechanistic and structural insights provided by the crystal structure of the hydrophilic domain. *Biochem* 46: 2275–2288.

Schrader M, Francini HD. (2006) Peroxisomes and oxidative stress. *Biochim Biophys Acta* 1763: 1755–1766.

Singh R, Singh S, Parihar P, *et al.* (2016) Reactive oxygen species (ROS): beneficial companions of plants' developmental processes. *Front Plant Sci* 7: 1299.

Singh S, Khan NA, Nazar R, Anjum NA. (2008) Photosynthetic traits and activities of antioxidant enzymes in blackgram (*Vigna mungo* L. Hepper) under cadmium stress. *Am J Plant Physiol* 3: 25–32.

Singh S, Srivastava PK, Kumar D, Tripathi DK, Chauhan DK, Prasad SM. (2015) Morpho-anatomical and biochemical adapting strategies of maize (*Zea mays* L.) seedlings against lead and chromium stresses. *Biocatal Agric Biotechnol* 4: 286–295.

Singh S, Tripathi DK, Singh S, *et al.* (2017) Toxicity of aluminium on various levels of plant cells and organism: a review. *Environ Exp Bot* 137: 177–193.

Singh VP, Tripathi DK, Kumar D, Chauhan DK. (2011) Influence of exogenous silicon addition on aluminium tolerance in rice seedlings. *Biol Trace Elem Res* 144: 1260–1274.

Siraki AG, Pourahmad J, Chan TS, *et al.* (2002) Endogenous and endobiotic induced reactive oxygen species formation by isolated hepatocytes. *Free Radical Biol Med* 32: 2–10.

Skovsen E, Snyder JW, Lambert JD, Ogilby PR. (2005) Lifetime and diffusion of singlet oxygen in a cell. *J Phys Chem B* 109: 8570–8573.
Stolz DB, Zamora R, Vodovotz Y, *et al.* (2002) Peroxisomal localization of inducible nitric oxide synthase in hepatocytes. *Hepatol* 36: 81–93.
Sweetlove LJ, Fell D, Fernie AR. (2008) Getting to grips with the plant metabolic network. *Biochem J* 409: 27–41.
Sweetlove LJ, Foyer CH. (2004) Roles for reactive oxygen species and antioxidants in plant mitochondria. In: Day DA, Millar AH, Whelan J (eds), *Plant Mitochondria: From Genome to Function, Advances in Photosynthesis and Respiration*, vol. 1. Kluwer Academic Press, Dordrecht, pp. 307–320.
Takahashi M, Shiraishi T, Asada K. (1998) *Arch Biochem Biophys* 267: 714–722.
Tarantino D, Vannini C, Bracale M, Campa M, Soave C, Murgia I. (2005) Antisense reduction of thylakoidal ascorbate peroxidase in Arabidopsis enhances paraquat-induced photooxidative stress and nitric oxide induced cell death. *Planta* 221: 757–765.
Telfer A, Frolov D, Barber J, *et al.* (2003) Oxidation of the two β-carotene molecules in the photosystem II reaction center. *Biochemistry* 42: 1008–1015.
Temple MD, Perrone GG, Dawes IW. (2005) Complex cellular responses to reactive oxygen species. *Trends Cell Biol* 15: 319–326.
Torres MA, Dangl JL, Jones JD. (2002) Arabidopsis gp91phox homologues AtrbohD and AtrbohF are required for accumulation of reactive oxygen intermediates in the plant defense response. *Proc Natl Acad Sci USA* 99: 517–522.
Trebst A. (2003) Function of β-carotene and tocopherol in photosystem II. *Z Naturforsch C* 58: 609–620.
Tripathi DK, Singh VP, Kumar D, Chauhan DK. (2012a) Impact of exogenous silicon addition on chromium uptake, growth, mineral elements, oxidative stress, antioxidant capacity, and leaf and root structures in rice seedlings exposed to hexavalent chromium. *Acta Physiol Plant* 34: 279–289.
Tripathi DK, Singh VP, Kumar D, Chauhan DK. (2012b) Rice seedlings under cadmium stress: effect of silicon on growth, cadmium uptake, oxidative stress, antioxidant capacity and root and leaf structures. *Chem Ecol* 28: 281–291.
Tripathi DK, Singh S, Singh VP, Prasad SM, Chauhan DK, Dubey NK. (2016a) Silicon nanoparticles more efficiently alleviate arsenate toxicity than silicon in maize cultivar and hybrid differing in arsenate tolerance. *Front Environ Sci* 4: 46.
Tripathi DK, Singh VP, Prasad SM, Dubey NK, Chauhan DK, Rai AK. (2016b) LIB spectroscopic and biochemical analysis to characterize lead toxicity alleviative nature of silicon in wheat (Triticum aestivum L.) seedlings. *J Photochem Photobiol B: Biol* 154: 89–98.
Tripathi DK, Singh S, Singh VP, Prasad SM, Dubey NK, Chauhan DK. (2017a) Silicon nanoparticles more effectively alleviated UV-B stress than silicon in wheat (Triticum aestivum) seedlings. *Plant Physiol Biochem* 110: 70–81.
Tripathi DK, Singh S, Singh S, *et al.* (2017b) Nitric oxide alleviates silver nanoparticles (AgNps)-induced phytotoxicity in Pisum sativum seedlings. *Plant Physiol Biochem* 110: 167–177.
Tripathi DK, Mishra RK, Singh S, *et al.* (2017c) Nitric oxide ameliorates zinc oxide nanoparticles phytotoxicity in wheat seedlings: implication of the ascorbate–glutathione cycle. *Front Plant Sci* 8: 1.
Tripathy BC, Oelmüller R. (2012) Reactive oxygen species generation and signaling in plants. *Plant Signal Behav* 712: 1621–1633.
Tripathy BC, Mohapatra A, Gupta I. (2007) Impairment of the photosynthetic apparatus by oxidative stress induced by photosensitization reaction of protoporphyrin IX. *Biochim Biophys Acta* 1767: 860–868.

Turrens JF. (2003) Mitochondrial formation of reactive oxygen species. *J Physiol* 552: 335–344.
Tuteja N. (2007) Mechanisms of high salinity tolerance in plants. *Meth Enzymol: Osmosens Osmosignal* 428: 419–438.
Tuteja N. (2010) Cold, salt and drought stress. In: Hirt H (ed.), *Plant Stress Biology: From Genomics towards System Biology*. Wiley-Blackwell, Weinheim, pp. 137–159.
Ushioda R, Hoseki J, Araki K, *et al.* (2008) ERdj5 is required as a disulfide reductase for degradation of misfolded proteins in the ER. *Science* 321: 569–572.
Valko M, Morris H, Cronin MTD. (2005) Metals, toxicity and oxidative stress. *Curr Med Chem* 12: 1161–1208.
Van der Vlies D, Makkinje M, Jansens, A *et al.* (2003) Oxidation of ER resident proteins upon oxidative stress: Effects of altering cellular redox/antioxidant status and implications for protein maturation. *Antioxid Redox Signal* 5: 381–387.
Votyakova TV, Reynolds IJ. (2001) ψm-Dependent and -independent production of reactive oxygen species by rat brain mitochondria. *J Neurochem* 79: 266–277.
Yabuta Y, Motoki T, Yoshimura K, *et al.* (2002) Thylakoid membrane-bound ascorbate peroxidase is a limiting factor of antioxidative systems under photo-oxidative stress. *Plant J* 32: 915–925.
Yoda H, Hiro Y, Sano H. (2006) Polyamine oxidase is one of the key elements for oxidative burst to induce programmed cell death in tobacco cultured cells. *Plant Physiol* 142: 193–206.

11

Quenching of Reactive Oxygen Species Inside the Cell: Physiological, Biochemical, and Molecular Mechanisms

Shivam Yadav and Neelam Atri

Banaras Hindu University, Varanasi, India

Introduction

In plants, reactive oxygen species (ROS) are continuously produced either as by-products of various metabolic pathways localized in different cellular compartments, or due to various biotic (e.g., pathogen attack) or abiotic (e.g., high light, drought, low temperature, high temperature, salt, UV-B, heavy metal, mechanical stress) constraints (Malan *et al.*, 1990; Elstner, 1991; Prasad *et al.*, 1994; Tsugane *et al.*, 1999; V.P. Singh *et al.*, 2011; Tripathi *et al.*, 2012a,b, 2016a,b, 2017a–c; S. Singh *et al.*, 2015, 2017; R. Singh *et al.*, 2016). Being very transient species and having high chemical reactivity, ROS are very likely to cause lipid peroxidation, and oxidation and degradation of enzymes and proteins (Matés *et al.*, 1999). ROS formation initiates due to univalent reduction of molecular oxygen using electrons and generating superoxides, hydrogen peroxide (H_2O_2), or hydroxyl radical (·OH) respectively, or with the generation of singlet oxygen ($^1O_2$) through transfer of excess excitation energy to O_2 (Alscher *et al.*, 2002; Wu *et al.*, 2008).

Among ROS, hydroxyl radicals are much more reactive than superoxide anion and hydrogen peroxide. As there are no known scavengers of hydroxyl radicals, the only way to avoid hydroxyl radical is to control the reactions that cause its generation. Therefore, various sophisticated strategies have been evolved by cells to keep the concentration of superoxide anion and hydrogen peroxide under tight control.

It is well known that in response to elevated ROS levels, plants activate their defense system (Foyer *et al.*,1994). The defense system of plants includes a complex antioxidative system composed of enzymatic as well as non-enzymatic components (Smirnoff, 1995; Ali *et al.*, 2008). The enzymatic defense system comprises of superoxide dismutase (SOD), ascorbate peroxidase (APX), glutathione reductase (GR), catalase (CAT), and peroxidases (POD). The non-enzymatic component of the defense system includes antioxidants divided in two general classes: (i) water-soluble reductants (ascorbate, glutathione, etc.); and (ii) lipid-soluble antioxidants (e.g., α-tocopherol and β-carotenes). The activities of the above-mentioned antioxidants are reported to be upregulated in various abiotic stresses such as salinity (Jaleel *et al.*, 2007a–d; Rai *et al.*, 2014), drought (Sankar *et al.*, 2007), UV-B (Shrivastava *et al.*, 2015), and temperature (Ali *et al.*, 2005).

In view of the above the present chapter is focused on mechanical quenching of ROS including its physiological, biochemical, and molecular aspects.

Reactive Oxygen Species in Plants: Boon Or Bane - Revisiting the Role of ROS, First Edition. Edited by Vijay Pratap Singh, Samiksha Singh, Durgesh Kumar Tripathi, Sheo Mohan Prasad, and Devendra Kumar Chauhan.

ROS Quenching: Physiological and Biochemical Aspects

ROS are known to cause oxidative damage to pigments, proteins, and lipids of different cellular compartments. In order to prevent the damaging effects of ROS on various cellular components, non-enzymatic and enzymatic ROS quenching is involved, discussed in the present section.

Non-Enzymatic ROS Quenching: Mechanism and Components

Non-enzymatic ROS scavengers are hydrophobic antioxidants, divided into the following categories:

1) Prenyl lipids (comprised of carotenoids – neoxanthin, zeaxanthin, and β-carotene).
2) Prenyl quinols (α-tocopherol, plastoquinol).
3) Cellular redox buffers (ascorbate, glutathione).
4) Phenols.

Carotenoids

Carotenoids are lipid-soluble antioxidants that perform various functions in plants, including oxidative stress tolerance. The ground triplet state of molecular oxygen ($^3O_2$) and a triplet excited state of carotenoids ($^3Car^*$) are formed after transfer of excitation energy from $^1O_2$ to carotenoid; $^3Car^*$ converts to ground state carotenoid while the triplet excitation energy is converted into heat (Equation 11.1).

$$^1O_2 + Car \rightarrow {}^3O_2 + {}^3Car^*$$

$$^3Car^* \rightarrow Car + heat \qquad (11.1)$$

The singlet-triplet energy transfer from singlet oxygen ($O_2(^1\Delta g)$) to carotenoid proceeds through the formation of $1(^1\Delta.S_0)$ encounter complex. This complex deactivates irreversibly by the internal conversion to ($T^{1.3}\Sigma$) complex, thus producing molecular oxygen and triplet carotenoid (Edge and Truscott, 1999).

$$O_2\left(^1\Delta g\right) + Car \rightarrow {}^1\left(^1\Delta \bullet S_0\right) \rightarrow {}^1\left(T^{1.3}\Sigma\right) \rightarrow O_2\left({}^3\sum_g\right) + {}^3Car^*$$

Xanthophylls

Xanthophylls can very effectively quench singlet oxygen formed in light-harvesting complex II (LHCII). α-Xanthophyll (lutein) and β-xanthophyll (zeaxanthin and neoxanthin) quench triplet chlorophyll and $^1O_2$, respectively (Dall'Osto *et al.*, 2007). Violaxanthin converts to zeaxanthin by removal of epoxy groups in the presence of violaxanthin de-epoxidase. This conversion is reversible, and the zeaxanthin to violaxanthin conversion takes place in the presence of zeaxanthin epoxidase.

β-Carotene

β-Carotene is the effective quencher of singlet oxygen, formed in the photosystem II (PSII) core antenna complex and the PSII reaction center (Trebst, 2003; Telfer, 2005).

α-Tocopherol

Tocopherols are lipophilic antioxidants and an important component of biological membranes. They are known to not only interact with polyunsaturated acyl groups of lipids and stabilize membranes but also quench various ROS and by-products of oxidative stress (Wang and Quinn, 2000). α-Tocopherol is a chain-breaking antioxidant, that is, it possesses the ability to repair oxidizing radicals directly thus inhibiting the chain propagation step during lipid auto-oxidation. Irreversible oxidation of α-tocopherols by O_2 is as follows:

$$^{1}O_2 + \alpha\text{-Toc} \rightarrow \alpha\text{-TOOH}\,(\text{8-hydroperoxy-tocopherone intermediate}) \rightarrow \alpha\text{-TQ}$$

The chromanol ring of α-tocopherol is expected to be located close to the membrane edge; therefore α-tocopherol is effective in quenching those O_2. that escape from the carotenoid antioxidant barrier formed in the interior of the thylakoid membrane (Kruk *et al.*, 1992; Jemioła-Rzemińska *et al.*, 2003).

Plastoquinols

There are various recent reports of ROS quenching by plastoquinols (PQH_2) (Kruk and Trebst, 2008; Yadav *et al.*, 2010). Oxidation of PQH_2 by O_2 leads to the generation of plastoquinone (PQ), which is subsequently oxidized by O_2 to the plastoquinone derivatives PQ(OH) and $PQ(OH)_3$, which contain one and three hydroxyl groups, respectively. Although PQH_2 possesses a lower quenching rate constant than α-tocopherol, PQH_2 has a large capability to scavenge O_2. In contrast to α-tocopherol, the plastoquinone ring is located in the membrane within the hydrophobic interior of the thylakoid membrane where $^{1}O_2$ is generated.

Ascorbic Acid

Ascorbic acid is one of the crucial components of the non-enzymatic antioxidant system (Smirnoff and Wheeler, 2000). Ascorbic acid is known to scavenge ROS formed from both photosynthetic and respiratory processes (Guo *et al.*, 2005), and it is also required for regeneration of α-tocopherol (Smirnoff, 2000) from tocopheroxyl radical. It sustains dissipation of excess excitation energy because it acts as a cofactor for violaxanthin de-epoxidase (Smirnoff, 2000). Ascorbic acid quenches O_2^{-} and $^{1}O_2$, scavenges H_2O_2, and reduces thiol radicals via conversion of monodehydroascorbate (MDHA) or dehydroascorbate (DHA) to ascorbate by the action of GR in a NADPH-dependent reaction (Tanaka *et al.*, 1999; Hamada, 2000).

Glutathione (GSH)

Glutathione, a tripeptide present in nearly all cell compartments, is the major source of non-protein thiols in most plant cells. Glutathione plays a major role in controlling H_2O_2 levels (Alscher *et al.*, 2002). The chemical reactivity of the thiol group of GSH along with relative stability and high water solubility make it ideal to protect organisms against oxidative stress (Noctor *et al.*, 2002). The central role of GSH in antioxidative defense relies on its ability to regenerate another powerful water-soluble antioxidant, ascorbic acid (AsA), via the AsA-GSH cycle (Noctor and Foyer, 1998; Halliwell, 2006). The roles of glutathione in ROS quenching have been reported by several authors. Pietrini *et al.* (2003) reported that enhanced antioxidant activity in the leaves and chloroplasts of *Phragmites australis* was associated with a large pool of GSH. *Arabidopsis* with low levels of GSH was highly sensitive to even low levels of Cd^{2+} (Xiang *et al.*, 2001).

Phenols

Polyphenolic structures possess the ability to eliminate radical species and also function as metal chelators (Ksouri *et al.*, 2008). A cycle was hypothesized by Santiago *et al.* (2000) whereby H_2O_2 is quenched by phenolics via a peroxidase; further phenolics are oxidized to phenoxyl radicals, which can be reduced by ascorbic acid. Reports suggest that antioxidant activity correlates well with total phenolics content and anthocyanin content in blueberry (Prior *et al.*, 1998; Kalt *et al.*, 2000; Ehlenfeldt and Prior, 2001). Salicylic acid, a phenolic hormone, plays a crucial role in stress resistance in plants (Alvarez, 2000).

Enzymatic ROS Quenching

Key enzymes involved in the detoxification of ROS are SOD, CAT, PODs, APX, GR, dehydroascorbate reductase (DHAR), and a few others involved in the Halliwell and Asada cycle. Under stress conditions the enhanced accumulation of almost all these enzymes has been reported (Zaka *et al.*, 2002; Zhao *et al.*, 2008; Pandey *et al.*, 2012; Agrawal *et al.*, 2014; Rai *et al.*, 2014; Srivastava *et al.*, 2015). This section incorporates biochemical and physiological aspects of ROS quenching by enzymatic antioxidants.

Superoxide Dismutase (EC: 1.15.1.1)

Superoxide dismutase (SOD) is a metal-containing enzyme that catalyzes the dismutation of superoxide radical to O_2 and H_2O_2:

$$O_2^{\bar{}} + O_2^{\bar{}} + 2H^+ \rightarrow H_2O_2 + O_2$$

SOD and CAT together stop the formation of most toxic and highly reactive hydroxyl radicals (·OH), which can react with all macromolecules. Moreover, there are no known direct scavengers of singlet oxygen $^1O_2$ or ·OH. SOD is believed to function in their elimination by chemical reaction (Alvarez, 2000). SODs are classified based on their metal cofactor into three types, namely Fe-SOD (chloroplasts), Mn-SOD (mitochondria), and Cu/Zn-SOD (chloroplasts, peroxisomes, and cytosol) (del Rio *et al.*, 1996). SOD activity under various abiotic stresses such as salinity, extreme temperature, drought, heavy metals, and waterlogging suggests execution of different mechanisms in different stress injuries (Jebara *et al.*, 2005; Karuppanapandian *et al.*, 2006a–c, 2009; Moussa and Abdel-Aziz, 2008). SOD constitutes the first line of defense inside a cell against ROS (Alscher *et al.*, 2002). ROS quenching through SOD is very complex and involves some other peroxidases (Scandalios, 1997).

Catalase (EC: 1.11.1.6)

Catalase (H_2O_2 oxidoreductase) is a tetrameric heme-containing enzyme that catalyzes the conversion of H_2O_2 into water and O_2. Various isoforms of CAT have been described in several plant species (Dat *et al.*, 2003; Vandenabeele *et al.*, 2004). Reports suggest the presence of three isoforms in *Arabidopsis*, namely cat-1, cat-2, and cat-3 (Frugoli *et al.*, 1996).

The enzyme is abundant in the glyoxisomes of lipid-storing tissues in germinating barley, where it degrades H_2O_2 (Fazeli *et al.*, 2007). It also has marked abundance in the peroxisomes of the leaves of C_3 plants, where it removes H_2O_2 produced during photorespiration by conversion of glycolate into glyoxylate (Foyer and Fletcher, 2001; Jayakumar *et al.*, 2008).

Ascorbate Peroxidase (EC: 1.11.1.11)

Whereas CAT scavenges H_2O_2 in peroxisomes, APX performs the same function in chloroplasts and cytosol of plant cells with the help of a hydrogen donor (ascorbate). APX has two cytosolic forms and a membrane-bound form. It converts H_2O_2 to H_2O and MDHA (monodehydroascorbate) (Asada, 2000), MDHA disproportionates to DHA non-enzymatically. Monodehydroascorbate reductase (MDHAR) produces ascorbic acid using NAD(P)H, and DHAR regenerates ascorbic acid utilizing GSH to form oxidized glutathione (GSSG). APX and CAT represent the key enzymes of H_2O_2 detoxification (Vanacker *et al.*, 1998). Various studies have shown APX involvement in protecting plants against oxidative damages due to biotic and abiotic stresses (S. Singh *et al.*, 2015, 2017; R. Singh *et al.*, 2016; Tripathi *et al.*, 2016a,b, 2017a–c). Naya *et al.* (2007) reported upregulated mRNA of cytosolic APX under drought stress in the alfalfa nodule. Furthermore, transgenic *Arabidopsis* plants overexpressing cytosolic APXs exhibited increased tolerance to salt stress (Lu *et al.*, 2007).

Monodehydroascorbate Reductase (EC: 1.6.5.4)

MDHAR is involved in regeneration of reduced ascorbate using NAD(P)H directly. MDHA is reduced directly to ascorbate using electrons derived from the photosynthetic electron transport chain. Schutzendübel *et al.* (2007) reported increased MDHAR activity in *Pinus sylvestris* under cadmium stress. Overexpression of MDHAR in transgenic tobacco increased tolerance against salt and osmotic stress (Eltayeb *et al.*, 2007).

Dehydroascorbate Reductase (EC: 1.8.5.1)

DHAR catalyzes regeneration of ascorbic acid from its oxidized state in a reaction that requires GSH (Eltayeb *et al.*, 2007). It contributes significantly to the regulation of the symplastic and apoplastic ascorbic acid pool size, and is expressed in rate-limiting amounts. DHAR overexpression has been shown to enhance stress tolerance in *Arabidopsis* and tobacco (Chen and Gallie, 2006; Eltayeb *et al.*, 2007).

Glutathione Reductase (EC: 1.6.4.2)

GR catalyzes the NADPH-dependent formation of a disulfide bond in glutathione disulfide (GSSG) and thus maintains the pool of reduced GSH. Its role in the Halliwell–Asada pathway in H_2O_2 quenching is well established (Noctor and Foyer, 1998; Asada, 2000) and it catalyzes the rate-limiting step. Upregulation of GR results in the accumulation of GSH thus conferring stress tolerance in plants.

Guaiacol Peroxidase (EC: 1.11.1.7)

Guaiacol peroxidase (GPX) is a heme-containing enzyme that catalyzes oxidation of certain substrates utilizing H_2O_2, thus helping cells to eliminate excess peroxides. GPX degrades indole-3-acetic acid (IAA) by consuming H_2O_2 in the cytosol, vacuoles, and cell wall as well as in extracellular spaces. Increased GPX activity has been reported in *Phaseolus vulgaris* under salinity stress conditions (Jebara *et al.*, 2005). Similarly Aravind and Prasad (2003) reported an increase in GPX activity in Cd-exposed plants of *Ceratophyllum demersum*.

Molecular Mechanism of ROS Quenching

Various studies indicate that cells have the means to sense ROS, leading to induction of specific responses; however, the underlying mechanism is as yet poorly understood (Scandalios, 1997; Dalton *et al.*, 1999). Although the transcriptional machinery that responds to ROS in eukaryotes is currently being deciphered, the prokaryotic system is better understood. Numerous studies have established that H_2O_2 induces ~30 proteins in bacteria (Christman *et al.*, 1985), of which around nine are regulated by the OxyR regulatory protein. Similarly, SoxRS regulatory proteins are known to regulate expression of $O_2^{\cdot-}$-responsive protein in bacteria. However, no obvious homologs of OxyR, SoxR, or SoxS have been found in eukaryotes (Storz and Imlay, 1999), although a number of other transcription factors have been found to play a role in regulating the expression of antioxidant genes in eukaryotes. In yeast, transcription regulators of antioxidant genes include ACE1, MAC1, YAP1, YAP2, HAP1, and HAP2/3/4 (Doke, 1997).

In mammals, NF-κB and AP-1 are responsible for the regulation of oxidative stress responses. NF-κB, AP-1, and ARF have also been found in the promoter regions of antioxidant genes in higher plants (Scandalios, 1997).

Numerous reports have shown that ROS serve as common factors in regulating various signaling pathways (Scandalios, 1997; Somssich and Halbrock, 1998). They activate the mitogen-activated protein kinase (MAPK) pathway leading to activation of antioxidant defense genes. In addition, exogenous application of several plant hormones and toxins has been shown to induce $O_2^{\cdot-}$ and H_2O_2 synthesis, leading to differential induction of some antioxidant genes and isozymes (Figure 11.1) (Williamson and Scandalios, 1992; Scandalios, 1997, 2001; Guan *et al.*, 2000).

Large-scale cDNA microarray analysis of the *Arabidopsis* transcriptome under oxidative stress leads to the identification of 175 expressed sequence tags (ESTs) from a sample of 11,000 regulated by H_2O_2. Out of the 175, 62 are repressed whereas 113 are induced. However, a considerable portion of these genes have predicted functions in defense responses, cell rescue, and signaling and transcription, thus emphasizing the multiple roles of H_2O_2 in plants' responses to stress. Similar to other organisms, expression of the MAPKs in *Arabidopsis* is induced by ROS, which ultimately mediate the induction of oxidative stress-responsive genes (Kovtun *et al.*, 2000).

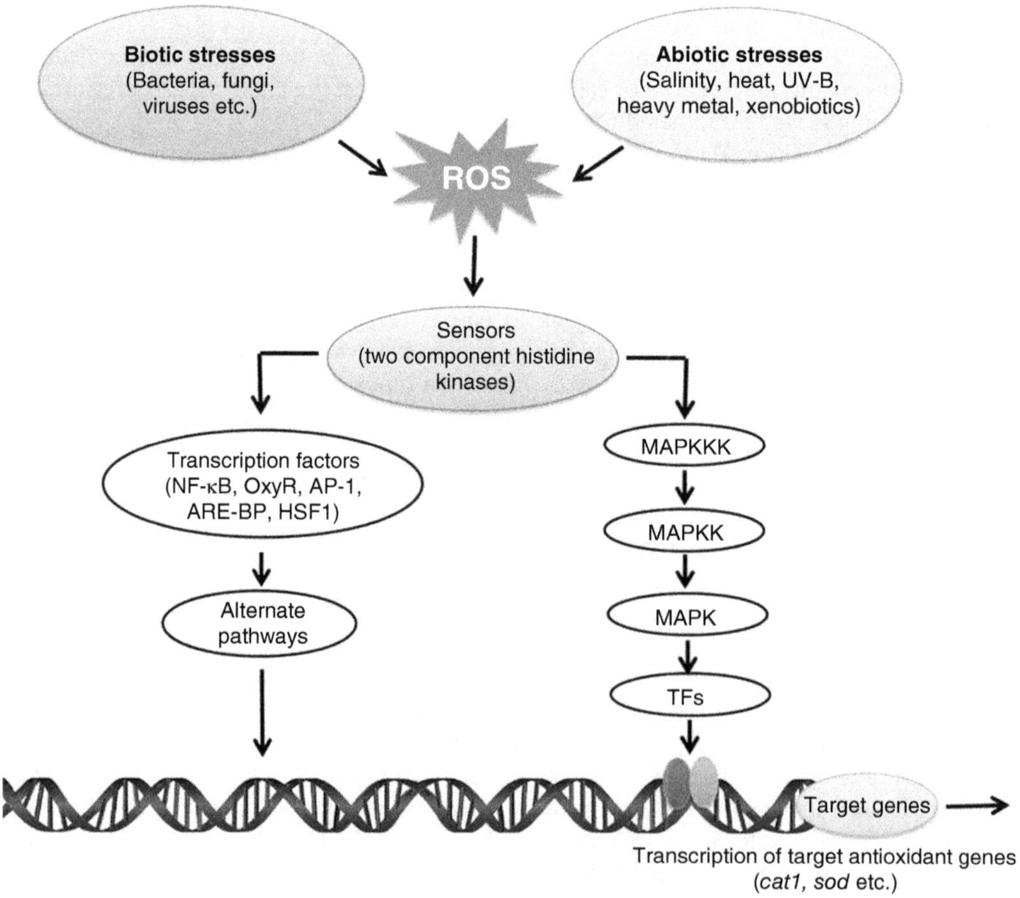

Figure 11.1 The major signaling pathways activated in response to reactive oxygen species (ROS). MAPK, mitogen-activated protein kinase; MAPKK, MAPK kinase; MAPKKK, MAPKK kinase; TFs, transcription factors.

In summary, ROS can activate gene transcription in two ways: (i) via transcription factors such as NF-κB, AP-1, and ARE-BP (ARE-binding proteins) that can interact directly with DNA promoters or motifs of target genes, or (ii) via activation of MAPK cascades, which in turn activate transcription factors that triggers target gene transcription.

Conclusion

Oxidative stress occurs due to an imbalance between the generation and elimination of ROS, inflicting indiscriminate damage on virtually all biomolecules. ROS pose a serious threat because unquenched species react spontaneously with organic molecules and cause membrane lipid peroxidation, protein oxidation, and enzyme inhibition ultimately threatening the organism's survival. The destructive impact of ROS is counteracted by the action of both enzymatic and non-enzymatic antioxidants. And yet, ROS also serve as ubiquitous signal molecules in higher plants. Genomic and proteomic tools are expanding our knowledge of ROS responses and the pleiotropic roles of ROS. Although much work has been done in recent years, there remain many uncertainties and gaps in our understanding of the molecular mechanism of ROS scavenging. Further studies at the gene level will elucidate the detailed mechanisms underlying ROS quenching in all aspects.

References

Agrawal C, Sen S, Singh S, *et al.* (2014) Comparative proteomics reveals association of early responsive protein in conferring butachlor tolerance in N_2-fixing cyanobacterium *Anabaena. J Proteomics* 96: 271–290.

Ali B, Hasan SA, Hayat S, *et al.* (2008) A role for brassinosteroids in the amelioration of aluminium stress through antioxidant system in mung bean (*Vigna radiata* L. Wilczek). *Environ Exp Bot* 62: 153–159.

Ali MB, Hahn EJ, Paek KY. (2005) Effects of temperature on oxidative stress defense systems, lipid peroxidation and lipoxygenase activity in *Phalaenopsis. Plant Physiol Biochem* 43: 213–223.

Alscher RG, Erturk N, Heath LS. (2002) Role of superoxide dismutases (SODs) in controlling oxidative stress in plants. *J Exp Biol* 53: 1331–1341.

Alvarez ME. (2000) Salicylic acid in the machinery of hypersensitive cell death and disease resistance. *Plant Mol Biol* 44: 429–442.

Aravind P, Prasad MNV. (2003) Zinc alleviates cadmium-induced oxidative stress in *Ceratophyllum demersum* L.: a free-floating freshwater macrophyte. *Plant Physiol Biochem* 41: 391–397.

Asada K. (2000) The water-water cycle as alternative photon and electron sinks. *Phil Trans R Soc Lond B Biol Sci* 355: 1419–1431.

Chen Z, Gallie DR. (2006) Dehydroascorbate reductase affects leaf growth, development, and function. *Plant Physiol* 142: 775–787.

Christman MF, Morgan RW, Jacobson FS, Ames BN. (1985) Positive control of a regulon for defenses against oxidative stress and some heat-shock proteins in *Salmonella typhimurium. Cell* 41: 753–762.

Dall'Osto L, Fiore A, Cazzaniga S, Giuliano G, Bassi R. (2007) Different roles of alpha and beta-branch xanthophylls in photosystem assembly and photoprotection. *J Biol Chem* 282: 35056–35068.

Dalton TP, Shertzer HG, Puga A. (1999) Regulation of gene expression by reactive oxygen. *Annu Rev Pharmacol Toxicol* 39: 67–101.

Dat JF, Pellinen R, Beeckman T, *et al.* (2003) Changes in hydrogen peroxide homeostasis trigger an active cell death process in tobacco. *Plant J* 33: 621–632.
del Rio LA, Palma JM, Sandalio LM, *et al.* (1996) Peroxisomes as a source of superoxide and hydrogen peroxide in stressed plants. *Biochem Soc Trans* 24: 434–438.
Doke N. (1997) The oxidative burst: roles in signal transduction and plant stress. In: Scandalios JG (ed.), *Oxidative Stress and the Molecular Biology of Antioxidant Defenses*. Cold Spring Harbor Laboratory Press, Plainview, NY, pp. 785–813.
Edge R, Truscott TG. (1999) Carotenoid radicals and the interaction of carotenoids with active oxygen species. In: Frank HA, Young AJ, Britton D, Cogdell RJ (eds), *Advances in Photosynthesis: the Photochemistry of Carotenoids*, vol. 8. Kluwer Academic Publishers, Dordrecht, pp. 223–234.
Ehlenfeldt MK, Prior RL. (2001) Oxygen radical absorbance capacity (ORAC) and phenolic and anthocyanin concentrations in fruit and leaf of high bush blueberry. *J Agric Food Chem* 49: 2222–2227.
Elstner EF. (1991) Mechanisms of oxygen activation in different compartments of plant cells. In: Pelland EJ, Steffen KL (eds), *Active Oxygen/Oxidative Stress in Plant Metabolism*. American Society of Plant Physiology, Rockville, MD, pp. 13–25.
Eltayeb AE, Kawano N, Badawi GH, *et al.* (2007) Overexpression of monodehydroascorbate reductase in transgenic tobacco confers enhanced tolerance to ozone, salt and polyethylene glycol stresses. *Planta* 225: 1255–1264.
Fazeli F, Ghorbanli M, Niknam V. (2007) Effect of drought on biomass, protein content, lipid peroxidation and antioxidant enzymes in two sesame cultivars. *Biol Plant* 51: 98–103.
Foyer CH, Descourvières P, Kunert KJ. (1994) Protection against oxygen radicals: an important defence mechanism studied in transgenic plants. *Plant Cell Environ* 17: 507–523.
Foyer CH, Fletcher JM. (2001) Plant antioxidants: colour me healthy. *Biologist* 48: 115–120.
Frugoli JA, Zhong HH, Nuccio ML, *et al.* (1996) Catalase is encoded by a multigene family in *Arabidopsis thaliana* (L.) Heynh. *Plant Physiol* 112: 327–336.
Guan LM, Zhao J, Scandalios JG. (2000) Cis-elements and transacting factors that regulate expression of the maize Cat1 antioxidant gene in response to ABA and osmotic stress: H_2O_2 is the likely intermediary signaling molecule for the response. *Plant J* 22: 87–95.
Guo Z, Tan H, Zhu Z, Lu S, Zhou B. (2005) Effect of intermediates on ascorbic acid and oxalate biosynthesis of rice and in relation to its stress resistance. *Plant Physiol Biochem* 43: 955–962.
Halliwell B. (2006) Oxidative stress and neurodegeneration: where are we now? *J Neurochem* 97: 1634–1658.
Hamada AM. (2000) Amelioration of drought stress by AA, thiamine or aspirin in wheat plants. *Indian J Plant Physiol* 5: 358–364.
Jayakumar K, Vijayarengan P, Zhao CX, Jaleel CA. (2008) Soil applied cobalt alters the nodulation, leg-haemoglobin content and antioxidant status of *Glycine max* (L.) Merr. *Colloids Surf B Biointerfaces* 67: 272–275.
Jaleel CA, Manivannan P, Kishorekumar A, Sankar B, Panneerselvam R. (2007a) Calcium chloride effects on salinity induced oxidative stress, proline metabolism and indole alkaloid accumulation in Catharanthus roseus. *CR Biol* 330: 674–683.
Jaleel CA, Gopi R, Manivannan P, Panneerselvam R. (2007b) Responses of antioxidant defense system of *Catharanthus roseus* (L.) G. Don. to paclobutrazol treatment under salinity. *Acta Physiol Plant* 29: 205–209.
Jaleel CA, Manivannan P, Lakshmanan GMA, Sridharan R, Panneerselvam R. (2007c) NaCl as a physiological modulator of proline metabolism and antioxidant potential in *Phyllanthus amarus*. *CR Biol* 330: 806–813.

Jaleel CA, Gopi R, Manivannan P, Panneerselvam R. (2007d) Antioxidative potentials as a protective mechanism in Catharanthus roseus (L.) G. Don. plants under salinity stress. *Turk J Bot* 31: 245–251.

Jebara S, Jebara M, Limam F, Aouani ME. (2005) Changes in ascorbate peroxidase, catalase, guaiacol peroxidase and superoxide dismutase activities in common bean (*Phaseolus vulgaris*) nodules under salt stress. *J Plant Physiol* 162: 929–936.

Jemioła-Rzemińska M, Kruk J, Strzałka K. (2003) Anisotropy measurements of intrinsic fluorescence of prenyl lipids reveal much higher mobility of plastoquinol than α-tocopherol in model membranes. *Chem Phys Lipids* 123: 233–243.

Kalt W, McDonald JE, Donner H. (2000) Anthocyanins, phenolics and antioxidant capacity of processed lowbush blueberry products. *J Food Sci* 65: 390–393.

Karuppanapandian T, Sinha PB, Kamarul Haniya A, Manoharan K. (2006a) Differential antioxidative responses of ascorbate-glutathione cycle enzymes and metabolites to chromium stress in green gram (*Vigna radiata* L. Wilczek) leaves. *J Plant Biol* 49: 440–447.

Karuppanapandian T, Sinha PB, KamarulHaniya A, Premkumar G, Manoharan K. (2006b) Aluminium-induced changes in antioxidative enzyme activities, hydrogen peroxide content and cell wall peroxidase activity in green gram (*Vigna radiata* L. cv. Wilczek) roots. *J Plant Biol* 33: 241–246.

Karuppanapandian T, Sinha PB, Premkumar G, Manoharan K. (2006c) Chromium toxicity: Correlated with increase in degradation of photosynthetic pigments and total soluble protein and increased peroxidase activity in green gram (*Vigna radiata* L.) seedlings. *J Swamy Bot-Cl* 23: 117–122.

Karuppanapandian T, Sinha PB, Kamarul Haniya A, Manoharan K. (2009) Chromium-induced accumulation of peroxide content, stimulation of antioxidative enzymes and lipid peroxidation in green gram (*Vigna radiata* L. cv. Wilczek) leaves. *Afr J Biotechnol* 8: 475–479.

Kovtun Y, Chiu WL, Tena G, Sheen J. (2000) Functional analysis of oxidative stress-activated mitogen-activated protein kinases cascade in plants. *Proc Natl Acad Sci USA* 97: 2940–2945.

Kruk J, Trebst A. (2008) Plastoquinol as a singlet oxygen scavenger in photosystem II. *Biochim Biophys Acta* 1777: 154–162.

Kruk J, Strzałka K, Leblanc RM. (1992) Monolayer study of plastoquinones, α-tocopherol quinone, their hydroquinone forms and their interaction with monogalactosyl diacylglycerol. Charge-transfer complexes in a mixed monolayer. *Biochim Biophys Acta* 1112: 19–26.

Ksouri R, Megdiche W, Falleh H, *et al.* (2008) Influence of biological, environmental and technical factors on phenolic content and antioxidant activities of Tunisian halophytes. *CR Biol* 331: 865–873.

Lu Z, Liu D, Liu S. (2007) Two rice cytosolic ascorbate peroxidases differentially improve salt tolerance in transgenic *Arabidopsis*. *Plant Cell Rep* 26: 1909–1917.

Malan C, Gregling MM, Gressel J. (1990) Correlation between CuZn superoxide dismutase and glutathione reductase and environmental and xenobiotic stress tolerance in maize inbreds. *Plant Sci* 69:157–166.

Matés JM, Pérez-Gómez C, Núñez de Castro I. (1999) Antioxidant enzymes and human diseases. *Clin Biochem* 32: 595–603.

Moussa HR, Abdel-Aziz SM. (2008) Comparative response of drought tolerant and drought sensitive maize genotypes to water stress. *Aust J Crop Sci* 1: 31–36.

Naya L, Ladrera R, Ramos J, *et al.* (2007) The response of carbon metabolism and antioxidant defenses of alfalfa nodules to drought stress and to the subsequent recovery of plants. *Plant Physiol* 144: 1104–1114.

Noctor G, Foyer CH. (1998) Ascorbate and glutathione: keeping active oxygen under control. *Annu Rev Plant Physiol Plant Mol Biol* 49: 249–270.

Noctor G, Leonardo G, Vanacker H, Foyer CH. (2002) Interactions between biosynthesis, compartmentation and transport in the control of glutathione homeostasis and signaling. *J Exp Bot* 53: 1283–1304.

Pandey S, Rai R, Rai LC. (2012) Proteomics combines morphological, physiological and biochemical attributes to unravel the survival strategy of *Anabaena* sp. PCC 7120 under arsenic stress. *J Proteomics* 75: 921–937.

Pietrini F, Iannelli MA, Pasqualini S, Massacci A. (2003) Interaction of cadmium with glutathione and photosynthesis in developing leaves and chloroplasts of *Phragmites australis* (Cav.) Trin. ex Steudel. *Plant Physiol* 133: 829–837.

Prasad TK, Anderson MD, Martin BA, Stewart CR. (1994) Evidence for chilling-induced oxidative stress in maize seedlings and a regulatory role for hydrogen peroxide. *Plant Cell* 6: 65–74.

Prior RL, Cao G, Martin A, *et al.* (1998) Antioxidant capacity as influenced by total phenolic and anthocyanin content, maturity and variety of vaccinium species. *J Agric Food Chem* 46: 2686–2693.

Rai S, Agrawal C, Shrivastava AK, Singh PK, Rai LC. (2014) Comparative proteomics unveils cross species variations in *Anabaena* under salt stress. *J Proteomics* 98: 254–270.

Sankar B, Jaleel CA, Manivannan P, Kishorekumar A, Somasundaram R, Panneerselvam R. (2007) Effect of paclobutrazol on water stress amelioration through antioxidants and free radical scavenging enzymes in *Arachis hypogaea* L. *Colloids Surf B Biointerfaces* 60: 229–235.

Santiago LJM, Louro RP, De Oliveira DE. (2000) Compartmentation of phenolic compounds and phenylalanine ammonia lyase in leaves of *Phyllanthus tenellus* Roxb. and their induction by copper sulphate. *Ann Bot* (Lond) 86: 1023–1032.

Scandalios JG. (1997) *Oxidative Stress and the Molecular Biology of Antioxidant Defenses*. Cold Spring Harbor Laboratory Press, Plainview, NY.

Scandalios JG. (2001) Molecular responses to oxidative stress. In: Hawkesford MJ, Buchner P (eds), *Molecular Analysis of Plant Adaptation to the Environment*. Kluwer Academic Publishers, Dordrecht, pp. 181–208.

Schutzendübel A, Schwanz P, Teichmann T, *et al.* (2007) Cadmium-induced changes in antioxidative systems, H_2O_2 content and differentiation in pine (*Pinus sylvestris*) roots. *Plant Physiol* 127: 887–898.

Shrivastava AK, Chatterjee A, Yadava S, Singh PK, Singh S, Rai LC. (2015) UV-B stress induced metabolic rearrangements explored with comparative proteomics in three *Anabaena* species. *J Proteomics* 127: 122–133.

Singh R, Singh S, Parihar P, *et al.* (2016) Reactive oxygen species (ROS): beneficial companions of plants' developmental processes. *Front Plant Sci* 7: 1299.

Singh S, Srivastava PK, Kumar D, Tripathi DK, Chauhan DK, Prasad SM. (2015) Morpho-anatomical and biochemical adapting strategies of maize (*Zea mays* L.) seedlings against lead and chromium stresses. *Biocatal Agric Biotechnol* 4: 286–295.

Singh S, Tripathi DK, Singh S, *et al.* (2017) Toxicity of aluminium on various levels of plant cells and organism: a review. *Environ Exp Bot* 137: 177–193.

Singh VP, Tripathi DK, Kumar D, Chauhan DK. (2011) Influence of exogenous silicon addition on aluminium tolerance in rice seedlings. *Biol Trace Elem Res* 144: 1260–1274.

Smirnoff M. (2000) AA: metabolism and functions of a multi-faceted molecule. *Curr Opin Plant Biol* 3: 229–235.

Smirnoff N. (1995) Antioxidant systems and plant response to the environment. In: Smirnoff V (ed.) *Environment and Plant Metabolism: Flexibility and Acclimation*. BIOS Scientific Publishers, Oxford, pp. 217–243.

Smirnoff N, Wheeler GL. (2000) AA in plants: biosynthesis and function. *Crit Rev Plant Sci* 19: 267–290.

Somssich IE, Halbrock K. (1998) Pathogen defence in plants – a paradigm of biological complexity. *Trends Plant Sci* 3: 86–90.

Storz G, Imlay JA. (1999) Oxidative stress. *Curr Opin Microbiol* 2: 188–194.

Tanaka R, Oster U, Kruse E, Rüdiger W, Grimm B. (1999) Reduced activity of geranyl geranyl reductase leads to loss of chlorophyll and tocopherol and to partially geranyl geranylated chlorophyll in transgenic tobacco plants expressing antisense RNA for geranyl geranyl reductase. *Plant Physiol* 120: 695–704.

Telfer A. (2005) Too much light? How β-carotene protects the photosystem II reaction centre. *Photochem Photobiol Sci* 4: 950–956.

Trebst A. (2003) Function of beta-carotene and tocopherol in photosystem II. *Z Naturforsch C* 58: 609–620.

Tripathi DK, Singh VP, Kumar D, Chauhan DK. (2012a) Impact of exogenous silicon addition on chromium uptake, growth, mineral elements, oxidative stress, antioxidant capacity, and leaf and root structures in rice seedlings exposed to hexavalent chromium. *Acta Physiol Plant* 34: 279–289.

Tripathi DK, Singh VP, Kumar D, Chauhan DK. (2012b) Rice seedlings under cadmium stress: effect of silicon on growth, cadmium uptake, oxidative stress, antioxidant capacity and root and leaf structures. *Chem Ecol* 28: 281–291.

Tripathi DK, Singh S, Singh VP, Prasad SM, Chauhan DK, Dubey NK. (2016a) Silicon nanoparticles more efficiently alleviate arsenate toxicity than silicon in maize cultivar and hybrid differing in arsenate tolerance. *Front Environ Sci* 4: 46.

Tripathi DK, Singh VP, Prasad SM, Dubey NK, Chauhan DK, Rai AK. (2016b) LIB spectroscopic and biochemical analysis to characterize lead toxicity alleviative nature of silicon in wheat (Triticum aestivum L.) seedlings. *J Photochem Photobiol B: Biol* 154: 89–98.

Tripathi DK, Singh S, Singh VP, Prasad SM, Dubey NK, Chauhan DK. (2017a) Silicon nanoparticles more effectively alleviated UV-B stress than silicon in wheat (Triticum aestivum) seedlings. *Plant Physiol Biochem* 110: 70–81.

Tripathi DK, Singh S, Singh S, *et al.* (2017b) Nitric oxide alleviates silver nanoparticles (AgNps)-induced phytotoxicity in Pisum sativum seedlings. *Plant Physiol Biochem* 110: 167–177.

Tripathi DK, Mishra RK, Singh S, *et al.* (2017c) Nitric oxide ameliorates zinc oxide nanoparticles phytotoxicity in wheat seedlings: implication of the ascorbate–glutathione cycle. *Front Plant Sci* 8: 1.

Tsugane K, Kobayashi K, Niwa Y, Ohba Y, Wada K, Kobayashi H. (1999) A recessive *Arabidopsis* mutant that grows enhanced active oxygen detoxification. *Plant Cell* 11: 1195–1206.

Vanacker H, Carver TLW, Foyer CH. (1998) Pathogen-induced changes in the antioxidant status of the apoplast in barley leaves. *Plant Physiol* 117: 1103–1114.

Vandenabeele S, Vanderauwera S, Vuylsteke M, *et al.* (2004) Catalase deficiency drastically affects gene expression induced by high light in *Arabidopsis thaliana*. *Plant J* 39: 45–58.

Wang X, Quinn PJ. (2000) The location and function of vitamin E in membranes. *Mol Membr Biol* 17: 143–156.

Williamson J, Scandalios JG. (1992) Differential response of maize catalases and superoxide dismutases to the photoactivated fungal toxin cercosporin. *Plant J* 2: 351–358.

Wu W, Wu J, Hu C. (2008) Signal cross talks for sustained MAPK activation and cell migration: the potential role of reactive oxygen species. *Cancer Metastasis Rev* 27: 303–314.

Xiang C, Werner BL, Christensen EM, Oliver DJ. (2001) The biological function of glutathione revisited in Arabidopsis transgenic plants with altered glutathione levels. *Plant Physiol* 126: 564–574.

Yadav DK, Kruk J, Sinha RK, Pospíšil P. (2010) Singlet oxygen scavenging activity of plastoquinol in photosystem II of higher plants: electron paramagnetic resonance spin-trapping study. *Biochim Biophys Acta* 1797: 1807–1811.

Zaka R, Vandecasteele CM, Misset T. (2002) Abscisic acid increase in non-growing and paclobutrazol-treated fruit of seedless mandarins. *Plant Physiol* 95: 613–619.

Zhao CX, Guo LY, Jaleel CA, Hong-Bo S, Yang HB. (2008) Prospects for dissecting plant-adaptive molecular mechanisms to improve wheat cultivars in drought environments. *CR Biol* 331: 579–586.

12

Adjustment of Plant Metabolism Against Reactive Oxygen Species: Past and Present

Gajendra S. Jeena, Ujjal J. Phukan, and Rakesh K. Shukla

CSIR-Central Institute of Medicinal and Aromatic Plants, Lucknow (UP), India

Introduction

Overproduction of reactive oxygen species (ROS) occurs in plants under various abiotic stresses. ROS are immensely reactive and toxic to cellular components such as proteins, lipids, carbohydrates, and DNA, which ultimately results in oxidative stress. The ROS generally include hydrogen peroxide (H_2O_2), superoxide anion ($O_2^{\cdot-}$), hydroxyl radical (·OH), and singlet oxygen ($^1O_2$) as the by-products of plant physiological metabolism, and are specifically restrained by potent enzymatic and non-enzymatic antioxidant defense systems. ROS act as second messengers involved in the stress response signal transduction pathway (Figure 12.1). ROS also control the expression of several genes and consequently govern many processes like growth, the cell cycle, programmed cell death, abiotic stress responses, pathogen defense, systemic signaling, and development. Plants can receive and decipher ROS signals to initiate relevant cellular responses via redox-sensitive enzymes, calcium mobilization, protein phosphorylation, and gene expression (Xiong *et al.*, 2002). The function of ROS in signal transduction signifies that there must be combined function of a regulation system to maintain ROS at non-toxic levels in a subtle balancing act between ROS generation – comprising ROS-generating enzymes and unavoidable production of ROS during basic cellular metabolism – and ROS scavenging pathways (You and Chan, 2015). To survive adverse environmental conditions, plants invoke a wide range of physiological and metabolic responses by activation of stress-responsive genes and synthesis of functional proteins through a complex signal transduction network, so as to confer tolerance under various stress conditions (Hirayama and Shinozaki, 2010; V.P. Singh *et al.*, 2011; Tripathi *et al.*, 2012a,b; S. Singh *et al.*, 2015). Stress-induced ROS production is neutralized by enzymatic antioxidant systems that encompass a variety of scavengers, such as superoxide dismutase (SOD), ascorbate peroxidase (APX), glutathione peroxidase (GPX), glutathione *S*-transferase (GST), and catalase (CAT), and non-enzymatic low molecular metabolites, such as ascorbic acid (AsA), glutathione (GSH), α-tocopherol, carotenoids, and flavonoids (Mittler *et al.*, 2004; V.P. Singh *et al.*, 2011; Tripathi *et al.*, 2012a,b; S. Singh *et al.*, 2015). Antioxidants are found in almost all cellular compartments, demonstrating the importance of ROS detoxification for cellular survival. Plants also accumulate various important metabolites that play protective roles such as aliphatic polyamines, including putrescine, spermidine, and spermine, which function in stress tolerance largely by modulating

Reactive Oxygen Species in Plants: Boon Or Bane - Revisiting the Role of ROS, First Edition. Edited by Vijay Pratap Singh, Samiksha Singh, Durgesh Kumar Tripathi, Sheo Mohan Prasad, and Devendra Kumar Chauhan.

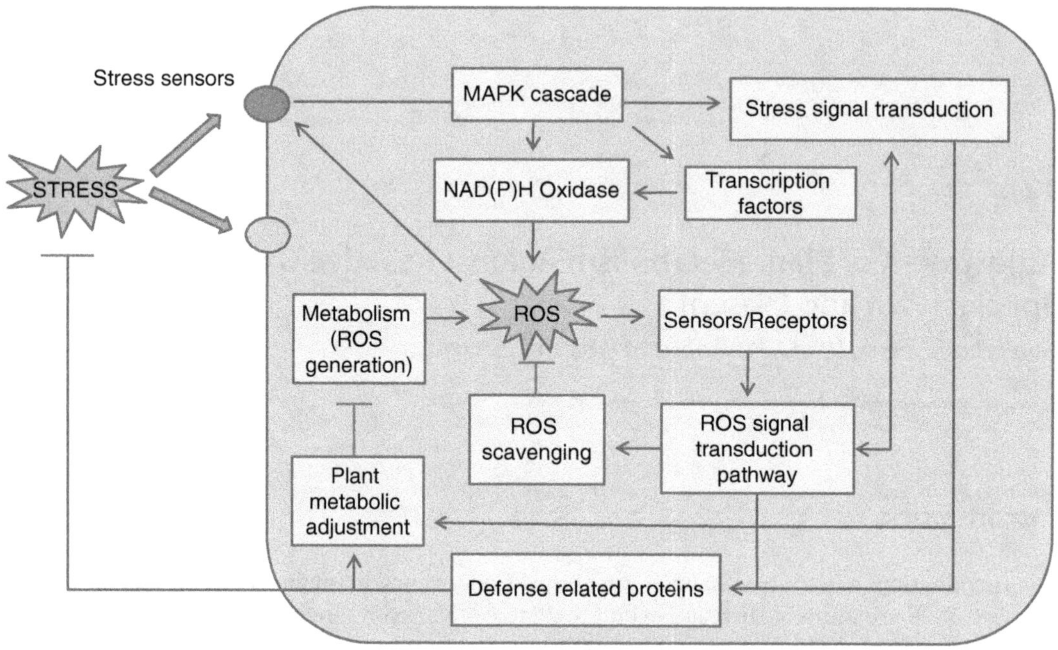

Figure 12.1 Reactive oxygen species (ROS) interplay during biotic and abiotic stress in plants. Stress stimulus received by plant sensors is mediated by various stress signal transduction pathways, such as the mitogen-activated protein kinase (MAPK) cascade, which activates defense-related proteins and modulates plant metabolism by reducing ROS generation. Increased ROS generation by NADPH oxidase under stress activates the plant's antioxidant defense system, which is involved in ROS scavenging.

the homeostasis of ROS (J.H. Liu *et al.*, 2015). In this chapter we highlight the role of plant metabolites in regulation of ROS homeostasis. The origin of ROS, the source and mode of their generation in response to different stresses, as well as metabolic processes are discussed. Endogenous ROS production under normal growth and development facilitates plants to mediate various positive adaptive signaling responses. Following stress stimuli, ROS signal the activation of stress-responsive genes to counteract the adverse effects. Mostly, the stress-specific triggers of ROS induction and response do not act alone but are often involved in cross-talk with other signaling molecules, like plant hormones, mitogen-activated protein kinase (MAPK), and calcium. Only when ROS are in excess do they exert a detrimental effect on plant cellular metabolism and physiology. Therefore particular attention is given to the moieties having potential antioxidant properties, especially those belonging to the different classes of enzymatic and non-enzymatic radical scavengers. These ROS scavengers help to protect against oxidative damage and maintain cellular homeostasis. We also focus on how these metabolites are regulated at the upstream transcription level. How transcription factors are activated, how they regulate antioxidant responsive genes directly or indirectly, and how they provide oxidative stress tolerance are described briefly. Although the mode and mechanism of ROS detoxification described in the past is considered, we highlight recent developments and progress in this area to the best of our knowledge. Certain aspects of the signaling pathway involved in ROS-induced oxidative damage are still under investigation but we know that biosynthesis, transport, and metabolism of ROS influence plant response mechanisms. So our aim is to present a comprehensive chapter to show how plants regulate the contrasting roles (beneficial as well as detrimental) of ROS, to attain a balance between development and survival. However, regardless of recent available data and advances in this field of signaling cascades, many questions remain unanswered, and further studies concentrating on the signaling and behavior of these molecules in plants are urgently required. So we also emphasize those aspects that future researchers should take into consideration to understand the entire system of ROS homeostasis.

ROS Homeostasis in Plants

The evolution of metabolic processes such as respiration and photosynthesis led to the generation of ROS in mitochondria, chloroplasts, and peroxisomes. Apart from these organelles, plasma membrane and apoplast are the main sites for ROS generation under various stress signals including exogenous environmental stimuli. Cell surface or apoplast localized enzymes, such as NADPH oxidases, amine oxidases, polyamine oxidases, oxalate oxidases, and a large family of class III peroxidases mainly contribute to production of apoplastic ROS (Gill and Tuteja, 2010). Intracellular ROS are mainly produced at low levels in organelles under normal growth conditions. However, ROS production is dramatically increased during stress. Abiotic stress-induced stomatal closure limits CO_2 uptake, increases the rate of photorespiratory H_2O_2 generation in peroxisomes, and produces superoxide free radical and H_2O_2 by decreasing the rate of the photosynthetic electron transport chain (Noctor *et al.*, 2014). Under stress conditions, elevated ROS production is also thought to act as a signal for the activation of stress-related genes (Baxter *et al.*, 2014). To protect themselves against oxidative damage, plants have evolved a potent enzymatic and non-enzymatic antioxidant system and fine inflection of low levels of ROS for signal transduction. Along with conventional enzymatic and non-enzymatic antioxidants, there is increasing evidence that soluble sugars, including disaccharides, raffinose family oligosaccharides, and fructans have a dual role with respect to ROS (Keunen *et al.*, 2013). Total soluble sugars involved in the regulation of ROS production by regulating the metabolic pathways involved in ROS generation, such as mitochondrial respiration or photosynthesis. Contrarily, they also feed NADPH-producing antioxidant metabolism (Couee *et al.*, 2006). Moreover, the antioxidant system regulates ROS production; mitigating the effect of stresses on plant metabolism is also important for maintaining ROS homeostasis. Alternative oxidases can avoid the excess production of ROS in mitochondrial electron transport chain (Maxwell *et al.*, 1999). Alternative oxidases can decrease the chance of electron leakage to O_2 to generate $O_2^{\cdot-}$ by diverting the flow of electrons through the electron transport chains. Other plant mechanisms such as leaf movement and curling, and rearranging the photosynthetic apparatus, may also avoid the over-reduction of ROS by balancing the amount of energy absorbed by the plant with the availability of CO_2 (Mittler, 2002). Cyclophilins are ubiquitous proteins located in subcellular compartments and involved in the regulation of a wide array of cellular responses. OsCyp2-P in rice is involved in positive regulation of multiple abiotic stress responses by regulating ROS and ion homeostasis (Kumari *et al.*, 2015). Normally lectin-like receptor kinases are present in plasma membranes and are involved in signal transduction, but lately it has been reported that they are involved in ROS homeostasis and stress tolerance. For example, overexpression of a pea lectin-like receptor kinase in transgenic *Nicotiana* resulted in enhanced salt tolerance by reducing the effect of both ionic and osmotic components of the concerned stress. This led to better tissue compartmentalization of Na^+, induced ROS-scavenging activity, alleviated membrane damage, enhanced growth and improved yield even after salinity stress (Vaid *et al.*, 2015). Along with ROS, reactive nitrogen species (RNS) also confer oxidative stress. *S*-nitrosoglutathione reductase is an important enzyme that regulates RNS-mediated responses by modulating NO metabolism and *S*-nitrosothiol homeostasis. Overexpression of *S*-nitrosoglutathione reductase in tomato led to enhanced tolerance to alkali stress by regulating redox signaling including ROS and RNS (Gong *et al.*, 2015). Other examples of antioxidant regulation involve glutaredoxin, 12-oxo-phytodienoic acid reductase, and SUV3 helicases. GRXS13 provided enhanced tolerance to photo-oxidative stress in *Arabidopsis*, while TaOPR1 overexpression in wheat and *Arabidopsis* imparted salt stress tolerance by modulating abscisic acid (ABA) and ROS signaling pathways (Laporte *et al.*, 2012; Dong *et al.*, 2013). The NTP-dependent RNA/DNA helicase OsSUV3 in rice gave rise to decreased H_2O_2 production and lipid peroxidation. OsSUV3 transgenic lines also showed increased antioxidant enzyme activities and induced tolerance to salt stress (Tuteja *et al.*, 2013).

Regulation of the Antioxidant System in Plants

The antioxidant system in plants consists of various enzymatic and non-enzymatic antioxidant components that work together with the ROS-generating pathway to maintain ROS homeostasis (Figure 12.2). Several studies showed that antioxidant components have important roles in maintaining ROS homeostasis in crop plants. During high-light photo-oxidative stress, ROS production is increased and primary antioxidants like isoprenoid quinones and chromanols are oxidized to other compounds that are indicators of singlet oxygen stress (Kruk *et al.*, 2016). Some important ROS-scavenging enzymes, such as GST (Dixon and Edwards, 2010), thioredoxin, and glutaredoxin (Meyer *et al.*, 2012), comprise large multigene families

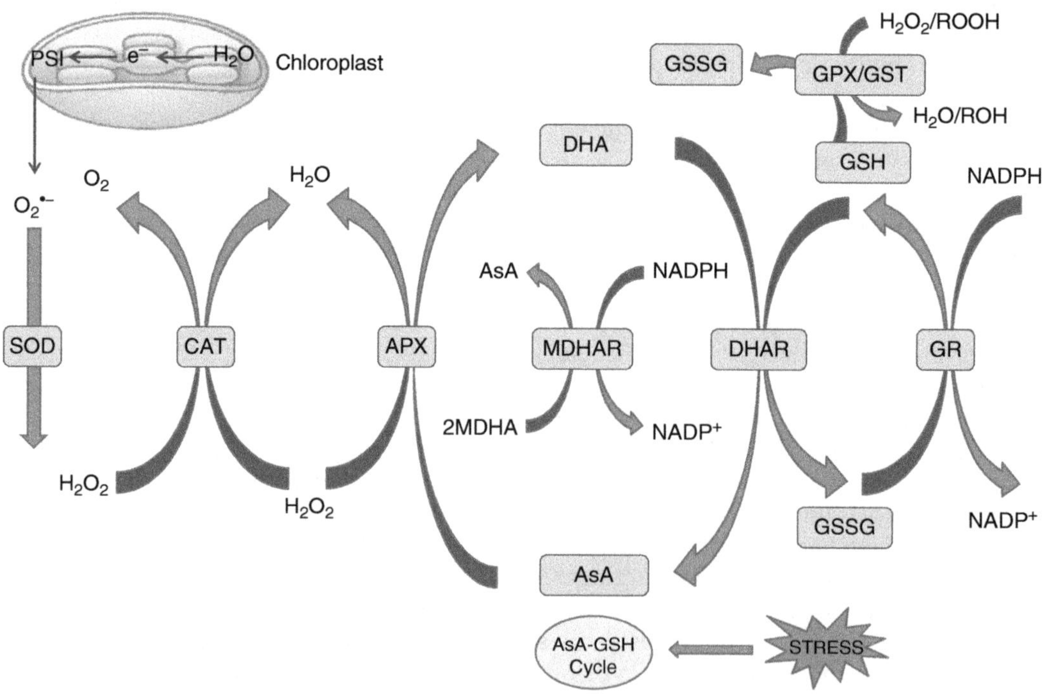

Figure 12.2 Enzymatic and non-enzymatic antioxidants in plants. Plants have an efficient antioxidant system that includes both enzymatic and non-enzymatic antioxidants. Enzymatic antioxidants include SOD (superoxide dismutase), dehydroascorbate reductase (DHAR), monodehydroascorbate reductase (MDHAR), catalase (CAT), and ascorbate peroxidase (APX). Non-enzymatic antioxidants include glutathione (GSH), glutathione *S*-transferases (GST), and ascorbic acid (AsA). DHA, dehydroascorbate; GPX, glutathione peroxidase; GR, glutathione reductase; GSSG, oxidized glutathione; PSI, photosystem I.

with distinct roles that deal with an array of adverse environmental conditions. OsTRXh1 from rice encodes h-type thioredoxin which regulates the redox state of apoplast and participates in plant development and stress responses (Zhang *et al.*, 2011). OsTRXh1 protein has reduction activity and is secreted into the extracellular space. Overexpression of OsTRXh1 in rice lowers H_2O_2 content under salt stress by reducing the expression of the salt-responsive genes responsible for a salt-sensitive phenotype. Another study showed that NADPH thioredoxin reductase in rice exploits NADPH to reduce the chloroplast 2-Cys PRX BAS1, which protects chloroplast against oxidative damage by reducing H_2O_2 (Perez-Ruiz *et al.*, 2006). There are three NADPH-dependent thioredoxin reductases found in *Arabidopsis* and rice that transfer reducing power to the thioredoxin/peroxiredoxin system for scavenging ROS and providing stress tolerance (Cha *et al.*, 2015). The plant stress hormone ABA is the key regulator of abiotic stress resistance in plants; it regulates a large number of stress-responsive genes by a complex regulatory network and provides environmental stress tolerance (Cutler *et al.*, 2010). ABA induces stress tolerance by partially activating plant antioxidant defense systems, including both enzymatic and non-enzymatic components, thus protecting plant cells from oxidative damage (Huang *et al.*, 2012). ABA accumulation under water stress and exogenous ABA treatment trigger the production of ROS, which leads to the activation of the antioxidant system in crops (Ye *et al.*, 2011). Molecules such as Ca^{2+}, calmodulin, NO, and ROS have been shown to play vital roles in ABA-induced antioxidant defense (Hu *et al.*, 2007). A recent study showed that histidine kinase OsHK3 in rice functions upstream of OsDMI3 and OsMPK1, and is necessary for the ABA-induced antioxidant defense system (Wen *et al.*, 2015). The maize calcium-dependent protein kinase (CDPK) gene, ZmCPK11, acts upstream of ZmMPK5, upregulates the expression and activities of SOD and APX, and hence the production of H_2O_2 in maize leaves in an ABA-dependent manner (Ding *et al.*, 2013). ZmMPK5 is required for NADPH oxidase-dependent self-propagation of ROS in brassinosteroid-induced antioxidant defense systems in maize (Zhang *et al.*, 2010). A further study found that a ZmMPK5 directly phosphorylates 65 kDa microtubule associated protein ZmMAP65-1a, which is required for induced antioxidant defense (Zhu *et al.*, 2013). Recently, Ca^{2+} and the maize CCaMK gene, ZmCCaMK, were shown to be required for the brassinosteroid-induced antioxidant defense system (Yan *et al.*, 2015).

ROS-Scavenging Enzymatic Antioxidants

The generation of ROS rapidly causes cell damage by triggering a chain reaction. Cells have evolved an elaborate system of enzymatic antioxidants against ROS that help them to scavenge indigenously generated free radicals. Various enzymes involved in ROS scavenging have been manipulated, overexpressed, or downregulated to add to the present knowledge and understanding of the role of the antioxidant systems. Plants possess very efficient enzymatic antioxidant defense systems, as discussed in detail below.

Superoxide Dismutase (SOD)

It is well known that various environmental stresses often lead to the increased generation of ROS, and superoxide formation, which may induce both protective responses and cellular damage. The scavenging of superoxide is achieved through SODs, which catalyze the dismutation of superoxide to H_2O_2 at rates 10^3-fold faster than spontaneous dismutation (Bowler and Montagu, 1992). The *Arabidopsis thaliana* genome consists of multiples genes encoding different SODs, which include three Fe-SOD genes (FSD1, FSD2, and FSD3), three Cu/Zn-SOD genes (CSD1, CSD2, and CSD3), and one Mn-SOD gene (MSD1) (Kliebenstein *et al.*, 1999). The rice genome has eight genes that encode putative SODs, including two cytosolic copper-zinc SODs, one putative Cu/Zn-SOD-like, one plastid SOD, two iron SODs (Fe-SOD2 and Fe-SOD3), and one manganese SOD (Nath *et al.*, 2014). A unique mitochondrial protein in *Arabidopsis*, AtMTM1, is also found to activate Mn-SOD under stress (Su *et al.*, 2007). The subcellular localization of SOD isozymes is also found to be peculiar. Mn-SOD is localized in the mitochondria and peroxisomes of eukaryotic cells (Rio *et al.*, 2003). Some Cu/Zn-SOD isozymes are found in the cytosolic fractions as well as in the chloroplasts of higher plants (Diaz Vivancos *et al.*, 2008). Peroxisomes and glyoxysomes of *Citrullus vulgaris* contain both Cu/Zn and Mn-SOD activity, but there are no reports of extracellular SOD enzymes in plants (Sandalio and Rio 1988). The upregulation of SODs is involved in avoiding oxidative stress caused by biotic and abiotic factors, and performs an important role in the plants' survival under environmental stress. Over-expression of Mn-SOD increases the level of H2O2 and also increases the activity of SOD and GR activities under photo-oxidative stress (Melchiorre *et al.*, 2009). *Arabidopsis* has three types of Fe-SODs but only FeSOD2 and FeSOD3 play essential roles in early chloroplast development. FeSOD2 and FeSOD3 act as ROS scavengers in the maintenance of early chloroplast development by protecting the chloroplast nucleoids from ROS (Myouga *et al.*, 2008). Cytoplasmic Cu/Zn-SOD1 scavenges intracellular superoxide radicals and affects fungal growth; SOD1-mediated deoxynivalenol production contributes to the virulence of *Fusarium graminearum* in wheat head infection (Yao *et al.*, 2016). Under water-deficit conditions, ROS play an important regulatory role in the process of *Cassava* leaf abscission. Co-overexpression of the ROS-scavenging proteins SOD and CAT1 in *Cassava* lowers the production of H_2O_2 in the abscission zone. Delayed co-overexpression of ROS-scavenging proteins SOD and CAT1 in plants results in cell separation of the pulvinus abscission zone both *in vitro* and at the plant level (Liao *et al.*, 2016). A wheat SOD gene, *TaSOD2*, increases salt resistance through regulating redox homeostasis by stimulating NADPH oxidase activity (M. Wang *et al.*, 2016). Necrotrophic pathogens improve the colonization process by producing ROS under stress. Absence of the Cu/Zn-SOD, BcSOD1, decreases *Botrytis cinerea* pathogenesis in *Arabidopsis* and tomato plants, which plays a role between ROS, callose, and signaling pathways (Lopez-Cruz *et al.*, 2017). The PaSOD gene from *Potentilla atrosanguinea* in single and double transgenic plants positively regulates expression of lignin biosynthesis pathway genes, leading to increased and ectopic accumulation of lignin in vascular tissues with larger xylem fibres under salt stress (Shafi *et al.*, 2015).

Catalase (CAT)

CAT constitutes another important class of antioxidant enzymes that help in scavenging the H_2O_2 produced in peroxisomes by oxidases involved in β-oxidation of fatty acids, photorespiration, and purine catabolism. Detoxification of H_2O_2 by CAT, mainly located in peroxisomes, is highly energy efficient, as it does not utilize cellular reducing equivalents. Different isoforms of CAT have been extensively studied in higher plants (Polidoros and Scandalios, 1999). Maize has three CAT isoforms (CAT1, CAT2, and CAT3), genes for which are located on separate chromosomes and are differentially expressed and independently regulated (Scandalios *et al.*, 1990). It has also been reported that apart from scavenging H_2O_2, CAT also reacts with some hydroperoxides like methyl hydrogen peroxide (Ali and Alqurainy *et al.*, 2006). Transgenic rice plants overexpressing OsMT1a showed enhanced CAT activity and thus enhanced tolerance to drought (Yang *et al.*, 2009). Overexpression of the CAT3 gene from *Brassica juncea* in tobacco provides enhanced tolerance to Cd stress in the transgenic plants (Guan *et al.*, 2009). The differential expression pattern of the NnCAT gene from *Nelumbo nucifera* indicates its vital role in the initial stages of oxidative stress, regulating ROS levels by extracellular stimuli such as short-time mechanical wounding (Dong *et al.*, 2015). The physical interaction of glycolate oxidase with catalase in rice leaves can be deregulated by the induction of salicylic acid. Moreover, the glycolate oxidase-mediated H_2O_2 accumulation is synergistically enhanced by salicylic acid (Zhang *et al.*, 2016).

Ascorbate Peroxidase (APX)

Ascorbate peroxidase, or scavenging peroxidase, is a heme protein with protoporphyrin IX as its prosthetic group and a molecular size similar to the plant peroxidase guaiacol peroxidase. Its primary function is rapid removal of H_2O_2 at the site of generation (Asada, 1992). The APX family of antioxidant enzymes consists of at least five different isoforms in thylakoid membrane, glyoxisome membrane, chloroplast stroma, and cytosol (Noctor and Foyer, 1998). Under mild drought stress, plants show elevated chloroplastic-APX (c-APX) activity, but c-APX activity declines under high drought stress (Sharma and Duby, 2005). The expression patterns of differentially localized isoforms of APX were analyzed in roots of etiolated *Oryza sativa* seedlings under salt stress, and it was found that chloroplastic OsAPX8 has enhanced activity compared to cytoslic APX (Hong *et al.*, 2007). Overexpression of APX-like 1 gene from *Capsicum annuum* in transgenic tobacco plants led to enhanced tolerance to oxidative stress, and to the oomycete pathogen *Phytophthora nicotianae*. It was suggested that the overproduction of APX increased the peroxidase (POD) activity, which strengthens the ROS-scavenging system and leads to oxidative stress tolerance and oomycete pathogen resistance (Sarowar *et al.*, 2005). The APX-6 gene in *Arabidopsis* protects against desiccation and helps in germination of seeds by mediating cross-talk between ROS, ABA, and auxin signaling (Chen *et al.*, 2014). A peroxisomal APX from *Puccinellia tenuiflora* improves abiotic stress tolerance in transgenic *Arabidopsis* by lowering the accumulation of H_2O_2 (Guan *et al.*, 2015). Increased expression of the native cytosolic form of Cu/Zn-SOD and APX in *Cassava* promotes tolerance to oxidative and chilling stresses (Xu *et al.*, 2014). Enhanced activity of chloroplastic APX of transgenic plants increased salt tolerance by enhancing ROS-scavenging activity under short-term NaCl stress conditions. Overexpression of APX from *Jatropha curcas* in tobacco enhanced tolerance under salt stress during seedling establishment and growth. Moreover, transgenic plants showed higher chlorophyll content and APX activity, which results in decreased H_2O_2 content compared to wild type plants (Liu *et al.*, 2013).

Guaiacol Peroxidase (GPOX)

GPOX can be distinguished from APX by differences in amino acid sequence and physiological functions. GPOX breaks down indole-3-acetic acid and performs a significant role in the biosynthesis of lignin and defense against biotic stresses by utilizing H_2O_2. GPOX promotes production of aromatic electron donors like guaiacol and pyragallol through oxidation of ascorbate (Asada, 1999). GPOX activity generally varies depending upon plant species and environmental stresses. GOPX activity increased in *Triticum aestivum* under cadmium stress to enhance scavenging of ROS (Milone *et al.*, 2003). A synergistic increase in GPOX activity in both leaf and root tissues of *O. sativa* has also been shown under salinity stress (Koji *et al.*, 2009). In *Vicia faba* plants selenium elevates GPOX activity in roots under oxidative stress induced by heavy metal lead (Pb), which may be suggestive of its potent antioxidant properties (Mroczek and Wojcik, 2012).

Glutathione Reductase (GR)

Glutathione reductase is an important antioxidant enzyme of the AsA-GSH cycle and performs a critical role in defense against ROS by maintaining the reduced status of GSH. It is localized primarily in chloroplasts, but is found to some extent in mitochondria and cytosol (Creissen *et al.*, 1994). GR catalyzes the reduction of GSH, which is involved in various metabolic and antioxidant processes in plants. GR is important in maintaining the GSH pool by catalyzing the NADPH-dependent reduction of oxidized glutathione (GSSG) (Chalapathi Rao and Reddy, 2008). Transgenic plants exhibiting low levels of GR activity showed increased sensitivity to oxidative stress. It was proposed that GR plays a crucial role in regenerating GSH and protects plants against oxidative stress by maintaining the AsA pool (Ding *et al.*, 2009). Under stress conditions, GR also has a role in upholding the reduced glutathione pool. In *Pennisetum glaucum*, the transcript level of PgGR was differentially governed by heat, cold, salinity, and methyl viologen induced oxidative stress (Achary *et al.*, 2015). The evolutionary relationship of GR genes from rice and *Arabidopsis* was evaluated using the multiple sequence alignment tool and phylogenetic tree, which determined evolutionary conserved pyridine nucleotide disulfide oxidoreductases class-I active sites among the GR proteins (Trivedi *et al.*, 2013).

Monodehydroascorbate Reductase (MDHAR)

MDHAR catalyzes the regeneration of ascorbate from MDHA. MDHARs are flavin nucleotide (FAD)-containing enzymes found in chloroplasts, cytosol, mitochondria, and peroxisomes. MDHAR shows high specificity for MDHA as the electron acceptor, promoting NADH rather than NADPH as the electron donor (Asada, 1999). It is well studied that disproportionation occurs in the thylakoids by photoreduced ferrodoxin. Photoreduced ferredoxin can reduce MDHA more adequately than $NADP^+$. MDHAR is unable to catalyze the reduction of MDHA in the thylakoidal scavenging system because MDHAR requires NAD(P)H to function unlike photoreduced ferredoxin that require NADP+ to reduce MDHA. (Asada, 1999). Concomitant APX and MDHAR is also localized in mitochondria and peroxisomes, where it shows H_2O_2-scavenging activity (Rio *et al.*, 2002). Differential transcript levels of the MDHAR gene in *Brassica rapa* have been detected in response to distinct oxidative stresses, in addition to H_2O_2, salicylic acid, paraquat, and ozone (Yoon *et al.*, 2004). When MDHAR gene of Arabidopsis was over-expressed in tobacco, it conferred enhanced tolerance to salt and osmotic stress in the transgenic lines (Eltayeb *et al.*, 2007). TaMDAR6 in wheat acts as a negative regulator of plant cell death; it functions indirectly in stomatal regulation during infection with of wheat stripe rust fungus. TaMDAR6 knock down plants displayed enhanced resistance to wheat-Puccinia striiformis through increased ROS production and PCD events. Infection leads to development of substomatal vesicles which is greatly reduced in TaMDAR6 knocked-down plants. (Abou-Attia *et al.*, 2016). Co-expression of MDHAR and DHAR from *Brassica rapa* adequately provides tolerance to freezing-induced oxidative stress (Shin *et al.*, 2013). Overexpression of AeMDHAR from a mangrove plant confers enhanced salt tolerance in rice (Sultana *et al.*, 2012).

Dehydroascorbate Reductase (DHAR)

DHAR is a key enzyme in the recycling of ascorbate, catalyzing the glutathione-dependent reduction of oxidized ascorbate. As a result, DHAR regenerates a pool of reduced ascorbate and detoxifies ROS. DHAR has an ascorbate-binding site that overlaps with the glutathione-binding site, suggesting a ping-pong kinetic mechanism for electron transfer at the common Cys20 active site (Do *et al.*, 2016). Plant cells contain cytosolic and chloroplastic DHARs, indicating that dehydroascorbate can be directly reduced in the cytoplasm and chloroplast by DHARs. A novel DHAR was found in the vacuole of *Zea mays* (Zhang *et al.*, 2015). Overexpression of the DHAR gene enhanced tolerance for salt in *Arabidopsis* (Ushimaru *et al.*, 2006) and for ozone toxicity in tobacco (Chen and Gallie, 2005). Overexpression of *Arabidopsis* cytosolic DHAR in transgenic tobacco increased the level of ascorbate, which provided enhanced tolerance to plants under aluminum (Al) stress (Yin *et al.*, 2010). Tobacco chloroplast transformants expressing genes encoding DHAR exhibited altered antioxidant metabolism and improved abiotic stress tolerance (Le Martret *et al.*, 2011). The chloroplast localized DHAR3 gene regulates the redox state of both ascorbate and glutathione acclimated to high light tolerance in *Arabidopsis* (Noshi *et al.*, 2016).

Glutathione *S*-Transferases (GSTs)

Plant GSTs are the key enzymes involved in detoxification of herbicides, hormone homeostasis, vacuolar sequestration of anthocyanin, tyrosine metabolism, hydroxyperoxide detoxification, and regulation of apoptosis in plant responses to biotic and abiotic stresses (Dixon *et al.*, 2010). GST/GPX overexpression in transgenic tobacco plants sustained increased scavenging of glutathione-dependent peroxide and adjustment in glutathione and ascorbate metabolism that led to reduced oxidative damage (Roxas *et al.*, 2000). The induction of OsGSTu3 and OsGSTu4, encoding the tau class of GSTs, occurred under environmental stress in rice seedlings (Moons, 2003). Transgenic plants overexpressing GSTcr1 exhibit significant increases in the activities of GST and GPX, which enhance the antioxidant defense of the transgenic plants to combat oxidative stress (Yu *et al.*, 2003). The GST Nt107 gene expressed in transgenic *Gossypium hirsutum* lines was used to find out the tolerance range under distinct stresses like chilling, salinity, and herbicides. It was found that transgenic plants exhibit enhanced GST activity under both control conditions and salt stress (Light *et al.*, 2005). Another study found that overexpression of *Prosopis juliflora* GST in tobacco gave a better survival rate than control plants under 15% PEG stress. Further, GFP fusion studies revealed the presence of PjGSTU1 in the chloroplasts of transgenic plants, which was correlated with the role of GST in ROS scavenging (George *et al.*, 2010). Allyl isothiocyanate degrades glutathione and upregulates the expression of GST in *Arabidopsis thaliana*. Allyl isothiocyanate is involved in the onset of defense-related mechanisms such as the regulation of stomatal aperture. It also led to enhanced ROS accumulation in the roots and loss of chlorophyll during oxidative stress (Overby *et al.*, 2015). Overexpression of the GST gene from *Pyrus pyrifolia* fruit improves tolerance to abiotic stress in transgenic tobacco plants (D. Liu *et al.*, 2013).

Glutathione Peroxidase (GPX)

Glutathione peroxidases are thiol-based enzymes that catalyze the reduction of H_2O_2 and hydroperoxides to H_2O or alcohols. They mitigate the toxicity of these compounds to the cell mainly by using thioredoxin as an electron donor (Passaia and Margis 2015). Plant GPXs perform a crucial role in redox signal transduction coupled with their peroxidase activity. Introduction of the radish phospholipid hydroperoxide GPX gene (PHGPx3) into a yeast deletion mutant significantly rescued the growth of the recombinant cell exposed to linolenic acid, indicating a similar role to the yeast PHGPx3 gene in protection of membranes against lipid peroxidation. Transcriptome analysis of salt stress tolerant and susceptible foxtail millet lines showed that PHGPx gene is highly expressed in the salt tolerant lines. They also found higher PHGPx protein content in the tolerant lines (Sreenivasulu *et al.*, 2004). GPXh showed transcriptional upregulation by $^1O_2$ photosensitizers when Gpxh promoter was fused with the arylsulfatase reporter gene (Leisinger *et al.*, 2001). Different members of the GPX gene family were coordinately regulated under specific environmental stress conditions, supporting the important roles of TsGPXs in salt and drought stress responses in *Thellungiella salsuginea* (Gao *et al.*, 2014). Glutathione peroxidase of *Pennisetum glaucum* (PgGPx) is a functional Cd^{2+}-dependent peroxiredoxin; it has a significant role in cellular redox homeostasis under stress conditions, helping to maintain membrane integrity and enhance tolerance toward oxidative stress (Islam *et al.*, 2015).

Phospholipid Hydroperoxide Glutathione Peroxidase (PHGPX)

PHGPX constitutes an important class of antioxidant enzymes that play an important role in the protection of biological membranes exposed to oxidative stress. The PHGPX protein and its encoding gene, CSA, have been isolated and characterized in *Citrus*. It has been found that CSA is directly induced by the substrate of PHGPX under heat, cold, and salt stresses, and this induction occurs mainly via the production of ROS (Avsian-Kretchmer *et al.*, 1999). The role of plant PHGPX has so far remained largely unexplored. Investigations on their possible induction under a wide range of abiotic and biotic stresses are likely to highlight their specific role in ROS scavenging, such as the removal of lipid hydroperoxides. Jain and Bhatla (2014) reported the sensing of NaCl stress in sunflower seedlings in relation to accumulation of lipid hydroperoxide, and its scavenging through upregulation of PHGPX activity in the cotyledons. Overexpression of the antioxidant enzyme PHGPX in *Citrus* interferes with shoot organogenesis and suggests the involvement of ROS during the early stages of development (Faltin *et al.*, 2010).

Non-enzymatic Antioxidants

Glutathione (GSH)

Fluctuations in environmental conditions can hamper normal regulation of cell metabolism. The consequent redox signals enable the cell to regain homeostasis in different subcellular compartments (Scheibe and Dietz, 2012). ROS generation and redox signals, generated under various conditions, regulate different essential metabolic processes that convert and distribute energy and metabolic fluxes in cells, maintain different cellular functions, initiate adaptive responses through retrograde signaling, and optimize and control different systemic signaling pathways (Suzuki *et al.*, 2012). GSH, along with its oxidized form (GSSG), mediates the cellular redox balance and GSH signaling under normal conditions and in stress responses. In chloroplasts GSH helps to protect the photosynthetic apparatus from severe oxidative damage. Therefore, they can be targeted for improved varietal development because proteins with oxidizable thiols like glutathione and thioredoxin-1 are important for various functions including transcription regulation, stability of chromatin, anterograde trafficking, and DNA replication/repair (Go and Jones 2010; Lukosz *et al.*, 2010). But an increase in GSH content does not necessarily benefit the plant. For example, overexpression of a chloroplast-targeted g-glutamyl cysteine synthetase (g-ECS) in transgenic tobacco resulted in a three-fold increase in the GSH level and increased oxidative damage (Creissen *et al.*, 1999). Not only GSH but also various second messengers like kinases, phosphatases, phytohormones, and calcium deeply influence cellular redox homeostasis. They interact with each other to maintain the intracellular redox state, which determines various signaling events. It is highly probable that ROS–antioxidant interactions act as a metabolic interface that lies between environmental changes and the accompanying signaling responses (Foyer and Noctor, 2012). For example, modulation of the redox state by ROS–antioxidant interactions regulates many transcription factors (TFs) directly or indirectly, like TGA, Athb-9, and RAP2, and their downstream genes (Dietz, 2008). One of the most toxic mycotoxins that promotes oxidative stress is ochratoxin A, and glutathione-associated metabolism is closely related with ochratoxin A stress. Also *Arabidopsis* plants subjected to ochratoxin A showed changes in biosynthesis and metabolism of glutathione resulting in induced resistance to the toxin (Wang *et al.*, 2014).

Ascorbic Acid (AsA)

Ascorbic acid also protects plants from severe oxidative damage. It can lead to direct scavenging of superoxide, hydroxyl radicals, and singlet oxygen species; it also reduces H_2O_2 to water via the ascorbate peroxidase reaction (Noctor and Foyer, 1998). AsA can donate electrons in a number of enzymatic and non-enzymatic reactions, acting as one of the most potent ROS scavengers. It can provide protection to membranes by directly scavenging O_2 and OH· and by regenerating α-tocopherol from tocopheroxyl radical. In chloroplasts, AsA acts as a cofactor of violaxanthin de-epoxidase leading to sustained dissipation of excess excitation energy (Smirnoff, 2000). AsA biosynthesis and metabolism occurs mainly in mitochondria, where AsA is synthesized by L-galactono-g-lactone dehydrogenase, and it is also regenerated from its oxidized forms (Szarka *et al.*, 2007). Exogenous application of AsA to root treatment under water stress increased non-enzymatic antioxidant production and decreased ROS toxicity. It also led to the induction of genes involved in cell-wall loosening, leading to cell expansion and enhanced root elongation (Xu *et al.*, 2015). AsA concentration might regulate lipid peroxidation in foliar tissues as there was a negative correlation observed under ozone stress and zinc deficiency, but no change observed in iron toxicity. Lack of AsA led to reduced photosynthetic efficiency, induced carboxylation rate of rubisco and increased electron transport rate (Holler *et al.*, 2015). Overproduction of L-ascorbic acid by D-galacturonic acid reductase in transgenic potato resulted in decreased methylglyoxal concentration under salinity stress by maintaining a higher reduced GSH level and glyoxalase enzyme activity. It also led to reduced oxidative stress-mediated damage (Upadhyaya *et al.*, 2011).

Tocopherol (Vitamin E)

Tocopherols are lipid-soluble antioxidants that mostly act in biomembranes and are considered potential scavengers of ROS and lipid radicals (Hollander-Czytko *et al.*, 2005). They protect membrane stability by quenching or scavenging ROS, though they have some non-antioxidant functions. In plants tocopherols are mainly localized in the thylakoid membrane of chloroplasts (Kamal-Eldin and Appelqvist 1999). Chloroplast membranes of higher plants contain α-tocopherol, which protects them against photo-oxidative damage (Fryer, 1992). Vitamin E has a very important role after anoxia stress because it can directly repair oxidizing radicals and prevent the chain propagation step during lipid auto-oxidation (Serbinova and Packer, 1994). It is reported that increased levels of α-tocopherol and AsA in tomato after trizole treatment protects plant from oxidative damage and cold stress (Shao *et al.*, 2007). A vital enzyme catalyzing the key methylation step in both α/γ-tocopherol and plastoquinone biosynthetic pathway is 2-methyl-6-phytyl-1,4-benzoquinol methyltransferase (MPBQ-MT). When Ls-MPBQ-MT was overexpressed in *Lactuca sativa* it led to increases in α-/γ-tocopherol and plastoquinone contents, providing enhanced tolerance and photosynthesis under high light stress. This helps the plant to direct excessive light energy toward photosynthetic activity rather than toward the generation of more photo-oxidative damage (Tang *et al.*, 2016). Transgenic *Brassica juncea* having increased α-tocopherol content displayed better tolerance against heavy metal, salt, and osmotic stresses. α-Tocopherol might possibly be involved in interplay with other antioxidant enzymes like SOD, CAT, APX, and GR (Kumar *et al.*, 2013). The prenylation step in tocopherol biosynthesis is catalyzed by VTE2.1, which encodes the homogentisate phytyltransferase enzyme. ScVTE2.1, when expressed under the control of a stress-inducible promoter in transgenic tobacco plants, increased α-tocopherol content under drought conditions. It led to higher water content, increased photosynthetic activity, reduced oxidative damage, and delayed leaf senescence (Espinoza *et al.*, 2013).

Proline

Proline acts as an osmolyte, a potent antioxidant, and a potential cell death inhibitor in microbes, animals, and plants (Chen and Dickman, 2005). Free proline is also considered as a protein stabilizer, a metal chelator, an inhibitor of lipid peroxidation (LPO), and a $\cdot OH$ and $O_2^{\cdot -}$ scavenger (Trovato *et al.*, 2008). It is reported that increased proline content is correlated with increased tolerance to different abiotic stresses like salt and drought. Increased synthesis and accumulation of proline under different stresses actually help reduce cytoplasmic acidosis and maintain $NADP^+$:NADPH levels that are compatible with normal metabolism (Hare and Cress, 1997). Ornithine δ-aminotransferase (OAT) is a pyridoxal-5′-phosphate-dependent enzyme that has been proposed to be involved in proline and arginine metabolism. OsOAT, which is a downstream target of SNAC2, when overexpressed in rice resulted in enhanced δ-OAT activity, increased proline accumulation, and better stress tolerance by enhancing ROS-scavenging capacity (You *et al.*, 2012). Upon Cu stress ABA-mediated proline accumulation is also observed, which may be because of proteolysis and increased activities of Δ^1-pyrroline-5-carboxylate reductase or ornithine-δ-aminotransferase (Chen *et al.*, 2001). Salt stress significantly inhibited growth BY2 cells. This growth inhibition was reduced upon exogenous application of proline. It also alleviated the reduction of antioxidant enzymes CAT and POD during the stress (Hoque *et al.*, 2007). In cultured tobacco cells proline and betaine conferred cadmium stress tolerance by increasing the activities of SOD and CAT while decreasing the lipid peroxidation rate (Islam *et al.*, 2009).

Carotenoids

In all photosynthetic organisms carotenoids like β-carotene, zeaxanthin, and tocopherols provide an important photoprotective feature. They may dissipate excess excitation energy as heat or scavenge ROS or suppress LPO to protect plants from oxidative stress. When CsCCD4b (carotenoid cleavage dioxygenase) from *Crocus sativus* was overexpressed in *Arabidopsis*, it imparted dehydration, salt, and oxidative stress tolerance by regulating ROS machinery (Baba *et al.*, 2015). Carotenoid derivatives, such as volatile β-cyclocitral, derived from the oxidation of β-carotene, are reactive electrophile species that are bioactive and can induce changes in gene expression leading to acclimation to stress conditions. In many plant species apocarotenoids like β-ionone and β-cyclocitral, which are enzymatic products of CsCCD4b, act as stress signal molecules during high light stress and regulate expression of stress-responsive genes leading to plant defense responses (Ramel *et al.*, 2012). Lutein also plays an important role in dissipating excess excitation energy and removing ROS so that normal physiological function of cells can be maintained. Overexpression of carotenoid epsilon-ring hydroxylase gene (LeLUT1) in tomato leads to photo-inhibition and photo-oxidation reduction, ultimately decreasing sensitivity to chilling stress (Zhou *et al.*, 2013). CYP97A4 is a rice carotenoid β-ring hydroxylase involved in lutein biosynthesis; it regulates α-carotene hydroxylation in rice. Knock-out CYP97A4 mutant lines showed reduced lutein content but increased α-carotene, making the plants more sensitive to intense light (Lv *et al.*, 2012).

Flavonoids

Flavonoids are considered as potent bioactive plant secondary metabolites. They have important roles as ROS scavengers, locating and neutralizing free radicals before they affect cellular homeostasis under hazardous environmental conditions (Lovdal *et al.*, 2010). Flavonoids are distinguished by the number and arrangement of their hydroxyl groups present on the ring structures. Their ability to act as antioxidants depends on the reduction potentials of their radicals and accessibility of the radicals. Flavonoids as well as other phenolic compounds absorb UV light and therefore plants containing higher amounts of these compounds can tolerate UV irradiation better than mutants having impaired flavonoid pathways (Cle *et al.*, 2008). When PRODUCTION OF ANTHOCYANIN PIGMENT 1-Dominant (pap1-D) was overexpressed in *Arabidopsis*, it led to accumulation of higher endogenous anthocyanin levels and better tolerance to oxidative and drought stress compared to wild-type plants (Nakabayashi *et al.*, 2014). Similarly FLAVONOL SYNTHASE 1 (FLS1), which negatively regulates anthocyanin levels, could be manipulated to provide stress tolerance in *Arabidopsis* (Nguyen *et al.*, 2016).

Role of Polyamines

A characteristic property of polyamines is to regulate transcription and translation; an attribute assigned to these polycationic substances because of their capacity to bind to anionic macromolecules like nucleic acids and proteins (Tiburcio *et al.*, 2014). They also help in maintaining stability of membranes, modulating antioxidant systems, and stress tolerance under adverse conditions (Liu *et al.*, 2007). Polyamines modulate ROS homeostasis under stress in two ways: either they lead to the inhibition of auto-oxidation of metals that finally impairs the supply of electrons for ROS generation, or they activate antioxidant systems (Shi *et al.*, 2010). It is reported that priming of plants with polyamines eventually increased antioxidant enzyme activity and tolerance to abiotic stresses like drought, cold, and heat. For example, exogenous application of spermine to *Citrus reticulata* led to enhanced SOD, POD, and CAT activities, significantly alleviating ROS levels under dehydration stress. Also spermine prevented relative loss of water by significantly decreasing the size of stomatal apertures (Shi *et al.*, 2010). Similarly, an exogenous supply of spermidine to rice seedlings reduced the impact of heat-induced damage. Spermidine led to decreased H_2O_2 content by regulating antioxidant and glyoxalase systems (Mostofa *et al.*, 2014). Overexpression of arginine decarboxylase from *Poncirus trifoliata* in tobacco and tomato conferred increased drought tolerance by substantially repressing ROS generation and inducing expression of stress-responsive genes (Wang *et al.*, 2011). ICE1 (Inducer of CBF Expression 1), a bHLH TF of *Poncirus trifoliata*, positively regulated cold stress tolerance by regulating polyamine levels through physically interacting with arginine decarboxylase. ICE1 overexpressing in tobacco and lemon transgenic lines enhanced the activity of antioxidant enzymes and reduced ROS accumulation (Huang *et al.*, 2015). Overexpression of *S*-adenosyl-L-methionine (SAM), a common precursor for polyamines and ethylene in tomato, increased alkali stress tolerance, maintained nutrient balance, enhanced photosynthetic capacity, and lowered oxidative stress. Transgenic lines displayed polyamine metabolism (increased spermine and spermidine content) and a stronger root system (Gong *et al.*, 2014).

NO Metabolism

Nitric oxide (NO) plays an important signaling role in growth, development, and defense responses of plants (Moreau *et al.*, 2010). NO is generated in different stress conditions and its signaling regulates various downstream responses. NO biosynthesis is also regulated by H_2O_2-mediated activation of MAP kinase 6 in *Arabidopsis* (Wang *et al.*, 2010). The probable mechanism of action of NO revolves around the transmittance of its signal by regulating the activity of proteins. NO modulates protein activity either by nitrosylation and probably tyrosine nitration, or acting as a Ca^{2+}-mobilizing messenger (Besson-Bard *et al.*, 2008). These NO-mediated post-translational alterations then regulate cell signaling under different physiological and stress responses (Corpas *et al.*, 2015). The role of NO as an antioxidant molecule emerged when it was observed that after NaCl application, the cleavage of DNA strands by the hydroxyl radical·OH and oxidative damage by other ROS molecules was inhibited by NO in *Citrus* leaves (Tanou *et al.*, 2009). It was speculated that NO could protect plants from·OH-mediated oxidative damage in two possible ways. NO either directly scavenge ROS or indirectly enhance NO-mediated induction of ferritin proteins within the organelles, which diminish free Fe^{2+} concentration and Fenton-type reactions (Martin *et al.*, 2009). NO may also indirectly regulate epigenetic modifications like DNA methylation because·OH effectively targets DNA methylation and regulatory genes (Shen *et al.*, 2014). Also it is reported that NO modulates superoxide formation and inhibits lipid peroxidation to regulate positive oxidative stress tolerance responses (Caro and Puntarulo, 1998; Boveris *et al.*, 2000). ROS and NO interaction under drought stress induces ABA biosynthesis and leads to stomatal closure to reduce the effect of stress (Neill *et al.*, 2003). When exogenous NO was applied to sorghum embryos it led to increased Fe availability, which could be an important aspect for reducing oxidative stress and maintaining normal plant physiology (Jasid *et al.*, 2008).

Transcription Factors

Stress responses in plants are regulated through multiple signaling pathways that lead to activation of gene transcription and its downstream cascade. Analysis of ROS-related microarray data has revealed the involvement of 1500 TFs in response to different ROS in *Arabidopsis* (Gadjev *et al.*, 2006). AtMYB12 from *Arabidopsis* regulates the expression of key enzymes involved in flavonoid biosynthesis and leads to increased accumulation of flavonoids (F. Wang *et al.*, 2016). Overexpression of another TF, bHLH1 from *Pyrus ussuriensis*, induced expression of many stress-responsive genes and provided increased tolerance to cold stress in tobacco (Jin *et al.*, 2016). BpERF11 inhibits the expression of an AtMYB61 homolog and negatively regulates salt and osmotic stress tolerance by increasing stomatal aperture and elevating transpiration rate. Also BpERF11 downregulates the expression of LEA proteins, dehydrin, and antioxidant-responsive genes like SOD and POD. BpERF11-overexpressing lines displayed reduced proline content but increased ROS accumulation (Zhang *et al.*, 2016). Similarly, ROS-mediated TF ZAT12 directly interacts with bHLH protein FIT, which negatively regulates iron deficiency responses in *Arabidopsis* (Brumbarova *et al.*, 2016). GhWRKY41 enhances salt and drought tolerance in transgenic tobacco by increasing stomatal closure and reducing ROS levels through induced expression of antioxidant responsive genes (Chu *et al.*, 2015). ROS have an important role in plant defense responses against pathogens. OxyR is an oxidative stress-regulated TF in *Pseudomonas syringae* pv. tomato DC3000, which regulates antioxidant-responsive genes like *KatB* and *KatG*, and has an important role in the virulence of this pathogen. oxyR-regulated ROS-detoxifying enzymes help in the establishment of increased bacterial population and counteract the plant's ROS defense system in *Arabidopsis* and tomato (Ishiga and Ichinose, 2016). Abscisic acid stress ripening (ASR) proteins containing ABA/WDS domains form a major TF family in plants; these have important roles in growth, development, and abiotic stress responses. For example, overexpression of ASR1 from foxtail millet enhanced drought and oxidative tolerance in *Nicotiana* through regulation of ROS homeostasis and antioxidant-responsive genes (Feng *et al.*, 2016). Similarly wheat ASR gene TaASR1 positively regulated plant osmotic stress tolerance by modulating ROS homeostasis and activating expression of antioxidant- and stress-responsive genes (Hu *et al.*, 2013). Rice SNAC3 modulates ROS homeostasis in overexpressed plants showing significantly lower H_2O_2 content, reduced lipid peroxidation, and relatively reduction of electrolyte leakage compared to wild-type plants under heat stress (Fang *et al.*, 2015). A NAM, ATAF1/2, CUC2 transcription factor (NAC TF) from soybean, GmNAC2, when overexpressed in tobacco negatively regulated abiotic stress by involving ROS signaling pathways and modulating ROS-scavenging genes (Jin *et al.*, 2013). AtbHLH112 positively and negatively regulated expression of P5CS and P5CDH/ProDH, respectively, to enhance proline content and mediate a stress response in *Arabidopsis*. AtbHLH112 also positively regulated ROS homeostasis by inducing expression of antioxidant-responsive genes like POD and SOD (Y. Liu *et al.*, 2015). TabHLH1 when overexpressed in wheat regulated expression of phosphate/nitrate transporters and antioxidant-responsive genes to mediate adaptation to inorganic phosphate (Pi) and nitrogen (N) starvation stresses. Nutrient stress was normalized by increasing uptake of Pi/N and maintaining ROS cellular homeostasis (Yang *et al.*, 2016).

Conclusion and Future Perspectives

ROS are continuously generated as by-products of plant cellular metabolism. ROS act as second messengers and are often associated with positive signaling responses like programmed cell death, pathogen defense, and stomatal conductance. But under severe environmental fluctuations excessive ROS are generated, which causes oxidative stress resulting in cell death. During the course of evolution plants have developed several endogenous systems of ROS detoxification through enzymatic or non-enzymatic mechanisms. Maintenance of the delicate equilibrium between ROS production and their scavenging is governed at different levels through different processes. Antioxidant enzymes like SOD, CAT, APX, GPOX, GR, MDHAR, DHAR, GST, GPX, and PHGPX act together, probably in an interconnected system, to remove excess ROS so that cellular homeostasis can be maintained. Besides enzymatic modes of detoxification, plants recruit non-enzymatic molecules to regain normal physiology and metabolism within the cellular environment. GSH, ascorbic acid, vitamin E, prolines, carotenoids, flavonoids, and polyamines assist in that ROS-scavenging process. Apart from some harmful regulation of NO, they provide antioxidant properties under different stress conditions. The entire mechanism involved in the scavenging process is still not fully elucidated as a complex interplay of different factors affect the ROS concentrations. This vast web of cross-talk involves many TFs, like ERF, WRKY, MYB, bHLH, and NAC, which regulate their downstream targets of antioxidant-responsive genes to initiate the oxidative stress response. Endogenous regulation of oxidative stress is effective, but under simultaneous stresses plants often are unable to cope with their environment. Therefore, the transgenic approach to varietal development has surfaced recently to provide tolerance to plants under adverse conditions. Overexpression or silencing of TFs rather than antioxidant-responsive genes would yield better results in generating oxidative stress-tolerant plants because TFs can regulate multiple downstream targets simultaneously. They not only activate antioxidant-responsive genes but also induce respective stress-responsive genes to attain normal homeostasis. More thorough study and analysis requires our full attention on the underlying regulation involved in oxidative stress responses. We emphasize that for proper growth, development, and yield under various stresses, antioxidant responses should be monitored properly along with other factors. Particular attention should be paid to the regulation and interplay within different scavenging molecules, their individual mode of action, and downstream as well as upstream components. The following key points should be addressed for a proper understanding of how plants adjust their metabolism against reactive oxygen species:

1) Analysis at the molecular level is necessary. Homologs, isoforms, and spliced forms of antioxidant-responsive genes in different plant species should be studied. Interaction at the DNA/protein level, promoter analysis, and epigenetic regulation of the respective genes should be monitored. Large-scale high-throughput technologies would assist in this analysis.

2) Modified plants with recombinant DNA technology should be analyzed for any negative feedback or unwanted traits. Proper field trials and serious scrutiny of the affected traits are necessary. Newly emerging CRISPR/cas9 technology could be applied to generate mutants in plants so that the exact functions of antioxidant-responsive genes can be determined. Not only transgenic approaches but also breeding techniques could be employed to get better varieties.
3) The mode of action and entire operating cascade of different scavengers, whether enzymatic or non-enzymatic, should be deciphered not only in model plants but also in other commercially important plants. Cross-talk and indirect ROS generation from other stresses should be monitored.

References

Abou-Attia MA, Wang X, Nashaat Al-Attala M, Xu Q, Zhan G, Kang Z. (2016) TaMDAR6 acts as a negative regulator of plant cell death and participates indirectly in stomatal regulation during the wheat stripe rust–fungus interaction. *Physiol Plant* 156: 262–277.

Achary VM, Reddy CS, Pandey P, Islam T, Kaul T, Reddy MK. (2015) Glutathione reductase a unique enzyme: molecular cloning, expression and biochemical characterization from the stress adapted C4 plant, *Pennisetum glaucum* (L.) R. Br. *Mol Biol Rep* 42: 947–962.

Ali A, Alqurainy F. (2006) Activities of antioxidants in plants under environmental stress. In: Motohashi N (ed.), *The Lutein – Prevention and Treatment for Diseases.* Transworld Research Network, India, pp. 187–256.

Asada K. (1992) Ascorbate peroxidase – a hydrogen peroxide scavenging enzyme in plants. *Physiol Plant* 85: 235–241.

Asada K. (1999) The water cycle in chloroplasts: scavenging of active oxygen and dissipation of excess photons. *Plant Mol Biol* 50: 601–639.

Avsian-Kretchmer O, Eshdat Y, Gueta-Dahan Y, Ben-Hayyim G. (1999) Regulation of stress induced phospholipid hydroperoxide glutathione peroxidase expression in citrus. *Planta* 209: 469–477.

Baba SA, Jain D, Abbas N, Ashraf N. (2015) Overexpression of *Crocus* carotenoid cleavage dioxygenase, CsCCD4b, in *Arabidopsis* imparts tolerance to dehydration, salt and oxidative stresses by modulating ROS machinery. *J Plant Physiol* 189: 114–125.

Baxter A, Mittler R, Suzuki N. (2014) ROS as key players in plant stress signalling. *J Exp Bot* 65: 1229–1240.

Besson-Bard A, Pugin A, Wendehenne D. (2008) New insights into nitric oxide signaling in plants. *Annu Rev Plant Biol* 59: 21–39.

Boveris A, Costa LE, Poderoso JJ, Carreras MC, Cadenas E. (2000) Regulation of mitochondrial respiration by oxygen and nitric oxide. *Ann N Y Acad Sci* 899: 121–135.

Bowler C, Montagu MV, Inze D. (1992) Superoxide dismutase and stress tolerance. *Annu Rev Plant Phys* 43: 83–116.

Brumbarova T, Le CT, Ivanov R, Bauer P. (2016) Regulation of ZAT12 protein stability: The role of hydrogen peroxide. *Plant Signal Behav* 11: e1137408.

Caro A, Puntarulo S. (1998) Nitric oxide decreases superoxide anion generation by microsomes from soybean embryonic axes. *Physiol Plant* 104: 357–364.

Cha JY, Barman DN, Kim MG, Kim WY. (2015) Stress defense mechanisms of NADPH-dependent thioredoxin reductases (NTRs) in plants. *Plant Signal Behav* 10: e1017698.

Chalapathi Rao ASV, Reddy AR. (2008) Glutathione reductase: a putative redox regulatory system in plant cells. In: Khan NA, Singh S, Umar S (eds), *Sulfur Assimilation and Abiotic Stresses in Plants.* Springer, The Netherlands, pp. 111–147.

Chen C, Dickman MB. (2005) Proline suppresses apoptosis in the fungal pathogen Colletotrichum trifolii. *Proc Natl Acad Sci USA* 102: 3459–3464.
Chen C, Letnik I, Hacham Y, *et al.* (2014) ASCORBATE PEROXIDASE6 protects Arabidopsis desiccating and germinating seeds from stress and mediates cross talk between reactive oxygen species, abscisic acid, and auxin. *Plant Physiol* 166: 370–383.
Chen CT, Chen L, Lin CC, Kao CH. (2001) Regulation of proline accumulation in detached rice leaves exposed to excess copper. *Plant Sci* 160: 283–290.
Chen Z, Gallie DR. (2005) Increasing tolerance to ozone by elevating folia ascorbic acid confers greater protection against ozone than increasing avoidance. *Plant Physiol* 138: 1673–1689.
Chu X, Wang C, Chen X, *et al.* (2015) The cotton WRKY gene GhWRKY41 positively regulates salt and drought stress tolerance in transgenic *Nicotiana benthamiana*. *PLoS One* 10(11): e0143022.
Cle C, Hill LM, Niggeweg R, *et al.* (2008) Modulation of chlorogenic acid biosynthesis in Solanum lycopersicum; consequences for phenolic accumulation and UV-tolerance. *Phytochemistry* 69: 2149–2156.
Corpas FJ, Begara-Morales JC, Sanchez-Calvo B, Chaki M, Barroso JB. (2015) Nitration and S-nitrosylation: two post-translational modifications (PTMs) mediated by reactive nitrogen species (RNS) and their role in signalling processes of plant cells. In: Gupta KJ, Igamberdiev AU (eds), *Reactive Oxygen and Nitrogen Species Signaling and Communication in Plants*. Springer, Berlin, pp. 267–281.
Couee I, Sulmon C, Gouesbet G, Amrani EI. (2006) Involvement of soluble sugars in reactive oxygen species balance and responses to oxidative stress in plants. *J Exp Bot* 57: 449–459.
Creissen GP, Broadbent P, Kular B, Reynolds H, Wellburn AR, Mullineaux PM. (1994) Manipulation of glutathione reductase in transgenic plants: implications for plant responses to environmental stress. *Proc R Soc Edinb* 102B: 167–175.
Creissen G, Firmin J, Fryer M, *et al.* (1999) Elevated glutathione biosynthetic capacity in the chloroplasts of transgenic tobacco plants paradoxically causes increased oxidative stress. *The Plant Cell* 11: 1277–1291.
Cutler SR, Rodriguez PL, Finkelstein RR, Abrams SR. (2010) Abscisic acid: emergence of a core signaling network. *Annu Rev Plant Biol* 61: 651–679.
Diaz-Vivancos P, Clemente-Moreno MJ, Rubio M, *et al.* (2008) Alteration in the chloroplastic metabolism leads to ROS accumulation in pea plants in response to plum pox virus. *J Exp Bot* 59: 2147–2160.
Dietz KJ. (2008) Redox signal integration: from stimulus to networks and genes. *Physiol Plant* 133: 459–468.
Ding S, Lu Q, Zhang Y, Yang Z, Wen X, Zhang L, Lu C. (2009) Enhanced sensitivity to oxidative stress in transgenic tobacco plants with decreased glutathione reductase activity leads to a decrease in ascorbate pool and ascorbate redox state. *Plant Mol Biol* 69(5): 577–592.
Ding Y, Cao J, Ni L, *et al.* (2013) ZmCPK11 is involved in abscisic acid-induced antioxidant defence and functions upstream of ZmMPK5 in abscisic acid signalling in maize. *J Exp Bot* 64: 871–884.
Dixon DP, Edwards R. (2010) Glutathione transferases. *Arabidopsis Book* 8: e0131.
Dixon DP, Skipsey M, Edwards R. (2010) Roles for glutathione transferases in plant secondary metabolism. *Phytochemistry* 71: 338–350.
Do H, Kim IS, Jeon BW, *et al.* (2016) Structural understanding of the recycling of oxidized ascorbate by dehydroascorbate reductase (OsDHAR) from Oryza sativa L. japonica. *Sci Rep* 18: 19498.
Dong C, Zheng X, Diao Y, Wang Y, Zhou M, Hu Z. (2015) Molecular cloning and expression analysis of a catalase gene (NnCAT) from Nelumbo nucifera. *Appl Biochem Biotechnol* 177: 1216–1228.

Dong W, Wang M, Xu F, *et al.* (2013) Wheat oxophytodienoate reductase gene TaOPR1 confers salinity tolerance via enhancement of abscisic acid signaling and reactive oxygen species scavenging. *Plant Physiol* 161: 1217–1228.

Eltayeb AE, Kawano N, Badawi GH, *et al.* (2007) Overexpression of monodehydroascorbate reductase in transgenic tobacco confers enhanced tolerance to ozone, salt and polyethylene glycol stresses. *Planta* 225: 1255–1264.

Espinoza A, San Martin A, Lopez-Climent M, Ruiz-Lara S, Gomez-Cadenas A, Casaretto JA. (2013) Engineered drought-induced biosynthesis of α-tocopherol alleviates stress-induced leaf damage in tobacco. *J Plant Physiol* 170: 1285–1294.

Faltin Z, Holland D, Velcheva M, *et al.* (2010) Glutathione peroxidase regulation of reactive oxygen species level is crucial for in vitro plant differentiation. *Plant Cell Physiol* 51: 1151–1162.

Fang Y, Liao K, Du H, *et al.* (2015) A stress-responsive NAC transcription factor SNAC3 confers heat and drought tolerance through modulation of reactive oxygen species in rice. *J Exp Bot* 66: 6803–6817.

Feng ZJ, Xu ZS, Sun J, *et al.* (2016) Investigation of the ASR family in foxtail millet and the role of ASR1 in drought/oxidative stress tolerance. *Plant Cell Rep* 35: 115–128.

Foyer CH, Noctor G. (2012) Managing the cellular redox hub in photosynthetic organisms. *Plant Cell Environ* 35: 199–201.

Fryer MJ. (1992) The antioxidant effects of thylakoid vitamin E (α-tocopherol). *Plant Cell Environ* 15: 381–392.

Gadjev I, Vanderauwera S, Gechev TS, *et al.* (2006) Transcriptomic footprints disclose specificity of reactive oxygen species signaling in Arabidopsis. *Plant Physiol* 141: 436–445.

Gao F, Chen J, Ma T, *et al.* (2014) The glutathione peroxidase gene family in Thellungiella salsuginea: genome-wide identification, classification, and gene and protein expression analysis under stress conditions. *Int J Mol Sci* 15: 3319–3335.

George S, Venkataraman G, Parida A. (2010) A chloroplast localized and auxin induced glutathione S-transferase from phreatophyte Prosopis juliflora confer drought tolerance on tobacco. *J Plant Physiol* 167: 311–318.

Gill SS, Tuteja N. (2010) Reactive oxygen species and antioxidant machinery in abiotic stress tolerance in crop plants. *Plant Physiol Biochem* 48: 909–930.

Go YM, Jones DP. (2010) Redox control systems in the nucleus: mechanisms and functions. *Antioxid Redox Signal* 13: 489–509.

Gong B, Li X, Vanden-Langenberg KM, *et al.* (2014) Overexpression of S-adenosyl-L-methionine synthetase increased tomato tolerance to alkali stress through polyamine metabolism. *Plant Biotechnol J* 12: 694–708.

Gong B, Wen D, Wang X, *et al.* (2015) S-nitrosoglutathione reductase-modulated redox signaling controls sodic alkaline stress responses in Solanum lycopersicum L. *Plant Cell Physiol* 56: 790–802.

Guan Q, Wang Z, Wang X, Takano T, Liu S. (2015) A peroxisomal APX from Puccinellia tenuiflora improves the abiotic stress tolerance of transgenic Arabidopsis thaliana through decreasing of H_2O_2 accumulation. *J Plant Physiol* 175: 183–191.

Guan ZQ, Chai TY, Zhang YX, Xu J, Wei W. (2009) Enhancement of Cd tolerance in transgenic tobacco plants overexpressing a Cd-induced catalase Cdna. *Chemosphere* 76: 623–630.

Hare PD, Cress WA. (1997) Metabolic implications of stress-induced proline accumulation in plants. *Plant Growth Regul* 21: 79–102.

Hirayama T, Shinozaki K. (2010) Research on plant abiotic stress responses in the post genome era: past, present and future. *Plant J* 61: 1041–1052.

Hollander-Czytko H, Grabowski J, Sandorf I, Weckermann K, Weiler EW. (2005) Tocopherol content and activities of tyrosine aminotransferase and cystine lyase in Arabidopsis under stress conditions. *J Plant Physiol* 162: 767–770.

Holler S, Ueda Y, Wu L, *et al.* (2015) Ascorbate biosynthesis and its involvement in stress tolerance and plant development in rice (Oryza sativa L.). *Plant Mol Biol* 88: 545–560.
Hong CY, Hsu YT, Tsai YC, Kao CH. (2007) Expression of ASCORBATE PEROXIDASE 8 in roots of rice (Oryza sativa L.) seedlings in response to NaCl. *J Exp Bot* 58: 3273–3283.
Hoque MA, Okuma E, Banu MNA, Nakamura Y, Shimoishi V, Murata Y. (2007) Exogenous proline mitigates the detrimental effects of salt stress more than exogenous betaine by increasing antioxidant enzyme activities. *J Plant Physiol* 164: 553–561.
Hu W, Huang C, Deng X, *et al.* (2013) TaASR1, a transcription factor gene in wheat, confers drought stress tolerance in transgenic tobacco. *Plant Cell Environ* 36: 1449–1464.
Hu X, Jiang M, Zhang J, Zhang A, Lin F, Tan M. (2007) Calcium-calmodulin is required for abscisic acid-induced antioxidant defense and functions both upstream and downstream of H_2O_2 production in leaves of maize (Zea mays) plants. *New Phytol* 173: 27–38.
Huang J, Sun S, Xu D, *et al.* (2012) A TFIIIA-type zinc finger protein confers multiple abiotic stress tolerances in transgenic rice (Oryza sativa L.). *Plant Mol Biol* 80: 337–350.
Huang XS, Zhang Q, Zhu D, *et al.* (2015) ICE1 of Poncirus trifoliata functions in cold tolerance by modulating polyamine levels through interacting with arginine decarboxylase. *J Exp Bot* 66: 3259–3274.
Ishiga Y, Ichinose Y. (2016) Pseudomonas syringae pv. tomato OxyR is required for virulence in tomato and Arabidopsis. *Mol Plant Microbe Interact* 29: 119–131.
Islam MM, Hoque MA, Okuma E, *et al.* (2009) Exogenous proline and glycinebetaine increase antioxidant enzyme activities and confer tolerance to cadmium stress in cultured tobacco cells. *J Plant Physiol* 166: 1587–1597.
Islam T, Manna M, Reddy MK. (2015) Glutathione peroxidase of Pennisetum glaucum (PgGPx) is a functional Cd^{2+} dependent peroxiredoxin that enhances tolerance against salinity and drought stress. *PLoS One* 10(11): e0143344.
Jain P, Bhatla SC. (2014) Signaling role of phospholipid hydroperoxide glutathione peroxidase (PHGPX) accompanying sensing of NaCl stress in etiolated sunflower seedling cotyledons. *Plant Signal Behav* 9: e977746.
Jasid S, Simontacchi M, Puntarulo S. (2008) Exposure to nitric oxide protects against oxidative damage but increases the labile iron pool in sorghum embryonic axes. *J Exp Bot* 59: 3953–3962.
Jin C, Huang XS, Li KQ, *et al.* (2016) Overexpression of a bHLH1 transcription factor of Pyrus ussuriensis confers enhanced cold tolerance and increases expression of stress-responsive genes. *Front Plant Sci* 7: 441.
Jin HX, Huang F, Cheng H, Song HN, Yu DY. (2013) Overexpression of the GmNAC2 gene, a NAC transcription factor, reduces abiotic stress tolerance in tobacco. *Plant Mol Biol Rep* 31: 435–442.
Kamal-Eldin A, Appelqvist LA. (1996) The chemistry and antioxidant properties of tocopherols and tocotrienols. *Lipids* 31: 671–701.
Keunen E, Peshev D, Vangronsveld J, Van Den Ende W, Cuypers A. (2013) Plant sugars are crucial players in the oxidative challenge during abiotic stress: extending the traditional concept. *Plant Cell Environ* 36: 1242–1255.
Kliebenstein DJ, Dietrich RA, Martin AC, Last RL, Dangl JL. (1999) LSD1 regulates salicylic acid induction of copper zinc superoxide dismutase in Arabidopsis thaliana. *Mol Plant Microbe Interact* 12: 1022–1026.
Koji Y, Shiro M, Michio K, Mitsutaka T, Hiroshi M. (2009) Antioxidant capacity and damages caused by salinity stress in apical and basal regions of rice leaf. *Plant Prod Sci* 12: 319–326.
Kruk J, Szymanska R, Nowicka B, Dłuzewska J. (2016) Function of isoprenoid quinones and chromanols during oxidative stress in plants. *N Biotechnol* 33: 636–643.
Kumar D, Yusuf MA, Singh P, Sardar M, Sarin NB. (2013) Modulation of antioxidant machinery in α-tocopherol-enriched transgenic Brassica juncea plants tolerant to abiotic stress conditions. *Protoplasma* 250: 1079–1089.

Kumari S, Joshi R, Singh K, *et al.* (2015) Expression of a cyclophilin OsCyp2-P isolated from a salt-tolerant landrace of rice in tobacco alleviates stress via ion homeostasis and limiting ROS accumulation. *Funct Integr Genomics* 15: 395–412.

Laporte D, Olate E, Salinas P, Salazar M, Jordana X, Holuigue L. (2012) Glutaredoxin GRXS13 plays a key role in protection against photooxidative stress in Arabidopsis. *J Exp Bot* 63: 503–515.

Leisinger U, Rufenacht K, Fischer B, *et al.* (2001) The glutathione peroxidase homologous gene from Chlamydomonas reinhardtii is transcriptionally up-regulated by singlet oxygen. *Plant Mol Biol* 46: 395–408.

Le Martret B, Poage M, Shiel K, Nugent GD, Dix PJ. (2011) Tobacco chloroplast transformants expressing genes encoding dehydroascorbate reductase, glutathione reductase, and glutathione-S-transferase, exhibit altered antioxidant metabolism and improved abiotic stress tolerance. *Plant Biotechnol J* 9: 661–673.

Liao W, Wang G, Li Y, Wang B, Zhang P, Peng M. (2016) Reactive oxygen species regulate leaf pulvinus abscission zone cell separation in response to water-deficit stress in cassava. *Sci Rep* 6: 21542.

Light GG, Mahan JR, Roxas VP, Allen RD. (2005) Transgenic cotton (Gossypium hirsutum L.) seedlings expressing a tobacco glutathione S-transferase fail to provide improved stress tolerance. *Planta* 222: 346–354.

Liu D, Liu Y, Rao J, *et al.* (2013) Overexpression of the glutathione S-transferase gene from Pyrus pyrifolia fruit improves tolerance to abiotic stress in transgenic tobacco plants. *Mol Biol (Mosk)* 47: 591–601.

Liu JH, Kitashiba H, Wang J, Ban Y, Moriguchi T. (2007) Polyamines and their ability to provide environmental stress tolerance to plants. *Plant Biotechnol* 24: 117–126.

Liu JH, Wang W, Wu H, Gong X, Moriguchi T. (2015) Polyamines function in stress tolerance: from synthesis to regulation. *Front Plant Sci* 6: 827.

Liu Y, Ji X, Nie X, *et al.* (2015) Arabidopsis AtbHLH112 regulates the expression of genes involved in abiotic stress tolerance by binding to their E-box and GCG-box motifs. *New Phytol* 207: 692–709.

Liu Z, Bao H, Cai J, Han J, Zhou L. (2013) A novel thylakoid ascorbate peroxidase from Jatropha curcas enhances salt tolerance in transgenic tobacco. *Int J Mol Sci* 15: 171–185.

Lopez-Cruz J, Crespo-Salvador O, Fernandez-Crespo E, Garcia-Agustin P, Gonzalez-Bosch C. (2017) Absence of Cu-Zn-superoxide dismutase BCSOD1 reduces Botrytis cinerea virulence in Arabidopsis and in tomato plants, which reveals interplay among ROS, callose and signaling pathways. *Mol Plant Pathol* 18: 16–31.

Lovdal T, Olsen KM, Slimestad R, Verheul M, Lillo C. (2010) Synergetic effects of nitrogen depletion, temperature, and light on the content of phenolic compounds and gene expression in leaves of tomato. *Phytochemistry* 71: 605–613.

Lukosz M, Jakob S, Buchner N, Zschauer TC, Altschmied J, Haendeler J. (2010) Nuclear redox signalling. *Antioxid Redox Signal* 12: 713–742.

Lv MZ, Chao DY, Shan JX, *et al.* (2012) Rice carotenoid β-ring hydroxylase CYP97A4 is involved in lutein biosynthesis. *Plant Cell Physiol* 53: 987–1002.

Martin M, Colman MJR, Gomez-Casati DF, Lamattina L, Zabaleta EJ. (2009) Nitric oxide accumulation is required to protect against iron-mediated oxidative stress in frataxin-deficient Arabidopsis plants. *FEBS Lett* 583: 542–548.

Maxwell DP, Wang Y, Mcintosh L. (1999) The alternative oxidase lowers mitochondrial reactive oxygen production in plant cells. *Proc Natl Acad Sci USA* 96: 8271–8276.

Melchiorre M, Robert G, Trippi V, Racca R, Lascano HR. (2009) Superoxide dismutase and glutathione reductase overexpression in wheat protoplast: photooxidative stress tolerance and changes in cellular redox state. *Plant Growth Regul* 57: 57–68.

Meyer Y, Belin C, Delorme-Hinoux V, Reichheld JP, Riondet C. (2012) Thioredoxin and glutaredoxin systems in plants: molecular mechanisms, crosstalks, and functional significance. *Antioxid Redox Signal* 17: 1124–1160.

Milone MT, Sgherri C, Clijters H, Navari-Izzo F. (2003) Antioxidative responses of wheat treated with realistic concentrations of cadmium. *Env. Exp. Bot* 50:265e273.

Mittler R. (2002) Oxidative stress, antioxidants and stress tolerance. *Trends Plant Sci* 7: 405–410.

Mittler R, Vanderauwera S, Gollery M, Van Breusegem F. (2004) Reactive oxygen gene network of plants. *Trends Plant Sci* 9: 490–498.

Moons A. (2003) Osgstu3 and osgtu4, encoding tau class glutathione S-transferases, are heavy metal- and hypoxic stress-induced and differentially salt stress responsive in rice roots. *FEBS Lett* 553: 427–432.

Moreau M, Lindermayr C, Durner J, Klessig DF. (2010) NO synthesis and signaling in plants – where do we stand? *Physiol Plant* 138: 372–383.

Mostofa MG, Yoshida N, Fujita M. (2014) Spermidine pretreatment enhances heat tolerance in rice seedlings through modulating antioxidative and glyoxalase systems. *Plant Growth Regul* 73: 31–44.

Mroczek-Zdyrska M, Wojcik M. (2012) The influence of selenium on root growth and oxidative stress induced by lead in Vicia faba L. minor plants. *Biol Trace Elem Res* 147(1–3): 320–328.

Myouga F, Hosoda C, Umezawa T, *et al.* (2008) A heterocomplex of iron superoxide dismutases defends chloroplast nucleoids against oxidative stress and is essential for chloroplast development in Arabidopsis. *The Plant Cell* 20: 3148–3162.

Nakabayashi R, Yonekura-Sakakibara K, Urano K, *et al.* (2014) Enhancement of oxidative and drought tolerance in Arabidopsis by overaccumulation of antioxidant flavonoids. *Plant J* 77: 367–379.

Nath K, Kumar S, Poudyal RS, *et al.* (2014) Developmental stage-dependent differential gene expression of superoxide dismutase isoenzymes and their localization and physical interaction network in rice (Oryza sativa L.). *Genes Genom* 36: 45–55.

Neill SJ, Desikan R, Hancock JT. (2003) Nitric oxide signalling in plants. *New Phytol* 159: 11–35.

Nguyen NH, Kim JH, Kwon J, *et al.* (2016) Characterization of Arabidopsis thaliana FLAVONOL SYNTHASE 1 (FLS1) overexpression plants in response to abiotic stress. *Plant Physiol Biochem* 103: 133–142.

Noctor G, Foyer CH. (1998) Ascorbate and glutathione: keeping active oxygen under control. *Annu Rev Plant Phys* 49: 249–279.

Noctor G, Gomez L, Vanacker H, Foyer CH. (2002) Interactions between biosynthesis, compartmentation, and transport in the control of glutathione homeostasis and signalling. *J Exp Bot* 53: 1283–1304.

Noctor G, Mhamdi A, Foyer CH. (2014) The roles of reactive oxygen metabolism in drought: not so cut and dried. *Plant Physiol* 164: 1636–1648.

Noshi M, Hatanaka R, Tanabe N, Terai Y, Maruta T, Shigeoka S. (2016) Redox regulation of ascorbate and glutathione by a chloroplastic dehydroascorbate reductase is required for high-light stress tolerance in Arabidopsis. *Biosci Biotechnol Biochem* 80: 870–877.

Overby A, Stokland RA, Asberg SE, Sporsheim B, Bones AM. (2015) Allyl isothiocyanate depletes glutathione and upregulates expression of glutathione S-transferases in Arabidopsis thaliana. *Front Plant Sci* 6: 277.

Passaia G, Margis-Pinheiro M. (2015) Glutathione peroxidases as redox sensor proteins in plant cells. *Plant Sci* 234: 22–26.

Perez-Ruiz JM, Spinola MC, Kirchsteiger K, Moreno J, Sahrawy M, Cejudo FJ. (2006) Rice NTRC is a high-efficiency redox system for chloroplast protection against oxidative damage. *Plant Cell* 18: 2356–2368.

Polidoros NA, Scandalios JG. (1999) Role of hydrogen peroxide and different classes of antioxidants in the regulation of catalase and glutathione S-transferase gene expression in maize (Zea mays L.). *Physiol Plant* 106: 112–120.
Ramel F, Birtic S, Ginies C, Soubigou-Taconnat L, Triantaphylides C, Havaux M. (2012) Carotenoid oxidation products are stress signals that mediate gene responses to singlet oxygen in plants. *Proc Natl Acad Sci USA* 109: 5535–5540.
Rio LA, Corpas FJ, Sandalio LM, Palma JM, Gomez M, Barroso JB. (2002) Reactive oxygen species, antioxidant systems and nitric oxide in peroxisomes. *J Exp Bot* 53: 1255–1272.
Rio LA, Sandalio LM, Altomare DA, Zilinskas BA. (2003) Mitochondrial and peroxisomal manganese superoxide dismutase: differential expression during leaf senescence. *J Exp Bot* 54: 923–933.
Roxas VP, Lodhi SA, Garrett DK, Mahan JR, Allen RD. (2000) Stress tolerance in transgenic tobacco seedlings that overexpress glutathione S-transferase/glutathione peroxidase. *Plant Cell Physiol* 41: 1229–1234.
Sarowar S, Kim EN, Kim YJ, *et al.* (2005) Overexpression of a pepper ascorbate peroxidase-like 1 gene in tobacco plants enhances tolerance to oxidative stress and pathogens. *Plant Sci* 169: 55–63.
Scandalios JG. (1990) Response of plant antioxidant defense genes to environmental stress. *Adv Genet* 28: 1–41.
Scandalios M, Rio LA. (1988) Intra-organellar distribution of superoxide dismutase in plant peroxisomes (glyoxysomes and leaf peroxisomes). *Plant Physiol* 88: 1215–1218.
Scheibe R, Dietz KJ. (2012) Reduction-oxidation network for flexible adjustment of cellular metabolism in photoautotrophic cells. *Plant Cell Environ* 35: 202–216.
Serbinova EA, Packer L. (1994) Antioxidant properties of α-tocopherol and α-tocotrienol. *Methods Enzymol* 234: 354–366.
Shafi A, Chauhan R, Gill T, *et al.* (2015) Expression of SOD and APX genes positively regulates secondary cell wall biosynthesis and promotes plant growth and yield in Arabidopsis under salt stress. *Plant Mol Biol* 87: 615–631.
Shao HB, Chu LY, Wu G, Zhang JH, Lu ZH, Hu YC. (2007) Changes of some antioxidative physiological indices under soil water deficits among 10 wheat (Triticum aestivum L.) genotypes at tillering stage. *Colloid Surf B Biointerfaces* 54: 143–149.
Sharma P, Dubey RS. (2005) Modulation of nitrate reductase activity in rice seedlings under aluminium toxicity and water stress: role of osmolytes as enzyme protectant. *J Plant Physiol* 162: 854–864.
Shen L, Song CX, He C, Zhang Y. (2014) Mechanism and function of oxidative reversal of DNA and RNA methylation. *Annu Rev Biochem* 83: 585–614.
Shi J, Fu XZ, Peng T, Huang XS, Fan QJ, Liu JH. (2010) Spermine pretreatment confers dehydration tolerance of citrus in vitro plants via modulation of antioxidative capacity and stomatal response. *Tree Physiol* 30: 914–922.
Shin SY, Kim MH, Kim YH, Park HM, Yoon HS. (2013) Co-expression of monodehydroascorbate reductase and dehydroascorbate reductase from Brassica rapa effectively confers tolerance to freezing-induced oxidative stress. *Mol Cells* 36: 304–315.
Singh R, Singh S, Parihar P, *et al.* (2016) Reactive oxygen species (ROS): beneficial companions of plants' developmental processes. *Front Plant Sci* 7: 1299.
Singh S, Srivastava PK, Kumar D, Tripathi DK, Chauhan DK, Prasad SM. (2015) Morpho-anatomical and biochemical adapting strategies of maize (*Zea mays* L.) seedlings against lead and chromium stresses. *Biocatal Agric Biotechnol* 4: 286–295.
Singh S, Tripathi DK, Singh S, *et al.* (2017) Toxicity of aluminium on various levels of plant cells and organism: a review. *Environ Exp Bot* 137: 177–193.

Singh VP, Tripathi DK, Kumar D, Chauhan DK. (2011) Influence of exogenous silicon addition on aluminium tolerance in rice seedlings. *Biol Trace Elem Res* 144: 1260–1274.

Smirnoff N. (2000) Ascorbic acid: metabolism and functions of a multi-facetted molecule. *Curr Opin Plant Biol* 3: 229–235.

Sreenivasulu N, Miranda M, Prakash HS, Wobus U, Weschke W. (2004) Transcriptome changes in foxtail millet genotypes at high salinity: identification and characterization of a PHGPx gene specifically upregulated by NaCl in a salt-tolerant line. *J Plant Physiol* 161: 467–477.

Su Z, Chai M, Lu P, An R, Chen J, Wang X. (2007) AtMTM1, a novel mitochondrial protein, may be involved in activation of the manganese-containing superoxide dismutase in Arabidopsis. *Planta* 226: 1031–1039.

Sultana S, Khew CY, Morshed MM, Namasivayam P, Napis S, Ho CL. (2012) Overexpression of monodehydroascorbate reductase from a mangrove plant (AeMDHAR) confers salt tolerance on rice. *J Plant Physiol* 169: 311–318.

Suzuki N, Koussevitzky S, Mittler R, Miller G. (2012) ROS and redox signaling in the response of plants to abiotic stress. *Plant Cell Environ* 35: 259–270.

Szarka A, Horemans N, Kovacs Z, Grof P, Mayer M, Banhegyi G. (2007) Dehydroascorbate reduction in plant mitochondria is coupled to the respiratory electron transfer chain. *Physiol Plant* 129: 225–232.

Tang Y, Fu X, Shen Q, Tang K. (2016) Roles of MPBQ-MT in promoting α/γ-tocopherol production and photosynthesis under high light in lettuce. *PLoS One* 11(2): e0148490.

Tanou G, Molassiotis A, Diamantidis GR. (2009) Hydrogen peroxide and nitric oxide induced systemic antioxidant prime like activity under NaCl-stress and stress-free conditions in citrus plants. *J Plant Physiol* 166: 1904–1913.

Tiburcio AF, Altabella T, Bitrian M, Alcazar R. (2014) The roles of polyamines during the lifespan of plants: from development to stress. *Planta* 240: 1–18.

Tripathi DK, Singh VP, Kumar D, Chauhan DK. (2012a) Impact of exogenous silicon addition on chromium uptake, growth, mineral elements, oxidative stress, antioxidant capacity, and leaf and root structures in rice seedlings exposed to hexavalent chromium. *Acta Physiol Plant* 34: 279–289.

Tripathi DK, Singh VP, Kumar D, Chauhan DK. (2012b) Rice seedlings under cadmium stress: effect of silicon on growth, cadmium uptake, oxidative stress, antioxidant capacity and root and leaf structures. *Chem Ecol* 28: 281–291.

Tripathi DK, Singh S, Singh VP, Prasad SM, Chauhan DK, Dubey NK. (2016a) Silicon nanoparticles more efficiently alleviate arsenate toxicity than silicon in maize cultivar and hybrid differing in arsenate tolerance. *Front Environ Sci* 4: 46.

Tripathi DK, Singh VP, Prasad SM, Dubey NK, Chauhan DK, Rai AK. (2016b) LIB spectroscopic and biochemical analysis to characterize lead toxicity alleviative nature of silicon in wheat (Triticum aestivum L.) seedlings. *J Photochem Photobiol B: Biol* 154: 89–98.

Tripathi DK, Singh S, Singh VP, Prasad SM, Dubey NK, Chauhan DK. (2017a) Silicon nanoparticles more effectively alleviated UV-B stress than silicon in wheat (Triticum aestivum) seedlings. *Plant Physiol Biochem* 110: 70–81.

Tripathi DK, Singh S, Singh S, *et al.* (2017b) Nitric oxide alleviates silver nanoparticles (AgNps)-induced phytotoxicity in Pisum sativum seedlings. *Plant Physiol Biochem* 110: 167–177.

Tripathi DK, Mishra RK, Singh S, *et al.* (2017c) Nitric oxide ameliorates zinc oxide nanoparticles phytotoxicity in wheat seedlings: implication of the ascorbate–glutathione cycle. *Front Plant Sci* 8: 1.

Trivedi DK, Gill SS, Yadav S, Tuteja N. (2013) Genome-wide analysis of glutathione reductase (GR) genes from rice and Arabidopsis. *Plant Signal Behav* 8(2): e23021.

Trovato M, Mattioli R, Costantino P. (2008) Multiple roles of proline in plant stress tolerance and development. *Rendiconti Lincei* 19: 325–346.
Tuteja N, Sahoo RK, Garg B, Tuteja R. (2013) OsSUV3 dual helicase functions in salinity stress tolerance by maintaining photosynthesis and antioxidant machinery in rice (Oryza sativa L. cv. IR64). *Plant J* 76: 115–127.
Upadhyaya CP, Venkatesh J, Gururani MA, *et al.* (2011) Transgenic potato overproducing L-ascorbic acid resisted an increase in methylglyoxal under salinity stress via maintaining higher reduced glutathione level and glyoxalase enzyme activity. *Biotechnol Lett* 33: 2297–2307.
Ushimaru T, Nakagawa T, Fujioka Y, *et al.* (2006) Transgenic Arabidopsis plants expressing the rice dehydroascorbate reductase gene are resistant to salt stress. *J Plant Physiol* 163: 1179–1184.
Vaid N, Pandey P, Srivastava VK, Tuteja N. (2015) Pea lectin receptor-like kinase functions in salinity adaptation without yield penalty, by alleviating osmotic and ionic stresses and upregulating stress-responsive genes. *Plant Mol Biol* 88: 193–206.
Wang BQ, Zhang QF, Liu JH, Li GH. (2011) Overexpression of PtADC confers enhanced dehydration and drought tolerance in transgenic tobacco and tomato: effect on ROS elimination. *Biochem Biophys Res Commun* 413: 10–16.
Wang F, Kong W, Wong G, *et al.* (2016) AtMYB12 regulates flavonoids accumulation and abiotic stress tolerance in transgenic Arabidopsis thaliana. *Mol Genet Genomics* 291: 1549–1559.
Wang M, Zhao X, Xiao Z, Yin X, Xing T, Xia G. (2016) A wheat superoxide dismutase gene TaSOD2 enhances salt resistance through modulating redox homeostasis by promoting NADPH oxidase activity. *Plant Mol Biol* 91: 115–130.
Wang P, Du Y, Li Y, Ren D, Song CP. (2010) Hydrogen peroxide-mediated activation of MAP kinase 6 modulates nitric oxide biosynthesis and signal transduction in Arabidopsis. *Plant Cell* 22: 2981–2998.
Wang Y, Zhao W, Hao J, *et al.* (2014) Changes in biosynthesis and metabolism of glutathione upon ochratoxin A stress in Arabidopsis thaliana. *Plant Physiol Biochem* 79: 10–18.
Wen F, Qin T, Wang Y, *et al.* (2015) OsHK3 is a crucial regulator of abscisic acid signaling involved in antioxidant defense in rice. *J Integr Plant Biol* 57: 213–228.
Xiong L, Schumaker KS, Zhu JK. (2002) Cell signaling during cold, drought, and salt stress. *Plant Cell* 14: S165–S183.
Xu J, Yang J, Duan X, Jiang Y, Zhang P. (2014) Increased expression of native cytosolic Cu/Zn superoxide dismutase and ascorbate peroxidase improves tolerance to oxidative and chilling stresses in cassava (*Manihot esculenta* Crantz). *BMC Plant Biol* 14: 208.
Xu Y, Xu Q, Huang B. (2015) Ascorbic acid mitigation of water stress-inhibition of root growth in association with oxidative defense in tall fescue (*Festuca arundinacea* Schreb.). *Front Plant Sci* 6: 807.
Yan J, Guan L, Sun Y, *et al.* (2015) Calcium and ZmCCaMK are involved in brassinosteroid-induced antioxidant defense in maize leaves. *Plant Cell Physiol* 56: 883–896.
Yang T, Hao L, Yao S, Zhao Y, Lu W, Xiao K. (2016) TabHLH1, a bHLH-type transcription factor gene in wheat, improves plant tolerance to Pi and N deprivation via regulation of nutrient transporter gene transcription and ROS homeostasis. *Plant Physiol Biochem* 104: 99–113.
Yang Z, Wu Y, Li Y, Ling HQ, Chu C. (2009) OsMT1a, a type 1 metallothionein, plays the pivotal role in zinc homeostasis and drought tolerance in rice. *Plant Mol Biol* 70: 219–229.
Yao SH, Guo Y, Wang YZ, Zhang D, Xu L, Tang WH. (2016) A cytoplasmic Cu-Zn superoxide dismutase SOD1 contributes to hyphal growth and virulence of Fusarium graminearum. *Fungal Genet Biol* 91: 32–42.
Ye N, Zhu G, Liu Y, Li Y, Zhang J. (2011) ABA controls H_2O_2 accumulation through the induction of OsCATB in rice leaves under water stress. *Plant Cell Physiol* 52: 689–698.

Yin L, Wang S, Eltayeb AE, *et al.* (2010) Overexpression of dehydroascorbate reductase, but not monodehydroascorbate reductase, confers tolerance to aluminium stress in transgenic tobacco. *Planta* 231: 609–621.

Yoon HS, Lee H, Lee I, Kim KY, Jo J. (2004) Molecular cloning of the monodehydroascorbate reductase gene from Brassica campestris and analysis of its mRNA level in response to oxidative stress. *Biochim Biophys Acta* 1658: 181–186.

You J, Chan Z. (2015) ROS regulation during abiotic stress responses in crop plants. *Front Plant Sci* 6: 1092.

You J, Hu H, Xiong L. (2012) An ornithine δ-aminotransferase gene OsOAT confers drought and oxidative stress tolerance in rice. *Plant Sci* 197: 59–69.

Yu T, Li YS, Chen XF, Hu J, Chang X, Zhu YG. (2003) Transgenic tobacco plants overexpressing cotton glutathione S-transferase (GST) show enhanced resistance to methyl viologen. *J Plant Physiol* 160: 1305–1311.

Zhang A, Zhang J, Ye N, Cao J, Tan M, Jiang M. (2010) ZmMPK5 is required for the NADPH oxidase-mediated self-propagation of apoplastic H_2O_2 in brassinosteroid-induced antioxidant defence in leaves of maize. *J Exp Bot* 61: 4399–4411.

Zhang CJ, Zhao BC, Ge WN, *et al.* (2011) An apoplastic h-type thioredoxin is involved in the stress response through regulation of the apoplastic reactive oxygen species in rice. *Plant Physiol* 157: 1884–1899.

Zhang W, Yang G, Mu D, *et al.* (2016) An Ethylene-responsive factor BpERF11 negatively modulates salt and osmotic tolerance in Betula platyphylla. *Sci Rep* 6: 23085.

Zhang YJ, Wang W, Yang HL, *et al.* (2015) Molecular properties and functional divergence of the dehydroascorbate reductase gene family in lower and higher plants. *PLoS One* 10(12): e0145038.

Zhang Z, Xu Y, Xie Z, Li X, He ZH, Peng XX. (2016) Association-dissociation of glycolate oxidase with catalase in rice: a potential switch to modulate intracellular H_2O_2 levels. *Mol Plant* 9: 737–748.

Zhou B, Deng YS, Kong FY, Li B, Meng QW. (2013) Overexpression of a tomato carotenoid ε-hydroxylase gene alleviates sensitivity to chilling stress in transgenic tobacco. *Plant Physiol Biochem* 70: 235–245.

Zhu Y, Zuo M, Liang Y, *et al.* (2013) MAP65-1a positively regulates H_2O_2 amplification and enhances brassinosteroid-induced antioxidant defence in maize. *J Exp Bot* 64: 3787–3802.

13

The Regulation of Plant Development: Cross-talk of Reactive Oxygen Species and Plant Hormones

Sonal Mishra[1], Vikas Srivastava[2], Shakti Mehrotra[3], and Syed Naved Quadri[4]

[1] *University of Jammu, Jammu, India*
[2] *Central University of Jammu, Jammu (J & K), India*
[3] *CSIR-Central Institute of Medicinal and Aromatic Plants, Lucknow, India*
[4] *Jamia Hamdard University, New Delhi, India*

Introduction

Reactive oxygen species (ROS) appeared as unwanted by-products in the Earth's atmosphere around 2.7 billion years ago through oxygen-evolving photosynthetic bacteria (Halliwell, 2006). ROS play significant roles in cell signaling in organisms that ranges from prokaryotes to eukaryotes. Initially it was believed that ROS are solely toxic by-products of aerobic metabolism. However, further investigation suggested their crucial participation in complex signaling networks of cells (Mittler *et al.*, 2011). Chemically, ROS are versatile reactive molecules (peroxides, superoxide anion, hydroxyl radical, and singlet oxygen) containing oxygen as a highly reactive component. These ROS are formed during normal metabolism of oxygen during several endogenous metabolic activities. Additionally, certain exogenous physical and chemical stimulatory factors like radiation, drought, temperature extremes, heavy metals, other pollutants, and so forth, also play significant roles in cellular ROS formation (Singh *et al.*, 2011; Tripathi *et al.*, 2012a,b, 2016a,b, 2017a–c; Singh *et al.*, 2015, 2017; Singh *et al.*, 2016).

Plants, being sessile, respond to external cues through their cellular perception at cell boundaries followed by activation of molecular signaling networks of the cell. Showing an analogy to electronic transducers, these molecular signals involve a complex network of molecules/ions and enzymes whose interplay results in transmission of information at several cellular or subcellular locations. The ultimate result of these molecular events not only includes plants' responses to the environmental conditions in which they grow but also their overall growth and development. Out of several signaling molecules, ROS are considered important and ubiquitous participants of these networks. Emerging evidence reveals that ROS regulate a variety of physiological parameters, including responses related to stress and growth factor stimulation. Thus, cellular ROS formation and ROS signaling may contribute to a range of growth and developmental processes, such as root and shoot genesis, various physiological activities during fruit and seed development, programmed cell death, and plant responses to environmental stresses (Figure 13.1).

In most instances, the cross-talk between hormonal signaling pathways and ROS production provides reciprocal control and also includes Ca^{2+}- and mitogen-activated protein kinase (MAPK)-dependent cascades (Xia *et al.*, 2015). The concentration of ROS has a decisive impact

Reactive Oxygen Species in Plants: Boon Or Bane - Revisiting the Role of ROS, First Edition. Edited by Vijay Pratap Singh, Samiksha Singh, Durgesh Kumar Tripathi, Sheo Mohan Prasad, and Devendra Kumar Chauhan.

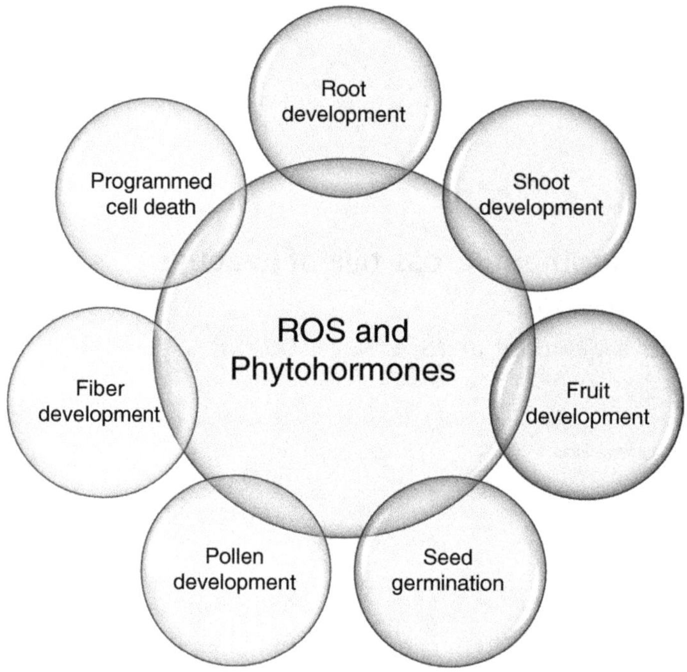

Figure 13.1 Reactive oxygen species (ROS) and phytohormones are required for different growth and development processes in plant life.

on the nature of the response. At low or moderate concentrations, ROS are ubiquitously recognized as signaling components of various processes in different organisms. Sometimes, a rise in ROS concentration (oxidative burst) occurs during a hypersensitive response leading to the death of plant tissues in the immediate vicinity to prevent further invasion of pathogens (Mur *et al.*, 2008). Additionally, ROS have recently been found to be effective regulators of hormonal interplay and associated developmental changes (Potikha *et al.*, 1999; Schopfer *et al.*, 2001; Pedreira *et al.*, 2004; Joo *et al.*, 2005; Kwak *et al.*, 2006; Müller *et al.*, 2009). It has become apparent that a number of plant processes are controlled by ROS, such as growth, development, responses to biotic and abiotic environmental stimuli, and programmed cell death. The ROS production sites involve various sub-cellular compartments like mitochondria, chloroplasts, peroxisomes, and so forth, which are highly coordinated due to the presence of various signaling pathways like calcium signaling and protein phosphorylation (Wrzaczek *et al.*, 2013).

In this chapter we discuss insights into ROS-mediated developmental changes in relation to plant hormones. We also take into account how these two important players (ROS and plant hormones) stimulate or suppress the biosynthesis or function of each other to attain directed development. Succinctly, the regulatory effect of ROS and phytohormones in determining various development processes like seed germination, root development, shoot development, fruit development, and programmed cell death (PCD) are discussed. The text also highlights the stress-induced developmental effects and participation of ROS during such processes.

ROS Producers and ROS Scavengers

ROS, chemically reactive oxygen-containing molecules, are produced after the partial reduction of molecular oxygen. These molecules can be further subcategorized into free radical forms (superoxide anion, peroxide, and hydroxyl radical) or non-radical forms (singlet oxygen and hydrogen peroxide). Importantly, the localized and temporal production of ROS is crucial for

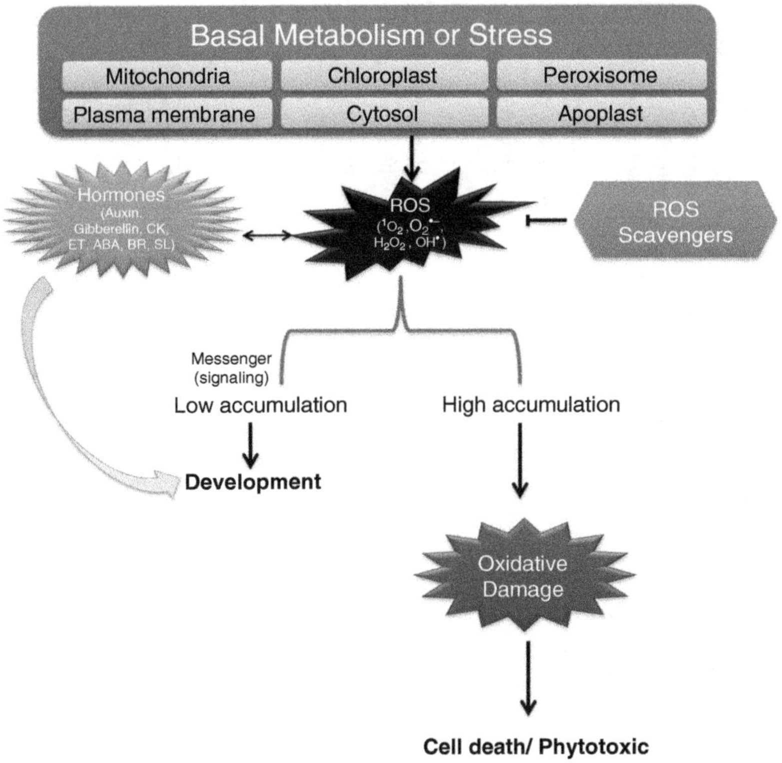

Figure 13.2 The impact of cellular reactive oxygen species (ROS) homeostasis and interplay with plant hormones. ABA, abscisic acid; BR, brassinosteroids; CK, cytokinin; ET, ethylene; SL, strigolactones.

cellular and intracellular ROS signal transduction. The interplay of ROS production and the enzymatic or non-enzymatic antioxidant systems, with which each cell is well equipped, results in regulation of ROS concentrations within the cell. Therefore, in contrast to Ca^{2+} signaling, which mainly depends on its storage and release, ROS dynamics involves their production and scavenging.

In cellular compartments, various ROS species such as $O_2^{\cdot-}$ and H_2O_2, and also perhaps $\cdot OH$, are generated during different metabolic activities. The ROS-generating systems in plants comprise many cellular locations such as plastids, mitochondria, peroxisomes (particularly in C_3 plants), cell membrane, and apoplast (cell wall) (Figure 13.2; Table 13.1). These are produced mainly as by-products during metabolic processes such as photosynthesis, photorespiration, and electron transport during respiratory metabolism. The major production of singlet oxygen, $O_2^{\cdot-}$, and H_2O_2 takes place during photosynthesis in chloroplasts. Photosynthetic carbon metabolism through photorespiration also supports H_2O_2 production in peroxisomes. Other sites of $O_2^{\cdot-}$ and H_2O_2 production are mitochondria through respiratory electron transport. Further, any physical, chemical, or metabolic shock stimulates the plasma membrane NADPH oxidases (plant homologs of mammalian respiratory-burst NADPH oxidases) and apoplastic peroxidases, leading to the rapid generation and accumulation of ROS (Table 13.1). As a result of exogenous stimulus, a rapid and transient production of huge amounts of ROS by cells causes an oxidative burst. An oxidative burst is characterized by the presence of ROS at elevated concentrations in cells leading to detrimental effects on proteins, lipids, carbohydrates, and DNA in the cell. Therefore, the cell needs a tight control system that involves ROS homeostasis to reduce the toxic consequences of ROS. Plants have developed effective antioxidant systems

Table 13.1 Major sites for reactive oxygen species (ROS) production and scavenging in the cell.

Cellular location/source	ROS production	ROS types	ROS scavengers (enzymatic)	ROS scavengers (non-enzymatic)
Chloroplast	Photosynthetic electron transport chain (PSI and PSII); singlet oxygen by triplet state of chlorophyll	Superoxide ($O_2^{\cdot-}$), hydroxyl radical ($\cdot OH$), hydrogen peroxide (H_2O_2), singlet oxygen ($^1O_2$)	Superoxide dismutase (SOD), ascorbate peroxidase (APX), monodehydroascorbate reductase (MDHAR), dehydroascorbate reductase (DHAR), glutathione reductase (GR), guaiacol peroxidase (GPX)	Ascorbic acid (AA), glutathione, carotenoids, tocopherol, proline
Mitochondrion	Respiratory electron transport chain (complex I, II, and III)	$O_2^{\cdot-}$, $\cdot OH$, H_2O_2, $^1O_2$	SOD, APX, MDHAR, DHAR, GR, GPX	AA, glutathione, tocopherol, proline
Peroxisome	Photorespiration (glycolate oxidase), fatty acid β-oxidation, xanthine oxidase, flavone oxidase	$O_2^{\cdot-}$, H_2O_2	SOD, APX, catalase (CAT)	AA, glutathione
Cytosol	Plasma membrane associated quinone oxidase	H_2O_2	SOD, APX, MDHAR, DHAR, GR, GPX	AA, glutathione, proline
Apoplast/plasma membrane/ cell wall	Plasma membrane associated NADPH-oxidase (in apoplast), cell wall associated oxalate oxidase, amine oxidase	$O_2^{\cdot-}$, $\cdot OH$, H_2O_2, $^1O_2$	SOD, APX	AA, tocopherol, glutathione
Endoplasmic reticulum	NAD(P)H-dependent electron transport involving Cyt P450	$O_2^{\cdot-}$	GPX	–

distributed at the sites of ROS production that control ROS concentrations in the cell up to desired levels. This antioxidant machinery comprises very efficient enzymatic components like superoxide dismutase (SOD), catalase (CAT), glutathione peroxidase (GPX), ascorbate peroxidase (APX), monodehydroascorbate reductase (MDHAR), dehydroascorbate reductase (DHAR), and so forth. Moreover, certain non-enzymatic components like non-protein amino acids, phenolic compounds, ascorbate, glutathione, among others, also function in antioxidant systems of the cell. These components keep a check on uncontrolled oxidation flow in the cell to prevent additional accumulation of ROS (Table 13.1). They also protect the cell from oxidative bursts by scavenging ROS (Gill and Tuteja, 2010).

Cross-talk of ROS With Phytohormones is Required for Plant Development

The morphological or anatomical changes that lead to growth, organogenesis, and consequently to complete structural development include cell growth and differentiation. Cell growth is characterized by meristematic cell division and cell expansion followed by organogenesis (differentiation). Topical insights suggest the involvement of ROS in the structural development of plants by regulating cell growth (Sugimoto-Shirasu and Roberts, 2003; Gapper and Dollan, 2006). During overall growth and developmental processes, the sources of ROS production are metabolic activities like respiration, photosynthesis, and photorespiration. However, to manage the amounts of ROS species during these cellular activities the antioxidant system also plays an important role through redox homeostasis during mitosis and cell elongation (Foyer and Noctor, 2005; Considine and Foyer, 2014). The degree of ROS accumulation also determines the response, which ranges from enhancement of gene expression to genetically programmed cell suicide pathways (Foyer and Noctor, 2005). The following explores the essential and prospectives roles of ROS and their interplay with the biosynthesis and function of various growth hormones during different growth and developmental phenomena of plants' life.

Seed Germination

ROS are reported to be continuously generated at various stages of seed development (Bailly *et al.*, 2008). At the appropriate concentrations, ROS are essential and major components of cell signaling pathways of various physiological activities during embryogenesis, seed germination, seed dormancy, and dormancy release. A crucial role of ROS during the transition from a quiescent seed to a metabolically active embryo has also been reported (Bailly *et al.*, 2008). Many species have been shown to accumulate hydrogen peroxide, nitric oxide, hydroxyl radicals, and superoxide radicals during the period of early seed imbibition (El-Maarouf-Bouteau and Bailly, 2008). The metabolically active seed's cells continuously produce ROS, which apparently play roles in various events like weakening of endosperm, seed reserve mobilization, protection against pathogens, and PCD due to adverse conditions during storage or germination (Gomes and Garcia, 2013). The internal ROS contents and the activities of ROS-scavenging systems determine the success of seed germination. Besides, ROS also act as transmitters of environmental cues that determine the timing of seed germination and hence dormancy release.

The sites of ROS production within the seed change during the course of seed development and depend on state of cell hydration. In dry seeds, ROS are mainly produced due to lipid peroxidation, which requires low moisture; however, in well-hydrated seeds metabolic activities in most of the cell organelles like glyoxysomes, peroxisomes, mitochondria, chloroplasts, and plasma membranes results in ROS accumulation (Bailly, 2004; Gomes and Garcia, 2013).

In contrast, the accumulation of surplus ROS causes oxidative damage, which ultimately reduces germination rates and even causes seed deterioration (Bailly *et al.*, 2008). However, seed cells endowed with antioxidant mechanisms strike a balance between ROS production and scavenging. This dual role of ROS might result from ROS interaction with growth hormone signaling.

Various studies have been conducted on the cross-talk between ROS and plant hormones during seed germination (Oracz and Karpiński, 2016). Gibberellic acid (GA)/abscisic acid (ABA) metabolism and signaling were observed to assess the role of ROS in barley seed germination (Ishibashi *et al.*, 2015). Seeds treated with diphenylene iodonium chloride (DPI), an NADPH oxidase inhibitor, exhibited delayed germination. The NADPH oxidases are the main producers of ROS during seed germination. However, the delayed germination effect was reverted by exogenous supplementation of H_2O_2. The application of exogenous H_2O_2 promoted gibberellin biosynthetic genes (*HvKAO1* and *HvGA3ox1*) in barley embryos. It has also been observed that ROS produced by embryonic NADPH oxidase stimulate GA biosynthesis, which activates ROS production from aleurone NADPH oxidase, which in turn activates α-amylase in aleurone cells (Ishibashi *et al.*, 2015). It was concluded that DPI suppressed GA biosynthesis in embryos resulting in delayed germination.

Moreover, the mRNAs encoding for ROS scavengers such as catalase, superoxide dismutase, and ascorbate peroxidase are untraceable in aleurone layers after GA treatment and only showed upregulation after ABA treatment. Bethke *et al.* (2002) have reported that GA is responsible for ROS-mediated death in aleurone cells but ABA maintained cell viability by keeping control over ROS production. Thus, the GA/ABA metabolism and signaling in embryo and aleurone cells regulate seed germination. This was also supported by studies of the role of ROS and other hormones in barley seed germination (Ishibashi *et al.*, 2012). In another case, a decrease in total antioxidant activity was observed in *Chenopodium rubrum* after GA treatment (Mitrović and Bogdanović, 2009). The application of GA induces upregulation of α-amylase and ROS-mediated PCD in barley aleurone cells (Fath *et al.*, 2001). The suppression of ROS-scavenging genes by GA sensitizes aleurone cells to ROS, which leads to synchronized cell death (Bethke and Jones, 2001). In dicotyledonous seeds also, ROS generation was found to be increased due to activation of NADPH oxidases (Schopfer *et al.*, 2001; Leymarie *et al.*, 2012). During germination, ROS enhance GA biosynthesis and signaling, but the regulation of localized ROS production due to GA signaling is not yet defined (Liu *et al.*, 2010; Bahin *et al.*, 2011; Leymarie *et al.*, 2012). Thus, there is also the possibility of feedback control during seed development by interplay of ROS and GA signaling. It has also been observed that nitric oxide (NO) can enhance antioxidant activity and thus suppresses the ROS-mediated PCD in GA-treated aleurone cells (Beligni *et al.*, 2002). This defines another line of control of ROS by NO during interactions with GA.

In contrast to the effect of GA, ABA inhibited seed germination due to suppressed accumulation of ROS, predominantly in the embryo of imbibed seeds (Ye and Zhang, 2012). This was due to enhancement of antioxidant enzyme activity after ABA treatment. Moreover, it has also been observed that ROS negatively regulate ABA signaling. Enhanced catabolism of ABA by H_2O_2 was noted during imbibition of *Arabidopsis* seeds through NO mediation (Liu *et al.*, 2010). Though, at the same time ROS upregulates GA biosynthesis. Furthermore, H_2O_2-mediated suppression of ABA-responsive protein kinase (PKABA) was also observed (Ishibashi *et al.*, 2012). The PKABA is known to regulate expression of the GA-regulated Myb transcription factor (GAMyb), which itself regulates α-amylase.

Apart from cross-talk of GA and ABA, ethylene has also been observed to play a role in seed germination. The inhibitors of ethylene biosynthesis and ethylene signaling mutants (*etr1, ein2, ain1, etr1,* and *erf1*) demonstrate the role of ethylene in the regulation of germination and dormancy (Corbineau *et al.*, 2014). At effective concentrations, ethylene releases dormancy. The ethylene-insensitive mutants of *Arabidopsis* are ABA sensitive and hence show improved dormancy. It has been recently noted that ethylene enhances ROS accumulation in dormant sunflower embryonic axes and probably utilizes NADPH oxidase (El-Maarouf-Bouteau *et al.*, 2015). Clear evidence of ROS, ethylene, and ABA interaction was observed during sunflower seed germination (El-Maarouf-Bouteau *et al.*, 2015). In contrast, in germinating seeds of *Arabidopsis* ethylene decreased ROS content under salinity stress (Lin *et al.*, 2012, 2013).

Root Development

The root system provides essential structural support to anchor the plant in soil and a means of nutrient and water uptake. The root system interacts with its surrounding environment and is capable of multiple responses as per requirement. The growth of the primary root, root hair, root tip, vascular tissue, and lateral root depends on root cell proliferation and differentiation. Various intrinsic and external factors, like external stimuli, hormones, secondary messengers, kinases, and transcription factors, regulate the balance between proliferation and differentiation in root cells. Under the influence of such factors root cell proliferation and differentiation work in fine coordination to execute root growth and development.

Though the development of roots is mainly associated with plant hormones, recent research has shown evidence of the involvement of ROS. Moreover, the two molecular entities, namely, ROS and plant growth hormones, work synergistically in an interdependent manner. ROS in roots are mainly produced by plasma membrane-localized NADPH oxidase, respiratory burst oxidases (RBOH), and by respiration in mitochondria. In the apoplast NADPH oxidase reduces oxygen (O_2) to superoxide anion ($O_2^{\cdot-}$), which is dismutated into H_2O_2 and O_2 spontaneously or enzymatically by the catalytic activity of SOD, apoplastic oxalate oxidase, diamine oxidase, or peroxidase. In a study of a specific cell-sorting data set for developing lateral root, the expression patterns of *Arabidopsis* S-phase kinase-associated protein 2 (SKP2B), revealed a number of genes actively participating in root growth or lateral root development. Of these genes the greatest number comprised those responsible for the activity of peroxidases. The study concluded that peroxidase activity and ROS signaling are specifically required for lateral root emergence, but not for primordial specification (Manzano *et al.*, 2014). The ROS play a fundamental role in tip-growing parts of plants such as root hairs and pollen growth (Cheung and Wu, 2008; Swanson and Gilroy, 2010). In recent years, ROS have emerged as significant molecules involved in various aspect of root growth via interaction with many hormones. The following text summarizes some of the literature reports of ROS as regulators of root growth and development at the molecular and cellular level, with or without interplay with hormones.

Auxin, an important plant hormone, plays a critical role in root growth and development by coordinating several signaling pathways and related factors involved in root development. The fate of root growth depends on the auxin gradient, which is established by active transport mediated by auxin transporters, PIN-FORMED (PIN) proteins, in cell membranes (Grunewald and Friml, 2010). ROS and auxin are closely associated as auxin-induced ROS cause cell wall loosening and play a role in cell elongation (Schopfer *et al.*, 2001). It is well established that ROS act as second messengers, thus auxin-induced ROS production causes various changes in cellular redox status, which are required as signals for development (Kocsy *et al.*, 2013). An auxin-induced ROS gradient is also responsible for auxin-mediated root gravitropism (Joo *et al.*, 2001). In a discussion about the regulation of cell proliferation and differentiation, it is relevant to state that the plant cell cycle is highly regulated by cellular redox status (Vernoux *et al.*, 2000; Vivancos *et al.*, 2010). Therefore, it is judicious to hypothesize that an increased auxin level is linked with an increased cellular oxidation state, which is further responsible for the maintenance of root meristem by arresting the cell cycle in the quiescent center (QC) (Jiang *et al.*, 2003). Hence, there might be some role for this redox gradient in association with an auxin gradient between QC cells and the surrounding stem/root cell initials. It seems that in the root tip, the differentiation and proliferation balance in root meristem and elongation zones is under the control of accumulated $O_2^{\cdot-}$ and H_2O_2 and their equilibrium (Tsukagosh *et al.*, 2010). Although little is known about auxin-mediated ROS production, in the past few years a plant-specific family of Rho-like small GTPases RAC/ROPs has been discovered. This group of proteins, activated by auxin, in turn mediates auxin signaling, auxin-regulated gene expression, and also auxin transport (Wu *et al.*, 2011). In rice, a study has shown the interaction of RAC with the N-terminal region of NADPH oxidase as part of the ROS-mediated defense response, suggesting direct regulation of NADPH oxidase by RAC/ROPs during auxin-mediated developmental processes (Wong *et al.*, 2007). However, in contrast, it is reported that under oxidative stress ROS inhibit auxin signaling, thus suppressing developmental processes to escape the stressed conditions (Potters *et al.*, 2007; Iglesias *et al.*, 2010). ROS-mediated auxin degradation and downregulation of expression of genes involved in polar auxin transport has also been reported (Blomster *et al.*, 2011; Peer *et al.*, 2013).

On the other hand, the relation between ABA and ROS is different in the case of root growth, as ABA inhibits primary root growth by activating the NADPH oxidases (Jiao *et al.*, 2013). This activation increases the level of ROS, which ultimately deactivate or degrade auxin required for primary root growth (He *et al.*, 2012). Gibberellins (GAs) are known to regulate various developmental processes like stem elongation, seed germination, dormancy, and flowering. They also maintain the redox homeostasis in cells under stressed conditions. The literature indicates the probable involvement of GA in root development besides ROS. GA signaling controls the expression of antioxidant genes mediated by DELLA proteins. DELLA proteins are an important component of GA signaling and are negatively regulated by the level of GA. In absence or lack of GA, DELLA protein activates and decreases the ROS level by upregulating expression of ROS scavengers (Achard *et al.*, 2008). ROS are involved in root development in a GA-dependent manner because in the absence of GA the activity of DELLA proteins increases followed by a reduction in ROS level, ultimately regulating root growth.

There are plenty of data showing a central role of ethylene (ET) and ROS in adventitious root development. The ET is also responsible for aerenchyma formation in a ROS-dependent manner. ROS help in lysogenous aerenchyma formation by activating PCD under submerged or water-logged conditions to increase gaseous exchange. ET-treated rice shows PCD in epidermal cells near adventitious root regions (Steffens and Sauter, 2009). So it might be possible that initially ET triggers ROS as it is also a stress hormone. On the other hand, ROS are reported to increase the biosynthesis of ET by regulating gene expression for the ET pathway (Wang *et al.*, 2002). ROS also play a significant role in root hair elongation under low ionic ambience. Low ammonium (NH_4^+) nutrient conditions resulted in strong root hair elongation in *Arabidopsis* seedlings. Treatment with auxin transport inhibitor (2,3,5-triiodobenzoic acid, TIBA; 1-naphthylphthalamic acid, NPA), auxin action inhibitor (*p*-chloro-phenoxyisobutyric acid, PCIB), ethylene biosynthesis inhibitor (amino-ethoxyvinylglycine, AVG), or ethylene perception antagonist (Ag^+) under low NH_4^+ conditions evidenced that root hair elongation is dependent on the ethylene pathway, but not related to the auxin pathway. Further studies have also suggested that the elongation is sensitive to low NH_4^+ treatment in auxin-insensitive mutants, but comparatively low in ethylene-insensitive mutants compared to wild-type plants. It has also been observed that low NH_4^+ conditions enhance the ROS levels in both root tip and root hair zone. Furthermore, the application of the NADPH oxidase inhibitor DPI and a ROS scavenger (dimethylthiourea, DMTU) inhibited low NH_4^+-induced root hair elongation, suggesting possible cross-talk of ethylene and ROS (Zhu *et al.*, 2016).

The brassinosteroids (BRs) are important growth regulators, found to play crucial roles in various cellular developmental processes like cell division and cell elongation, vascular differentiation, photomorphogenesis, photosynthesis, and stress tolerance (Haubrick and Assmann, 2006). In a mutual interaction with auxin, BRs regulate the growth and expression of genes that are involved in growth and development, directly or indirectly (Nemhauser *et al.*, 2004). The BR signaling cascade utilizes various receptors, including MAP kinases, and also requires second messengers in the form Ca^{2+} and ROS. The BR-induced elevated Ca^{2+} activates RBOH protein thus increasing the ROS level (Zhao *et al.*, 2013), indicating BR-mediated ROS production via a Ca^{2+} signaling cascade. As mentioned above that auxin promotes ROS production and root growth, BRs might also be involved in a similar manner as they enhance polar transport of auxin by controlling PIN2 (PIN-formed), auxin efflux transporter localization (Lanza *et al.*, 2012).

Shoot Development

The shoot apical meristem (SAM) is formed during embryogenesis and produces aerial plant parts. The functioning of SAM and outgrowth of lateral organs is regulated by many factors like plant hormones, peptide signals, and local transcriptional networks. The participation of ROS and their signaling has also been observed during development (Schmidt and Schippers, 2015). ROS homeostasis plays key roles in the regulation of shoot and root development with the help of ROS-generating and antioxidant components like thioredoxins, glutathione, glutaredoxins, and so forth. However, the precise role of ROS-producing and -scavenging components in shoot growth and development is still largely unexplored.

The typical SAM comprises a central zone (CZ), organizing center (OC), and peripheral zone (PZ). The high energy demand during cell division in SAM substantiates the contribution of mitochondrial metabolism, which in turn consumes high amounts of oxygen and thus increases mitochondrial ROS production. The respiratory electron transport chain is the main source of mitochondrial ROS. But the plasma membrane localized NADPH oxidase also contributes to ROS production. This was evidenced by the work done in tomato where apical dominance was significantly affected by lowering of NADPH oxidase activity. Moreover, the differential expression patterns of ten NADPH oxidase genes of *Arabidopsis* in different SAM tissue sites also suggested the role of NADPH oxidase generated ROS in physiological activities of the cells of apical meristem (Sagi *et al.*, 2004).

It was observed that RBOH-mediated H_2O_2 production negatively regulates axillary bud outgrowth. This was marked in the RBOH antisense-transformed lines of *Lycopersicon esculentum*, which displayed decreased apical dominance (Sagi *et al.*, 2004). Such transformed phenotypes exhibited highly branched and bushy forms in contrast with the wild type, which had intermediate growth with minor secondary branching. This behavior of RBOH is in contrast to many developmental processes like the polarized growth of pollen grains (*Nicotiana tabacum*), root hair growth (*Arabidopsis thaliana*), and lateral root growth and development (*Phaseolus vulgaris*), which require intense RBOH activity (Foreman *et al.*, 2003; Potocký *et al.*, 2007; Montiel *et al.*, 2013).

Plant hormones such as auxin, cytokinin, and strigolactones (SLs) are the chief modulators of plant architecture (Chen *et al.*, 2016). However, the role of ROS during regulation of axillary bud outgrowth and their relation with phytohormones is not much explored. In this context, to investigate the signaling pathway for apoplastic H_2O_2-induced axillary bud outgrowth, *RBOH1* and *WFI1* (respiratory burst oxidase homolog 1 and whitefly induce 1 – two important genes responsible for apoplastic H_2O_2 production) silenced tomato lines were developed (Chen *et al.*, 2016). This resulted in enhanced bud outgrowth along with indole acetic acid (IAA) and cytokinin accumulation in apex and stem node, respectively. This resulted due to decreased transcript of *FZY*, the rate limiting gene of IAA biosynthesis and increased transcript of *IPT2*, the cytokinin biosynthetic gene. However, exogenously supplemented H_2O_2 prevented these responses and supported IAA biosynthesis in the apex. Thus, it was concluded that RBOH-dependent apoplastic H_2O_2 promotes auxin biosynthesis in apical cells that restrain cytokinin biosynthesis and further bud outgrowth.

Senescence and Cell Death

Senescence in plants is a developmental program required for nutrient remobilization that leads to death of tissue and may also involve intrinsic or extrinsic factors (Guo and Gan, 2012). Programmed cell death (PCD) is one of the decisive events of a plant's development and defense mechanism (Reape *et al.*, 2007). The PCD involves sequential processes leading to controlled and organized destruction of cells (Lockshin and Zakeri, 2004). Cell death in plants has a developmental role as it involves synchronized degradation followed by mobilization of nutrient molecules from senescing tissues to the other plant parts active in various developmental processes (Zimmermann and Zentgraf, 2005). In the life of a plant PCD takes part in diverse ways, ranging from tissue development (xylem formation) to end of life (senescence) (Rogers, 2005). This process is also intended to remove redundant, misplaced, or damaged cells, which is vital for maintenance of multicellular organisms (Palavan-Unsal *et al.*, 2005). In plants, this process has been mainly described during the formation of tracheary elements, seed germination, and during senescence (Souter and Lindsey, 2000; Young and Gallie, 2000; Kuriyama and Fukuda, 2002; Gunawardena *et al.*, 2004).

In the proper development of a plant, PCD responds to various endogenous and exogenous stimulating factors (Lam, 2004). From a plant defense perspective, this approach allows plant cells to restrict the spread of pathogens. Fundamentally, plant PCD differs from apoptosis (cell shrinkage, nuclear condensation, and fragmentation) in animals due to the lack of engulfment by neighboring cells in plants (Lam, 2004). Furthermore, this phenomenon also offers versatility, for example during pathogen invasion (hypersensitive response) or a stress-related developmental response like aerenchyma formation, a response that involve ethylene (van Doorn and Woltering, 2005).

Senescence, which also acts as a plant defense mechanism, is promoted by major hormones including ethylene, jasmonic acid, salicylic acid, abscisic acid, and brassinosteroids. ROS, depending on their cellular concentration, may take part in signaling (at low levels), programmed cell death (at moderate levels), or even uncontrolled killing resulting in necrosis (at high levels) (Van Breusegem and Dat, 2006). ROS work in coordination with other signaling molecules, particularly phytohormones, during plant PCD (Overmyer *et al.*, 2005; Krystyna and Karpiński, 2016). This complex networking of transduction pathways, particularly in plant defense or the hypersensitive response, involves ROS, nitric oxide, SA, JA, and ethylene (Kunkel and Brooks, 2002). The increase in levels of SA and ethylene has been observed during ROS-dependent PCD (de Jong *et al.*, 2002). JA may either induce or suppress PCD, depending upon the ROS species available (singlet oxygen or superoxide), and the antagonistic effect of JA may involve suppression of SA biosynthesis or attenuation of ET sensitivity (Overmyer *et al.*, 2003). Moreover, ROS along with NO, either independently or synergistically, influence PCD (Wang *et al.*, 2013).

In a study where cell death was induced by camptothecin, it was noted that camptothecin-induced H_2O_2 production and PCD were disrupted due to inhibition of either ethylene synthesis or ethylene perception (de Jong *et al.*, 2002). This suggested the stimulatory effect of exogenous ethylene in camptothecin-induced H_2O_2 production and cell death. ROS also functions in hormone-regulated PCD of aleurone cells. Here, GA stimulates ROS accumulation (which is toxic and causes cell death) by suppressing antioxidant systems, and ABA reduces ROS concentration by utilizing the alternative oxidase pathway and the ROS-scavenging systems (Bethke *et al.*, 2007).

Fruit Ripening

Fruit ripening is an important phenomenon required for the completion of a plant's life cycle. The process includes various physiological, molecular, and biochemical events that ultimately contribute to color, flavor, and quality of fruit. The interplay between ROS and various phytohormones is responsible for the proper execution of such events during fruit ripening. Along with ROS, NO also plays a significant role in these events. Earlier, it was hypothesized that the gaseous hormone ethylene (ET) is the prime regulator of the fruit ripening process; however, later studies have also revealed the involvement of NO, and auxin. Nevertheless, at the signaling level, the process is rather complex and depends upon ROS. Therefore, the whole process involves complex cross-talk between ROS, NO, ET, auxin, and various other hormones.

Some of the evidence indicates a role of ROS in ethylene biosynthesis and related signaling during fruit ripening (Corbineau *et al.*, 2014). During ripening, the respiration rate in fruits increases due to ethylene induction, leading to an enhancement in ROS level (Hurr *et al.*, 2013). This overproduction of ROS causes deterioration of membranes followed by leakage and finally PCD. In a study exploring the role of different ROS species and antioxidant genes, it was observed that various ROS species are involved in signaling events at different stages of the ripening process. For example, at middle stages of fruit development, $O_2^{\cdot-}$ and H_2O_2 both function as effective signaling molecules, while at later stages H_2O_2 is the main ROS species to induce lipid peroxidation and oxidative stress (Huan *et al.*, 2016). This report further suggested that under low temperature conditions (4 °C) there is a decrease in respiration rate and ethylene level, which leads to a rise in catalase and glutathione peroxidases (antioxidant) activity. These enzymatic activities further upregulate key genes involved in easing of oxidative stress during late ripening stages (Huan *et al.*, 2016). In a study of apple fruits, Zermiani *et al.* (2015) reported that during prolonged cold exposure ethylene downregulates the expression of transcripts of *RBOH* and ROP-GAP machinery required for maintenance of H_2O_2 homeostasis under these conditions. Recently, NO has been shown to be an important molecule in fruit ripening. NO acts as a competitive inhibitor to lipoxygenase (LOX), a key enzyme in lipid peroxidation, and plays an important role in maintenance of the ROS level required for fruit ripening (Zhu *et al.*, 2006). It also upregulates the activities of SOD, CAT, and peroxidases (enzymes of the ROS scavenging system) in peach and kiwi fruits (Flores *et al.*, 2008; Zhu *et al.*, 2008). This suggests that NO negatively regulates ROS production and thus helps delay fruit ripening. But in contrast to this, NO biosynthesis and ROS production seem to be closely related as the nitric oxide synthase (NOS), an enzyme required for NO biosynthesis under arginine (substrate for NO biosynthesis) depletion, produces H_2O_2 (Stamler *et al.*, 1992). Also if NADPH oxidase activity is blocked by its inhibitor DPI then inhibition in NOS activity is also reported (Bolwell, 1999; Zhang and Zhao, 2008).

Along with ethylene, ABA and auxin have also emerged as novel entities involved in fruit ripening. For example, exogenous application of ethylene to certain tomato mutants, like ripening inhibitor (*rin*) and non-ripening (*nor*), was unable to promote ripening in spite of upregulation of ethylene biosynthetic pathway genes (Tigchelaar *et al.*, 1978; Yen *et al.*, 1995). These findings point to ABA functioning as another fruit-ripening hormone. More work is needed to explore the molecular mechanisms involved, which might include ROS. Auxin is another important component in the regulation of fruit ripening (Miller *et al.*, 1987; Gillaspy *et al.*, 1993; El-Sharkawy *et al.*, 2008, 2009, 2010, 2014). The ripening process in peach fruit requires coordinated signaling of auxin and ethylene (Trainotti *et al.*, 2007). Later studies by various groups suggested that auxin accelerates ethylene production in plum and peach (El-Sharkawy *et al.*, 2008, 2009, 2010; Tatsuki *et al.*, 2013). In contrast to the action of ethylene, auxin delays ripening when applied to tomato fruits by inhibiting lycopene accumulation. So like ABA signaling, auxin may engage in cross-talk with ROS species as it does in other developmental processes, or there is a complex network of auxin, ABA, and ethylene signaling that also incorporates ROS as an important agent for cross-talk between these signaling pathways.

Conclusion

Our understanding of the interplay between plant hormones and ROS during structural and physiological development of plants has been briefly addressed. The literature provides sufficient evidence about the participation of ROS during hormone-mediated signaling in plants. ROS-mediated responses were also observed in most of the stress conditions linked either to physiological or developmental outcomes. ROS-mediated plant behavior follows two routes. One originates from internal developmental cues followed by non-stress hormonal signaling. The other commences with a stress trigger, which in turn is perceived by the appropriate receptor and also involves participation of stress-related responses. Moreover, besides ROS this complex signaling also involves participation of important components such as Ca^{2+}, kinase, NO, and so forth. ROS influence the functioning of plant hormone-mediated responses at many tiers, like biosynthesis, transport, localization, and signaling.

In normal conditions the production and scavenging of ROS maintain homeostasis, and thus ROS do not have any adverse impact on plant cells. An array of candidate components work both for production and destruction of different ROS. Occasionally, due to any kind of unfavorable condition, this homeostasis is disrupted and ROS reach toxic levels, which can lead to an oxidative burst followed by cell death (Figure 13.2). Hence, an effective antioxidant control system is essential to avoid such circumstances. Moreover, the interdependency of ROS production, destruction, and signaling with different plant hormones plays a decisive role in various physiological events during plant growth.

Phytohormones may regulate ROS concentrations or may themselves be controlled by ROS. ROS again work in the biosynthesis and signaling of important phytohormones. Though this tight regulation has not been well established for many aspects of development, it can be clearly stated particularly in relation to GA/ABA-mediated seed germination, auxin/cytokinin (CK) in axillary bud proliferation, and so forth. The PCD finds its routes in the development particularly in seed germination, organ development or stress induced anatomical adaptation. However, this synchronized interdependent system has not been well established and requires more investigation. Many developmental features are reportedly influenced by plant hormones and ROS, but extensive work is required to understand the underlying mechanism of their cross-talk.

References

Achard P, Renou JP, Berthomé R, Harberd NP, Genschik P. (2008) Plant DELLAs restrain growth and promote survival of adversity by reducing the levels of reactive oxygen species. *Curr Biol* 18: 656–660.

Bahin E, Bailly C, Sotta B, Kranner I, Corbineau F, Leymarie J. (2011) Crosstalk between reactive oxygen species and hormonal signalling pathways regulates grain dormancy in barley. *Plant Cell Environ* 34: 980–993.

Bailly C. (2004) Active oxygen species and antioxidants in seed biology. *Seed Sci Res* 14: 93–107.

Bailly C, El-Maarouf-Bouteau H, Corbineau F. (2008) From intracellular signaling networks to cell death: the dual role of reactive oxygen species in seed physiology. *CR Biol* 331: 806–814.
Beligni MV, Fath A, Bethke PC, Lamattina L, Jones RL. (2002) Nitric oxide acts as an antioxidant and delays programmed cell death in barley aleurone layers. *Plant Physiol* 129: 1642–1650.
Bethke PC, Fath A, Spiegel YN, Hwang YS, Jones RL. (2002) Abscisic acid, gibberellin and cell viability in cereal aleurone. *Euphytica* 126: 3–11.
Bethke PC, Jones RL. (2001) Cell death of barley aleurone protoplasts is mediated by reactive oxygen species. *The Plant J* 25: 19–29.
Bethke PC, Libourel IG, Aoyama N, Chung YY, Still DW, Jones RL. (2007) The Arabidopsis aleurone layer responds to nitric oxide, gibberellin, and abscisic acid and is sufficient and necessary for seed dormancy. *Plant Physiol* 143: 1173–1188.
Blomster T, Salojärvi J, Sipari N, *et al.* (2011) Apoplastic reactive oxygen species transiently decrease auxin signaling and cause stress-induced morphogenic response in Arabidopsis. *Plant Physiol* 157: 1866–1883.
Bolwell GP. (1999) Role of active oxygen species and NO in plant defence. *Curr Opin Plant Biol* 2: 287–294.
Chen XJ, Xia XJ, Guo X, *et al.* (2016) Apoplastic H_2O_2 plays a critical role in axillary bud outgrowth by altering auxin and cytokinin homeostasis in tomato plants. *New Phytol* 211: 1266–1278.
Cheung AY, Wu HM. (2008) Structural and signaling networks for the polar cell growth machinery in pollen tubes. *Annu Rev Plant Biol* 59: 547–572.
Considine MJ, Foyer CH. (2014) Redox regulation of plant development. *Antioxid Redox Sign* 21: 1305–1326.
Corbineau F, Qiong Xia, Bailly C, El-Maarouf-Bouteau H. (2014) Ethylene, a key factor in the regulation of seed dormancy. *Front Plant Sci* 5: 539.
de Jong AJ, Yakimova ET, Kapchina VM, Woltering EJ. (2002) A critical role for ethylene in hydrogen peroxide release during programmed cell death in tomato suspension cells. *Planta* 214: 537–545.
De Sen K, Zheng Xun W. (2009) The role of reactive oxygen species in the ethylene biosynthesis in plants. *J Wuhan Bot Res* 27: 327–331.
El-Maarouf-Bouteau H, Bailly C. (2008) Oxidative signaling in seed germination and dormancy. *Plant Sign Behav* 3: 175–182.
El-Maarouf-Bouteau H, Sajjad Y, Bazin J, *et al.* (2015) Reactive oxygen species, abscisic acid and ethylene interact to regulate sunflower seed germination. *Plant Cell Environ* 38: 364–374.
El-Sharkawy I, Kim WS, Jayasankar S, Svircev AM, Brown DCW. (2008) Differential regulation of four members of *ACC* synthase gene family in plum. *J Exp Bot* 59: 2009–2027.
El-Sharkawy I, Sherif S, Mila I, Bouzayen M, Jayasankar S. (2009) Molecular characterization of seven genes encoding ethylene-responsive transcriptional factors during plum fruit development and ripening. *J Exp Bot* 60: 907–922.
El-Sharkawy I, Mila I, Bouzayen M, Jayasankar S. (2010) Regulation of two germin-like protein genes during plum fruit development. *J Exp Bot* 61: 1761–1770.
El-Sharkawy I, Sherif MS, Brian J, *et al.* (2014) TIR1-like auxin-receptors are involved in the regulation of plum fruit development. *J Exp Bot* 65: 5205–5215.
Fath A, Bethke PC, Jones RL. (2001) Enzymes that scavenge reactive oxygen species are down-regulated prior to gibberellic acid-induced programmed cell death in barley aleurone. *Plant Physiol* 126: 156–166.
Flores FB, Sánchez-Bel P, Valdenegro M, Romojaro F, Martínez-Madrid MC, Egea MI. (2008) Effects of a pretreatment with nitric oxide on peach (*Prunus persica* L.) storage at room temperature. *Eur Food Res Tech* 227: 1599–1611.

Foreman J, Demidchik V, Bothwell JH, *et al.* (2003) Reactive oxygen species produced by NADPH oxidase regulate plant cell growth. *Nature* 422: 442–446.

Foyer CH, Noctor G. (2005) Redox homeostasis and antioxidant signaling: a metabolic interface between stress perception and physiological responses. *The Plant Cell* 17: 1866–1875.

Gapper C, Dolan L. (2006) Control of plant development by reactive oxygen species. *Plant Physiol* 141: 341–345.

Gill SS, Tuteja N. (2010) Reactive oxygen species and antioxidant machinery in abiotic stress tolerance in crop plants. *Plant Physiol Biochem* 48: 909–930.

Gillaspy G, Ben-David H, Gruissem W. (1993) Fruits: a developmental perspective. *The Plant Cell* 5: 1439–1451.

Gomes MP, Garcia QS. (2013) Reactive oxygen species and seed germination. *Biologia* 68: 351–357.

Grunewald W, Friml J. (2010) The march of the PINs: developmental plasticity by dynamic polar targeting in plant cells. *EMBO J* 29: 2700–2714.

Gunawardena AHLAN, Greenwood JS, Dengler NG. (2004) Programmed cell death remodels lace plant leaf shape during development. *Plant Cell* 16: 60–73.

Guo Y, Gan SS. (2012) Convergence and divergence in gene expression profiles induced by leaf senescence and 27 senescence-promoting hormonal, pathological and environmental stress treatments. *Plant Cell Environ* 35: 644–655.

Halliwell B. (2006) Reactive species and antioxidants. Redox biology is a fundamental theme of aerobic life. *Plant Physiol* 141: 312–322.

Haubrick LL, Assmann SM. (2006) Brassinosteroids and plant function: some clues, more puzzles. *Plant Cell Environ* 29: 446–457.

He J, Duan Y, Hua D, *et al.* (2012) DEXH box RNA helicase-mediated mitochondrial reactive oxygen species production in Arabidopsis mediates crosstalk between abscisic acid and auxin signaling. *The Plant Cell* 24: 1815–1833.

Huan C, Jiang L, An X, *et al.* (2016) Potential role of reactive oxygen species and antioxidant genes in the regulation of peach fruit development and ripening. *Plant Physiol Biochem* 104: 294–303.

Hurr BM, Huber DJ, Vallejos CE, Lee E, Sargent SA. (2013) Ethylene-induced overproduction of reactive oxygen species is responsible for the development of watersoaking in immature cucumber fruit. *J Plant Physiol* 170: 56–62.

Iglesias MJ, Terrile MC, Bartoli CG, D'Ippólito S, Casalongué CA. (2010) Auxin signaling participates in the adaptative response against oxidative stress and salinity by interacting with redox metabolism in Arabidopsis. *Plant Mol Biol* 74: 215–222.

Ishibashi Y, Tawaratsumida T, Kondo K, *et al.* (2012) Reactive oxygen species are involved in gibberellin/abscisic acid signaling in barley aleurone cells. *Plant Physiol* 158: 1705–1714.

Ishibashi Y, Kasa S, Sakamoto M, *et al.* (2015) A role for reactive oxygen species produced by NADPH oxidases in the embryo and aleurone cells in barley seed germination. *PloS One* 10(11): pe0143173.

Jiang K, Meng YL, Feldman LJ. (2003) Quiescent center formation in maize roots is associated with an auxin-regulated oxidizing environment. *Development* 130: 1429–1438.

Jiao Y, Sun L, Song Y, *et al.* (2013) AtrbohD and AtrbohF positively regulate abscisic acid-inhibited primary root growth by affecting Ca^{2+} signalling and auxin response of roots in *Arabidopsis*. *J Exp Bot* 64: 4183–4192.

Joo JH, Bae YS, Lee JS. (2001) Role of auxin-induced reactive oxygen species in root gravitropism. *Plant Physiol* 126: 1055–1060.

Joo JH, Yoo HJ, Hwang I, Lee JS, Nam KH, Bae YS. (2005) Auxin-induced reactive oxygen species production requires the activation of phosphatidylinositol 3-kinase. *FEBS Lett* 579: 1243–1248.

Kai K, Kasa S, Sakamoto M, *et al.* (2016) Role of reactive oxygen species produced by NADPH oxidase in gibberellin biosynthesis during barley seed germination. *Plant Signal Behav* 11: e1180492.
Kocsy G, Tari I, Vanková R, *et al.* (2013) Redox control of plant growth and development. *Plant Sci* 211: 77–91.
Krystyna O, Karpiński S. (2016) Phytohormones, signaling pathways and ROS involvement in seed germination. *Front Plant Sci* 7: 864.
Kunkel BN, Brooks DM. (2002) Cross talk between signaling pathways in pathogen defense. *Curr Opin Plant Biol* 5: 325–331.
Kuriyama H, Fukuda H. (2002) Developmental programmed cell death in plants. *Curr Opin Plant Biol* 5: 568–573.
Kwak JM, Vinh N, Schroeder JI. (2006) The role of reactive oxygen species in hormonal responses. *Plant Physiol* 141: 323–329.
Lam E. (2004) Controlled cell death, plant survival and development. *Nat Rev Mol Cell Biol* 5: 305–315.
Lanza M, Garcia-Ponce B, Castrillo G, *et al.* (2012) Role of actin cytoskeleton in brassinosteroid signaling and in its integration with the auxin response in plants. *Dev Cell* 22: 1275–1285.
Leymarie J, Vitkauskaité G, Hoang HH, *et al.* (2012) Role of reactive oxygen species in the regulation of Arabidopsis seed dormancy. *Plant Cell Physiol* 53: 96–106.
Lin Y, Wang J, Zu Y, Tang Z. (2012) Ethylene antagonizes the inhibition of germination in Arabidopsis induced by salinity by modulating the concentration of hydrogen peroxide. *Acta Physiol Plant* 34: 1895–1904.
Lin Y, Yang L, Paul M, Zu Y, Tang Z. (2013) Ethylene promotes germination of Arabidopsis seed under salinity by decreasing reactive oxygen species: evidence for the involvement of nitric oxide simulated by sodium nitroprusside. *Plant Physiol Biochem* 73: 211–218.
Liu Y, Ye N, Liu R, Chen M, Zhang J. (2010) H_2O_2 mediates the regulation of ABA catabolism and GA biosynthesis in Arabidopsis seed dormancy and germination. *J Exp Bot* 61: 2979–2990.
Lockshin RA, Zakeri Z. (2004) Caspase-independent cell death? *Oncogene* 23: 2766–2773.
Manzano C, Pallero-Baena M, Casimiro I, *et al.* (2014) The emerging role of ROS signalling during lateral root development. *Plant Physiol* 165: 1105–1119.
Miller AN, Walsh CS, Cohen JD. (1987) Measurement of indole-3-acetic acid in peach fruits (*Prunus persica* L. Batsch cv. Redhaven) during development. *Plant Physiol* 84: 491–494.
Mitrović A, Bogdanović J. (2009) Effect of gibberellic acid on total antioxidant activity during *Chenopodium rubrum* L. ontogenesis in vitro. *Arch Biol Sci* 61: 49–55.
Mittler R, Vanderauwera S, Suzuki N, *et al.* (2011) ROS signaling: the new wave? *Trends Plant Sci* 16: 300–309.
Montiel J, Arthikala MK, Quinto C. (2013) Phaseolus vulgaris RbohB functions in lateral root development. *Plant Signal Behav* 8: e22694.
Müller K, Linkies A, Vreeburg RA, Fry SC, Krieger-Liszkay A, Leubner-Metzger G. (2009) In vivo cell wall loosening by hydroxyl radicals during cress seed germination and elongation growth. *Plant Physiol* 150: 1855–1865.
Mur LA, Kenton P, Lloyd AJ, Ougham H, Prats E. (2008) The hypersensitive response; the centenary is upon us but how much do we know? *J Exp Bot* 59: 501–520.
Nemhauser JL, Mockler TC, Chory J. (2004) Interdependency of brassinosteroid and auxin signaling in Arabidopsis. *PLoS Biol* 2(9): e258.
Oracz K, Karpiński S. (2016) Phytohormones signaling pathways and ROS involvement in seed germination. *Front Plant Sci* 7: 864.
Overmyer K, Brosché M, Kangasjärvi J. (2003) Reactive oxygen species and hormonal control of cell death. *Trends Plant Sci* 8: 335–342.

Overmyer K, Brosché M, Pellinen R, *et al.* (2005) Ozone-induced programmed cell death in the Arabidopsis radical-induced cell death1 mutant. *Plant Physiol* 137: 1092–1104.

Palavan-Unsal N, Buyuktuncer ED, Tufekci MA. (2005) Programmed cell death in plants. *J Cell Mol Biol* 4: 9–23.

Pedreira J, Sanz N, Pena MJ, *et al.* (2004) Role of apoplastic ascorbate and hydrogen peroxide in the control of cell growth in pine hypocotyls. *Plant Cell Physiol* 45: 530–534.

Peer WA, Cheng Y, Murphy AS. (2013) Evidence of oxidative attenuation of auxin signalling. *J Exp Bot* 64: 2629–2639.

Potikha TS, Collins CC, Johnson DI, Delmer DP, Levine A. (1999) The involvement of hydrogen peroxide in the differentiation of secondary walls in cotton fibers. *Plant Physiol* 119: 849–858.

Potocký M, Jones MA, Bezvoda R, Smirnoff N, Žárský V. (2007) Reactive oxygen species produced by NADPH oxidase are involved in pollen tube growth. *New Phytol* 174: 742–751.

Potters G, Pasternak TP, Guisez Y, Palme KJ, Jansen MA. (2007) Stress-induced morphogenic responses: growing out of trouble? *Trends Plant Sci* 12: 98–105.

Reape TJ, Molony EM, McCabe PF. (2008) Programmed cell death in plants: distinguishing between different modes. *J Exp Bot* 59: 435–444.

Rogers HJ. (2005) Cell death and organ development in plants. *Curr Top Dev Biol* 71: 225–261.

Sagi M, Davydov O, Orazova S, *et al.* (2004) Plant respiratory burst oxidase homologs impinge on wound responsiveness and development in *Lycopersicon esculentum*. *The Plant Cell* 16: 616–628.

Schmidt R, Schippers JH. (2015) ROS-mediated redox signaling during cell differentiation in plants. *BBA Gen Subjects* 1850: 1497–1508.

Schopfer P, Plachy C, Frahry G. (2001) Release of reactive oxygen intermediates (superoxide radicals, hydrogen peroxide, and hydroxyl radicals) and peroxidase in germinating radish seeds controlled by light, gibberellin, and abscisic acid. *Plant Physiol* 125: 1591–1602.

Singh R, Singh S, Parihar P, *et al.* (2016) Reactive oxygen species (ROS): beneficial companions of plants' developmental processes. *Front Plant Sci* 7: 1299.

Singh S, Srivastava PK, Kumar D, Tripathi DK, Chauhan DK, Prasad SM. (2015) Morpho-anatomical and biochemical adapting strategies of maize (*Zea mays* L.) seedlings against lead and chromium stresses. *Biocatal Agric Biotechnol* 4: 286–295.

Singh S, Tripathi DK, Singh S, *et al.* (2017) Toxicity of aluminium on various levels of plant cells and organism: a review. *Environ Exp Bot* 137: 177–193.

Singh VP, Tripathi DK, Kumar D, Chauhan DK. (2011) Influence of exogenous silicon addition on aluminium tolerance in rice seedlings. *Biol Trace Elem Res* 144: 1260–1274.

Souter M, Lindsey K. (2000) Polarity and signalling in plant embryogenesis. *J Exp Bot* 51: 971–983.

Stamler JS, Singel DJ, Loscalzo J. (1992) Biochemistry of nitric oxide and its redox-activated forms. *Science* 18: 258: 1898–1902.

Steffens B, Sauter M. (2009) Epidermal cell death in rice is confined to cells with a distinct molecular identity and is mediated by ethylene and H_2O_2 through an autoamplified signal pathway. *The Plant Cell* 21: 184–196.

Sugimoto-Shirasu K, Roberts K. (2003) "Big it up": Endoreduplication and cell-size control in plants. *Curr Opin Plant Biol* 6: 544–553.

Swanson S, Gilroy S. (2010) ROS in plant development. *Physiol Plant* 138(4): 384–392.

Tatsuki M, Nakajima N, Fujii H, *et al.* (2013) Increased levels of IAA are required for system 2 ethylene synthesis causing fruit softening in peach (*Prunus persica* L. Batsch). *J Exp Bot* 64: 1049–1059.

Tigchelaar EC, McGlasson WB, Buescher RW. (1978) Genetic regulation of tomato fruit ripening. *Hort Sci* 13: 508–513.

Trainotti L, Alice T, Giorgio C. (2007) The involvement of auxin in the ripening of climacteric fruits comes of age: the hormone plays a role of its own and has an intense interplay with ethylene in ripening peaches. *J Exp Bot* 58: 3299–3308.

Tripathi DK, Singh VP, Kumar D, Chauhan DK. (2012a) Impact of exogenous silicon addition on chromium uptake, growth, mineral elements, oxidative stress, antioxidant capacity, and leaf and root structures in rice seedlings exposed to hexavalent chromium. *Acta Physiol Plant* 34: 279–289.
Tripathi DK, Singh VP, Kumar D, Chauhan DK. (2012b) Rice seedlings under cadmium stress: effect of silicon on growth, cadmium uptake, oxidative stress, antioxidant capacity and root and leaf structures. *Chem Ecol* 28: 281–291.
Tripathi DK, Singh S, Singh VP, Prasad SM, Chauhan DK, Dubey NK. (2016a) Silicon nanoparticles more efficiently alleviate arsenate toxicity than silicon in maize cultivar and hybrid differing in arsenate tolerance. *Front Environ Sci* 4: 46.
Tripathi DK, Singh VP, Prasad SM, Dubey NK, Chauhan DK, Rai AK. (2016b) LIB spectroscopic and biochemical analysis to characterize lead toxicity alleviative nature of silicon in wheat (Triticum aestivum L.) seedlings. *J Photochem Photobiol B: Biol* 154: 89–98.
Tripathi DK, Singh S, Singh VP, Prasad SM, Dubey NK, Chauhan DK. (2017a) Silicon nanoparticles more effectively alleviated UV-B stress than silicon in wheat (Triticum aestivum) seedlings. *Plant Physiol Biochem* 110: 70–81.
Tripathi DK, Singh S, Singh S, *et al.* (2017b) Nitric oxide alleviates silver nanoparticles (AgNps)-induced phytotoxicity in Pisum sativum seedlings. *Plant Physiol Biochem* 110: 167–177.
Tripathi DK, Mishra RK, Singh S, *et al.* (2017c) Nitric oxide ameliorates zinc oxide nanoparticles phytotoxicity in wheat seedlings: implication of the ascorbate–glutathione cycle. *Front Plant Sci* 8: 1.
Tsukagosh, H, Busch W, Benfey PN. (2010) Transcriptional regulation of ROS controls transition from proliferation to differentiation in the root. *Cell* 143: 606–616.
Van Breusegem F, Dat JF. (2006) Reactive oxygen species in plant cell death. *Plant Physiol* 141: 384–390.
van Doorn WG, Woltering EJ. (2005) Many ways to exit? Cell death categories in plants. *Trends Plant Sci* 10: 117–122.
Vernoux T, Wilson RC, Seeley KA, *et al.* (2000) The ROOT MERISTEMLESS1/CADMIUM SENSITIVE2 gene defines a glutathione-dependent pathway involved in initiation and maintenance of cell division during postembryonic root development. *The Plant Cell* 12: 97–109.
Vivancos PD, Wolff T, Markovic J, Pallardó FV, Foyer CH. (2010) A nuclear glutathione cycle within the cell cycle. *Biochem J* 431: 169–178.
Wang KLC, Li H, Ecker JR. (2002) Ethylene biosynthesis and signaling networks. *The plant cell* 14 (1): S131–S151.
Wang Y, Loake GJ, Chu C. (2013) Cross-talk of nitric oxide and reactive oxygen species in plant programed cell death. *Front Plant Sci* 4: 314.
Wong HL, Pinontoan R, Hayashi K, *et al.* (2007) Regulation of rice NADPH oxidase by binding of RacGTPase to its N-terminal extension. *The Plant Cell* 19: 4022–4034.
Wrzaczek M, Brosché M, Kangasjärvi J. (2013) ROS signaling loops—production, perception, regulation. *Curr Opin Plant Biol* 16: 575–582.
Wu HM, Hazak O, Cheung AY, Yalovsky S. (2011) RAC/ROP GTPases and auxin signaling. *The Plant Cell* 23: 1208–1218.
Xia XJ, Zhou YH, Shi K, Zhou J, Foyer CH, Yu JQ. (2015) Interplay between reactive oxygen species and hormones in the control of plant development and stress tolerance. *J Exp Bot* 66: 2839–2856.
Ye N, Zhang J. (2012) Antagonism between abscisic acid and gibberellins is partially mediated by ascorbic acid during seed germination in rice. *Plant Signal Behav* 7: 563–565.

Yen HC, Lee S, Tanksley SD, Lanahan MB, Klee HJ, Giovannoni JJ. (1995) The tomato Never-ripe locus regulates ethylene-inducible gene expression and is linked to a homolog of the Arabidopsis ETR1 gene. *Plant Physiol* 107: 1343–1353.

Young TE, Gallie DR. (2000) Programmed cell death during endosperm development. *Plant Mol Biol* 44: 283–301.

Zermiani M, Zonin E, Nonis A, *et al.* (2015) Ethylene negatively regulates transcript abundance of ROP-GAP rheostat-encoding genes and affects apoplastic reactive oxygen species homeostasis in epicarps of cold stored apple fruits. *J Exp Bot* 66: 7255–7270.

Zhang L, Zhao L. (2008) Production of nitric oxide under ultraviolet-B irradiation is mediated by hydrogen peroxide through activation of nitric oxide synthase. *J Plant Biol* 51: 395–400.

Zhao Y, Qi Z, Berkowitz GA. (2013) Teaching an old hormone new tricks: cytosolic Ca^{2+} elevation involvement in plant brassinosteroid signal transduction cascades. *Plant Physiol* 163: 555–565.

Zhu C, Yang N, Guo Z, Qian M, Gan L. (2016) An ethylene and ROS-dependent pathway is involved in low ammonium-induced root hair elongation in Arabidopsis seedlings. *Plant Physiol Biochem* 105: 37–44.

Zhu JY, Sae-Seaw J, Wang ZY. (2013) Brassinosteroid signalling. *Development* 140: 1615–1620.

Zhu S, Liu M, Zhou J. (2006) Inhibition by nitric oxide of ethylene biosynthesis and lipoxygenase activity in peach fruit during storage. *Postharvest Biol Tec* 42: 41–48.

Zhu S, Sun L, Liu M, Zhou J. (2008) Effect of nitric oxide on reactive oxygen species and antioxidant enzymes in kiwifruit during storage. *J Sci Food Agric* 88: 2324–2331.

Zimmermann P, Zentgraf U. (2005) The correlation between oxidative stress and leaf senescence during plant development. *Cell Mol Biol Lett* 10: 515.

14

Cross-talk of Reactive Oxygen Species and Nitric Oxide in Various Processes of Plant Development: Past and Present

Zsuzsanna Kolbert and Gábor Feigl

University of Szeged, Szeged, Hungary

Nitric Oxide (NO) and Related Intermediates: Reactive Nitrogen Species (RNS)

Nitric oxide (NO) was first characterized as a cardiovascular signal molecule in 1987 (Furchgott, 1988), some 25 years after the first demonstration of reactive oxygen species (ROS) (Iyer *et al.*, 1961). Thanks to intensive research, it has been found that living organisms contain a group of NO-originated molecules, which are called reactive nitrogen species (RNS) by analogy to ROS. Both groups of intermediates are diverse in the sense that they include radical and non-radical forms with different physico-chemical properties and reactivity. Moreover, both ROS and RNS are redox active, thus contributing to the maintenance of the cellular redox homeostasis (Potters *et al.*, 2010). The ROS are generated by the one-, two-, or three-electron oxidation of molecular oxygen, while RNS intermediates are nitrogenous products and derive usually from the oxidation of the nitric oxide radical (NO·). Among RNS, we can distinguish the redox forms of NO radical (nitroxyl anion, NO^-; nitrosonium cation, NO^+), the higher oxides of nitrogen – dinitrogen pentoxide (N_2O_5), dinitrogen tetroxide (N_2O_4), dinitrogen trioxide (N_2O_3) – peroxynitrite ($ONOO^-$), *S*-nitrosothiol compounds (RSNOs), and dinitrosyl iron complexes (Arasimowicz-Jelonek *et al.*, 2015). Nitric oxide serves as a precursor for the other RNS molecules; at the same time NO itself is the major biologically active reactive nitrogen intermediate, therefore our chapter focuses on it.

Physico-chemical Properties of NO: A Comparison With ROS

Nitric oxide is a small diatomic gas molecule with an unpaired electron in its π orbital, which assigns a radical character for it. Moreover, because nitric oxide is a redox-active molecule it exists in three interchangeable forms in biological systems. Oxidation yields the nitrosonium cation (NO^+), while reduction of the NO radical leads to the formation of nitroxyl anion (NO^-) (Stamler *et al.*, 1992). It is worth emphasizing that NO is a peculiar radical with a surprisingly low reactivity. It only shows reactivity toward molecules with a free radical character, and toward transition metals like heme iron (Beckman and Koppenol, 1996). Therefore, NO has a relatively long half-life (in the order of seconds; Table 14.1), which is, however, influenced by its

Reactive Oxygen Species in Plants: Boon Or Bane - Revisiting the Role of ROS, First Edition. Edited by Vijay Pratap Singh, Samiksha Singh, Durgesh Kumar Tripathi, Sheo Mohan Prasad, and Devendra Kumar Chauhan.

Table 14.1 Some relevant physico-chemical properties of reactive oxygen and nitrogen species.

	Hydrogen peroxide (H_2O_2)	Superoxide anion ($O_2^{\cdot-}$)	Peroxynitrite ($ONOO^-$)	Nitric oxide (NO)
Radical character	No	Yes	No	Yes
Charge	Non-charged	Negatively charged	Negatively charged	Non-charged
Half-life (ms)	~1	0.002–0.004	<10	5000–15,000
Diffusion distance (μm)	~ × 10	~ × 10	4	100–200

own concentration (Wink and Mitchell, 1998). Besides its relative stability, its good diffusion capability (Lancaster, 1997), owing to its lipophilic character, helps nitric oxide to be an excellent signal molecule within and between cells.

In comparison, H_2O_2, which likewise is a non-charged signal molecule, possesses a half-life two orders of magnitude shorter (Vranova *et al.*, 2002), consequently its diffusibility is less (one-tenth that of NO) (Krumova and Cosa, 2016). The other relevant reactive oxygen intermediate is superoxide radical anion ($O_2^{\cdot-}$), which has a remarkably shorter biological half-life compared to NO (Table 14.1) (Vranova *et al.*, 2002) and, because of its negative charge at physiological pH, its diffusion across membranes depends on the presence of anion channels (Denicola *et al.*, 1998). The fast reaction between superoxide and nitric oxide results in the formation of a non-radical anion, peroxynitrite ($ONOO^-$). The different diffusion properties of $O_2^{\cdot-}$ and NO suggest that in biological systems, peroxynitrite generates close to the sites of $O_2^{\cdot-}$ formation where NO produced at distant cellular spaces arrives (Denicola *et al.*, 1998). Peroxynitrite itself has a longer half-life compared to the other discussed ROS (Siegel *et al.*, 2015) (Table 14.1), but it is more reactive than NO. Regarding the diffusion distance of peroxynitrite, it is similar to that of H_2O_2 and $O_2^{\cdot-}$, but it is shorter compared to NO (Denicola *et al.*, 1998).

Common Sites of ROS and NO Production Within Plant Cells

Regarding ROS–RNS cross-talk, it is important to have information about the relationships between their production in particular cellular organelles and areas. In general, ROS are formed by electron transport activities of chloroplasts, mitochondria, and plasma membranes or as a by-product in various metabolic pathways (Bashir and Jan, 2015). In contrast, the endogenous production of NO in plant cells is more complex, since several oxidative and reductive pathways are implicated; furthermore, NO can be synthesized with or without enzymatic catalysis (Mur *et al.*, 2013). The fact that there are uncertainties regarding some NO synthesis pathways further complicates the matter. Plant NO can derive from the oxidative catabolism of polyamines (Wimalasekera *et al.*, 2011) or the oxidation of hydroxylamines (Rümer *et al.*, 2009), although these mechanisms are barely documented.

A similarly obscure field of plant NO research is the questionable existence of nitric oxide synthase (NOS) enzymes, the well-characterized NO synthesizing enzyme family of animals. These enzymes use L-arginine as substrate to produce NO (Stamler, 1994).To this day, no enzyme protein or coding sequence for these enzymes has been identified in the genome of higher plants. Interestingly, though, NOS-like activities were detected in some green algae (e.g., *Ulva compressa*, *Chatonella marina*; Gonzalez *et al.*, 2012; Kim *et al.*, 2006, respectively), and the NOS-like enzyme of *Ostreococcus tauri* was both structurally and functionally characterized (Foresi *et al.*, 2010; Gupta *et al.*, 2016). These results indicate that during evolution, higher plants lost their capability to generate NO via the NOS-system (Fröhlich and Durner, 2011), since during their highly nitrate-dependent metabolism, NO can be produced more thriftily via nitrate reductase activity, without maintaining the extra enzyme system. Therefore, the main enzyme for reductive NO formation is considered to be nitrate reductase (NR) (Mur *et al.*, 2013). This cytosolic enzyme undergoes a regulatory switch from its preferential substrate, nitrate, to nitrite, and produces NO by the one-electron reduction of nitrite (Yamasaki and Sakihama, 2000). Another organelle-linked reductive mechanism is when nitrite acts as a terminal electron acceptor for the mitochondrial cytochrome coxidase/reductase (Castello *et al.*, 2006) in case of low oxygen availability (Planchet *et al.*, 2005). Mitochondria represent a major source also for intracellular ROS production. The superoxide radicals primarily generate in the electron transport chain at NADH dehydrogenase (complex I) and at ubiquinone-cytochrome *c* reductase (complex III). Furthermore, superoxide dismutase (SOD) and other antioxidants are responsible for the generation of H_2O_2 from $O_2^{\cdot-}$ in the mitochondrion (Krumova and Cosa, 2016). This means that in certain physiological conditions, NO and $O_2^{\cdot-}$ may generate at the same time in the electron transport chain, which can result in the formation of mitochondrial peroxynitrite (Noctor *et al.*, 2007). The peroxisomally localized xanthine oxidoreductase (XOR) has to be considered as a reductive enzymatic source of NO, since it reduces nitrite to NO using NADH as electron donor under low-oxygen conditions (Corpas *et al.*, 2008). When the oxygen tension is higher, XOR catalyzes the formation of superoxide anion, therefore it can be considered as a relevant source for ROS and RNS (del Río, 2011). Moreover, in their early work, Barroso *et al.* (1999) determined and biochemically characterized L-arginine-dependent NOS activity in isolated peroxisomes, suggesting that NO may be generated by at least two independent pathways (L-arginine-dependent and XOR-mediated) in peroxisomes. More recently, Corpas and Barroso (2013) detected the common generation of NO, $O_2^{\cdot-}$, and $ONOO^-$ in the root and stomatal peroxisomes of *Arabidopsis* expressing the CFP-PTS1 reporter gene construct.

The plasma membrane has a prominent role in intracellular ROS synthesis, since it binds the major superoxide-generating system, the NADPH oxidase enzyme complex (Torres *et al.*, 2002). At the same time, plasma membrane (PM) also provides space for NO generation. The PM-localized isoform of nitrate reductase supplies the nitrite substrate for nitrite-nitric oxide reductase (Ni-NOR) (Stöhr *et al.*, 2001), which reduces it and synthetizes NO. The Ni-NOR enzyme has a higher molecular weight and more acidic pH optimum than NR, it uses reduced cytochrome *c* as an *in vitro* electron donor, and it was found exclusively in root plasma membrane of tobacco. Although Ni-NOR has not been genetically characterized so far, its involvement in root development, response to anoxia, and symbiosis has been suggested (Stöhr *et al.*, 2001; Vandana *et al.*, 2012). Also, the plant-specific organelle, the chloroplast, provides a site for simultaneous ROS and RNS generation. Singlet oxygen ($^1O_2$) is a natural by-product of photosystem II (PSII) function, while $O_2^{\cdot-}$ and H_2O_2 are generated in the electron transport chain (Das and Roychoudhury, 2014). Regarding NO production, it was shown that the proper functionality of chloroplasts is significant for maintaining NO levels in soybean cotyledons (Galatro *et al.*, 2013). Earlier, the nitrite- and L-arginine-dependent pathways of chloroplastic NO synthesis had been described (Jasid *et al.*, 2006), while, according to others, NOS-like activity, but not nitrate reductase, is responsible for NO synthesis in chloroplasts of protoplasts (Tewari *et al.*, 2013).Additionally, different ROS types can be generated by different cell-wall-associated enzymes such as peroxidases (Figure 14.1), oxalate oxidases, or amine oxidases (Das and Roychoudhury, 2014). The cell wall is also the site of non-enzymatic NO generation, since nitrite can be chemically reduced by ascorbic acid at pH 3–6 yielding NO and dehydroascorbic

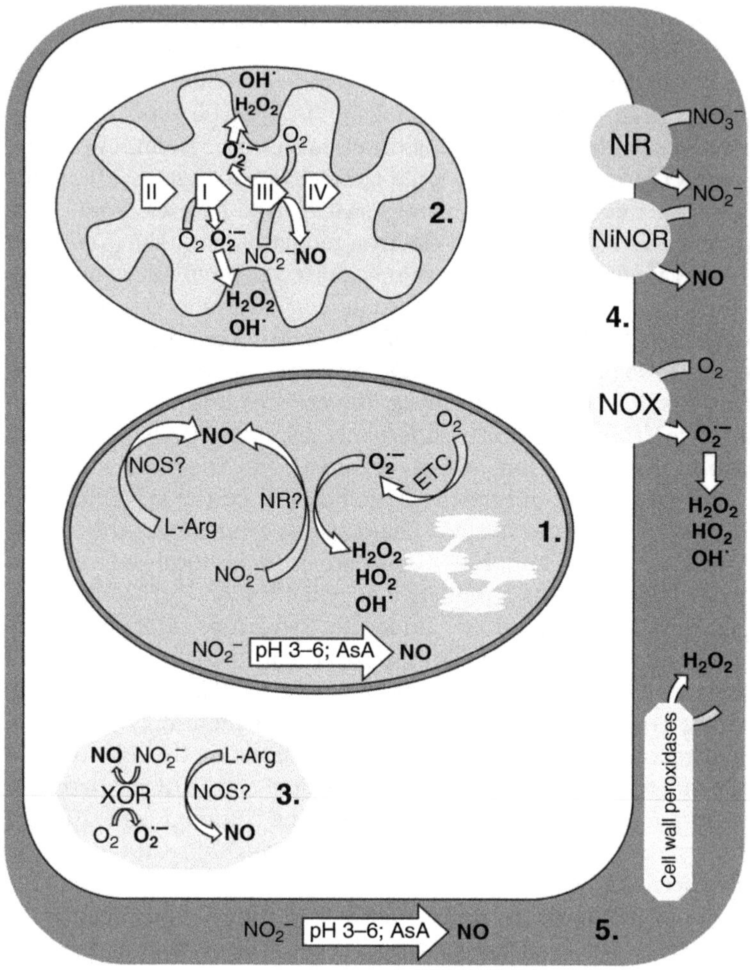

Figure 14.1 Schematic overview of the most relevant subcellular sites and mechanisms of reactive oxygen species (ROS) and nitric oxide (NO) generation in plant cells. **1.** Chloroplast: NO may originate from different sources such as nitrite or L-arginine, $O_2^{\bullet-}$ is generated in the electron transport chain (ETC), H_2O_2 and other forms are produced by enzymatic activities (e.g., superoxide dismutase, SOD). **2.** Mitochondrion: both NO and $O_2^{\bullet-}$ are produced in the ETC. SODs convert $O_2^{\bullet-}$ to H_2O_2. **3.** Peroxisome: the major source of NO is xanthine oxidoreductase (XOR), which produces also $O_2^{\bullet-}$. Another putative NO source in peroxisomes is the mammalian-like nitric oxide synthase (NOS) enzyme. **4.** Plasma membrane: NO is synthesized by nitrite-nitric oxide reductase (Ni-NOR) in the root cells, while NADPH oxidase enzyme complex (NOX) catalyzes the formation of $O_2^{\bullet-}$. Other ROS forms are produced enzymatically (e.g., SOD). **5.** Cell wall: NO is produced in a non-enzymatic reduction reaction at low pH in the presence of reductants such as ascorbic acid (AsA). ROS can be formed, *inter alia*, by the activity of cell wall-associated peroxidases.

acid (Beligni *et al.*, 2002). This reaction may occur under microlocalized pH conditions in the chloroplast and apoplastic space, where ascorbic acid is known to be present (Henry *et al.*, 1997).

Taken together, the mitochondrion, the chloroplast, and the peroxisome are organelles providing space for the common production of intracellular ROS and RNS. Furthermore, the plasma membrane and the cell wall contribute to the formation of both groups of reactive molecules (summarized in Figure 14.1). The simultaneous production of ROS (particularly $O_2^{\bullet-}$) and NO at the same subcellular location permits their reaction, which makes $ONOO^-$ generation probable in the above mentioned organelles.

Transfer of NO Action: NO-dependent Posttranslational Modifications

The formation of peroxynitrite represents a significant link between the actions of ROS and RNS, since this reactive species is able to perform both oxidative and nitrative modifications of macromolecules. The peroxynitrite-triggered protein tyrosine nitration (PTN) is an irreversible two-step posttranslational modification (PTM) during which a nitro group (-NO_2) attaches to the aromatic ring of tyrosine (Tyr) in the *ortho* position yielding 3-nitrotyrosine (Souza *et al.*, 2008). In plant cells, PTN mostly inhibits the activity of the particular enzyme protein (reviewed by Corpas *et al.*, 2013); however, in animal systems, PTN-triggered activation or no change of activity is also conceivable (Yeo *et al.*, 2015). Another exciting consequence of PTN is that it may either prevent or induce the tyrosine phosphorylation thus influencing cell signaling (Souza *et al.*, 2008). In a comprehensive study, 127 nitrated proteins were identified in wild-type *Arabidopsis thaliana* grown during control conditions (Lozano-Juste *et al.*, 2011).

The other amino acid modification triggered by NO/RNS ensuring the transfer of NO bioactivity, is *S*-nitrosylation. This reversible covalent chemical reaction results in the modification of the cysteine thiol group, and as a consequence generates *S*-nitrosothiol (-SNO). This PTM is catalyzed by, for example, the higher oxides of NO or nitrosonium cation (NO^+), by metal-NO complexes, by low molecular weight *S*-nitrosothiols (*S*-nitrosocysteine, CysNO) or *S*-nitrosoglutathione (GSNO) (Lamotte *et al.*, 2014). *S*-nitrosylation may induce conformational changes of proteins thus affecting their activities, subcellular localization, and interactions or binding activities. In plant systems, Hu *et al.* (2015) provided the most comprehensive dataset of *S*-nitrosylated proteins so far, since they identified 1195 endogenously *S*-nitrosylated peptides in 926 proteins in the *Arabidopsis* proteome.

A further possibility for NO to modify proteins is metal nitrosylation. This NO-dependent PTM involves the attachment of NO to a transition metal center in the metalloprotein resulting in the formation of a metal-nitrosyl complex. In animal systems the best characterized enzyme candidate for metal nitrosylation is guanylate cyclase (GC) (Russwurm and Koesling, 2004). The effect of NO on a plant's GC activity as well as the nature of this interaction is still unresolved.

NO-dependent Posttranslational Modifications of ROS-related Enzymes

The above discussed posttranslational modifications accomplish the transmission of the NO signal and also ensure cross-talk with ROS. Hence, NO is able to regulate the metabolism of reactive oxygen intermediates through PTMs.

Most importantly, NO has been shown to induce the *S*-nitrosylation of Arabidopsis NADPH oxidase (RBOHD) at Cys890 and consequently inhibit the superoxide-producing activity of this enzyme complex (Yun *et al.*, 2011). Similarly to NADPH oxidase, the main enzymatic source of peroxisomal H_2O_2, glycolate oxidase (GOX), suffers *S*-nitrosylation-triggered inactivation in pea leaf (Ortega-Galisteo *et al.*, 2012). Interestingly, GOX was identified among *in vivo* nitrated proteins in *Arabidopsis*as well (Lozano-Juste *et al.*, 2011), which raises the possibility of dual NO-dependent regulation. Besides the producing enzymes, several ROS-scavenging antioxidants proved to be targets for NO-dependent PTMs. According to our present knowledge, catalase (CAT) can be affected by both *S*-nitrosylation (Ortega-Galisteo *et al.*, 2012) and nitration (Lozano-Juste *et al.*, 2011), meaning that multiple NO-related pathways may control its action. Similarly, ascorbate peroxidase (APX) is under dual, NO-mediated regulation. The *S*-nitrosylation of APX was demonstrated, *inter alia*, in tobacco, *Arabidopsis*, and pea (Correa-Aragunde *et al.*, 2013; de Pinto *et al.*, 2013; Begara-Morales *et al.*, 2014). In case of pea APX, the nitrosylated site (Cys32) was found to be located at the ascorbate-binding site, and the modification caused functional induction (Begara-Morales *et al.*, 2014). Also in *Arabidopsis*, the *S*-nitrosylation at Cys32 in APX1 protein led to the activation of the enzyme (Yang *et al.*, 2015). More interestingly, *S*-nitrosylation of APX was shown to prevent carbonylation of the protein in *Antiaris toxicaria* seeds (Bai *et al.*, 2011). Moreover, Correa-Aragunde *et al.* (2013) demonstrated that denitrosylation causes partial inactivation of APX1 during root development. Regarding the nitration of APX, the modification of Tyr235 located near to the heme group resulted in functional loss (Begara-Morales *et al.*, 2014). Peroxynitrite is able to catalyze the tyrosine nitration of the mitochondrial manganase superoxide dismutase (Mn-SOD, MSD1), the peroxisomal copper/zinc SOD (CSD3), and the chloroplastic iron SOD3 (FSD3) (Holzmeister *et al.*, 2015). In case of the MSD1 protein, Tyr69 was identified as the site of nitration, suggesting that nitration is able to influence the substrate-binding activity of Mn-SOD. The functional consequence of the peroxynitrite-mediated nitration is the loss of Mn-SOD activity (Holzmeister *et al.*, 2015). Moreover, SODs can be candidates also for NO-triggered *S*-nitrosylation, as it was revealed in different plant systems (Lindermayr *et al.*, 2005; Tanou *et al.*, 2009; Sehrawat *et al.*, 2013), although the functional consequence has not been determined so far.

Based on the experimental evidence available so far, NO-related posttranslational modifications significantly affect relevant enzymes involved in, for example, ROS production or the ascorbate-glutathione cycle (summarized in Table 14.2). This suggests that NO keeps ROS homeostasis under tight control. However, the impact of NO goes beyond the above mentioned processes, an assertion supported by the fact that NO is able to regulate the hydrogen peroxide-reducing and peroxynitrite-detoxifying activities of peroxiredoxin II E via *S*-nitrosylation as well (Romero-Puertas *et al.*, 2007).

ROS-regulated NO Production in Plant Systems

It is easy to see that the interplay between ROS and NO is not unidirectional, but that mutual relationships between them are conceivable. Accordingly, ROS are also able to influence the metabolism of endogenous NO in biological systems.

The guard cell is an excellent example to study ROS–NO signal interactions, supported by the large number of experimental data in this system. For example, Srivastava *et al.* (2009) detected fast ROS production followed by slower NO formation in stomata of chitosan-treated pea plants, which suggests that NO participates in chitosan-induced stomatal movements downstream of ROS production. Likewise, in guard cells, hydrogen-rich water-promoted NO generation was shown to be dependent on RbohF-related H_2O_2 production. In this case, nitrate reductase proved to be responsible for ROS-triggered NO generation (Xie *et al.*, 2014). Similarly, a direct link between ROS and NO was observed in NO-overproducing *noe1* rice plants, where the accumulated ROS induced the activity of nitrate reductase consequently resulting in NO liberation (Lin *et al.*, 2012). Another example of the signal link between ROS and NO was related to the thermotolerance of *Arabidopsis* seedlings, where NO proved to be involved in H_2O_2 signaling as a downstream factor (Wang *et al.*, 2014). However, these studies often lack an explanation of the exact mechanism of the ROS–NO interaction at the molecular level. To address this shortcoming, P. Wang *et al.* (2010) reported that the H_2O_2-induced MAP kinase 6 phosphorylates NIA2 protein, whereupon the activity of nitrate reductase is promoted and NO is generated.

It is clear that the relationships between reactive oxygen and nitrogen species are mutual and diverse. At the same time, the nature of this cross-talk can also vary depending on several factors such as the plant species, tissue or cell type, and more importantly the presence or absence of stress. Regarding the interactions of ROS and NO during biotic and abiotic stress, we have extensive knowledge as evidenced by the numerous review papers and book chapters recently published on this topic (Scheler *et al.*, 2013; Hebelstrup and Møller, 2015; Locato *et al.*, 2016; Pucciariello and Perata, 2017). Beyond stress responses, ROS and NO are known to be involved

Table 14.2 List of antioxidant enzyme proteins modulated by *S*-nitrosylation or tyrosine nitration.

Protein	Plant species	Reference
S-nitrosylation		
RBOHD	*Arabidopsis thaliana*	Yun *et al.*, 2011
GOX	*Citrus*	Tanou *et al.*, 2009
	Pisum sativum	Ortega-Galisteo *et al.*, 2012
APX	*Arabidopsis thaliana*	Correa-Aragunde *et al.*, 2013
	Pisum sativum	Begara-Morales *et al.*, 2014
	Citrus aurantium	Tanou *et al.*, 2009
	Anticaris toxicaria	Bai *et al.*, 2011
	Arabidopsis thaliana	Yang *et al.*, 2015
	Nicotiana tabacum	de Pinto *et al.*, 2013
SODs	*Arabidopsis thaliana*	Lindermayr *et al.*, 2005
	Brassica juncea	Sehrawat *et al.*, 2013
	Citrus aurantium	Tanou *et al.*, 2009
CAT	*Pisum sativum*	Ortega-Galisteo *et al.*, 2011
GR	*Antiaris toxicaria*	Bai *et al.*, 2011
GPOX	*Arabidopsis thaliana*	Lindermayr *et al.*, 2005
GSTs	*Arabidopsis thaliana*	Lindermayr *et al.*, 2005
	Arabidopsis thaliana	Hu *et al.*, 2015
	Citrus aurantium	Tanou *et al.*, 2009
Thioredoxins	*Arabidopsis thaliana*	Hu *et al.*, 2015
	Brassica rapa	Sehrawat *et al.*, 2013
Peroxiredoxins	*Arabidopsis thaliana*	Lindermayr *et al.*, 2005
	Arabidopsis thaliana	Romero-Puertas *et al.*, 2007
Glutaredoxins	*Citrus aurantium*	Tanou *et al.*, 2009
	Arabidopsis thaliana	Lindermayr *et al.*, 2005
Tyrosine nitration		
GOX	*Arabidopsis*	Lozano-Juste *et al.*, 2011
APX	*Arabidopsis thaliana*	Lozano-Juste *et al.*, 2011
	guanylate cyclase; Pisum sativum	Begara-Morales *et al.*, 2014
SOD (Mn)	*Arabidopsis thaliana*	Holzmeister *et al.*, 2015
CAT	*Arabidopsis*	Lozano-Juste *et al.*, 2011
GR	*Helianthus annuus*	Chaki *et al.*, 2009
POX	*Arabidopsis thaliana*	Lozano-Juste *et al.*, 2011
MDHAR	*Pisum sativum*	Begara-Morales *et al.*, 2015

APX, ascorbate peroxidase; CAT, catalase; GOX, glycolate oxidase; GPOX, glutathione peroxidase; GR, glutathione reductase; GST, glutathione *S*-transferase; MDHAR, monodehydroascorbate reductase; RBOHD, respiratory burst homolog oxidase D; SOD, superoxide dismutase.

in signal transduction regulating plant growth and development. What's more, their signaling is linked at several points and the regulation of growth responses is thought to be realized through their tight cooperation. The known elements of ROS–NO cross-talk during growth and developmental processes will now be discussed in detail.

ROS–NO Interaction During Plant Growth and Development

The seed-to-seed life cycle of seed-bearing plants includes embryogenesis, and the vegetative and reproductive phases. According to the current state of knowledge, ROS as well as NO are both relevant regulator signals involved in all three developmental periods, as discussed below.

ROS–NO Interplay During Seed Embryogenesis

Embryogenesis in higher plants involves differentiation processes resulting in the formation of a mature embryo. Moreover, somatic cells can also be induced to develop into embryos in a process called somatic embryogenesis, which is an excellent model for understanding the mechanisms of plant embryogenesis (Harada *et al.*, 2010). It has been known for a while that exogenous H_2O_2 promotes embryogenesis, and elevated intracellular levels of H_2O_2 are associated with embryonic cell formation (Kairong *et al.*, 1999). In an early proteomic analysis, Kairong *et al.* (2002) observed several H_2O_2-induced translated proteins and concluded that H_2O_2 is an active messenger and induces gene expression at the mRNA level during somatic embryogenesis. The involvement of nitric oxide as a signal molecule in plant embryogenesis was evidenced a couple of years later. According to Ötvös *et al.* (2005) exogenous NO (sodium nitroprusside, SNP) promoted, and the inhibitor of the mammalian NOS enzymes (L-NMMA) delayed, the exogenous auxin-dependent formation of embryogenic cells. Moreover, similarly to H_2O_2, NO is generated endogenously in the embryonic cells of different plant species as the effect of polyamine treatments (Silveira *et al.*, 2006; Santa-Catarina *et al.*, 2007). Nitric oxide is produced in the endosperm during the embryogenesis of *Sechium edule* and its level could be reduced by NOS inhibitor, suggesting that a NOS-like enzyme may be responsible for NO synthesis during seed embryogenesis (Lombardi *et al.*, 2012). According to recent results, NO is an integral signal in glutathione-induced embryogenesis of *Araucaria angustifolia* (Vieira *et al.*, 2012). Furthermore, suppression of the *Arabidopsis* hemoglobin gene (Glb1) resulted in NO overproduction, which in turn induced ethylene synthesis and altered embryogenesis. The liberated ethylene negatively affected auxin synthesis thus inhibiting somatic embryogenesis (Mira *et al.*, 2015). Regarding the hormonal interactions of H_2O_2, the salicylic acid-triggered embryogenesis was accompanied by H_2O_2 production due to the inhibition of APX and CAT (Luo *et al.*, 2001). In a recent proteomic study, several ROS-scavenging enzyme proteins were found to be induced in somatic embryos of cotton (Zhou *et al.*, 2016). Suppressing the expression of specifically identified GhAPX proteins resulted in the inhibition of dedifferentiation, reflecting the involvement of ROS in the differentiation process. Moreover, disrupting redox homeostasis led to inhibited dedifferentiation and altered expression of genes related to auxin transport and signaling, suggesting a link between ROS and auxin during somatic embryogenesis.

Although the available literature does not reveal the exact nature of the cross-talk between ROS and NO during seed embryogenesis, similarities in their production and actions can be determined. For instance, it is apparent that both ROS (mainly H_2O_2) and NO applied exogenously are able to induce embryogenesis, and both of them are endogenously synthesized in embryonic cells (Figure 14.2). Anyway, the putative H_2O_2–NO interplay during embryogenesis could be an exciting topic for future research.

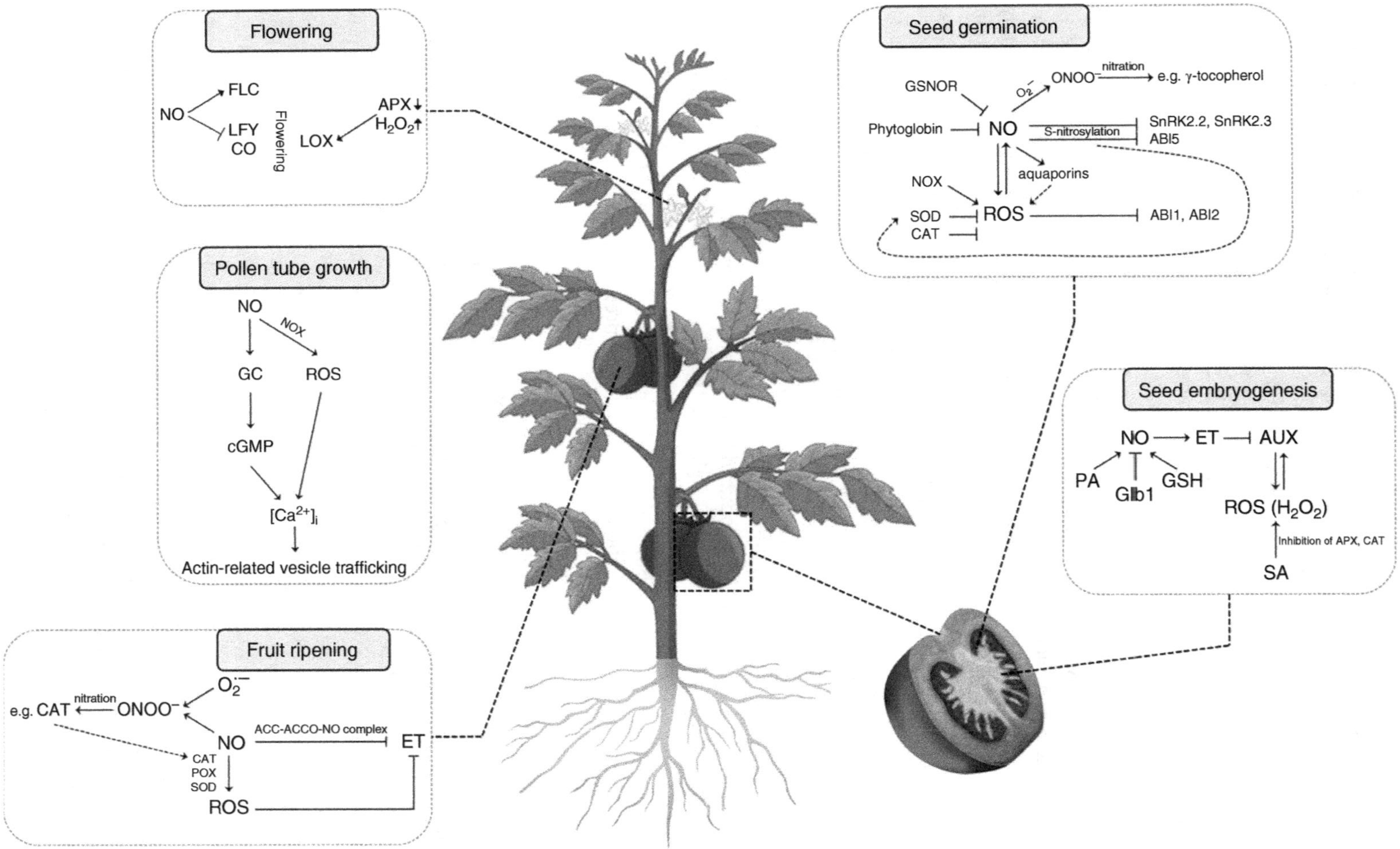

Figure 14.2 Signal interactions between reactive oxygen species (ROS) and nitric oxide (NO) during reproductive developmental processes such as flowering, pollen tube growth, and fruit ripening and during seed developmental events like embryogenesis and germination. Solid lines indicate experimentally verified links, while dashed lines represent putative interactions. ABI, ABA insensitive (transcription factor); APX, ascorbate peroxidase; AUX, auxin; CAT, catalase; CO, *CONSTANS*; ET, ethylene; FLC, *FLOWERING LOCUS* C; GC, guanylate cyclase; Glb1, class 1 hemoglobin gene; GSH, glutathione; GSNOR, *S*-nitrosoglutathione reductase; LFY, *LEAFY*; LOX, lipoxygenase; NOX, NADPH oxidase; PA, polyamine; POX, peroxidase; SA, salicylic acid; SnRK, sucrose nonfermenting 1-related protein kinase; SOD, superoxide dismutase.

ROS–NO Interplay in Mature Seeds

Within the dormant mature embryo, the shoot and root axis can be clearly recognized, and it may also contain several immature leaves. Germination is the process allowing seedling development through elongation of the embryonic axis of a seed. During the first phase (imbibition), the quiescent seed bloats as a consequence of water uptake. Then, the metabolic and molecular mechanisms are activated and germination takes place. In a germinating seed, not only does the size of the embryo increase but also internal cell differentiation occurs. Experimental results of the last 15 years have revealed that the transition of a dormant seed to a metabolically active plant involves the generation of both reactive oxygen and nitrogen species.

Regarding the formation of the main RNS, nitric oxide, in plant seeds, non-enzymatic processes seem to be primarily important. These include the spontaneous reduction of nitrite to NO at acidic pH in the apoplast of the aleurone cells, where the presence of reductants (e.g., phenolic compounds) may facilitate NO production (Bethke *et al.*, 2004). Moreover, the aleurone-localized, nitrate-inducible nitrate reductase (Ferrari and Varner, 1970) may also indirectly contribute to NO production via nitrite generation. The aleurone layer is also an important source of reactive oxygen intermediates, since during germination the generation of $O_2^{\cdot-}$, H_2O_2, and $\cdot OH$ were detected in the seed coat of radish (Schopfer *et al.*, 2001). The initiation of seed germination is controlled also by the mitochondrial production of NO. The seeds take up water during imbibition leading to the hypoxic conditions, during which mitochondrial nitrite reduction and the consequent production of NO can occur (Gupta and Igamberdiev, 2011). In hydrated seeds, all metabolically active compartments may be sources of ROS, such as mitochondria (Bailly *et al.*, 2008), which represent another common site for ROS and NO generation. Examining the kinetics of the production of reactive species, ROS formation occurs during imbibition, while the generation of NO is induced by oxygen depletion later, upon germination. The work of Ma *et al.* (2016) gives a deeper insight into the metabolism of NO during seed germination. The germinating seed produces NO, which consequently leads to the intensification of thiol nitrosylation. The phytoglobin and *S*-nitrosoglutathione reductase (GSNOR) genes involved in the regulation of NO metabolism were highly expressed in the germinating seed, suggesting their role in controlling the NO level (Ma *et al.*, 2016). At the same time, the redox state of glutathione is tightly controlled by NO via the synthesis and the GSNOR-mediated catabolism of GSNO, reflecting a link between ROS and NO metabolism during seed germination. The role of GSNOR activity in controlling both available NO and ROS pools during seed germination was supported by the delayed seed germination of AtGSNOR knockout *Arabidopsis* plants (Holzmeister *et al.*, 2011). According to Oracz *et al.* (2009) the intense formation of H_2O_2 and $O_2^{\cdot-}$ during germination is associated with the inhibition of CAT and SOD and with the induction of NADPH oxidase. As discussed earlier, these ROS-producing and antioxidant enzymes can be targets of NO-dependent posttranslational modifications (Table 14.2), thus it is conceivable that germination-induced NO regulates these enzyme proteins and consequently ROS levels; however, experimental data are lacking to support such cross-talk.

The promoting effect of NO on seed germination of different plant species has been known for a while (Beligni and Lamattina, 2000; Bethke *et al.*, 2004; Sarath *et al.*, 2007), and the aleurone layer seems to have a pivotal role in NO-perception, NO synthesis, and NO-regulated responses (Bethke *et al.*, 2007). Sarath *et al.* (2007) provided the first evidence for ROS–NO interaction in seeds, observing exogenous H_2O_2-triggered germination and concomitant NO liberation. Additionally, scavenging of NO completely inhibited the H_2O_2-induced seed germination of switchgrass. Interestingly, NO is involved in the regulation of aquaporin gene expression in germinating rice seeds, thus ensuring efficient water uptake and better germination performance (Liu *et al.*, 2007). Certain isoforms of aquaporins allow the movement of signal molecules (e.g., H_2O_2), and thus it is attractive to hypothesize that NO may result in intensified H_2O_2 signaling during germination.

Based on the published data, it is feasible that the signal transductions of H_2O_2 and NO in the germinating seed are linked at several points, although the molecular details about the majority of possible interactions are still obscure. One example of common targets of H_2O_2 and NO in the signaling cascade of germination is the regulation of ABI transcription factors. The NO-dependent *S*-nitrosylation of ABI5 at Cys153 results in its degradation through the activities of CULLIN4-based and KEEP ON GOING E3 ligases, inhibiting abscisic acid (ABA) signaling and promoting seed germination (Albertos *et al.*, 2015). Similarly, exogenous H_2O_2 arrests ABA signaling by inactivating ABI1 and ABI2 type 2C protein phosphatases (Meinhard and Grill, 2001; Meinhard *et al.*, 2002). Later, Liu *et al.* (2010) proposed that H_2O_2 mediates ABA catabolism through a NO signal and triggers gibberellic acid synthesis in germinating *Arabidopsis* seeds. However, multiple (at least two) NO-dependent pathways for the regulation of ABA signaling in germinating seed can be supposed, since *S*-nitrosylation of sucrose nonfermenting 1 (SNF1)-related protein kinases (SnRK2.2 and SnRK2.3) results in the inhibition of ABA-related signal transduction (Wang *et al.*, 2015).

The antioxidant property of NO means a further possibility for ROS and NO cross-talk. In the aleurone, the endogenously generated NO regulates programmed cell death by the maintenance of antioxidant enzyme activities (Beligni *et al.*, 2002). Furthermore, NO enhances the iron availability in germinating seeds, which is associated with the promotion of oxidative stress and ROS production (Jasid *et al.*, 2008). As was mentioned above, the NO-associated posttranslational regulation of ROS-detoxifying enzymes in the seed has not been examined yet. In germinating durum wheat, the upregulation of *S*-nitrosylation and at least 13 target proteins were observed using the biotin switch assay, although those proteins were not identified (Sen, 2010). The other NO-related macromolecule modification, nitration, requires the formation of peroxynitrite, which thus has a pivotal role in ROS–NO interactions. Similar to animal systems, the nitration of γ-tocopherol was evidenced in plant tissues (Desel *et al.*, 2007), and also its negative effect on NO content and on germination was observed (Desel and Krupinska, 2005), although the mechanism of its action remains unknown.

In general, both ROS and NO are formed during seed germination, and the aleurone layer represents a common site for their generation in the germinating seed.

Within the seed, the life of the endosperm is limited; it ends with germination, when it undergoes programmed cell death (PCD) (Ingram, 2010), which is a well-regulated process. In rice, a ROS burst due to mitochondrial membrane permeabilization accompanies PCD (Kobayashi *et al.*, 2013). According to Fath *et al.* (2000), aleurone PCD during germination might be initiated with the invovlement of ROS and NO as second messengers.

The available data suggest that ROS and NO mutually regulate each other's metabolism and activate common downstream targets in the signal transduction of ABA-regulated germination (Figure 14.2). However, further experiments are needed to complete our knowledge of germination-associated molecules modulated by ROS and NO.

ROS–NO Interplay During Vegetative Plant Development

Shoot Developmental Processes

Cotyledon and Leaf Growth

Both groups of reactive species play a role in the growth of aerial plant parts such as cotyledon/leaf and hypocotyl. Pioneer studies showed that at low concentrations, exogenous NO increases leaf expansion, but elevated doses lead to growth inhibition of pea (Leshem and Haramaty, 1996), tomato (Anderson and Mansfield, 1979), and lettuce (Hufton *et al.*, 1996). Interestingly, similarly to NO, H_2O_2 applied exogenously had a concentration-dependent regulating effect on the growth of pea seedlings (Barba-Espin *et al.*, 2010). Besides biochemical evidence, the altered morphology of NO and ROS mutants also supports the role of these reactive species in growth regulation. For instance, GSNOR-deficient plants containing elevated GSNO and NO levels have an increased number of shoots and altered shoot leaf morphology and trichome density compared to the wild-type (Holzmeister *et al.*, 2011). Similarly, the NO-overproducing *nox1/cue1* mutant possesses reduced shoot size (He *et al.*, 2004; Liu W-Z *et al.*, 2013; Frungillo *et al.*, 2014). At the same time, *nia1nia2* and *nia1nia2noa1-2 Arabidopsis* lines containing reduced NO levels also show growth defects resulting in highly reduced shoot (and root) development (Lozano-Juste and León 2010; Kolbert *et al.*, 2015). These findings suggest that tight control of NO levels is needed for normal shoot growth.

Greening

A peculiar process during shoot growth is greening and plastid development. Upon seedling greening, NO intensified the light responses, and promoted the synthesis of chlorophylls and chloroplast-related proteins (Beligni and Lamattina, 2000; Zhang *et al.*, 2006). Moreover, NO formation intensified in association with de-etiolation in *Arabidopsis* and wheat seedlings (Lozano-Juste and León, 2011; Liu Y *et al.*, 2013) reflecting the involvement of NO in the process. The recent work of Melo *et al.* (2016) revealed that exogenous NO reduces ethylene but increases auxin content during the de-etiolation of tomato. These hormonal changes accomplish negative feedback regulation of the NO level. Furthermore, NO can mimic transcriptional changes similar to those induced during phytochrome-dependent light perception, suggesting a regulatory link also between NO and phytochromes upon greening (Melo *et al.*, 2016). In greening cotyledons, the proteomic analysis of peroxisomal proteins revealed that the generated H_2O_2 is scavenged mainly by CAT3, suggesting the role of this ROS-detoxifying enzyme in regulating H_2O_2 levels during cotyledon greening (Fukao *et al.*, 2002). The possible interactions are summarized in Figure 14.3.

Hypocotyl Elongation

Besides the growth processes of the cotyledons and leaves, NO also participates in the regulation of hypocotyl elongation. Exogenous application of NO negatively impacts hypocotyl growth, as reported in potato, lettuce, and *Arabidopsis* (Beligni and Lamattina, 2000). Hypocotyls also produce superoxide anions and hydrogen peroxide in the xylem tissues during their growth, and the production of these ROS is associated with the sites of distribution of CuZn-SOD enzyme and lignin (Ogawa *et al.*, 1997). Cell wall stiffening, a necessary mechanism for efficient hypocotyl elongation, involves a precisely controlled ratio between ascorbate and H_2O_2 concentrations, evidencing the regulatory role of ROS and the redox state in hypocotyl growth (Pedreira *et al.*, 2004). In dark-grown *Arabidopsis* seedlings, extracellular ATP (eATP) caused hypocotyl elongation and the concomitant formation of NO and superoxide generation. Moreover, NO proved to be required for eATP action and NO generation is linked to that of superoxide in the *Arabidopsis* hypocotyls (Tonón *et al.*, 2010).

Based on the available data, physiological processes during shoot development seem to be regulated in parallel by NO and ROS. However, the mechanism of their interplay is mostly unknown (Figure 14.3).

Root Developmental Processes

Development of root system architecture includes growth of the primary root, lateral or adventitious roots, and root hairs and also gravitropic responses. The accumulated literature data clearly show that NO and ROS participate in all of these processes and in most cases their signaling pathways intersect.

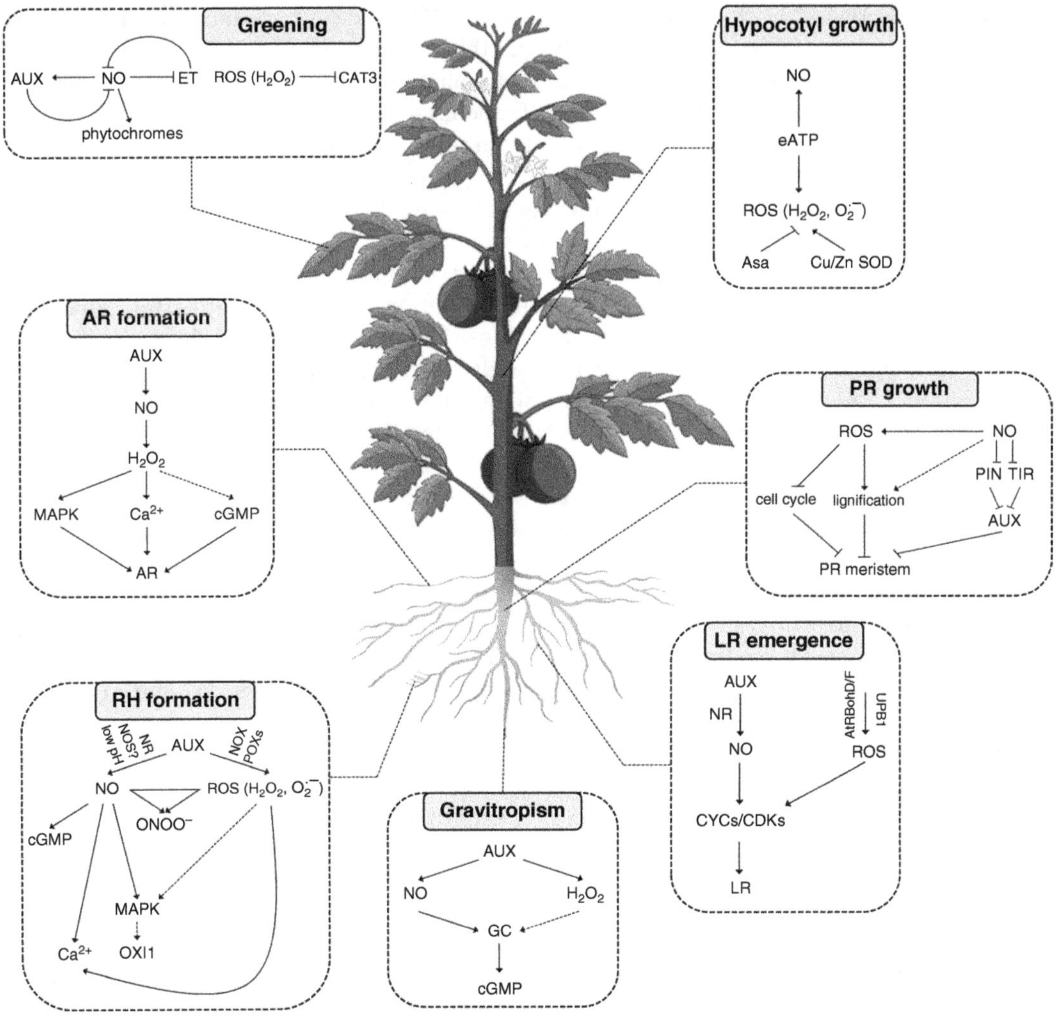

Figure 14.3 Summary of signal relationships between reactive oxygen species (ROS) and nitric oxide (NO) during different processes of vegetative plant development. Solid lines indicate experimentally verified links, while dashed lines represent hypothesized interactions. AR, adventitious root; Asa, ascorbic acid; AUX, auxin; CAT, catalase; CDK, cyclin-dependent kinase; CYC, cyclin; GC, guanylate cyclase; LR, lateral root; MAPK, mitogen-activated protein kinase; NOX, NADPH oxidase; NR, nitrate reductase; PIN, PIN-FORMED protein; POX, peroxidase; PR, primary root; RH, root hair; SOD, superoxide dismutase; TIR, TRANSPORT INHIBITOR RESPONSE protein.

Primary Root Growth

Analyzing NO and ROS mutants reveals that not only the signaling but also the metabolism of the two reactive species are connected. For instance, primary root (PR) tips of the NO-deficient *nia1nia2* and *nia1nia2noa1-2* mutants showed two-fold accumulation of superoxide anion as well as total ROS (Pető *et al.*, 2013; Kolbert *et al.*, 2015). In *gsnor1-3* plants with elevated NO content, the levels of superoxide and H_2O_2 were 50% lower relative to the wild-type (Pető *et al.*, 2013), indicating a negative relationship between ROS and NO metabolism in the root system. Similarly, the root meristems of *noa1, nia1nia2, nia1nia2noa1-2,* and wild-type plants treated with L-NMMA contained increased ROS levels, which were presumed to contribute to the reduced root growth properties of these plants (Sanz *et al.*, 2014).

As discussed previously, there are common sites of ROS and NO production, which in the root cells are the plasma membrane, the cell wall, the mitochondria, and the plastids. Due to the formation and the catabolism of ROS in the cell wall, the structure of the cell wall is directly affected through the induction of lignification (Francoz *et al.*, 2015). Similarly, NO and related molecules are thought to be involved in cell wall synthesis and modification (Correa-Aragunde *et al.*, 2008; Kolbert, 2016).

Interestingly, both ROS and NO were shown to decrease PR meristem size, but their modes of action seem to be different. Exogenous H_2O_2 treatment has been shown to reduce cell cycle gene expression (Tsukagoshi *et al.*, 2010), while NO donor disrupted auxin transport mediated by PINOID auxin efflux carrier (PIN) leading to the inhibition of meristem growth (Fernández-Marcos *et al.*, 2011). These results accord with those of Shi *et al.* (2015), where significantly reduced basipetal auxin transport and protein levels of PIN1 and PIN2 were measured in the *gsnor1-3* mutant. NO also modulates auxin signaling and sensitivity through the NO-dependent *S*-nitrosylation of auxin receptor TIR1 (TRANSPORT INHIBITOR RESPONSE 1) promoting its interaction with AUX/IAA (AUXIN/INDOL-3-ACETIC ACID) transcriptional co-repressor proteins (Terrile *et al.*, 2012). However, recently the degradation of AXR3NT-GUS (reporter for auxin-mediated degradation of AUX/IAA by TIR1) was found to be delayed in *gsnor1-3* plants compared with the wild-type, showing that the TIR1-mediated auxin signaling pathway was compromised in this mutant (Shi *et al.*, 2015). Other results imply the possibility also of an auxin-H_2O_2 regulatory link, since increased auxin levels were shown to mediate H_2O_2 production in the PR meristem leading to inhibition of root cell elongation and consequently root growth (Ivanchenko *et al.*, 2013). The possible interactions are summarized in Figure 14.3.

Lateral Root Emergence

Both ROS and NO were shown to be involved in lateral root (LR) development (Figure 14.3), and their accumulation has been detected in LR primordia. While NO generation proved to be linked to auxin and occurs through nitrate reductase activity (Kolbert *et al.*, 2008), the production of ROS is regulated by UP BEAT 1 (UPB1) transcription factors (Tsukagoshi *et al.*, 2010) and AtrbohD and F (Li *et al.*, 2015). Nitric oxide is a downstream element of auxin-mediated signaling during LR emergence (reviewed by Kolbert, 2016), but the ROS-regulated LR formation seems to be independent from auxin (Li *et al.*, 2015). These findings suggest that although both reactive species participate in the regulatory network of LR development their signaling pathways are not linked via auxin. At the same time, ROS are known to promote auxin-mediated cell cycle entry and thus may have a positive effect on the plant cell cycle (Fehér *et al.*, 2008).

Adventitious Root Formation

NO and H_2O_2 also regulate the emergence of adventitious roots (AR), since exogenous application of them resulted in enhanced AR number in species like cucumber and mung bean (Pagnussat *et al.*, 2003; Li *et al.*, 2007, 2009). Furthermore, both molecules accumulate during the induction of AR formation (Pagnussat *et al.*, 2003; Li *et al.*, 2007). In contrast to lateral root emergence, in ARs auxin induces NO and ROS generation, and both signals seem to be downstream elements of auxin-mediated signaling (Pagnussat *et al.*, 2003, Li *et al.*, 2009). Furthermore, NO was identified as an upstream signaling molecule for H_2O_2 production in auxin-induced AR formation in marigold (Liao *et al.*, 2011). As for NO, several pathways of its action have been demonstrated such as cGMP-dependent (Pagnussat *et al.*, 2003), mitogen-activated protein kinase (MAPK) cascade-linked (Pagnussat *et al.*, 2004), calcium-associated (Lanteri *et al.*, 2006), and phosphatide acid-mediated (Lanteri *et al.*, 2008). According to Liao *et al.* (2009), cGMP is not involved in H_2O_2-induced AR formation in marigold, which suggests that signaling pathways of ROS and NO are partly independent (Figure 14.3).

Root Hair Formation

Root hairs (RHs) represent both structurally and functionally relevant parts of the root system showing characteristic tip growth primarily regulated by auxin (Lee and Cho, 2013). Within the primary root, superoxide anion was detected in the apoplast of elongation zone cells, while the formation of H_2O_2 localized in the cell walls of root hairs in the differentiation zone (Dunand *et al.*, 2007) suggesting its regulatory role during root hair elongation. Peroxidases and a particular NADPH oxidase (RbohC) were held responsible for the auxin-induced generation of ROS in the root hair cell (Foreman *et al.*, 2003; Dunand *et al.*, 2007). Root hairs also produce NO during their normal development, but at different subcellular locations than for H_2O_2 production. In actively growing root hairs, NO was detected in the vacuole, whereas in more mature root hairs the cytoplasm proved to be the site of NO production (Lombardo and Lamattina, 2012). According to some hypotheses, auxin may promote the synthesis of NO in several ways, including nitrate reductase, NOS-like activity, and low pH-mediated nitrite reduction (Lombardo *et al.*, 2006).

Scavenging of ROS ($O_2^{\cdot-}$ or H_2O_2) or NO reduced the root hair formation capability of *Arabidopsis* and lettuce, respectively (Lombardo *et al.*, 2006; Dunand *et al.*, 2007). The *Arabidopsis* plants carrying a mutation in RHD2/AtrbohC showed reduced ROS accumulation and defective root hair formation (Foreman *et al.*, 2003). Moreover, the NO-deficient *nia1nia2* mutant had altered root hair growth (Lombardo and Lamattina, 2012), while in low SNO-containing *atgsnor1-1* mutants elongated root hairs were detected, and in SNO-overproducer *atgsnor1-3* mutants reduced root hairs were detected (Kwon *et al.*, 2012). These data collectively indicate that both ROS and NO are essential for normal root hair growth. On the other hand, exogenous application of NO donor was able to induce the formation of root hairs, and the participation of NO as a downstream element of auxin action was assumed (Lombardo *et al.*, 2006). More target effectors of NO signals during root cell growth are possible, such as the modulation of the redox-state via the formation of $ONOO^-$ and the activation of cGMP-, MAPK-, or calcium-dependent signaling cascades (Lombardo *et al.*, 2006). The primary role of NO in root hair growth is the regulation of vesicle formation and trafficking, as proposed by Lombardo and Lamattina (2012). Some of the putative signal pathways of NO can be linked with ROS signaling. Namely, ROS was shown to facilitate Ca^{2+} influx thus contributing to the formation of the calcium gradient within the root hair cell, which in turn regulates, *inter alia*, actin dynamics (Foreman *et al.*, 2003). Another downstream element of ROS signals is OXIDATIVE BURST INDUCIBLE1 (OXI1) kinase (Rentel *et al.*, 2004), which is activated by ROS likely via the modulation of the MAPK cascade.

The experimental data available so far strongly indicate that both NO and ROS are essential regulators of root hair development. However, their roles seem to be different. While ROS regulate cytoskeleton dynamics, NO participates in the modulation of vesicle formation and trafficking. Although the ROS- and NO-dependent signal transductions in the root hair cell are much less known, common effectors of ROS and NO are probable (Figure 14.3).

Root Gravitropism

Root gravitropism implements adaptation to the changing environment. Despite the relevance of gravitropism in plant stress physiology, only a limited number of publications deal with it in connection with ROS and NO.

Both ROS and NO were shown to accumulate on the lower side of the gravistimulated root, and both signals proved to be auxin-inducible (Joo *et al.*, 2001; Hu *et al.*, 2005). As for the role of NO in gravitropism, it appears that NO signaling is mediated via cGMP (Hu *et al.*, 2005); in other cell types ROS were shown to induce the GC-cGMP system (Dubovskaya *et al.*, 2011), implying the possibility of similar cross-talk in gravistimulated root (Figure 14.3). Several lines of evidence indicate that calcium and inositol trisphosphate (IP_3) as second messengers are also involved in the gravitropic response, but how ROS- and NO-mediated signals are integrated into a physiological response remains to be elucidated.

ROS–NO Interplay During Reproductive Plant Development

The survival of a plant species largely depends on sexual reproduction, which ensures the rise of new generations, and new populations. In flowering plants the adult plant produces flowers, which after fertilization develop seeds as the main method of dispersal (Arc *et al.*, 2013).

Floral Transition

The switch to flowering is a crucial developmental transition in the life of a plant, with a significant effect on the reproductive success of a species (Simpson and Dean, 2002; Henderson and Dean, 2004). In this process numerous exogenous and endogenous signals have a role, for instance vernalization, photoperiod, or gibberellins were shown to have a positive effect on floral transition (Mouradov *et al.*, 2002; Simpson and Dean, 2002).

During the vegetative/generative transition of the shoot apical meristem, notable NO production was detected in wheat and thus a regulatory function of this gaseous signal in the induction of flower development was supposed (Kolbert *et al.*, 2011). Earlier, He *et al.* (2004) found that the application of the NO donor SNP delayed the induction of flowering. They also reported that *Arabidopsis* mutants with high endogenous NO levels have delayed flowering, while lower endogenous NO levels are associated with earlier flowering, compared to the wild-type. High NO content suppresses the *LFY* (*LEAFY*) gene in the floral meristem and *CO* (*CONSTANS*) gene responsible for floral promotion, while the expression of the *FLC* (*FLOWERING LOCUS C*) floral repressor gene is induced by NO. The *nia1nia2* mutant possessing reduced NO content produced flowers earlier than the wild-type, confirming the negative effect of NO on flowering (Seligman *et al.*, 2008).

The effect of ROS on floral transition is mostly unknown. Similarly to NO, H_2O_2 also is generated in the shoot meristem during floral transition (Zimmermann *et al.*, 2006). The H_2O_2 formation was accompanied by reduced APX activity, but was not associated with ROS-related senescence; thus it might act as a signal for leaves for flower induction or for preparation of the development of reproductive structures (Bañuelos *et al.*, 2008). Moreover, the accumulated H_2O_2 proved to be associated with transient increase in leaf lipoxygenase (LOX) activity before/at the floral transition. These findings suggest that H_2O_2 activates plastid LOX, which plays a role in oxylipin synthesis, an important factor in the transient lipid peroxidation observed at floral transition (Bañuelos *et al.*, 2008). The possible mechanisms are visualized in Figure 14.2.

Pollination

For the success of the entire fertilization process, many factors have to match in time. For successful pollen germination, pollen tube development, and fertilization, the male pollen and the female pistil must continuously exchange several different signals, to assure the compatibility of the two partners (Heslop-Harrison, 1978, 2000; Herrero, 2003).

Several studies have considered the involvement of ROS and NO signaling in pollen germinaton and pollen tube growth (Prado *et al.*, 2004; Cárdenas *et al.*, 2006; Potocký *et al.*, 2007) or pollen–stigma interactions (McInnis *et al.*, 2006).

An interesting connection between ROS and NO in the process of pollination/pollen germination is hypothesized by Hiscock and Allen (2008). They observed that H_2O_2/ROS levels are high in stigmas, as a defense against pathogens, while pollens with high NO content may reduce the levels of ROS locally, making pollen germination possible. The interaction between ROS and NO might have a potential role in species recognition, to avoid non-compatible pollination; however, little is known regarding this (Hiscock *et al.*, 2007; Swanson *et al.*, 2004).

Another prospect is that both NO and ROS may be part of a basic recognition system that can distinguish pollen and microbes. Despite the constant secretion of sugars and lipids, receptive stigmas are highly resistant to microbial attack, because of the high concentrations of ROS here and in the nectar, which might be sufficiently toxic against pathogens (Heslop-Harrison and Shivanna, 1977; Carter and Thornburg, 2000, 2004). The source of the ROS in the stigma is mostly unknown; only one stigma-specific peroxidase has been detected so far (McInnis *et al.*, 2005).

It is now known that ROS and NO coordinate the self-incompatibility (SI) response in incompatible pollen (Serrano *et al.*, 2015). The ultimate result of the SI response is PCD, in which both reactive species have important roles (Bosch *et al.*, 2010; Wilkins *et al.*, 2011; Serrano *et al.*, 2012a,b; Jiang *et al.*, 2014). In the growing pollen tube tip of incompatible pollen of *Pyrus pyrifolia*, the inactivation of NADPH oxidase is responsible for the S-RNase-dependent disruption of the ROS content (Wang *et al.*, 2010). The lack of high ROS levels in the tip alters intracellular Ca^{2+} levels, leading to disruption of the actin cytoskeleton and to DNA degradation in the nucleus (Obara *et al.*, 2001; Thomas *et al.*, 2006). In *Papaver rhoeas* SI pollen tube, ROS and later NO accumulation lead to the reorganization of the actin cytoskeleton (Geitmann *et al.*, 2000; Bosch and Franklin-Tong, 2007; Wilkins *et al.*, 2011) through posttranscriptional modifications (carbonylation, *S*-nitrosylation) causing interference with actin polymerization, leading to PCD (Rodríguez-Serrano *et al.*, 2014). In *Olea europaea* SI pollens, NO and $O_2^{\cdot-}$ are generated and react to produce peroxynitrite, which promotes nitration in both papillar cells of the stigma and pollen undergoing PCD. It raises the possibility that tyrosine nitration triggered by ROS–NO interaction might have a role in SI PCD (Serrano *et al.*, 2012a).

If the ROS content of the stigma is deadly to microbes, why is it not toxic to pollen? It is now known that in stigmatic papillae in contact with pollen grains the ROS level is reduced (Hiscock *et al.*, 2007). Considering the high amounts of NO in pollen, it seems likely that NO acts as a signal by which the pollen "turns off" the ROS barrier (Hiscock *et al.*, 2007, McInnis *et al.*, 2006).

Numerous ROS-associated activities (catalase, superoxide dismutase, ascorbate peroxidase, monodehydroascorbate reductase, or GSH-dependent processes) were characterized in pollen grains (Acevedo and Scandalios, 1990; Dai *et al.*, 2006). Also, NADPH oxidase (NOX)-dependent ROS production is found at the tip of the growing pollen tube, which is thought to be important in maintaining polarized tip growth (Potocký *et al.*, 2007), just like in the tip-localized growth of root hairs.

Unlike ROS, NO is excluded from the tip of the growing pollen tube, but is generated in peroxisomes of subapical regions, and it is proposed to play an important role in directional growth of the pollen tube. This spatial distribution might allow the apical growth of the pollen tube, while on the other hand it might act as a positive feedback for elongation when the surrounding environment lacks an exogenous NO source. Also, external NO is able to cause the reduction and reorientation of pollen tube growth (Prado *et al.*, 2004). The effect of NO seems to be mediated by at least two pathways. Endogenous NO generated by the germinating pollen tube activates cGMP synthesis, which in turn increases Ca^{2+} uptake leading to the accumulation of intracellular calcium. At the same time NO can promote NOX activation and consequently ROS formation, also leading to calcium accumulation. As a signal, calcium regulates actin-related vesicle transport to the tip of the pollen tube. Nitric oxide also might have a role (as a "guide") in the further orientation of the growing pollen tube until it reaches the micopyle; however, the exact mechanisms are yet to be established (Hiscock and Allen, 2008).

When the pollen tube reaches the micropyle, several events happen close to each other in time. Accessory cells, like antipodial cells and synergids, undergo PCD around fertilization (Heydlauff and Groß-Hardt, 2014). The receptive synergid attracts the pollen tube to the female gametophyte, and then degenerates rapidly (Sandaklie-Nikolova *et al.*, 2007). During this PCD ROS and NO as signal molecules are suspected to play an important role in the communication between gametophytes before sperm release (Boisson-Dernier *et al.*, 2008).

As detailed above NO and ROS contribute to pollen–microbe distinction, to the self-incompatibility response, as well as to pollen germination. During the SI response, the importance of NO- and ROS-dependent posttranslational modifications have been demonstrated, while in the growing pollen tube NO-induced ROS production and a common downstream signal (Ca^{2+}) have been described (Figure 14.2).

Fruit Ripening

Fruit set, development, maturation, and ripening are controlled by several phytohormones such as ethylene, auxin, cytokinin, and gibberellin (McAtee *et al.*, 2013). The involvement of ROS- or NO-dependent signaling in the early stages after pollination, such as fruit set and development, is mainly unexplored. In contrast, the ripening of fruits is a well-characterized process in regard to NO and ROS actions.

Ripening of climacteric fruits, such as mango or papaya, is an oxidative phenomenon, in which ROS play an important role (Mondal *et al.*, 2004). Furthermore, it is known that ROS have a role in fruit development, but the available information is limited. In peach, two peaks of ROS content were observed by Huan *et al.* (2016) in the middle of fruit development, which proved to be important in the further development of the fruit, suggesting that ROS might act as positive signals in ripening. Hydrogen peroxide formation occurs during the ripening of tomato (Kumar *et al.*, 2016), grape (Pilati *et al.*, 2007), or guava (Mondal *et al.*, 2009).

As for NO, it has anti-ripening and senescence effects, and is able to extend postharvest shelf-life of crops (Manjunatha *et al.*, 2010). It is known that the NO level in unripe fruits is much higher than in ripe ones (Leshem *et al.*, 1998). Principally, NO inhibits the autocatalytic biosynthesis of ethylene through its binding to 1-aminocyclopropane-1-carboxylic acid (ACC) oxidase, resulting in the formation of the ACC-ACC oxidase-NO complex, which leads to the reduction of ethylene production (Rudell and Mattheis, 2006). Alternatively, NO is able to trigger ROS-associated redox changes, which may affect ethylene biosynthesis, but further research is needed to support these putative interactions (Manjunatha *et al.*, 2012). Moreover, the major antioxidant enzymes like CAT, peroxidase (POX), and SOD are regulated by NO in peaches (Flores *et al.*, 2008) or in kiwi fruit (Zhu *et al.*, 2008). Also, polyphenolase, phenylalanine ammonia lyase, or lipoxygenase activities were shown to be affected by NO in fruit tissues (Duan *et al.*, 2007; Zhu *et al.*, 2009). Another link between ROS and NO can be the accumulation of salicylic acid triggered by NO and suppressing the level of superoxide radical and other ROS (Zhang *et al.*, 2003), resulting in the maintenance of membrane integrity. During the ripening of pepper, protein nitration intensified and six proteins were identified to be nitrated (Chaki *et al.*, 2015). Among them catalase was the most abundant in both green and ripened fruits. In red pepper fruits, the decrease of catalase activity suggests diminished H_2O_2 detoxification, which presumably promotes lipid peroxidation (Chaki *et al.*, 2015).

The results available so far indicate that NO exerts its anti-ripening effect both through the modulation of ET biosynthesis and signaling, and through the modification of ROS metabolism. This also reveals an antagonistic relationship between NO and ROS during fruit ripening (Figure 14.2).

Conclusion

It is established that both ROS and NO are involved in the regulatory networks of all three phases of plant development from seed. However, in certain processes (e.g., seed or pollen tube germination, root system growth) their role and interplay are better known, whereas in others (e.g., shoot system development, flowering, fruit set) they are poorly characterized. Nevertheless, ROS–NO interaction may occur by direct chemical reaction between NO and $O_2^{\cdot -}$ (e.g., during seed germination, fruit ripening) or by the modifying effect of NO on ROS-detoxifying enzymes (e.g., during seed germination, pollen tube growth). Furthermore, there are common effector elements in their signal transduction, such as MAPK and cGMP during root hair or adventitious root formation, for instance, representing an additional level of ROS–NO cross-talk. So, thanks to intensive research, a lot of information is now available regarding the relevance of ROS and NO in plant growth and development, but many exciting questions remain unanswered.

Acknowledgements

Authors would like to thank the financial support for the János Bolyai Research Scholarship of the Hungarian Academy of Sciences (Grant no. BO/00751/16/8) and the National Research, Development and Innovation Fund (Grant no. NKFI-6, K120383).This research was supported by the EU-funded Hungarian grant EFOP-3.6.1-16-2016-00008. Authors are grateful to Dr. Attila Ördög (Department of Plant Biology, University of Szeged) for his kind help with digital figure editing.

References

Acevedo A, Scandalios JG. (1990) Expression of the catalase and superoxide dismutase genes in mature pollen in maize. *Theoret App Genet* 80: 705–711.

Albertos P, Romero-Puertas MC, Tatematsu K, *et al.* (2015) S-nitrosylation triggers ABI5 degradation to promote seed germination and seedling growth. *Nature Comm* 6: Art. no. 8669; doi: 10.1038/ncomms9669.

Anderson LS, Mansfield TA. (1979) The effects of nitric oxide pollution on the growth of tomato. *Environ Pollut* 20: 113–121.

Arasimowicz-Jelonek M, Floryszak-Wieczorek J, Abramowski D, Izbiańska K. (2015) Nitric oxide and reactive nitrogen species. In: Kahn MN, Mobin M, Mohammad F, Corpas FJ (eds), *Nitric Oxide in Plants: Metabolism and Role in Stress Physiology*. Springer, pp. 165–184.

Arc E, Sechet J, Corbineau F, Rajjou L, Marion-Poll A. (2013) ABA crosstalk with ethylene and nitric oxide in seed dormancy and germination. *Front Plant Sci* 4: 63.

Bai X, Yang L, Tian M, *et al.* (2011) Nitric oxide enhances desiccation tolerance of recalcitrant Antiaris toxicaria seeds via protein S-nitrosylation and carbonylation. *PLoS ONE* 6: e20714.

Bailly C, El-Maarouf-Bouteau H, Corbineau F. (2008) From intracellular signaling networks to cell death: the dual role of reactive oxygen species in seed physiology. *CR Biol* 331: 806–814.

Bañuelos GR, Argumedo R, Patel K, Ng V, Zhou F, Vellanoweth RL. (2008) The developmental transition to flowering in *Arabidopsis* is associated with an increase in leaf chloroplastic lipoxygenase activity. *Plant Sci* 174: 366–373.

Barba-Espin G, Diaz-Vivancos P, Clemente-Moreno MJ, *et al.* (2010) Interaction between hydrogen peroxide and plant hormones during germination and the early growth of pea seedlings. *Plant Cell Environ* 33: 981–994.

Barroso JB, Corpas FJ, Carreras A, *et al.* (1999) Localization of nitric-oxide synthase in plant peroxisomes. *J Biol Chem* 274: 36729–36733.

Bashir F, Jan S. (2015) Oxidative stress and antioxidant defence systems in response to pesticide stress. In: Azooz MM, Ahmad P (eds), *Legumes under Environmental Stress: Yield, Improvement, and Adaptations*. John Wiley & Sons, Ltd, pp. 103–123.

Beckman JS, Koppenol WH. (1996) Nitric oxide, superoxide and peroxynitrite: the good, the bad, and the ugly. *Am J Physiol* 271: C1424–C1437.

Begara-Morales JC, Sánchez-Calvo B, Chaki M. (2014) Dual regulation of cytosolic ascorbate peroxidase (APX) by tyrosine nitration and S-nitrosylation. *J Exp Bot* 65: 527–538.

Begara-Morales JC, Sánchez-Calvo B, Chaki M, *et al.* (2015) Differential molecular response of monodehydroascorbate reductase and glutathione reductase by nitration and S-nitrosylation. *J Exp Bot* 66: 5983–5996.

Beligni MV, Lamattina L. (2000) Nitric oxide stimulates seed germination and de-etiolation, and inhibits hypocotyl elongation, three light-inducible responses in plants. *Planta* 210: 215–221.

Beligni MV, Fath A, Bethke PC, Lamattina L, Jones RL. (2002) Nitric oxide acts as an antioxidant and delays programmed cell death in barley aleurone layers. *Plant Phys* 129: 1642–1650.

Bethke PC, Murray R. Badger MR, Jones RL. (2004) Apoplastic synthesis of nitric oxide by plant tissues. *Plant Cell* 16: 332–341.

Bethke PC, Libourel IG, Aoyama N, Chung YY, Still DW, Jones RL. (2007) The *Arabidopsis* aleurone layer responds to nitric oxide, gibberellin, and abscisic acid and is sufficient and necessary for seed dormancy. *Plant Phys* 143: 1173–1188.

Boisson-Dernier A, Frietsch S, Kim TH, Dizon MB, Schroeder JI. (2008) The peroxin loss-of-function mutation abstinence by mutual consent disrupts male-female gametophyte recognition. *Curr Biol* 18: 63–68.

Bosch M, Franklin-Tong VE. (2007) Temporal and spatial activation of caspase-like enzymes induced by self-incompatibility in *Papaver* pollen. *Proc Natl Acad Sci USA* 104: 18327–18332.

Bosch M, Poulter NS, Perry RM, Wilkins KA, Franklin-Tong VE. (2010) Characterization of a legumain/vacuolar processing enzyme and YVADase activity in *Papaver* pollen. *Plant Mol Biol* 74: 381–393.

Cárdenas L, McKenna ST, Kunkel JG, Hepler PK. (2006) NAD(P)H oscillates in pollen tubes and is correlated with tip growth. *Plant Phys* 142: 1460–1468.

Carter C, Thornburg RW. (2000) Tobacco nectarin I purification and characterization as a germin-like, manganese superoxide dismutase implicated in the defense of floral reproductive tissues. *J Biol Chem* 275: 36726–36733.

Carter C, Thornburg RW. (2004) Is the nectar redox cycle a floral defense against microbial attack? *Trends Plant Sci* 9: 320–324.

Castello PR, David PS, McClure T, Crook Z, Poyton RO. (2006) Mitochondrial cytochrome oxidase produces nitric oxide under hypoxic conditions: implications for oxygen sensing and hypoxic signaling in eukaryotes. *Cell Metab* 3: 277–287.

Chaki M, Valderrama R, Fernández-Ocaña A, *et al.* (2009) Protein targets of tyrosine nitration in sunflower (*Helianthus annuus* L.) hypocotyls. *J Exp Bot* 60: 4221–4234.

Chaki M, Álvarez de Morales P, Ruiz C, *et al.* (2015) Ripening of pepper (*Capsicum annuum*) fruit is characterized by an enhancement of protein tyrosine nitration. *Ann Bot* 116: 637–647.

Corpas FJ, Barroso JB. (2013) Peroxynitrite ($ONOO^-$) is endogenously produced in *Arabidopsis* peroxisomes and is overproduced under cadmium stress. *Ann Bot* 113: 87–96.

Corpas FJ, Palma JM, Sandalio LM, Valderrama R, Barroso JB, del Río LA. (2008) Peroxisomal xanthine oxidoreductase: Characterization of the enzyme from pea (*Pisum sativum* L.) leaves. *J Plant Phys* 165: 1319–1330.
Corpas FJ, Palma JM, del Río LA, Barroso JB. (2013) Protein tyrosine nitration in higher plants grown under natural and stress conditions. *Front Plant Sci* 4: 29.
Correa-Aragunde N, Lombardo C, Lamattina L. (2008) Nitric oxide: an active nitrogen molecule that modulates cellulose synthesis in tomato roots. *New Phytol* 179: 386–396.
Correa-Aragunde N, Foresi N, Delledonne M, Lamattina L. (2013) Auxin induces redox regulation of ascorbate peroxidase 1 activity by S-nitrosylation/denitrosylation balance resulting in changes of root growth pattern in *Arabidopsis*. *J Exp Bot* 64: 3339–3349.
Dai S, Li L, Chen T, Chong K, Xue Y, Wang T. (2006) Proteomic analyses of *Oryza sativa* mature pollen reveal novel proteins associated with pollen germination and tube growth. *Proteomics* 6: 2504–2529.
Das K, Roychoudhury A. (2014) Reactive oxygen species (ROS) and response of antioxidants as ROS-scavengers during environmental stress in plants. *Front Environ Sci* 2: 53.
del Río LA. (2011) Peroxisomes as a cellular source of reactive nitrogen species signal molecules. *Arch Biochem Biophys* 506: 1–11.
Denicola A, Souza JM, Radi R. (1998) Diffusion of peroxynitrite across erythrocyte membranes. *Proc Natl Acad Sci USA* 95: 3566–3571.
de Pinto MC, Locato V, Sgobba A, *et al.* (2013) S-Nitrosylation of ascorbate peroxidase is part of programmed cell death signaling in tobacco bright yellow-2 cells. *Plant Phys* 163: 1766–1775.
Desel C, Krupinska K. (2005) The impact of tocochromanols on early seedling development and NO release. *J Plant Phys* 162: 771–776.
Desel C, Hubbermann EM, Schwarz K, Krupinska K. (2007) Nitration of γ-tocopherol in plant tissues. *Planta* 226: 1311–1322.
Duan X, Su X, You Y, Qu H, Li Y, Jiang Y. (2007) Effect of nitric oxide on pericarp browning of harvested longan fruit in relation to phenolic metabolism. *Food Chem* 104: 571–576.
Dubovskaya LV, Bakakina YS, Kolesneva EV, *et al.* (2011) cGMP-dependent ABA-induced stomatal closure in the ABA-insensitive *Arabidopsis* mutant *abi1-1*. *New Phytol* 191: 57–69.
Dunand C, Crèvecoeur M, Penel C. (2007) Distribution of superoxide and hydrogen peroxide in *Arabidopsis* root and their influence on root development: possible interaction with peroxidases. *New Phytol* 174: 332–341.
Fath A, Bethke P, Lonsdale J, Meza-Romero R, Jones R. (2000) Programmed cell death in cereal aleurone. In: Lam E, Fukuda H, Greenberg J (eds), *Programmed Cell Death in Higher Plants*. Springer Netherlands, pp. 11–22.
Fehér A, Ötvös K, Pasternak TP, Szandtner AP. (2008) The involvement of reactive oxygen species (ROS) in the cell cycle activation (G(0)-to-G(1) transition) of plant cells. *Plant Signal Behav* 3: 823–826.
Fernández-Marcos M, Sanz L, Lewis DR, Muday GK, Lorenzo O. (2011) Nitric oxide causes root apical meristem defects and growth inhibition while reducing PIN-FORMED 1 (PIN1)-dependent acropetal auxin transport. *Proc Natl Acad Sci USA* 108: 18506–18511.
Ferrari TE, Varner JE. (1970) Control of nitrate reductase activity in barley aleurone layers. *Proc Natl Acad Sci USA* 65: 729–736.
Flores FB, Sánchez-Bel P, Valdenegro M, Romojaro F, Martínez-Madrid MC, Egea MI. (2008) Effects of a pretreatment with nitric oxide on peach (*Prunus persica* L.) storage at room temperature. *Eur Food Res Technol* 227: 1599–1611.
Foreman J, Demidchik V, Bothwell JH, *et al.* (2003) Reactive oxygen species produced by NADPH oxidase regulate plant cell growth. *Nature* 422: 442–446.
Foresi N, Correa-Aragundea N, Parisi G, Caló G, Salerno G, Lamattina L. (2010) Characterization of a nitric oxide synthase from the plant kingdom: NO generation from the green alga *Ostreococcus tauri* is light irradiance and growth phase dependent. *Plant Cell* 22: 3816–3830.

Francoz E, Ranocha P, Nguyen-Kim H, Jamet E, Burlat V, Dunand C. (2015) Roles of cell wall peroxidases in plant development. *Phytochemistry* 112: 15–21.
Fröhlich A, Durner J. (2011) The hunt for plant nitric oxide synthase (NOS): Is one really needed? *Plant Sci* 18: 401–404.
Frungillo L, Skelly MJ, Loake GJ, Spoel SH, Salgado I. (2014) S-nitrosothiols regulate nitric oxide production and storage in plants through the nitrogen assimilation pathway. *Nat Commun* 5: Art. no. 5401.
Fukao Y, Hayashi M, Nishimura M. (2002) Proteomic analysis of leaf peroxisomal proteins in greening cotyledons of *Arabidopsis thaliana*. *Plant Cell Physiol* 43: 689–696.
Furchgott RF. (1988) Studies on relaxation of rabbit aorta by sodium nitrite: the basis for the proposal that the acid-activatable inhibitory factor from retractor penis is inorganic nitrite and the endothelium-derived relaxing factor is nitric oxide. In: Vanhoutte PM (ed.), *Vasodilatation: Vascular Smooth Muscle, Peptides, Autonomic Nerves and Endothelium*. Raven Press, New York, pp. 401–414.
Galatro A, Puntarulo S, Guiamet JJ, Simontacchi M. (2013) Chloroplast functionality has a positive effect on nitric oxide level in soybean cotyledons. *Plant Phys Biochem* 66: 26–33.
Geitmann A, Snowman BN, Emons AMC, Franklin-Tong VE. (2000) Alterations in the actin cytoskeleton of pollen tubes are induced by the self-incompatibility reaction in *Papaver rhoeas*. *Plant Cell* 12: 1239–1251.
Gonzalez A, de los Angeles Cabrera M, Henriquez MJ, Contreras RA, Morales B, Moenne A. (2012) Cross talk among calcium, hydrogen peroxide, and nitric oxide and activation of gene expression involving calmodulins and calcium-dependent protein kinases in *Ulva compressa* exposed to copper excess. *Plant Phys* 158: 1451–1462.
Gupta AK, Kumari A, Mishra S, Wany A, Gupta KJ. (2016) The functional role of nitric oxide in plant mitochondrial metabolism. *Adv Bot Res* 77: 145–163.
Gupta KJ, Igamberdiev AU. (2011) The anoxic plant mitochondrion as a nitrite: NO reductase. *Mitochondrion* 11: 537–543.
Harada JJ, Belmonte MF, Kwong RW. (2010) Plant embryogenesis (zygotic and somatic). eLS; doi: 10.1002/9780470015902.a0002042.pub2.
He Y, Tang RH, Hao Y, *et al.* (2004) Nitric oxide represses the *Arabidopsis* floral transition. *Science* 305: 1968–1971.
Hebelstrup KH, Møller IM. (2015) Mitochondrial signaling in plants under hypoxia: use of reactive oxygen species (ROS) and reactive nitrogen species (RNS). In: Gupta KJ, Igamberdiev AU (eds), *Reactive Oxygen and Nitrogen Species Signaling and Communication in Plants*. Springer, pp. 63–77.
Henderson IR, Dean C. (2004) Control of *Arabidopsis* flowering: the chill before the bloom. *Development* 131: 3829–3838.
Henry YA, Ducastel B, Guissani A. (1997) Basic chemistry of nitric oxide and related nitrogen oxides. In: Henry YA, Guissani A, Ducastel B (eds), *Nitric Oxide Research from Chemistry to Biology*. Landes Company, pp. 15–46.
Herrero M. (2003) Male and female synchrony and the regulation of mating in flowering plants. *Phil Trans R Soc B Biol Sci* 358: 1019–1024.
Heslop-Harrison J. (1978) Recognition and response in the pollen-stigma interaction. *Sym Soc Exp Biol* 32: 121.
Heslop-Harrison Y. (2000) Control gates and micro-ecology: the pollen-stigma interaction in perspective. *Ann Bot* 85: 5–13.
Heslop-Harrison Y, Shivanna KR. (1977) The receptive surface of the angiosperm stigma. *Ann Bot* 41: 1233–1258.
Heydlauff J, Groß-Hardt R. (2014) Love is a battlefield: programmed cell death during fertilization. *J Exp Bot* 65: 1323.

Hiscock SJ, Allen AM. (2008) Diverse cell signalling pathways regulate pollen-stigma interactions: the search for consensus. *New Phytol* 179: 286–317.
Hiscock S, Bright J, McInnis SM, Desikan R, Hancock JT. (2007) Signaling on the stigma: potential new roles for ROS and NO in plant cell signaling. *Plant Signal Behav* 2: 23–24.
Holzmeister C, Fröhlich A, Sarioglu H, Bauer N, Durner J, Lindermayr C. (2011) Proteomic analysis of defense response of wildtype *Arabidopsis thaliana* and plants with impaired NO-homeostasis. *Proteomics* 11: 1664–1683.
Holzmeister C, Gaupels F, Geerlof A, *et al.* (2015) Differential inhibition of *Arabidopsis* superoxide dismutases by peroxynitrite-mediated tyrosine nitration. *J Exp Bot* 66: 989–999.
Hu J, Huang X, Chen L, *et al.* (2015) Site-specific nitrosoproteomic identification of endogenously S-nitrosylated proteins in *Arabidopsis. Plant Physiol* 167: 1731–1746.
Hu X, Neill SJ, Tang Z, Cai W. (2005) Nitric oxide mediates gravitropic bending in soybean roots. *Plant Phys* 137: 663–670.
Huan C, Jiang L, An X, *et al.* (2016) Potential role of reactive oxygen species and antioxidant genes in the regulation of peach fruit development and ripening. *Plant Physiol Biochem* 104: 294–303.
Hufton CA, Besford PT, Wellburn AR. (1996) Effects of NO (+ NO_2) pollution on growth, nitrate reductase activities and associated protein contents in glasshouse lettuce grown hydroponically in winter with CO_2 enrichment. *New Phytol* 133: 495–501.
Ingram GC. (2010) Family life at close quarters: communication and constraint in angiosperm seed development. *Protoplasma* 247: 195–214.
Ivanchenko MG, den Os D, Monshausen GB, Dubrovsky JG, Andrea Bednářová A, Natraj Krishnan N. (2013) Auxin increases the hydrogen peroxide (H_2O_2) concentration in tomato (*Solanum lycopersicum*) root tips while inhibiting root growth. *Ann Bot* 112: 1107–1116.
Iyer GYN, Islam MF, Quastel JH. (1961) Biochemical aspects of phagocytosis. *Nature* 11: 535–541.
Jasid S, Simontacchi M, Bartoli CG, Puntarulo S. (2006) Chloroplasts as a nitric oxide cellular source. Effect of reactive nitrogen species on chloroplastic lipids and proteins. *Plant Physiol* 142: 1246–1255.
Jasid S, Simontacchi M, Susana Puntarulo S. (2008) Exposure to nitric oxide protects against oxidative damage but increases the labile iron pool in sorghum embryonic axes. *J Exp Bot* 59: 3953–3962.
Jiang X, Gao Y, Zhou H, Chen J, Wu J, Zhang S. (2014) Apoplastic calmodulin promotes self-incompatibility pollen tube growth by enhancing calcium influx and reactive oxygen species concentration in *Pyrus pyrifolia. Plant Cell Rep* 33: 255–263.
Joo JH, Bae JS, Lee JS. (2001) Role of auxin-induced reactive oxygen species in root gravitropism. *Plant Physiol* 126: 1055–1060.
Kairong C, Gengsheng X, Xinmin L, Gengmei X, Yafu W. (1999) Effect of hydrogen peroxide on somatic embryogenesis of *Lycium barbarum* L. *Plant Sci* 146: 9–16.
Kairong C, Ji L, Gengmei X, Jianlong L, Lihong W, Yafu W. (2002) Effect of hydrogen peroxide on synthesis of proteins during somatic embryogenesis in *Lycium barbarum. Plant Cell Tiss Org Cult* 68: 187–193.
Kim D, Yamaguchi K, Oda T. (2006) Nitric oxide synthase-like enzyme mediated nitric oxide generation by harmful red tide phytoplankton, *Chattonella marina. J Plankton Res* 28: 613–620.
Kobayashi H, Ikeda TM, Nagata K. (2013) Spatial and temporal progress of programmed cell death in the developing starchy endosperm of rice. *Planta* 237: 1393–1400.
Kolbert Zs. (2016) Implication of nitric oxide (NO) in excess element-induced morphogenic responses of the root system. *Plant Physiol Biochem* 101: 149–161.
Kolbert Zs, Bartha B, Erdei L. (2008) Exogenous auxin-induced NO synthesis is nitrate reductase-associated in *Arabidopsis thaliana* root primordia. *J Plant Phys* 165: 967–975.

Kolbert Zs, Pető A, Szőllősi R, Erdei L, Tari I. (2011) Nitric oxide generation during vegetative/generative transition of the apical meristem in wheat. *Acta Biol Szeged* 55: 95–97.

Kolbert Zs, Pető A, Lehotai N, Feigl G, Erdei L. (2015) Copper sensitivity of *nia1nia2noa1-2* mutant is associated with its low NO level. *Plant Growth Regul* 77: 255–263.

Krumova K, Cosa G. (2016) Overview of reactive oxygen species. In: Nonell S, Flors C (eds), *Singlet Oxygen: Applications in Biosciences and Nanosciences*. Royal Society of Chemistry, pp. 3–21.

Kumar V, Irfan M, Ghosh S, Chakraborty N, Chakraborty S, Datta A. (2016) Fruit ripening mutants reveal cell metabolism and redox state during ripening. *Protoplasma* 253: 581–594.

Kwon E, Feechan A, Yun B-W, *et al.* (2012) AtGSNOR1 function is required for multiple developmental programs in *Arabidopsis*. *Planta* 236: 887–900.

Lamotte O, Bertoldo JB, Besson-Bard A, *et al.* (2014) Protein S-nitrosylation: specificity and identification strategies in plants. *Front Chem* 2: 114.

Lancaster Jr JR (1997) A tutorial on the diffusibility and reactivity of free nitric oxide. *Nitric Oxide* 1: 18–30.

Lanteri ML, Laxalt AM, Lamattina L. (2008) Nitric oxide triggers phosphatidic acid accumulation via phospholipase D during auxin-induced adventitious root formation in cucumber. *Plant Physiol* 147: 188–198.

Lanteri ML, Pagnussat GC, Lamattina L. (2006) Calcium and calcium-dependent protein kinases are involved in nitric oxide- and auxin-induced adventitious root formation in cucumber. *J Exp Bot* 57: 1341–1351.

Lee RDW, Cho HT. (2013) Auxin, the organizer of the hormonal/environmental signals for root hair growth. *Front Plant Sci* 4: 448.

Leshem YY, Haramaty E. (1996) The characterization and contrasting effects of the nitric oxide free radical in vegetative stress and senescence of *Pisum sativum* Linn. *J Plant Phys* 148: 258–263.

Leshem YY, Wills RBH, Ku VV. (1998) Evidence for the function of the free radical gas – nitric oxide (NO·) – as an endogenous maturation and senescence regulating factor in higher plants. *Plant Physiol Biochem* 36: 825–833.

Li N, Sun L, Zhang L, *et al.* (2015) AtrbohD and AtrbohF negatively regulate lateral root development by changing the localized accumulation of superoxide in primary roots of *Arabidopsis*. *Planta* 241: 591–602.

Li S, Xue L, Xu S, Feng H, An L. (2007) Hydrogen peroxide involvement in formation and development of adventitious roots in cucumber. *Plant Growth Regul* 52: 173–180.

Li S-W, Xue L, Xu S, Feng H, An L. (2009) Hydrogen peroxide acts as a signal molecule in the adventitious root formation of mung bean seedlings. *Environ Exp Bot* 65: 63–71.

Liao W, Xiao H, Zhang M. (2009) Role and relationship of nitric oxide and hydrogen peroxide in adventitious root development of marigold. *Acta Physiol Plant* 31: 1279–1289.

Liao W, Huang G, Yu J, Zhang M, Shi X. (2011) Nitric oxide and hydrogen peroxide are involved in indole-3-butyric acid-induced adventitious root development in marigold. *J Hort Sci Biotech* 86: 159–165.

Lin A, Wang Y, Tang J, *et al.* (2012) Nitric oxide and protein S-nitrosylation are integral to hydrogen peroxide-induced leaf cell death in rice. *Plant Physiol* 158: 451–464.

Lindermayr C, Saalbach G, Durner J. (2005) Proteomic identification of S-nitrosylated proteins in *Arabidopsis*. *Plant Phys* 137: 921–930.

Liu H-Y, Yu X, Cui D-Y, *et al.* (2007) The role of water channel proteins and nitric oxide signaling in rice seed germination. *Cell Res* 17: 638–649.

Liu W-Z, Kong D-D, Gu X-X, *et al.* (2013) Cytokinins can act as suppressors of nitric oxide in *Arabidopsis*. *Proc Natl Acad Sci USA* 110: 1548–1553.

Liu Y, Ye N, Liu R, Chen M, Zhang J. (2010) H_2O_2 mediates the regulation of ABA catabolism and GA biosynthesis in *Arabidopsis* seed dormancy and germination. *J Exp Bot* 61: 2979–2990.
Liu Y, Xinna Li X, Xu L, Shen W. (2013) De-etiolation of wheat seedling leaves: cross talk between heme oxygenase/carbon monoxide and nitric oxide. *PLoS ONE* 8: e81470.
Locato V, Paradiso A, Sabetta W, De Gara L, de Pinto MC. (2016) Nitric oxide and reactive oxygen species in PCD signaling. *Adv Bot Res* 77: 165–192.
Lombardi L, Mariotti L, Picciarelli P, Ceccarelli N, Lorenzi R. (2012) Ethylene produced by the endosperm is involved in the regulation of nucellus programmed cell death in *Sechium edule* Sw. *Plant Sci* 187: 31–38.
Lombardo MC, Lamattina L. (2012) Nitric oxide is essential for vesicle formation and trafficking in *Arabidopsis* root hair growth. *J Exp Bot* 63: 4875–4885.
Lombardo MC, Graziano M, Polacco JC, Lamattina L. (2006) Nitric oxide functions as a positive regulator of root hair development. *Plant Signal Behav* 1: 28–33.
Lozano-Juste J, León J. (2010) Enhanced abscisic acid-mediated responses in *nia1nia2noa1-2* triple mutant impaired in NIA/NR- and AtNOA1-dependent nitric oxide biosynthesis in *Arabidopsis*. *Plant Physiol* 152: 891–903.
Lozano-Juste J, Colom-Moreno R, León J. (2011) In vivo protein tyrosine nitration in Arabidopsis thaliana. *J Exp Bot* 62: 3501–3517.
Luo J-P, Jiang S-T, Pan L-J. (2001) Enhanced somatic embryogenesis by salicylic acid of *Astragalus adsurgens* Pall.: relationship with H_2O_2 production and H_2O_2-metabolizing enzyme activities. *Plant Sci* 161: 125–132.
Ma Z, Marsolais F, Bykova NV, Igamberdiev AU. (2016) Nitric oxide and reactive oxygen species mediate metabolic changes in barley seed embryo during germination. *Front Plant Sci* 7: 138.
Manjunatha G, Lokesh V, Neelwarne B (2010) Nitric oxide in fruit ripening: Trends and opportunities *Biotech Adv* 28: 489–499.
Manjunatha G, Gupta KJ, Lokesh V, Mur LAJ, Neelwarne B. (2012) Nitric oxide counters ethylene effects on ripening fruits. *Plant Signal Behav* 7: 476–483.
McAtee P, Karim S, Schaffer R, David K. (2013) A dynamic interplay between phytohormones is required for fruit development, maturation, and ripening. *Front Plant Sci* 4: 79.
McInnis SM, Costa LM, Gutiérrez-Marcos JF, Henderson CA, Hiscock SJ. (2005) Isolation and characterization of a polymorphic stigma-specific class III peroxidase gene from *Senecio squalidus* L. (Asteraceae). *Plant Mol Biol* 57: 659–677.
McInnis SM, Emery DC, Porter R, Desikan R, Hancock JT, Hiscock SJ. (2006) The role of stigma peroxidases in flowering plants: insights from further characterization of a stigma-specific peroxidase (SSP) from *Senecio squalidus* (Asteraceae). *J Exp Bot* 57: 1835–1846.
Meinhard M, Grill E. (2001) Hydrogen peroxide is a regulator of ABI1, a protein phosphatase 2C from *Arabidopsis*. *FEBS Lett* 508: 443–446.
Meinhard M, Rodriguez PL, Grill E. (2002) The sensitivity of ABI2 to hydrogen peroxide links the abscisic acid-response regulator to redox signalling *Planta* 214: 775–782.
Melo NKG, Bianchetti RE, Lira BS, *et al.* (2016) Nitric oxide, ethylene, and auxin cross talk mediates greening and plastid development in deetiolating tomato seedlings. *Plant Phys* 170: 2278–2294.
Mira MM, Adel E-S, Stasolla C. (2015) Ethylene is integrated into the nitric oxide regulation of Arabidopsis somatic embryogenesis. *J Genet Eng Biotech* 13: 7–17.
Mondal K, Sharma NS, Malhotra SP, Dhawan K, Singh R. (2004) Antioxidant systems in ripening tomato fruits. *Biol Plant* 48: 49–53.
Mondal K, Malhotra SP, Jain V, Singh R. (2009) Oxidative stress and antioxidant systems in Guava (*Psidium guajava* L.) fruits during ripening. *Physiol Molecul Biol Plants* 15: 327–334.

Mouradov A, Cremer F, Coupland G. (2002) Control of flowering time interacting pathways as a basis for diversity. *Plant Cell* 14: S111–S130.

Mur LAJ, Mandon J, Persijn S, *et al.* (2013) Nitric oxide in plants: an assessment of the current state of knowledge. *AoB PLANTS* 5: pls052; doi: 10.1093/aobpla/pls052.

Noctor G, De Paepe R, Foyer CH. (2007) Mitochondrial redox biology and homeostasis in plants. *Trends Plant Sci* 12: 1360–1385.

Obara K, Kuriyama H, Fukuda H. (2001) Direct evidence of active and rapid nuclear degradation triggered by vacuole rupture during programmed cell death in Zinnia. *Plant Phys* 125: 615–626.

Ogawa K, Kanematsu S, Asada K. (1997) Generation of superoxide anion and localization of CuZn-superoxide dismutase in the vascular tissue of spinach hypocotyls: their association with lignification. *Plant Cell Physiol* 38: 1118–1126.

Oracz K, El-Maarouf-Bouteau H, Kranner I, Bogatek R, Corbineau F, Bailly C. (2009) The mechanisms involved in seed dormancy alleviation by hydrogen cyanide unravel the role of reactive oxygen species as key factors of cellular signaling during germination. *Plant Phys* 150: 494–505.

Ortega-Galisteo AP, Rodríguez-Serrano M, Pazmiño DM, Gupta DK, Sandalio LM Romero-Puertas MC. (2012) S-Nitrosylated proteins in pea (*Pisum sativum* L.) leaf peroxisomes: changes under abiotic stress. *J Exp Bot* 63: 2089–2103.

Ötvös K, Pasternak TP, Miskolczi P, *et al.* (2005) Nitric oxide is required for, and promotes auxin-mediated activation of, cell division and embryogenic cell formation but does not influence cell cycle progression in alfalfa cell cultures. *Plant J* 43: 849–860.

Pagnussat GC, Lanteri ML, Lamattina L. (2003) Nitric oxide and cyclic GMP are messengers in the indole acetic acid-induced adventitious rooting process. *Plant Phys* 132: 1241–1248.

Pagnussat GC, Lanteri ML, Lombardo MC, Lamattina L. (2004) Nitric oxide mediates the indole acetic acid induction activation of a mitogen-activated protein kinase cascade involved in adventitious root development. *Plant Phys* 135: 279–286.

Pedreira J, Sanz N, Peña MJ, Sánchez M, Queijeiro E, Revilla G, Zarra I. (2004) Role of apoplastic ascorbate and hydrogen peroxide in the control of cell growth in pine hypocotyls. *Plant Cell Physiol* 45: 530–534.

Pető A, Lehotai N, Feigl G, *et al.* (2013) Nitric oxide contributes to copper tolerance by influencing ROS metabolism in *Arabidopsis*. *Plant Cell Rep* 32: 1913–1923.

Pilati S, Perazzolli M, Malossini A, *et al.* (2007) Genome-wide transcriptional analysis of grapevine berry ripening reveals a set of genes similarly modulated during three seasons and the occurrence of an oxidative burst at veraison. *BMC Gen* 8: 1.

Planchet E, Jagadis Gupta K, Sonoda M, Kaiser WM. (2005) Nitric oxide emission from tobacco leaves and cell suspensions: rate limiting factors and evidence for the involvement of mitochondrial electron transport. *Plant J* 41: 732–743.

Potocký M, Jones MA, Bezvoda R, Smirnoff N, Žárský V. (2007) Reactive oxygen species produced by NADPH oxidase are involved in pollen tube growth. *New Phytol* 174: 742–751.

Potters G, Horemans N, Jansen MAK. (2010) The cellular redox state in plant stress biology – a charging concept. *Plant Phys Biochem* 48: 292–300.

Prado AM, Porterfield DM, Feijó JA. (2004) Nitric oxide is involved in growth regulation and re-orientation of pollen tubes. *Development* 131: 2707–2714.

Pucciariello C, Perata P. (2017) New insights into reactive oxygen species and nitric oxide signalling under low oxygen in plants. *Plant Cell Environ* 40: 473–482.

Rentel MC, Lecourieux D, Ouaked F, *et al.* (2004) OXI1 kinase is necessary for oxidative burst-mediated signalling in *Arabidopsis*. *Nature* 427: 858–861.

Rodríguez-Serrano M, Pazmiño DM, Sparkes I, *et al.* (2014) 2,4-Dichlorophenoxyacetic acid promotes S-nitrosylation and oxidation of actin affecting cytoskeleton and peroxisomal dynamics. *J Exp Bot* 65: 4783–4793.

Romero-Puertas MC, Laxa M, Mattè A, *et al.* (2007) S-nitrosylation of peroxiredoxin II E promotes peroxynitrite mediated tyrosine nitration. *Plant Cell* 19: 4120–4130.
Rudell DR, Mattheis JP. (2006) Nitric oxide and nitrite treatments reduce ethylene evolution from apple fruit disks. *Hort Sci* 41: 1462–1465.
Rümer S, Gupta KJ, Kaiser WM. (2009) Plant cells oxidize hydroxylamines to NO. *J Exp Bot* 60: 2065–2072.
Russwurm M, Koesling D. (2004) NO activation of guanylyl cyclase. *EMBO J* 23: 4443–4450.
Sandaklie-Nikolova L, Palanivelu R, King EJ, Copenhaver GP, Drews GN. (2007) Synergid cell death in *Arabidopsis* is triggered following direct interaction with the pollen tube. *Plant Phys* 144: 1753–1762.
Santa-Catarina C, Silveira V, Scherer GFE, Floh EIS. (2007) Polyamine and nitric oxide levels relate with morphogenetic evolution in somatic embryogenesis of *Ocotea catharinensis. Plant Cell Tiss Organ Cult* 90: 93–101.
Sanz L, Fernández-Marcos M, Modrego A, *et al.* (2014) Nitric oxide plays a role in stem cell niche homeostasis through its interaction with auxin. *Plant Phys* 166: 1972–1984.
Sarath G, Hou G, Baird LM, Mitchell RB. (2007) Reactive oxygen species, ABA and nitric oxide interactions on the germination of warm-season C4-grasses. *Planta* 226: 697–708.
Scheler C, Durner J Astier J. (2013) Nitric oxide and reactive oxygen species in plant biotic interactions. *Curr Opin Plant Biol* 16: 534–539.
Schopfer P, Plachy C, Frahry G. (2001) Release of reactive oxygen intermediates (superoxide radicals, hydrogen peroxide, and hydroxyl radicals) and peroxidase in germinating radish seeds controlled by light, gibberellin, and abscisic acid. *Plant Phys* 125: 1591–1602.
Sehrawat A, Abat JK, Deswal R. (2013) RuBisCO depletion improved proteome coverage of cold responsive S-nitrosylated targets in *Brassica juncea. Front Plant Sci* 4: 342.
Seligman K, Saviani EE, Oliveira HC, Pinto-Maglio CAF, Salgado I. (2008) Floral transition and nitric oxide emission during flower development in *Arabidopsis thaliana* is affected in nitrate reductase-deficient plants. *Plant Cell Physiol* 49: 1112–1121.
Sen S. (2010) S-nitrosylation process acts as a regulatory switch for seed germination. *Am J Plant Phys* 5: 122–132.
Serrano I, Romero-Puertas MC, Rodrıguez-Serrano M, Sandalio LM, Olmedilla A. (2012a) Peroxynitrite mediates programmed cell death both in papillar cells and in self-incompatible pollen in the olive (*Olea europaea* L.). *J Exp Bot* 63: 1479–1493.
Serrano I, Romero-Puertas MC, Rodríguez Serrano M, Sandalio LM, Olmedilla A. (2012b) Role of peroxynitrite in programmed cell death induced in self-incompatible pollen. *Plant Signal Behav* 7: 779–781.
Serrano I, Romero-Puertas MC, Sandalio LM, Olmedilla A. (2015) The role of reactive oxygen species and nitric oxide in programmed cell death associated with self-incompatibility. *J Exp Bot* 66: 2869–2876.
Shi Y-F, Wang D-L, Wang C, *et al.* (2015) Loss of GSNOR1 function leads to compromised auxin signaling and polar auxin transport. *Mol Plant* 8: 1350–1365.
Siegel JM, De Campos RPS, Gunasekara DB, Da Silva JAF, Lunte SM. (2015) Electrophoretic methods for separation of peroxynitrite and related compounds. In: Peteu S, Szunerits S, Bayachou M (eds), *Peroxynitrite Detection in Biological Media: Challenges and Advances.* Royal Society of Chemistry, pp 121–150.
Silveira V, Santa-Catarina C, Tun NN, *et al.* (2006) Polyamine effects on the endogenous polyamine contents, nitric oxide release, growth and differentiation of embryogenic suspension cultures of *Araucaria angustifolia* (Bert.) O. Ktze. *Plant Sci* 171: 91–98.
Simpson GG, Dean C. (2002). *Arabidopsis*, the Rosetta stone of flowering time? *Science* 296: 285–289.

Souza JM, Peluffo G, Radi R. (2008) Protein tyrosine nitration – functional alteration or just a biomarker? *Free Rad Biol Med* 45: 357–366.

Srivastava N, Gonugunta VK, Puli MR, Raghavendra AS. (2009) Nitric oxide production occurs downstream of reactive oxygen species in guard cells during stomatal closure induced by chitosan in abaxial epidermis of *Pisum sativum*. *Planta* 229: 757–765.

Stamler JS. (1994) Redox signaling: nitrosylation and related target interactions of nitric oxide. *Cell* 78: 931–936.

Stamler JS, Singel DJ, Loscalzo J. (1992) Biochemistry of nitric oxide and its redox-activated form. *Science* 258: 1988–1902.

Stöhr C, Strube F, Marx G, Ullrich WR, Rockel P. (2001) A plasma membrane-bound enzyme of tobacco roots catalyses the formation of nitric oxide from nitrite. *Planta* 212: 835–841.

Swanson R, Edlund AF, Preuss D. (2004) Species specificity in pollen-pistil interactions. *Annu Rev Genet* 38: 793–818.

Tanou G, Job C, Rajjou L, *et al.* (2009). Proteomics reveals the overlapping roles of hydrogen peroxide and nitric oxide in the acclimation of citrus plants to salinity. *Plant J* 60: 795–804.

Terrile MC, Paris R, Calderon-Villalobos LI, *et al.* (2012) Nitric oxide influences auxin signalling through S-nitrosylation of the *Arabidopsis* TRANSPORT INHIBITOR RESPONSE 1 auxin receptor. *Plant J* 70: 492–500.

Tewari RK1, Prommer J, Watanabe M. (2013) Endogenous nitric oxide generation in protoplast chloroplasts. *Plant Cell Rep* 32: 31–44.

Thomas SG, Huang S, Li S, Staiger CJ, Franklin-Tong VE. (2006) Actin depolymerization is sufficient to induce programmed cell death in self-incompatible pollen. *J Cell Biol* 174: 221–229.

Tonón C, Terrile MC, Iglesias MJ, Lamattina L, Casalongue C. (2010) Extracellular ATP, nitric oxide and superoxide act coordinately to regulate hypocotyl growth in etiolated *Arabidopsis* seedlings. *J Plant Phys* 167: 540–546.

Torres MA, Dangl JL, Jones JD. (2002) *Arabidopsis gp91phox* homologues AtrbohD and AtrbohF are required for accumulation of reactive oxygen intermediates in the plant defence response. *Proc Natl Acad Sci USA* 99: 517–522.

Tsukagoshi H, Busch W, Benfey PN. (2010) Transcriptional regulation of ROS controls transition from proliferation to differentiation in the root. *Cell* 143: 606–616.

Tsukagoshi H. (2016) Control of root growth and development by reactive oxygen species. *Curr Opin Plant Biol* 29: 57–63.

Vandana S, Sustmann R, Rauen U, Stöhr C. (2012) FNOCT as a fluorescent probe for in vivo localization of nitric oxide distribution in tobacco roots. *Plant Physiol Biochem* 59: 80–89.

Vieira LN, Santa-Catarina C, Fraga HPF, *et al.* (2012) Glutathione improves early somatic embryogenesis in *Araucaria angustifolia* (Bert) O. Kuntze by alteration in nitric oxide emission. *Plant Sci* 195: 80–87.

Vranova E, Inze D, Van Breusegem F. (2002) Signal transduction during oxidative stress. *J Exp Bot* 53: 1227–1136.

Wang CL, Wu J, Xu GH, *et al.* (2010) S-RNase disrupts tip-localized reactive oxygen species and induces nuclear DNA degradation in incompatible pollen tubes of *Pyrus pyrifolia*. *J Cell Sci* 123: 4301–4309.

Wang L, Guo Y, Jia L, *et al.* (2014) Hydrogen peroxide acts upstream of nitric oxide in the heat shock pathway in *Arabidopsis* seedlings. *Plant Physiol* 164: 2184–2196.

Wang P, Du Y, Li Y, Ren D, Song C-P. (2010) Hydrogen peroxide-mediated activation of MAP kinase 6 modulates nitric oxide biosynthesis and signal transduction in Arabidopsis. *Plant Cell* 22: 2981–2998.

Wang P, Zhu J-K, Lang Z. (2015) Nitric oxide suppresses the inhibitory effect of abscisic acid on seed germination by S-nitrosylation of SnRK2 proteins. *Plant Signal Behav* 10(6): e1031939; doi: 10.1080/15592324.2015.1031939.
Wilkins KA, Bancroft J, Bosch M, Ings J, Smirnoff N, Franklin-Tong VE. (2011) Reactive oxygen species and nitric oxide mediate actin reorganization and programmed cell death in the self-incompatibility response of *Papaver*. *Plant Phys* 156: 404–416.
Wimalasekera R, Tebartz F, Scherer GF (2011) Polyamines, polyamine oxidases and nitric oxide in development, abiotic and biotic stresses. *Plant Sci* 181: 593–603.
Wink DA, Mitchell JB. (1998) Chemical biology of nitric oxide: Insights into regulatory, cytotoxic, and cytoprotective mechanisms of nitric oxide. *Free Radic Biol Med* 25: 434–456.
Xie Y, Mao Y, Zhang W, Lai D, Wang Q, Shen W. (2014) Reactive oxygen species-dependent nitric oxide production contributes to hydrogen-promoted stomatal closure in *Arabidopsis*. *Plant Phys* 165: 759–773.
Yamasaki H, Sakihama Y. (2000) Simultaneous production of nitric oxide and peroxynitrite by plant nitrate reductase: in vitro evidence for the NR-dependent formation of active nitrogen species. *FEBS Lett* 468: 89–92.
Yang H, Mu J, Chen L, *et al.* (2015) S-Nitrosylation positively regulates ascorbate peroxidase activity during plant stress responses. *Plant Phys* 167: 1604–1615.
Yeo W-S, Kim YJ, Kabir MH, Kang JW, Kim KP. (2015) Mass spectrometric analysis of protein tyrosine nitration in aging and neurodegenerative diseases. *Mass Spectrom Rev* 34: 166–183.
Yun BW, Feechan A, Yin M, *et al.* (2011) S-nitrosylation of NADPH oxidase regulates cell death in plant immunity. *Nature* 478: 264–268.
Zhang L, Wang Y, Zhao L, Shi S, Zhang L. (2006) Involvement of nitric oxide in light-mediated greening of barley seedlings. *J Plant Phys* 163: 818–826.
Zhang Y, Chen K, Zhang S, Ferguson I (2003) The role of salicylic acid in postharvest ripening of kiwifruit. *Postharvest Biol Technol* 28: 67–74.
Zhou T, Yang X, Guo K, *et al.* (2016) ROS homeostasis regulates somatic embryogenesis via the regulation of auxin signaling in cotton. *Mol Cell Proteom* 15: 2108–2124.
Zhu S, Sun L, Liu M, Zhou J. (2008) Effect of nitric oxide on reactive oxygen species and antioxidant enzymes in kiwifruit during storage. *J Sci Food Agric* 88: 2324–2331.
Zhu S, Sun L, Zhou J. (2009) Effects of nitric oxide fumigation on phenolic metabolism of postharvest Chinese winter jujube (*Zizyphus jujuba* Mill. cv. Dongzao) in relation to fruit quality. *Food Sci Technol* 42: 1009–1014.
Zimmermann P, Heinlein C, Orendi G, Zentgraf U. (2006) Senescence-specific regulation of catalases in *Arabidopsis thaliana* (L.) *Heynh. Plant Cell Environ* 29: 1049–1060.

15

Reactive Oxygen Species Signaling and Seed Germination: An Overview

Prabhakaran Soundararajan, Abinaya Manivannan, and Byoung Ryong Jeong

Gyeongsang National University, Jinju, South Korea

Introduction

In most angiosperms life starts with the germination of seeds and ends with the production of seeds. Reactive oxygen species (ROS) play roles in all stages of plant growth and development, but in seed development they play a major role in the fusion of male and female gametophytes, which ensures proper fertilization and seed dispersal (Serrano *et al.*, 2015). From the adhesion of pollen in the stigma surface, growth of the pollen tube and maintenance of its integrity, to rupture of the tube and guidance of sperm cells to fuse with the egg cell and central cells – all are directly or indirectly regulated by ROS-mediated mechanisms (Dresselhaus and Franklin-Tong, 2013). Among the various ROS, the production of hydroxyl radicals (·OH) in the entrance of gametophytes plays a major role in rupture of the pollen tube and sperm release (Müller *et al.*, 2013). The osmotic changes created by the $Ca^{2+}$and K^{+} direct the haploid cells to perform the double fertilization (Iwano *et al.*, 2004, 2012; Wu *et al.*, 2011). A series of events occurs during the polarization of pollen tube into the micropyle. Intercellular communication during gamete fusion involves guidance from proteins, ROS, nitric oxide (NO)/reactive nitrogen species (RNS), and Ca^{2+} (Dresselhaus and Franklin-Tong, 2013).

Angiosperms and gymnosperms are generally classified as seed-producing plants. In the gymnosperm ovules are unclosed and "naked seeds" are produced after fertilization. In angiosperms the ovule is enclosed by the endosperm, and seeds are produced by double fertilization (Bell *et al.*, 2010). Two sperm cells from the pollen are involved in the fertilization. One sperm cell fuses with the egg cell to produce a diploid (2n) zygote. The other fuses with the central haploid polar cell nuclei to produce a triploid (3n) cell. The former fusion product develops into the embryo and the latter one forms the endosperm (Bewley and Black, 1994). Pollen grains have evolved protection against dessication, pathogens, and wind (Doyle, 2012). Both spermatophytes and gametophytes have mechanisms for self or non-self recognition (Igic *et al.*, 2004). The complex and coordinative communication between the pollen and receptive pistil determines whether a pollen grain is compatible and accepted or incompatible and rejected (Serrano *et al.*, 2015). A study on *Arabidopsis* reported that a single pollen grain is guided properly to reach one of about 50 ovules (Dresselhaus and Franklin-Tong, 2013). Planivelu and Tsukamoto (2012) observed that to reach the ovule, the pollen tube must interact with at least seven different types of pistil cells.

Reactive Oxygen Species in Plants: Boon Or Bane - Revisiting the Role of ROS, First Edition. Edited by Vijay Pratap Singh, Samiksha Singh, Durgesh Kumar Tripathi, Sheo Mohan Prasad, and Devendra Kumar Chauhan.

After fertilization, development of the ovules into seeds and the germination of seeds is a complex process. Seed development after successful pollination starts with cell division, the formation of seed reserves, the dehydration during the post-germination, desiccation, and germination (upon imbibition). Successful germination is generally marked by protrusion of the radicle outside the endosperm layer (Bewley and Black, 1994). Although excess accumulations of ROS are quarantined as hazardous molecules that induce oxidative stress, elevated concentrations of ROS were identified in the cell-wall surface during germination (Maarouf-Bouteau and Bailly, 2008). Large amounts of ROS stimulate the endonucleases leading to the breakage of cell wall (Leymarie *et al.*, 2012). The hypersensitive reaction (HR) induced by ROS protects the newly emerging seedling from pathogens (Grant and Loake, 2000). On the other hand, essential cellular components and metabolism are effectively prevented by the oxidative stress usually created due to the excessive generation of ROS (Bailly, 2004). Proper antioxidant mechanisms tend to control the oxidative damage occurring due to the excessive accumulation of ROS (S. Singh *et al.*, 2015, 2017; R. Singh *et al.*, 2016; Tripathi *et al.*, 2016a,b, 2017a–c). For breakage of dormancy gibberellins (GAs) have been identified as essential plant growth regulators (PGR). Abscisic acid (ABA) is another important PGR for the induction of dormancy and/or prevention of germination before proper maturation, especially during the late embryonic stage (Seo *et al.*, 2006). The interactions of GA and ABA with ROS are considered as essential for the proper development and dispersal of seed (Kucera *et al.*, 2005). In addition, previous reports suggested that other PGRs such as auxins, ethylene, and brassinosteroids were also active participants in seed development (Ghassemian *et al.*, 2000; Bouquin *et al.*, 2001; Fu and Harberd, 2003; Liu *et al.*, 2013).

In this chapter, the role of ROS is described in relation to two different aspects of seed biology: (i) the involvement of ROS in the recognition process between pollen and pistil and guidance of the pollen tube components leading to fusion of cells; and (ii) the role of ROS in radicle protrusion and its efficient interaction with the PGRs during seed germination.

Influence of ROS on Male-female Cross-talk During Fertilization

Pollen–pistil Recognition

Initially pollen grains landing on the surface of the stigma undergo adhesion and hydration. The pollen tube emerges and navigates to the maternal tissues. Transduction cascades signal the pollen tube's arrival to the maternal tissues (Dresselhaus and Franklin-Tong, 2013). For example, studies conducted on the cross-pollination of *Brassica* and *Arabidopsis* failed due to poor pollen adhesion, whereas the success rate of interspecific crosses between *Brassica* species was directly correlated with the higher pollen adhesion rate (Luu *et al.*, 1998; Zinkl *et al.*, 1999, 2000). The specificity of pollen adhesion depends on the hydration of the pollen. Pollen coat proteins, pollen coat lipids, and plasma membrane stigmatic proteins are responsible for the rapid stickiness of proper pollen on the stigma as well as the non-adhesiveness of incompatible pollens (Takayama *et al.*, 2000; Takayama and Isogai, 2005). Following the adhesion of pollen grains, hydration of pollen is supported by the stigma. The protein exocyst subunit EXO70 family protein A1 (EXO70A1), a plasma membrane-localized stigmatic protein, is crucial for the hydration status during the interaction of pollen and pistil. A study showed that an *exo70a1* mutant was incapable of recognizing even compatible pollen. In contrast, overexpression of *exo70a1* partially overcomes the self-incompatibility in *Brassica*. Meanwhile *exo70A1* mediates the tethering of post-Golgi secretory vesicles on the plasma membrane of the stigma (Samuel *et al.*, 2009). An overview of the pollen–pistil interaction based on the hydration status

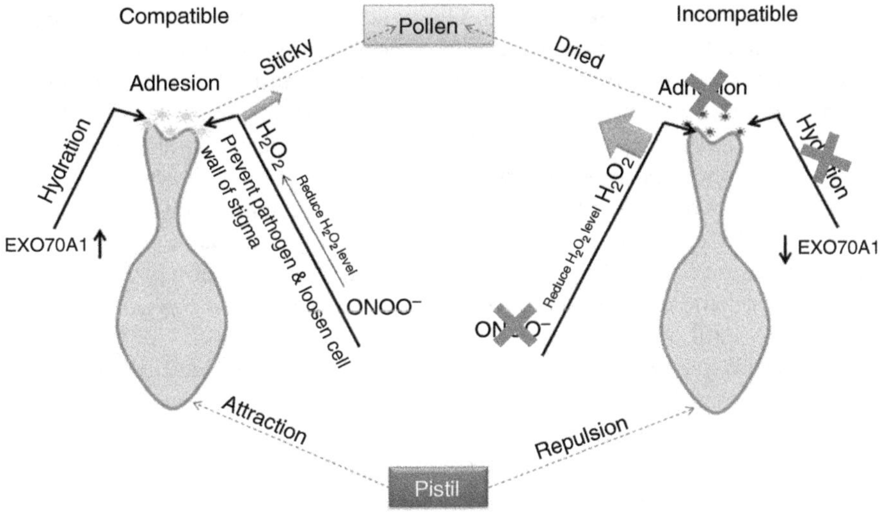

Figure 15.1 An heuristic overview of the pollen–pistil interaction based on the hydration status mediated by reactive oxygen species (ROS) and EXO70A1. The concept of this figure has been adapted from the following sources: Dresselhaus and Franklin-Tong (2013), Takayama and Isogai (2005), Takayama *et al.* (2000), Samuel *et al.* (2009), Serrano *et al.* (2015), Luu *et al.* (1998), Zinkl and Preuss (2014), and Zinkl *et al.* (1999).

mediated by ROS and EXO70A1 is described in Figure 15.1. Chapman and Goring (2011) reported that phosphatidylinositol-4-phosphate (PI4P) plays an important role during the later stages of pistil signaling the hydration of compatible pollen. The extracellular matrix (ECM) provides guidance to the pollen tube as it extends toward the ovule by means of chemoattractants (Palanivelu and Tsukamoto, 2012). Chemical gradients help to avoid mistargeting and insufficient growth. Arabinogalactan, transmitting tissue specific (TTS) protein stimulates the pollen tube growth (Cheung *et al.*, 1995). Sulfinylated azadecalins act in specific manner to ensure appropriate fertilization (Qin *et al.*, 2011). Chemocyanin directs the orientation of pollen tubes. Steep gradients of plantacyanin from the stigma to ovary support the idea of pistil chemoattractant gradients guiding the pollen tube (Dong *et al.*, 2005; Chae *et al.*, 2009). The expression of genes differs between the hydration and dry state of pollen. Qin *et al.* (2009) reported that an Affymetrix whole-genome array showed over 1000 genes were differentially expressed between the hydrated and dried pollen grains. Hydrogen peroxide (H_2O_2) was reported to play an important role in the development of the pollen–stigma interaction (Serrano *et al.*, 2015). The level of H_2O_2 was elevated when the stigma was unreceptive, but was lower in a stigma that supported pollen development. Under favorable conditions the H_2O_2 loosened the cell wall of the stigma to allow the pollen to penetrate, and also acted as a defense against pathogens. Serrano *et al.* (2012) reported that during pollen-pistil recognition the NO produced by the pollen reduced the H_2O_2 level. The NADPH oxidase level was also increased, indicating the accumulation of O_2, which subsequently produces peroxynitrite ($ONOO^-$)-dependent nitration. It helps the navigation of pollen through the pistil tissues. Exchange of signals between the stigma and pollen is regulated by the ROS and RNS (Serrano *et al.*, 2012; Dresselhaus and Franklin-Tong, 2013).

Navigation of the Pollen Tube to the Micropyle

Germination of a compatible pollen grain entails cytoskeletal changes involving actin assembly at the pollen-binding site (Iwano *et al.*, 2007). FERONIA (FER), an important protein involved in pollen tube guidance, is strongly mediated by ROS. The avoidance of polyspermy is correlated with FER expression. A mutant of *fer* failed to prevent other sperms entering the micropyle after the fusion of compatible pollen (Duan *et al.*, 2014). The double mutant *aux1aux2* had the defect on entering into ovule (Boisson-Dernier *et al.*, 2009). Several proteins, such as RAC/ROP GTPases (Duan *et al.*, 2010), nicotinamide adenine dinucleotide phosphate (NADPH) oxidase (Zhang *et al.*, 2014), and *Zea mays* embryo sac 4 (ZmES4) (Amien *et al.*, 2010), have been found to control the levels of ROS at the entrance to the female gametophyte. A reduced level of ROS at the micropylar region is observed in the *fer* mutant (Escobar-Restrepo *et al.*, 2007). Duan *et al.* (2014) reported that the involvement of FER in pollen tube growth is in a ROS- and Ca^{2+}-dependent manner. Gamma-aminobutyric acid (GABA) also has a critical role in pollen tube growth. The pistil of the pollen-pistil 2 (*pop2*) mutant (lack of GABA) fails to guide the pollen tube (Palanivelu *et al.*, 2003). A transcriptomics study revealed that in the *pop* mutant 60% of proteins encoding for secretion or secretory pathways showed reduced expression compared to the wild-type (Renault *et al.*, 2011). Most events during the stigma recognition process are signaled by cytosolic calcium concentration [Ca^{2+}]. Fluorescence resonance energy transfer (FRET) showed that Ca^{2+} is elevated in compatible pollen binding sites. Ca^{2+} concentration is also increased near the micropylar region during pollen tube protrusion and penetration into the ovary (Iwano *et al.*, 2004). In tomato, pollen receptor kinases (PRK) such as LePRK1 and LePRK2 form a complex and are phosphorylated in a favorable pollen interaction. Dephosphorylation was observed in pollen membranes incubated with style extracts (Wengier *et al.*, 2010). The major groups of proteins activated during the delivery of a sperm cell in the gametophyte are cysteine-rich proteins (CRPs), defensin-like proteins (DEFLs), lipid-transfer proteins (LTPs), proteinase inhibitors, thionins, snackins, and so on. In addition signals from cell wall-related proteins, extracellular signaling proteins, and pathogen defense-related proteins help to coordinate fertilization as well as protect tissues from microbial assault (Silverstein *et al.*, 2007; Dresselhaus and Franklin-Tong, 2013).

Ovular Guidance Mechanisms

The sperm cell is released after death of the synergid cell and rupture of the pollen tube. Other processes activated during fertilization are concerned with compatibility/incompatibility, avoidance of polyspermy, and/or recovery of fertilization failures (Kessler and Grossniklaus, 2011; Dresselhaus and Sprunck, 2012). Both the rupture of the pollen tube and death of the synergids involve programmed cell death (PCD), and are signaled by the ROS/NO molecules. The ovule guides the pollen tube by two mechanisms, namely, funicular and micropylar (Higashiyama *et al.*, 2003). In the funicular mechanism the pollen tube exits the transmitting track and grows on the surface of the septum and funiculars toward the ovule. In the micropylar guidance mechanism, the pollen tube navigates toward the egg apparatus through the micropyle (Lausser and Dresselhaus, 2010; Dresselhaus and Franklin-Tong, 2013). Gametophytic regulation of funicular and micropylar mechanisms involves the secretion and perception of either attraction or repulsion signals. In the former mechanism, GABA signals the tract growth (Palanivelu *et al.*, 2003). In the latter case, cyclic nucleotide-gated channel 18 (CNGC18) facilitates the micropylar guidance (Michard *et al.*, 2011). Another gene, SR1, which encodes the serine racemase that converts L-serine into D-serine, is highly expressed on the micropyle (Snyder and Ferris, 2000). Cation/proton exchangers (CHX) increase K^+ uptake in a pH-dependent manner; this plays a pivotal role in the growth of pollen toward the ovule. Pollen cells of the *chx21chx23* mutant are unable to exit the tract and move toward the ovule (Lu *et al.*, 2011). Egg Apparatus 1 (EA1), secreted by the gametophyte, attracts the pollen grain by secretion of small peptides (Márton *et al.*, 2012). In *Torenia* it was reported that LUREs, a subgroup of CRPs belonginig to the DEFL family, are generated by the synergid cells (Okuda *et al.*, 2009).

Pollen tube burst and receptive synergid degeneration occurred simultaneously along with several endonucleases. In this case the level of Ca^{2+} activates the PCD required for the arrest of pollen tube extension and induction of pollen tube rupture.

Previous studies indicate that FER encodes for the CrRLK1L receptor-like kinases, and ANXUR1 (ANX1) and ANXUR2 (ANX2) direct the pollen tube rupture (Lindner *et al.*, 2012; Boisson-Dernier *et al.*, 2009). Downstream of FER, pathogen-specific proteins such as NORTIA (NTA) are also identified as cell-surface indicators (Kesler *et al.*, 2010). In the FER mutant reduction of ROS accumulation, mainly by the downregulation of NADPH oxidase, was observed. A calmodulin (CaM) binding protein activates the Ca^{2+} to sense the polar localization. Capron *et al.* (2008) observed that mutant of *lre* showed overgrowth of the pollen tube and flaws in sperm cell release. Similarly, the mutant of *abstinence by mutual consent* (*AMC*), a peroxisomal protein involved in generation of ROS and NO, was unable to produce and import proteins to the peroxisomes (Boisson-Dernier *et al.*, 2008). The VERDANDI (VDD) transcription factor regulates the pollen tube rupture, which was defective in the *vdd* mutant. The gametophytic gene, ES1-4, attracts the pollen was correlated with the changes in the osmotic potential and concentrations of sucrose and Ca^{2+} (Amien *et al.*, 2010). Calmodulin-activated Ca^{2+} ATPase (ACA) regulates the Ca^{2+} concentration and helps to position/discharge the sperm cell across the major reproductive barrier (Schiott *et al.*, 2004). After the invasion of synergids, ROS-mediated mechanisms in female gametophytes ensure proper fertilization and prevent polyspermy.

ROS and NO in Repulsion or Attraction Between Sporophyte and Gametophyte

Most flowering plants have both male and female reproductive organs in a single individual. Two types of fertilization can occur in the plant, that is, self-fertilization or cross-fertilization. The inability of fertile hermaphrodite seed plants to produce zygotes after self-fertilization is defined as self-incompatibility. In some cases plants promote self-incompatibility to avoid self-fertilization (de Nettancourt, 1997). This intraspecific barrier involves the ability of the pistil to discriminate between self-pollen and non-self-pollen, thereby enabling rejection or acceptance of the pollen. Self-incompatibility is triggered by PCD, which is mediated by ROS and NO to avoid self-pollen (Serrano *et al.*, 2015). Van Doorn *et al.* (2011) described how PCD in pollen due to self-incompatibility involves characteristic vacuole enlargement as well as swelling of mitochondria. Morphological changes occurring during the PCD are alteration in the morphology of the nuclease, swelling of vacuole, mitochondria, and endoplasmic reticulum, shrinkage of protoplast, and reorganization of the cytoskeleton. Other molecular/biochemical events associated with PCD are DNA fragmentation, caspase-like activity, and ROS/RNS accumulation (Van Doorn *et al.*, 2011; Serrano *et al.*, 2015). PCD is a genetically controlled form of cell death and a fundamental cellular process occurring throughout a plant's life (Bozhkov and Lam, 2011). Through *S*-nitrosylation or nitration, NO and RNS regulate gene transcription, activating second messengers or directly regulating proteins (Rodriguez-Serrano *et al.*, 2009). Meanwhile to fine-tune the ROS/NO systems, ROS-producing antioxidant enzymes are the target of *S*-nitrosylation. ROS and RNS, key signaling molecules of the HR, accumulate in higher amounts during self-incompatibility (Grant and Loake, 2000). The enhanced accumulation of ROS and RNS in both the pollen and pistil signals PCD with self-incompatibility (SI), whereas a lesser accumulation gives the clue of self-compatibility. In the latter process, the generation of ROS triggers the signaling network for the further development of the embryo.

Pollen Tube Rupture

ROS and RNS signaling for the pollen tube have been identified as the copper-binding specific phytocyanins (Dong *et al.*, 2005). Among the various ROS, including superoxide anion ($O_2^{\cdot-}$),

H_2O_2, and·OH, ·OH was considered the most prevalent species (McInnis *et al.*, 2006). Indeed, both $O_2^{\cdot-}$ and H_2O_2 tend to convert to·OH by the Fenton reaction. Blockade of the Ca^{2+} transmembrane signals and K^+ influx (osmoregulation) inhibits the timely rupturing of the pollen tube (Simontacchi *et al.*, 2013). In the presence of ROS with in the short period of time, Ca^{2+} gradient is at maximum towards the distal end of the cytoplasm. Even with the addition of ROS, inhibition of Ca^{2+} influx delays the dissipation of pollen tube (Duan *et al.*, 2014). Another important protein to regulate ROS level, especially in the female gametophyte, is LORELEI (LRE) (Capron *et al.*, 2008). Tsukamoto *et al.* (2010) identified that plants lacking FER also showed diminished expression of LRE. Overgrowth of the pollen tube around the ovule could be caused by the tripartite complex. In addition to the LRE, FER is reported to interact with ROPGEFs and RAC/ROP GTPases (Duan *et al.*, 2010). Reduced expression of RAC/ROP GTPases leads to defective polysaccharide breakage of the pollen tube, and catastrophic failure of sperm discharge (Duan *et al.*, 2014). The proper balance of turgor and cell wall integrity mediated by ROS via Ca^{2+} is responsible for the penetration as well as fusion of gametes. The longevity of the pollen tube and timing of its rupturing is triggered by the Ca^{2+} signals via controlled PCD (Iwano *et al.*, 2009). In incompatible crosses, even between different species of the same genus, for example, in the case of *Arabidopsis lyrata* and *Arabidopsis thaliana*, incompatibility occurred due to failure in the priming of the pollen tube (Escobar-Restrepo *et al.*, 2007). In the compatible condition, timely bursting of the pollen tube ensures successful fertilization. In both the cases, the level of ROS was identified as a crucial factor. A simple model of the involvement of ROS and Ca^{2+} and their signaling in the fusion of male and female gametophytes is described in Figure 15.2.

Links Between ROS and Predicted Self-incompatibility Mechanisms

Previous studies (McClure and Franklin-Tong, 2006; Eaves *et al.*, 2014) reported that self-incompatibility (SI) is controlled by the single highly polymorphic *S* locus. There are two types of self-incompatibility, that is, sporophytic SI (SSI) and gametophytic SI (GSI). In SSI, pollen rejection involved interaction with the stigma-expressed S-locus receptor kinase (SRK). The SRK is located in the plasma membrane of the stigma. In case of pollen, S-locus cysteine rich protein (SCR) the pollen-coat localized ligand. The *SRK* gene encodes the serine/threonine receptor kinase, and *SRC* encodes the cysteine-rich protein. Germination will not occur if the pollen parent matches with either *S* alleles of the pistil parent (Dresselhaus and Franklin-Tong, 2013; Leducq *et al.*, 2014). The GSI type is reported only in the genus *Pyruspyrifolia*. The female determinant protein is characterized as PrsS and the male determinant is characterized as PrpS. The former is secreted in the surface of the pistil while the latter is located in the transmembrane region of the pollen grain. During the interaction of the PrsS and PrpS, intracellular Ca^{2+} levels are increased in the cell. This induces the multi-layered SI signaling cascade to culminate in PCD of incompatible pollen (Eaves *et al.*, 2014; Wilkins *et al.*, 2014). The other type of PCD is reported in the Solanaceae, Plantaginaceae, and Rosaceae and is associated with the *S-RNase* and *S-locus F-box* (*SLF* or *SFB*) genes. The pistil specificity is regulated by the *S-RNase* and the pollen specificity is regulated by the *SLF* or *SFB*. The S-RNase degrades the RNA from the self-pollen to reject the pollen or prevent the growth of pollen in the style (Luu *et al.*, 2000; Kubo *et al.*, 2010).In the SI process, S-RNase particularly disturbs the tip-localized ROS on the germinated self-pollen via mitochondrial alteration and decrease in the mitochondrial and cytosolic NADPH oxidases. The other events associated with the tip-localized ROS disrupting mechanism include changes in the intracellular level Ca^{2+}, actin cytoskeleton depolymerization, nuclear DNA degradation, and PCD (Thomas *et al.*, 2006). Jiang *et al.* (2014) reported that the PCD occurred in a calmodulin (CaM)-dependent manner. Depending on whether the pollen–pistil interaction is self-compatible or self-incompatible, the transient

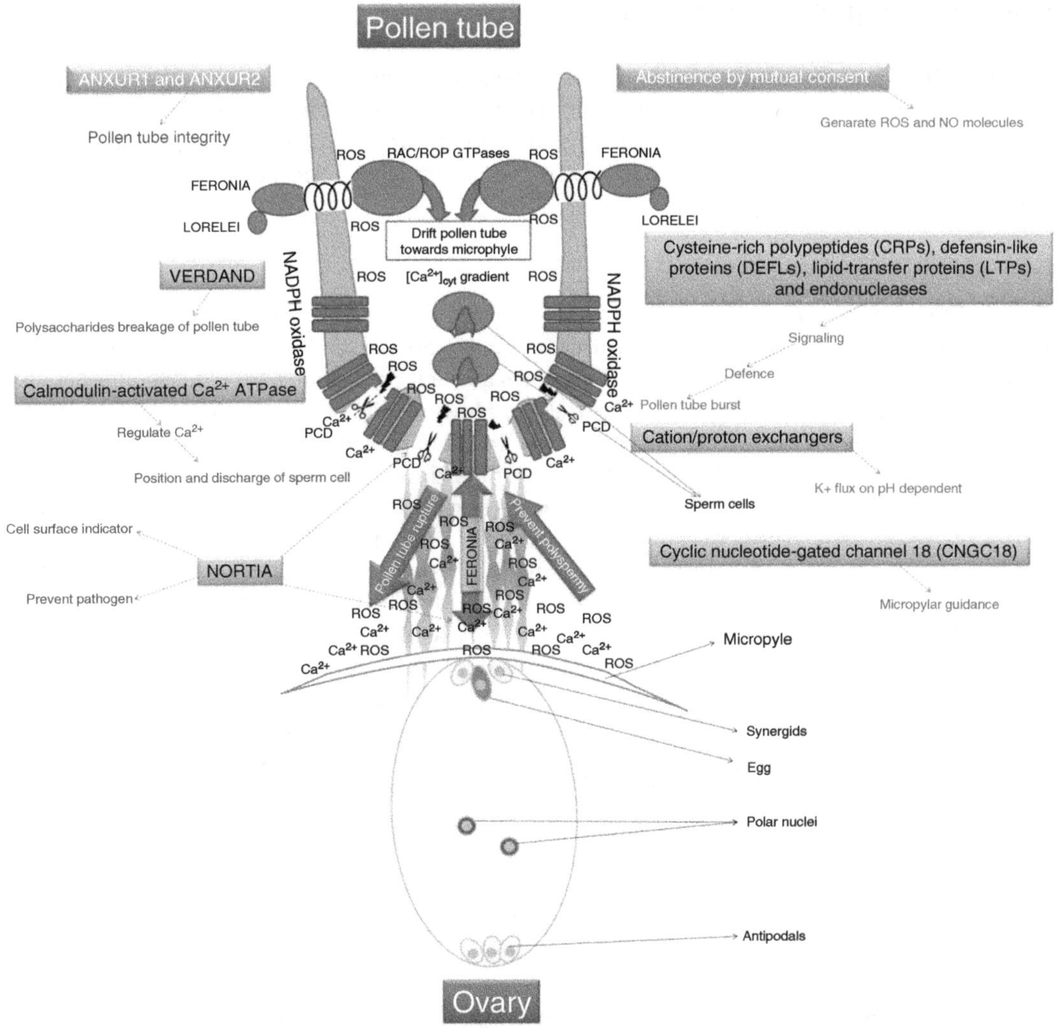

Figure 15.2 A model for the involvement of ROS- and Ca^{2+}-mediated mechanisms in the guidance of sperm cells toward the ovule. Signaling patterns have been conceived and adapted from various references cited in the text. However, the main concept has been adapted specially from the following: Boisson-Dernier *et al.* (2009), Capron *et al.* (2008), Duan *et al.* (2010, 2013, 2014), Dresselhaus and Franklin-Tong (2013), Iwano *et al.* (2009), Lausser and Dresselhaus (2010), Müller *et al.* (2013), Palanivelu *et al.* (2003), Palanivelu and Tsukamoto (2012), Rodriguez-Serrano *et al.* (2014), Serrano and Olmedilla (2012); Serano *et al.* (2015), Van Doorn *et al.* (2011), and Zhang *et al.* (2014). The model shows the general role of proteins involved in the signaling mediated by ROS and Ca^{2+}, during the fusion of male and female gametes. Detailed mechanism(s) of individual genes/proteins cannot be interpreted from this figure. ROS, reactive oxygen species; PCD, programmed cell death; Ca^{2+}, calcium.

accumulation of ROS and NO varies. Especially during SI, the cytosolic level of ROS is greatly increased followed by an increase in the NO level (Wilkins *et al.*, 2011). The increase in levels of ROS and NO changes the cytoskeletal actin content. In detail, posttranslational modification involving carbonylation and *S*-nitrosylation severely disturbs the polymerization. The consequent degradation of the cell wall polymers disturbs the structure and function of the cytoskeleton (Rodríguez-Serrano *et al.*, 2014).

From Embryogenesis to Endosperm Cleavage

After fertilization the ovule develops into a seed (egg cell fused with a sperm cell) and a triploid endosperm (polar cells fused with a sperm cell), and the integuments of the ovule sac develop into the seed coat. During embryogenesis, cell division is rapid initially, followed by active production of seed reserves (Cui *et al.*, 1999). In the later stages desiccation due to necessitate dehydration seed become quiescent.

Status of ROS During Desiccation

Though ROS are actively produced in all stages of embryogenesis, during the desiccation phase the seed produces less ROS compared with the imbibition phase, due to less water availability (El-Maarouf-Bouteau and Bailly, 2008). This plays a crucial role in maintaining seed longevity and triggering dormancy release upon hydration. However, in the desiccated seeds lipid peroxidation produces LOOH, ROOH-like peroxy reactive molecules via the Amadori and/or Maillard reactions (Leubner-Metzger, 2005). Improper desiccation as well as storage increased lipid peroxidation (LPO). In dormant seeds, inappropriate ROS accumulation affects the equilibrium between their production and detoxification. Damage during the dry state cannot be retained after hydration. Mobility and metabolic activities are almost completely arrested in the dry state (Bailly, 2004). Hence stimulation of antioxidant enzymes to scavenge the ROS produced by the breakdown of polyunsaturated fatty acid (PUFA) is not active at this stage (Lang, 1996). Peroxiredoxin and thioredoxin are reported to be involved in maintaining redox balance in the nucleus (Pulido *et al.*, 2009). However, during seed storage, the activity of glutathione *S*-hydrolase activity was reported to be increased. It induces apoptosis of its own cell by splitting thioredoxin. The signaling process mediated by the apoptosis-regulating kinase 1 (ASK1) induces mitogen-activated protein kinase (MAPK) (Chen *et al.*, 2013). Meanwhile, cytochrome *c* (apoptogenic proteins) expressed in the mitochondria activates the caspases and downstream proteolytic enzymes. This process is transmitted through the adenine nucleotide translocator (ANT) (Lam and del Pozo, 2000). Degradation of nuclear and structural proteins activates PCD. Either misfolding and/or unfolded proteins induce endoplasmic reticulum (ER) stress as most of the improperly folded/denatured proteins are stored in the lumen of ER (Chen *et al.*, 2013). Subsequently DNA laddering indicative of PCD and ubiquitin-proteasome pathways induce the upregulation of binding protein (BiP). BiP is a molecular chaperone mainly activated upon stress and involved in the restoration of proteins upon imbibition.

Resumption Into Active Form

Hydration activates the respiration of seeds, which stimulates the production of ROS from cell organelles such as mitochondria, glyoxysomes, and plasma membrane NADPH oxidase (McDonald, 1999). The level of ROS must be reduced for the survival of seeds during storage. In prolonged or unfavorable storage conditions, production of ROS cannot be controlled by the antioxidative mechanism as most of the cellular functions are arrested. Further, the reduced water content (5–15%) of dried seeds increases cytoplasm viscosity. As the availability of water is less, the activity of antioxidant enzymes is impeded, so they are less able to actively participate in detoxification. During seed hydration, higher accumulations of ROS are scavenged by the enzymatic and non-enzymatic antioxidant mechanisms (Bewley, 1997; Lang, 1996). Mitochondrial respiration shows a resurgence upon imbibition of the seed. The production of H_2O_2 as a by-product of mitochondrial metabolism reduces the electron pressure and increases protein oxidation rate. Activation of protein carbonylation effectively increases the protein turnover with lesser energy in the mitochondria (Barba-Espin *et al.*, 2011). Carbonylation of protein is the first event observed once the seed has retained moisture content.

The first major process activated upon hydration is DNA repair (Adamakis *et al.*, 2011). Biogenesis of mitochondria occurred rapidly immediately after imbibition. It provides the large amounts of ATP required for the cell-building activities (Law *et al.*, 2012). The energy required for biogenesis or import of mitochondrial proteins is comparatively less than the energy supplied by the mitochondria for essential cellular activities. During this process, enormous amounts of ROS are produced by the active cell organelles (Law *et al.*, 2014). The mitochondria alone are observed to consume around 2–3% of total oxygen (O_2) in the cell (Foyer and Noctor, 2003). The mitochondrial electron transport chain (ETC) was reported as the primary source for ROS. The chloroplast is an essential organelle for photosynthesis. The resumption of chlorophyll synthesis activates the photosystems I (PSI) and II (PSII), where $O_2^{\cdot-}$, $^1O_2$, and H_2O_2 are often generated. In addition, H_2O_2 is generated as a by-product of the breakdown of lipid reserves, catabolism of purines, and conversion of xanthine in the glyoxysomes and peroxisomes (Corpas *et al.*, 2001). For the cell-to-cell diffusion of ROS, cell wall-dependent peroxidases, amine oxidase, cytochrome P450, and germin-like oxidase are actively involved (Foyer and Noctor, 2003). Especially during the cleavage of cell wall, higher expression of cell wall-dependent peroxidase was observed (Corpas *et al.*, 2001). The mobility and viscosity of the cell environment determines the interaction of ROS with transcription factors (Walters, 1998). A reduced oxidation rate of pentose phosphate proteins and improper glycolysis are observed in dormant seeds. The *AtrbohD* mutant failed to express NADPH oxidase and showed a reduction in the cytosolic electron transfer. In the dry state, the oxidation rate of *AtrbohD* was less than in the wild-type, whereas during imbibition oxidation of proteins was higher than in the wild-type (Leymarie *et al.*, 2012). The *rbohD* involved in the NADPH oxidase catalyzes the conversion of H_2O_2 into $\cdot OH$ (Kumar *et al.*, 2015). In *Arabidopsis*, the *cat2-1* mutant and *vte1-1* mutant showed delayed germination. The former encodes the H_2O_2 detoxification and latter encodes tocopherol. These findings show that disturbance in the equilibrium of ROS leads to germination failures. However, exogenous application of H_2O_2 and menodione (to induce $O_2^{\cdot-}$ accumulation) enhanced chilling tolerance in maize (Prasad *et al.*, 1994). Applications of sodium benzoate, which tends to inhibit ROS production, strongly inhibited germination (Hung and Kao, 2004). Seeds incubated in diphenylene iodonium (DPI) failed to germinate, which was significantly correlated with reduced expression of NADPH oxidase (Leymari *et al.*, 2012).

Cross-talk with Plant Growth Regulators

Success of seed germination depends on the level of two important hormones, that is, gibberellins (GA) and abscisic acid (ABA). Relatively low endogenous levels of ABA and high levels of GA increase the success of germination. In contrast, elevated levels of ABA cause the seed to enter a dormant state. The application of GA is associated with the accumulation of $\cdot OH$ in the radicle region. In contrast, exogenous ABA inhibits the Fenton reaction, by which $\cdot OH$ is produced. In sunflower, H_2O_2 content was measured and compared between the germinated and non-germinated seeds. The total H_2O_2 of non-dormant seeds was higher than in the dormant seeds (Muller *et al.*, 2009). Bazin *et al.* (2011) suggested that in sunflower the oxidation level of mRNA of non-dormant seeds is more than in dormant seeds. ABA was reported to be possibly inhibiting the PCD. This result proved that $\cdot OH$ is much needed for controlled PCD during the emergence of the radicle. Amylase and GAMyb act as antagonists for the ABA (Ishibashi *et al.*, 2012). Though the exact mechanism of action of ROS varied between species, most reports suggest that higher accumulations of ROS with the proper guidance are necessary for successful germination. Often this mechanism is referred as the "oxidation window," that is, controlled usage of molecules (ROS) could deploy oxidation stress for the benefit (seed germination) of the plant (Bailly, 2008).

Emergence of the Radicle

Leymarie *et al.* (2012) observed that for the emergence of the radicle, proper spatial and temporal localization of ROS is necessary. The ROS are produced in the cytoplasm and move to the nucleus and to the cell wall. NADPH oxidase is involved in the cell-to-cell diffusion of ROS (Chandrakuntal *et al.*, 2011). Successful germination is signaled by protrusion of the radicle. The spatial and temporal localization of ROS, especially·OH, is responsible for the cleavage of cell-wall polymers (Leymarie *et al.*, 2012). For cldeavage of the polysaccharides ROS stimulate hydrolase enzymes. Weakening of the endosperm is an essential process for protrusion of the radicle. Most reports have suggested that·OH generated on the cell wall by NADPH oxidase is responsible for radicle emergence (Muller *et al.*, 2007; Leymarie *et al.*, 2012). Generally, production of ROS starts with $O_2^{\cdot-}$ and/or H_2O_2, which are partially converted into·OH by the Fenton reaction. Compared with the $O_2^{\cdot-}$ and H_2O_2, ·OH is extremely harmful (Pulido *et al.*, 1994). Without the proper control mechanisms, ·OH can easily affect the nucleus. Accumulation of·OH was reported to activate the hydrolysis (Muller *et al.*, 2007). Activation of endonucleases is involved in the cleavage of tightly bound polysaccharides of the cell wall. In dormant seeds, ROS were found to be dispersed whereas in non-dormant seeds they accumulated in a specific manner (Meitha *et al.*, 2015; Leymarie *et al.*, 2012). Especially near the cell wall, excessive ROS cleave the polymers (mannose, cellulose, etc.) directly as well as inducing enzymes such as glucanase (Muller *et al.*, 2007). Thus, the cleavage of polymers leads to emergence of the radicle from the endosperm.

Safeguarding the Newly Emerged Radicle

After emergence of the radicle ROS act as a safeguard against pathogen attack (Levine *et al.*, 1994). Chiefly H_2O_2 is involved in both directly attacking pathogens and induction of the hypersensitive reaction (HR), which triggers PCD. Higher rates of H_2O_2 accumulation induce systemic acquired resistance (SAR), which helps protect the newly emerging embryonic axis against pathogens. During radicle emergence, increase $O_2^{\cdot-}$ generation was efficiently converted into H_2O_2 by superoxide dismutase (SOD). This H_2O_2 tends to combat pathogens by killing infected cells to prevent the spread of pathogens (Dietz *et al.*, 2010). Liu *et al.* (2010) reported that for defense against pathogens, ROS induce numerous proteins. Class III peroxidases especially were increased in abundance during pathogen invasion (Cosio and Dunand, 2009). These enzymes are known to attack pathogens. It has been reported that in order to adapt to an uncertain environment, production of various ROS are induced by the class III peroxidases. In addition, formation of separate hydroxylic cycle produces the several ROS radicals (Passardi *et al.*, 2004). For avirulent pathogen interactions, subunits of NADPH-dependent oxidase such as *AtrbohD* and *AtrbohF* reduce ROS generation (Torres *et al.*, 2002). Meanwhile, enzymes such as germin-like oxalate oxidase (Hu *et al.*, 2003), amine oxidases (Walters, 2003), NADPH oxidase-like alternative oxidases (Apel and Hirt, 2004), glycolate oxidases, and class III peroxidases (Lui *et al.*, 2009) increase ROS production to kill the infected host cell as well as the pathogen. This signals the plant to activate other defense mechanisms to protect the emerging seedling from microbes. Figure 15.3 shows a simplified representation of the involvement of ROS and antioxidant enzymes for germination in dormant and non-dormant seeds.

Conclusions

The generation of ROS is an inevitable by-product of metabolic processes, and excessive generation of ROS is harmful to cellular components. Hence, excessive ROS are a sign of oxidative stress, and efficient antioxidant mechanisms are employed by plants to keep ROS under

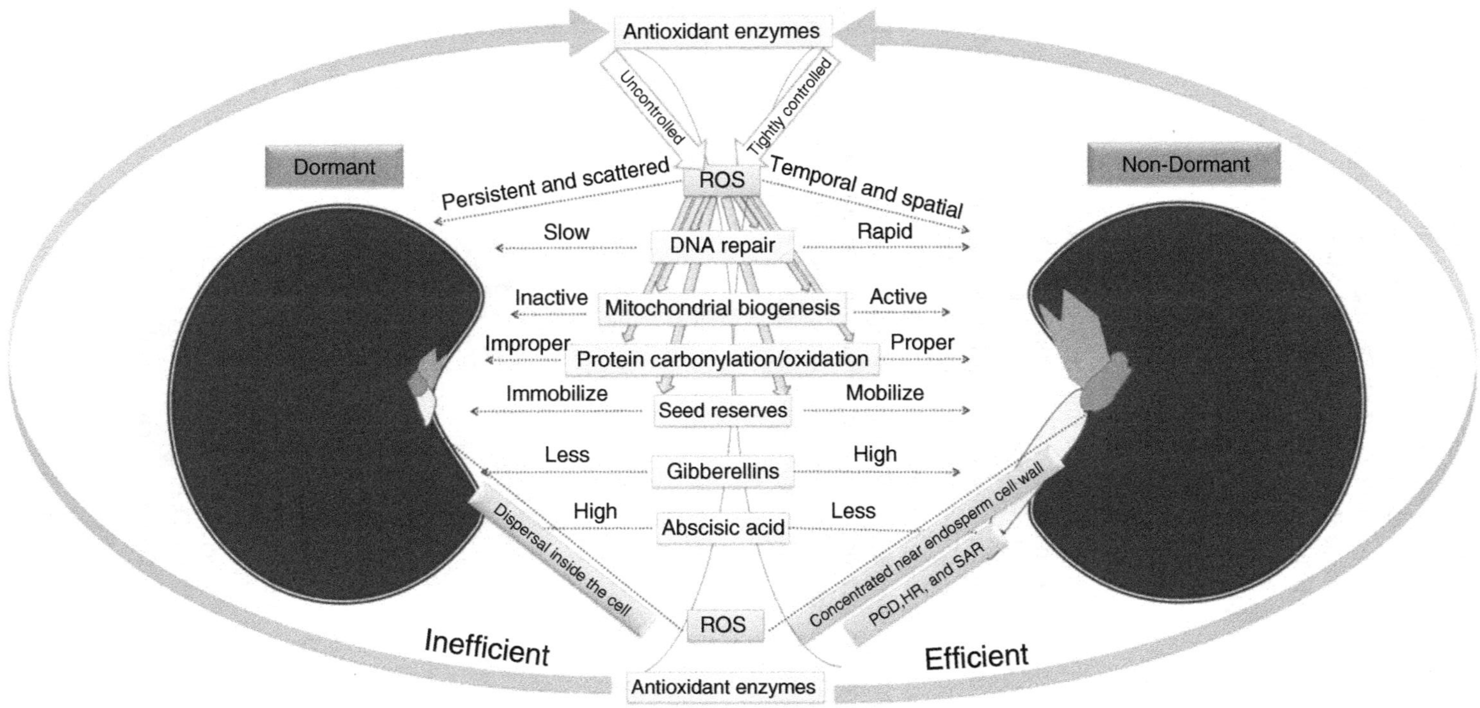

Figure 15.3 A simplified representation of the involvement of reactive oxygen species (ROS) and antioxidant enzymes in the germination of dormant and non-dormant seeds. HR, hypersensitive response; PCD, programmed cell death; SAR, systemic acquired resistance.

control. However, ROS play important roles in fertilization, starting with adhesion of the pollen to the stigma and fusion of sperm cells with the female gametes. For successful fertilization, the recognition of compatible pollen and guidance of the pollen tube to the ovule depend on an efficient ROS system. The functions of ROS are associated with PCD, NO, and Ca^{2+}/K^{+} influx. After imbibition, from the beginning of DNA repair, mitochondria biogenesis, and breakdown/movement of seed reserves, ROS play a pivotal role in seed germination until radical emergence. Protein oxidation/carbonylation especially ensures the proper development of the seedling, and is correlated with an organized role of ROS. During the cleavage of cell wall ROS are present in higher concentrations near the endosperm layer. Meanwhile, essential cellular components are effectively protected by the efficient antioxidant mechanism(s). Often this kind of mechanism is termed the "oxidative window." Finally, activation of SAR/PCD by ROS protects the newly emerged seedling from pathogen invasion.

Future Prospects

The various aspects of involvement of ROS in seed formation and seedling development gives new dimensions to redox biology. As the metabolism of ROS as well as the structure/germination mechanisms of seeds vary between species, the universal role of ROS in seed development is difficult to predict. However, several molecular and biochemical studies suggest that controlled elevations in ROS concentrations mediate the fusion of sperm cells with gametes, and radicle protrusion. Meanwhile, under non-favorable conditions, ROS function randomly or are more dispersed. This particular spatial and temporal localization plays a pivotal role in plant growth. Hence, in future further studies on tenacious of pollen with pistil and persistent of ROS in the development of plant could help to understand in detail the involvement of oxygen species in living processes of biological systems.

Acknowledgments

Prabhakaran Soundararajan was supported by a scholarship from the BK21 Plus Program, the Ministry of Education.

References

Adamakis IDS, Panteris E, Eleftheriou EP. (2011) The fatal effect of tungsten on *Pisum sativum* L. root cells: indications for endoplasmic reticulum stress-induced programmed cell death. *Planta* 234: 21–34.

Amien S, Kliwer I, Márton ML, *et al.* (2010) Defensin-like ZmES4 mediates pollen tube burst in maize via opening of the potassium channel KZM1. *PLoS Biol* 8(6): e1000388.

Apel K, Hirt H. (2004) Reactive oxygen species: metabolism, oxidative stress, and signal transduction. *Annu Rev Plant Biol* 55: 373–399.

Bailly C. (2004) Active oxygen species and antioxidants in seed biology. *Seed Sci Res* 14: 93–107.

Barba-Espín G, Diaz-Vivancos P, Job D, Belghazi M, Job C, Hernández JA. (2011) Understanding the role of H_2O_2 during pea seed germination: a combined proteomic and hormone profiling approach. *Plant Cell Environ* 34: 1907–1919.

Bazin J, Langlade N, Vincourt P, Arribat S, *et al.* (2011) Targeted mRNA oxidation regulates sunflower seed dormancy alleviation during dry after-ripening. *The Plant Cell* 23: 2196–2208.

Bell CD, Soltis DE, Soltis PS. (2010) The age and diversification of the angiosperms re-revisited. *Am J Bot* 97: 1296–1303.
Bewley JD, Black M. (1994) *Seeds: Physiology of Development and Germination*, 3rd edn. Plenum Press, New York.
Boisson-Dernier A, Frietsch S, Kim TH, Dizon MB, Schroeder JI. (2008) The peroxin loss-of-function mutation abstinence by mutual consent disrupts male-female gametophyte recognition. *Curr Biol* 18: 63–68.
Boisson-Dernier A, Roy S, Kritsas K, *et al.* (2009) Disruption of the pollen-expressed FERONIA homologs ANXUR1 and ANXUR2 triggers pollen tube discharge. *Development* 136: 3279–3288.
Bouquin T, Meier C, Foster R, Nielse ME, Mundy J. (2001) Control of specific gene expression by gibberellin and brassinosteroid. *Plant Physiol* 127: 450–458.
Bozhkov PV, Lam E. (2011) Green death: revealing programmed cell death in plants. *Cell Death Differ* 18: 1239–1240.
Capron A, Gourgues M, Neiva LS, *et al.* (2008) Maternal control of male-gamete delivery in Arabidopsis involves a putative GPI-anchored protein encoded by the LORELEI gene. *The Plant Cell* 20: 3038–3049.
Chae K, Kieslich CA, Morikis D, Kim SC, Lord EM. (2009) A gain-of-function mutation of Arabidopsis lipid transfer protein 5 disturbs pollen tube tip growth and fertilization. *The Plant Cell* 21: 3902–3914.
Chandrakuntal K, Shah AK, Thomas NM, *et al.* (2010) Blue light exposure targets NADPH oxidase to plasma membrane and nucleus in wheat coleoptiles. *J Plant Physiol* 29: 232–241.
Chapman LA, Goring DR. (2011) Misregulation of phosphoinositides in Arabidopsis thaliana decreases pollen hydration and maternal fertility. *Sex Plant Reprod* 24: 319–326.
Chen H, Osuna D, Colville L, *et al.* (2013) Transcriptome-wide mapping of pea seed ageing reveals a pivotal role for genes related to oxidative stress and programmed cell death. *PloS One* 8(10): e78471.
Cheung AY, Wang H, Wu HM. (1995) A floral transmitting tissue-specific glycoprotein attracts pollen tubes and stimulates their growth. *Cell* 82: 383–393.
Corpas FJ, Barroso JB, del Rio LA. (2001) Peroxisomes as a source of reactive oxygen species and nitric oxide signal molecules in plant cells. *Trends Plant Sci* 6:145–150.
Cosio C, Dunand C. (2009) Specific functions of individual class III peroxidase genes. *J Exp Bot* 60: 391–408.
de Nettancourt D. (1997) Incompatibility in angiosperms. *Sex Plant Reprod* 10: 185–199.
Dietz K, Jacquot J, Harris G. (2010) Hubs and bottlenecks in plant molecular signalling networks. *New Phytol* 188: 919–938.
Dong J, Kim ST, Lord EM. (2005) Plantacyanin plays a role in reproduction in Arabidopsis. *Plant Physiol* 138: 778–789.
Doyle JA. (2012) Molecular and fossil evidence on the origin of angiosperms. *Annu Rev Earth Pl Sci* 40: 301–326.
Dresselhaus T, Franklin-Tong N. (2013) Male-female crosstalk during pollen germination, tube growth and guidance, and double fertilization. *Mol Plant* 6: 1018–1036.
Dresselhaus T, Sprunck S. (2012) Plant fertilization: maximizing reproductive success. *Curr Biol* 22: R487–R489.
Duan Q, Kita D, Li C, Cheung AY, Wu HM. (2010) FERONIA receptor-like kinase regulates RHO GTPase signaling of root hair development. *Proc Natl Acad Sci USA* 107: 17821–17826.
Eaves DJ, Flores-Ortiz C, Haque T, Lin Z, Teng N, Franklin-Tong VE. (2014) Self-incompatibility in Papaver: advances in integrating the signalling network. *Biochem Soc Trans* 42: 370–376.
El-Maarouf-Bouteau H, Bailly C. (2008) Oxidative signaling seed germination and dormancy. *Plant Sign Behav* 3: 175–182.

Escobar-Restrepo JM, Huck N, Kessler S, *et al.* (2007) The FERONIA receptor-like kinase mediates male-female interactions during pollen tube reception. *Science* 317: 656–660.

Foyer CH, Noctor G. (2003) Redox sensing and signalling associated with reactive oxygen in chloroplasts, peroxisomes and mitochondria. *Physiol Plantarum* 119: 355–364.

Fu X, Harberd NP. (2003) Auxin promotes Arabidopsis root growth by modulating gibberellin response. *Nature* 421: 740–743.

Ghassemian M, Nambara E, Cutler S, Kawaide H, Kamiya Y, McCourt P. (2000) Regulation of abscisic acid signalling by the ethylene response pathway in Arabidopsis. *Plant Cell* 12: 1117–1126.

Grant JJ, Loake GJ. (2000) Role of reactive oxygen intermediates and cognate redox signaling in disease resistance. *Plant Physiol* 124: 21–30.

Higashiyama T, Kuroiwa H, Kuroiwa T. (2003) Pollen-tube guidance: beacons from the female gametophyte. *Curr Opin Plant Biol* 6: 36–41.

Hu X, Bidney DL, Yalpani N, *et al.* (2003) Overexpression of a gene encoding hydrogen peroxide-generating oxalate oxidase evokes defense responses in sunflower. *Plant Physiol* 133: 170–181.

Hung KT, Kao CH. (2004) Nitric oxide acts as an antioxidant and delays methyl jasmonate-induced senescence of rice leaves. *J Plant Physiol* 161: 43–52.

Igic B, Bohs L, Kohn JR. (2004) Historical inferences from the self-incompatibility locus. *New Phytol* 161: 97–105.

Ishibashi Y, Tawaratsumida T, Kondo K, *et al.* (2012) Reactive oxygen species are involved in gibberellin/abscisic acid signaling in barley aleurone cells. *Plant Physiol* 158: 1705–1714.

Iwano M, Shiba H, Miwa T, *et al.* (2004) Ca^{2+} dynamics in a pollen grain and papilla cell during pollination of Arabidopsis. *Plant Physiol* 136: 3562–3571.

Iwano M, Shiba H, Matoba K, *et al.* (2007) Actin dynamics in papilla cells of *Brassica rapa* during self- and cross-pollination. *Plant Physiol* 144: 72–81.

Iwano M, Entani T, Shiba H, *et al.* (2009) Fine-tuning of the cytoplasmic Ca^{2+} concentration is essential for pollen tube growth. *Plant Physiol* 150: 1322–1334.

Iwano M, Ngo QA, Entani T, *et al.* (2012) Cytoplasmic Ca^{2+} changes dynamically during the interaction of the pollen tube with synergid cells. *Development* 139: 4202–4209.

Kessler SA, Grossniklaus U. (2011) She's the boss: signaling in pollen tube reception. *Curr Opin Plant Biol* 14: 622–627.

Kessler SA, Shimosato-Asano H, Keinath NF, *et al.* (2010) Conserved molecular components for pollen tube reception and fungal invasion. *Science* 330: 968–971.

Lam E, Del Pozo O. (2000) Caspase-like protease involvement in the control of plant cell death. In: Lam E, Fukuda H, Greenberg J (eds), *Programmed Cell Death in Higher Plants*. Springer, Netherlands, pp. 173–184.

Lausser A, Dresselhaus T. (2010) Sporophytic control of pollen tube growth and guidance in grasses. *Biochem Soc Trans* 38: 631–634.

Leducq JB, Gosset CC, Gries R, *et al.* (2014) Self-incompatibility in Brassicaceae: identification and characterization of SRK-like sequences linked to the S-locus in the tribe Biscutelleae. *G3: Genes Genomes Genetics* 4: 983–992.

Leubner-Metzger G. (2005) β-1,3-Glucanase gene expression in low-hydrated seeds as a mechanism for dormancy release during tobacco after-ripening. *Plant J* 41: 133–145.

Levine A, Tenhaken R, Dixon R, Lamb C. (1994) H2O2 from the oxidative burst orchestrates the plant hypersensitive disease resistance response. *Cell* 79: 583–593.

Lindner H, Müller LM, Boisson-Dernier A, Grossniklaus U. (2012) CrRLK1L receptor-like kinases: not just another brick in the wall. *Curr Opin Plant Biol* 15: 659–669.

Liu X, Xing D, Li L, Zhang L. (2007) Rapid determination of seed vigor based on the level of superoxide generation during early imbibition. *Photochem Photobiol Sci* 6: 767–774.

Liu X, Zhang H, Zhao Y, *et al.* (2013) Auxin controls seed dormancy through stimulation of abscisic acid signaling by inducing ARF-mediated ABI3 activation in Arabidopsis. *Proc Natl Acad Sci USA* 110: 15485–15490.

Lu Y, Chanroj S, Zulkifli L, *et al.* (2011) Pollen tubes lacking a pair of K^+ transporters fail to target ovules in Arabidopsis. *The Plant Cell* 23: 81–93.

Luu DT, Passelègue E, Dumas C, Heizmann P. (1998) Pollen-stigma capture is not species discriminant within the Brassicaceae family. *CR Acad Sci III-Vie* 321: 747–755.

Márton ML, Fastner A, Uebler S, Dresselhaus T. (2012) Overcoming hybridization barriers by the secretion of the maize pollen tube attractant ZmEA1 from Arabidopsis ovules. *Curr Biol* 22: 1194–1198.

McDonald MB. (1999) Seed deterioration: physiology, repair and assessment. *Seed Sci Technol* 27: 177–237.

Meitha K, Konnerup D, Colmer TD, Considine JA, Foyer CH, Considine MJ. (2015) Spatio-temporal relief from hypoxia and production of reactive oxygen species during bud burst in grapevine (*Vitis vinifera*). *Ann Bot* 116: 703–711.

Michard E, Lima PT, Borges F, *et al.* (2011) Glutamate receptor-like genes form Ca^{2+} channels in pollen tubes and are regulated by pistil D-serine. *Science* 332: 434–437.

Muller K, Heß B, Leubner-Metzger G. (2007) A role for reactive oxygen species in endosperm weakening. In: Adkins S, Ashmore S, Navie S (eds), *Seeds: Biology, Development and Ecology*. CAB International, Wallingford, UK, pp. 287–295.

Müller K, Levesque-Tremblay G, Bartels S, *et al.* (2013) Demethylesterification of cell wall pectins in Arabidopsis plays a role in seed germination. *Plant Physiol* 161: 305–316.

Okuda S, Tsutsui H, Shiina K, *et al.* (2009) Defensin-like polypeptide LUREs are pollen tube attractants secreted from synergid cells. *Nature* 458: 357–361.

Palanivelu, Tsukamoto T. (2012) Pathfinding in angiosperm reproduction: pollen tube guidance by pistils ensures successful double fertilization. *WIRES Dev Biol* 1: 96–113.

Palanivelu R, Brass L, Edlund AF, Preuss D. (2003) Pollen tube growth and guidance is regulated by POP2, an Arabidopsis gene that controls GABA levels. *Cell* 114: 7–59.

Passardi F, Longet D, Penel C, Dunand C. (2004) The class III peroxidase multigenic family in rice and its evolution in land plants. *Phytochemistry* 65: 1879–1893.

Pulido P, Cazalis R, Cejudo FJ. (2009) An antioxidant redox system in the nucleus of wheat seed cells suffering oxidative stress. *Plant J* 57: 132–145.

Qin Y, Leydon AR, Manziello A, *et al.* (2009) Penetration of the stigma and style elicits a novel transcriptome in pollen tubes, pointing to genes critical for growth in a pistil. *PLoS Genet* 5(8): e1000621.

Qin Y, Wysocki RJ, Somogyi A, *et al.* (2011) Sulfinylated azadecalins act as functional mimics of a pollen germination stimulant in Arabidopsis pistils. *Plant J* 68: 800–815.

Renault H, El Amrani A, Palanivelu R, *et al.* (2011) GABA accumulation causes cell elongation defects and a decrease in expression of genes encoding secreted and cell wall-related proteins in Arabidopsis thaliana. *Plant Cell Physiol* 52: 894–908.

Rodríguez-Serrano M, Romero-Puertas MC, Sparkes I, Hawes C, Luis A, Sandalio LM. (2009) Peroxisome dynamics in Arabidopsis plants under oxidative stress induced by cadmium. *Free Radical Bio Med* 47: 1632–1639.

Samuel MA, Chong YT, Haasen KE, Aldea-Brydges MG, Stone SL, Goring DR. (2009) Cellular pathways regulating responses to compatible and self-incompatible pollen in Brassica and Arabidopsis stigmas intersect at Exo70A1, a putative component of the exocyst complex. *The Plant Cell* 21: 2655–2671.

Schiøtt M, Romanowsky SM, Bækgaard L, Jakobsen MK, Palmgren MG, Harper JF. (2004) A plant plasma membrane Ca^{2+} pump is required for normal pollen tube growth and fertilization. *Proc Natl Acad Sci USA* 101: 9502–9507.

Serrano I, Olmedilla A. (2012) Histochemical location of key enzyme activities involved in receptivity and self-incompatibility in the olive tree (*Olea europaea* L.). *Plant Sci* 197: 40–49.
Serrano I, Romero-Puertas MC, Sandalio LM, Olmedilla A. (2015) The role of reactive oxygen species and nitric oxide in programmed cell death associated with self-incompatibility. *J Exp Bot* 66: 2869–2876.
Silverstein KA, Moskal WA, Wu HC, *et al.* (2007) Small cysteine-rich peptides resembling antimicrobial peptides have been under-predicted in plants. *Plant J* 51: 262–280.
Singh R, Singh S, Parihar P, *et al.* (2016) Reactive oxygen species (ROS): beneficial companions of plants' developmental processes. *Front Plant Sci* 7: 1299.
Singh S, Srivastava PK, Kumar D, Tripathi DK, Chauhan DK, Prasad SM. (2015) Morpho-anatomical and biochemical adapting strategies of maize (*Zea mays* L.) seedlings against lead and chromium stresses. *Biocatal Agric Biotechnol* 4: 286–295.
Singh S, Tripathi DK, Singh S, *et al.* (2017) Toxicity of aluminium on various levels of plant cells and organism: a review. *Environ Exp Bot* 137: 177–193.
Snyder SH, Ferris CD. (2000) Novel neurotransmitters and their neuropsychiatric relevance. *Am J Psychiatry* 157: 1738–1751.
Takayama S, Isogai A. (2005) Self-incompatibility in plants. *Annu Rev Plant Biol* 56: 467–489.
Takayama S, Shiba H, Iwano M, *et al.* (2000) Isolation and characterization of pollen coat proteins of Brassica campestris that interact with S locus-related glycoprotein 1 involved in pollen-stigma adhesion. *Proc Natl Acad Sci USA* 97: 3765–3770.
Torres MA, Dangl JL, Jones JD. (2002) Arabidopsis gp91phox homologues AtrbohD and AtrbohF are required for accumulation of reactive oxygen intermediates in the plant defense response. *Proc Natl Acad Sci USA* 99: 517–522.
Tripathi DK, Singh S, Singh VP, Prasad SM, Chauhan DK, Dubey NK. (2016a) Silicon nanoparticles more efficiently alleviate arsenate toxicity than silicon in maize cultivar and hybrid differing in arsenate tolerance. *Front Environ Sci* 4: 46.
Tripathi DK, Singh VP, Prasad SM, Dubey NK, Chauhan DK, Rai AK. (2016b) LIB spectroscopic and biochemical analysis to characterize lead toxicity alleviative nature of silicon in wheat (Triticum aestivum L.) seedlings. *J Photochem Photobiol B: Biol* 154: 89–98.
Tripathi DK, Singh S, Singh VP, Prasad SM, Dubey NK, Chauhan DK. (2017a) Silicon nanoparticles more effectively alleviated UV-B stress than silicon in wheat (Triticum aestivum) seedlings. *Plant Physiol Biochem* 110: 70–81.
Tripathi DK, Singh S, Singh S, *et al.* (2017b) Nitric oxide alleviates silver nanoparticles (AgNps)-induced phytotoxicity in Pisum sativum seedlings. *Plant Physiol Biochem* 110: 167–177.
Walters C. (1998) Understanding the mechanisms and kinetics of seed aging. *Seed Sci Res* 8: 223–244.
Walters DR. (2003) Polyamines and plant disease. *Phytochemistry* 64: 97–107.
Wengier DL, Mazzella MA, Salem TM, McCormick S, Muschietti JP. (2010) STIL, a peculiar molecule from styles, specifically dephosphorylates the pollen receptor kinase LePRK2 and stimulates pollen tube growth in vitro. *BMC Plant Biol* 10: 1.
Wu J, Wang S, Gu Y, Zhang S, Publicover SJ, Franklin-Tong VE. (2011) Self-incompatibility in Papaver rhoeas activates nonspecific cation conductance permeable to Ca^{2+} and K^{+}. *Plant Physiol* 155: 963–973.
Zhang Y, Chen B, Xu Z, *et al.* (2014) Involvement of reactive oxygen species in endosperm cap weakening and embryo elongation growth during lettuce seed germination. *J Exp Bot* 65: 3189–3200.
Zinkl GM, Preuss D. (2000) Dissecting Arabidopsis pollen-stigma interactions reveals novel mechanisms that confer mating specificity. *Annals Bot* 85(Suppl. 1): 15–21.
Zinkl GM, Zwiebel BI, Grier DG, Preuss D. (1999) Pollen-stigma adhesion in Arabidopsis: a species-specific interaction mediated by lipophilic molecules in the pollen exine. *Development* 126: 5431–5440.

16

Reactive Oxygen Species Signaling and Root Hair Development

De-Jian Zhang[1], *Yu-Jie Yang*[1], *Chun-Yan Liu*[1], *and Qiang-Sheng Wu*[1,2]

[1] *Yangtze University, Jingzhou, Hubei, China*
[2] *University of Hradec Kralove, Hradec Kralove, Czech Republic*

Reactive Oxygen Species (ROS)

Reactive oxygen species (ROS) in plants are produced continuously as by-products of aerobic metabolism in mitochondria, chloroplasts, peroxysomes, and the apoplast (Miller *et al.*, 2009, Mittler *et al.*, 2011). In general, ROS are comprised of the superoxide anion ($O_2^{\cdot-}$), hydrogen peroxide (H_2O_2), hydroxyl radical (·OH), and so on (Miller *et al.*, 2009; Møller and Sweetlove, 2010; Tsukagoshi *et al.*, 2010; Yuan *et al.*, 2013). These ROS are highly reactive and toxic and often cause damage to proteins, lipids, carbohydrates, and DNA, resulting in nucleic acid damage, oxidizing of proteins, and lipid peroxidation (Costa *et al.*, 2010; Gill and Tuteja, 2010). Production and elimination of ROS are in a dynamic balance in cells under normal environments conditions (Gill and Tuteja, 2010). Furthermore, several abiotic stresses such as salinity, drought, temperature stress, nutrient deficiency, and pathogen attacks can strongly stimulate ROS accumulation in plant cells (Shoresh *et al.*, 2011; V.P. Singh *et al.*, 2011; Steffens, 2014; Tripathi *et al.*, 2012a,b, 2016a,b, 2017a–c; S. Singh *et al.*, 2015, 2017; R. Singh *et al.*, 2016).

Recently, widespread attention has been given to ROS as important signals in plants. In low concentrations ROS are seen as signaling molecules involved in controlling processes such as the cell cycle, programmed cell death, pathogen defense, hormone signaling, and biotic and abiotic stress responses (Apel and Hirt, 2004). However, plant cells contain large amounts of ROS during abiotic stresses, which cause damage to DNA, proteins, and other intracellular macromolecular components, thus impeding normal plant development and growth (Apel and Hirt, 2004). At the same time, plants also have two scavenging systems for ROS: enzymatic mechanisms (superoxide dismutase, catalase, peroxidase, etc.) and a non-enzymatic antioxidant system (ascorbic acid, glutathione, proline, etc.), which maintain ROS within a certain range to prevent the damage caused by high concentrations of ROS (Apel and Hirt, 2004; Gill and Tuteja, 2010). Therefore, like a double-edged sword, the dynamic balance of ROS production and elimination is critical to plant growth and stress responses.

ROS have also been proven to play a key role in regulating polar growth in roots and root hairs. Joo *et al.* (2001) revealed that ROS are asymmetrically generated in the superficial cortex of root tip during bending for gravitropism of maize root. However, the gravitropism of the root could disappear after treatment with ROS scavengers (Joo *et al.*, 2001). Recent studies

Reactive Oxygen Species in Plants: Boon Or Bane - Revisiting the Role of ROS, First Edition. Edited by Vijay Pratap Singh, Samiksha Singh, Durgesh Kumar Tripathi, Sheo Mohan Prasad, and Devendra Kumar Chauhan.

showed that ROS not only participates in root gravitropism, but also regulates root hair growth and development (Lohar *et al.*, 2007; Cárdenas *et al.*, 2008; Sundaravelpandian *et al.*, 2013a,b; Zhu *et al.*, 2016).

Root Hair

The root hair is a long tubular structure that emerges from a trichoblast (hair-forming epidermal cell). Their wide distribution among all major groups of vascular plants gives evidence for a long evolutionary history and their importance to the success of diverse plant groups in adapting to changing environmental conditions (Peterson and Farquhar, 1996). Originating from the root epidermis, root hairs greatly increase the root surface area, which is beneficial to water and nutrient uptake by roots, anchorage, and the interaction of plants and soil microorganisms (Peterson and Farquhar, 1996; Grierson and Schiefelbein, 2002; Tanaka *et al.*, 2014). Root hairs also provide access to immobile inorganic ions such as phosphate and ferrate, which are essential for plant development (Gahoonia *et al.*, 1997; Carol *et al.*, 2005). Furthermore, the root hair has significant implications for studying cell differentiation, cell development, and programmed cell death (Cao *et al.*, 2013; Li *et al.*, 2014).

The root hair has a development cycle lasting 2–20 days, and is typically 0.2–10 mm in length and 5–17 μm in diameter (Peterson andFarquhar, 1996; Grierson and Schiefelbein, 2002; Zhang *et al.*, 2013). The density of root hairs is dependent on plant species. For example, values for the root hair density of maize and citrus are 420 mm^{-2} and 81 mm^{-2}, respectively (Peterson and Farquhar, 1996; Grierson and Schiefelbein, 2002; Zhang *et al.*, 2013).

In general, development of a root hair is divided into four stages: cell fate specification, initiation, tip growth, and maturation (Lee and Cho, 2009; Zhang *et al.*, 2013). During tip growth, the cytoplasm and pectic substance of the hair are highly polarized, with secretory vesicles concentrated located behind the hair tip, followed by the organelles required for the production and secretion of new cell wall and plasma membrane materials (Galway *et al.*, 1997; Carol and Dolan, 2001, 2006). In the maturation stage, ribosomes, mitochondria, and endoplasmic reticulum concentrate at the root hair tip (Cormack, 1947; Nestler *et al.*, 2014). The cell wall includes two layers: an outer layer containing cellulose, pectic substances, and cutin; and an inner layer consisting of a dense microfibrous structure (Ma, 1999).

Recent studies have indicated that root hair development could be regulated by culture substrates (yellow soil, Hoagland solution, etc.), mineral nutrients (P, Ca, Mg, etc.), plant growth regulators (auxin, ethylene, jasmonic acid, strigolactone, etc.), soil edaphon (arbuscular mycorrhizal fungi), and so on (Lee and Cho, 2009; Peret *et al.*, 2011; Cao *et al.*, 2013; Zhang *et al.*, 2013; Wu *et al.*, 2016). However, we guess that such factors may work through interaction with ROS. Therefore, this review mainly elaborates the effect of ROS and their interactions with calcium (Ca), magnesium (Mg), and others on root hair development.

ROS and Root Hairs

Root wall loosening, which is essential for cell extension and root hair initiation, can be induced by hydroxyl radical (OH) generated from $O_2^{\cdot-}$ and H_2O_2 by a peroxidase-mediated reaction (Liszkay *et al.*, 2004; Shin *et al.*, 2005). It has been reported that the ROS level was increased in root hair tip growth, increasing the pH and changing the stiffness of cytoderm (Monshausen *et al.*, 2007). Visualizing ROS in roots stained by nitroblue tetrazolium (NBT), which reduces to a blue formazan deposit in the presence of superoxide, showed that ROS gathered in the dome of the trichoblast at the root hair initiation stage, then were tightly focused in the cell wall directly behind the growing root hair apex during the growth stage, only to disappear once tip growth ceased (Foreman *et al.*, 2003; Liszkay *et al.*, 2004; Carol *et al.*, 2005; Renew *et al.*, 2005; Carol and Dolan, 2006).

Screening and use of root hair mutants is the best way to study the regulatory mechanism of ROS on root hair development. For instance, the *Arabidopsis* mutant *root hair defective 2* (*rhd2*; AtrbohC) initiated root hairs normally, but tip growth was seriously hindered (Jones *et al.*, 2007; Monshausen *et al.*, 2007). The *rhd2* mutant has mutation of a gene that encodes a homolog of $gp91^{phox}$, the NADPH oxidase cytochrome, which is part of the respiratory burst oxidase that catalyzes the production of ROS as second messengers (Yoshioka *et al.*, 2001; Foreman *et al.*, 2003). ROS accumulation was observed in elongating root hair tips of wild-type plants, but not in those of the *rhd2* mutant (Yoshioka *et al.*, 2001; Foreman *et al.*, 2003; Bassil *et al.*, 2005). It was suggested that ROS production by *RHD2* is required for root-hair tip growth. To further prove this assumption, diphenylene iodonium chloride (DPI), which covertly binds in the reaction center of flavoproteins such as NADPH oxidases thus inhibiting their activity, was used (O'Donnell *et al.*, 1993; Carol and Dolan, 2006). The growth of wild-type root hairs is inhibited by DPI treatment, phenocopying the *rhd2* mutant hair (Foreman *et al.*, 2003). Equally, ROS accumulation in wild-type hairs is inhibited by DPI (Carol *et al.*, 2005). This evidence strongly links ROS production to root hair tip growth, and shows that an NADPH oxidase producing ROS is a fundamental requirement for tip growth. Hence, localized RHD2/AtrbohC NADPH oxidase produces ROS, which are necessary for root hair growth and development. Based on the important role of ROS in the root hair, Juárez *et al.* (2015) further exploited a protocol to detect ROS specifically in the root hair by oxidation of a derivative of fluorescein, 2′,7-dihydro-diclorofluorescein. In short, ROS production is crucial for tip growth of root hairs (Juárez *et al.*, 2015).

With regard to another gene, *SCN1* (*supercentipede1*), which is a RhoGTPase GDP dissociation inhibitor (RhoGDI), its mutant could spatially develop ectopic sites of growth on the outer face of the hair cell and multiple hairs from a single trichoblast in *Arabidopsis*, which is similar in part to the phenotype treated by exogenous ROS (·OH or H_2O_2) (Parker *et al.*, 2000; Carol *et al.*, 2005; Carol and Dolan, 2006). To test whether *SCN1* controls the site of outgrowth by spatially controlling ROS production, Carol *et al.* (2005) used NBT to reveal apoplastic ROS production in root hairs of *scn1* mutants. The results showed that a single trichoblast has multiple ROS accumulation sites, which resulted in multiple root hairs initiated in *scn1* mutants. To verify genetically that *SCN1* controls RHD2/AtrbohC activity during root hair initiation, the *scn1 rhd2* double mutant suppressed the root hair phenotype seen in the single *scn1* mutant (Carol *et al.*, 2005). When dyed with NBT, the development of ectopic sites of ROS accumulation in the *scn1* single mutant is absent from the *scn1 rhd2* double mutant (Carol *et al.*, 2005). It strongly suggests that *RHD2* activity is the source of ROS, which is delocalized in *scn1* mutants. Thus, Carol *et al.* (2005) thought that the ectopic sites of ROS accumulation were produced by the action of RHD2/AtrbohC, and RHD2 activity was required to develop the multiple root hairs from a single trichoblast in the *scn1* mutant.

In maize (*Zea mays* L.), the gene of *RTH5* (*roothairless5*) encoded a specific NADPH oxidase, and displayed a preferential expression in root hair (Nestler *et al.*, 2014). As compared to wild-type, *rth5* mutant had depressed levels of ROS in the tip of root hair and defects in root hair density and length (Nestler *et al.*, 2014). It suggests that *RTH5* is responsible for establishing the appropriate levels of ROS in root hair tip required for root hair initiation and elongation.

Transcriptional mediators, central co-regulators of transcription, have been identified as large protein complexes in eukaryotes and play important roles in several developmental processes in plants (Sundaravelpandian *et al.*, 2013a). The *PFT1* gene, a transcriptional mediator complex subunit, positively regulates cell wall remodeling and controls organ growth by restricting cell proliferation (Xu and Li, 2011; Sundaravelpandian *et al.*, 2013a). It is also essential for multiple stages of root hair development and interacts with ROS in that process (Xu and Li, 2011; Sundaravelpandian *et al.*, 2013a,b). In *Arabidopsis*, the *pft1* mutant has smaller and thinner root hairs than wild-type (Sundaravelpandian *et al.*, 2013a,b). Further studies indicated that it perturbed H_2O_2 distribution, suggesting that *PFT1* is critical to maintaining redox homeostasis (Sundaravelpandian *et al.*, 2013a,b). Distribution of ROS is essential to root hair initiation and elongation. Transcriptional profiling of *pft1* roots revealed that *PFT1* activated a subset of H_2O_2-producing class III peroxidases (Sundaravelpandian *et al.*, 2013a,b). Thus, *PFT1* maintains the ROS distribution that controls root hair development and growth. In short, ROS are important in root hair development and growth in plants.

ROS and Ca^{2+} in Root Hair Growth

Calcium ion (Ca^{2+}) is involved in most biological processes in the plant kingdom (Weinl and Kudla, 2009). It serves as a versatile second messenger and regulates a multitude of adaptation, developmental, and physiological processes in plants, including cell division, cell expansion, cytoplasmic streaming, responses to abiotic stress, adjustment of ion homeostasis, and pathogen defense (Bush, 1995; Sanders *et al.*, 2002; White and Broadley, 2003; Kim *et al.*, 2009). Cytoplasmic Ca^{2+} is known to regulate cytoskeletal rearrangements and vesicular trafficking in tip-growing systems such as root hair development (Hepler *et al.*, 2001).

With regard to the effect of Ca^{2+} on root hair growth, studies have shown that a growing root hair has a localized gradient of Ca^{2+} concentration toward the growing apex, and the intensity of this gradient correlates with the growth rate of the root hair (Schiefelbein *et al.*, 1992; Pierson *et al.*, 1996; Felle & Hepler, 1997; Wymer *et al.*, 1997; Takeda *et al.*, 2008; Li *et al.*, 2012). Further evidence for the role of Ca^{2+}, using the vibrating probe, suggested that Ca^{2+} influx is higher at the tip than at the base or sides of a root hair (Schiefelbein *et al.*1992; Herrmann and Felle, 1995; Jones *et al.* 1995). Furthermore, Ca^{2+} ionophores and channel blockers can inhibit root hair tip growth because they disrupt the Ca^{2+} concentration gradient (Miller *et al.*, 1992; Pierson *et al.*, 1994; Herrmann and Felle, 1995; Wymer *et al.*, 1997). Bibikova *et al.* (1997) revealed that a tip-focused Ca^{2+} gradient was centered at the site of active growth, based on confocal ratio imaging. If the root hair direction was changed, the Ca^{2+} gradient was also reoriented. Subsequently, if the root hair growth was returned to the original direction, the Ca^{2+} concentration gradient also did so. As a consequence, Ca^{2+} concentration gradient is a requirement in sustaining root hair tip growth.

With regard to the effect of ROS and Ca^{2+} on root hair development, Foreman *et al.* (2003) and Takeda *et al.* (2008) showed that ROS produced by NADPH oxidase activated Ca^{2+} channels in the apical plasma membrane, leading to the tip-focused Ca^{2+} concentration gradient required for root hair polar growth. Additional evidence from *Arabidopsis rhd2* mutant indicated that application of exogenous ROS to *in vivo* root apices stimulated cell growth and Ca^{2+} influxes in root hair, which induced root hair elongation (Foreman *et al.*, 2003). The effect of exogenous ROS could be blocked by pharmacological inhibitors of calcium channels (Foreman *et al.*, 2003). Artificially increasing cytosolic Ca^{2+} by treating with calcium ionophore A23187 (mixed calcium-magnesium salt) induced production of elevated levels of ROS around the root hair tip and inhibited its tip growth; these large amounts of ROS disrupted the Ca^{2+} concentration gradient.

To summarize, ROS can activate calcium channel-mediated influx of Ca^{2+}, which creates a Ca^{2+} concentration gradient and subsequently influences root hair elongation.

Influence of ROS and Mg Availability on Root Hair Development

Magnesium (Mg) is an essential mineral nutrient for plant metabolic processes and reactions, such as photophosphorylation, photo-oxidation in leaf tissues, chlorophyll formation, protein synthesis, photosynthetic CO_2 fixation, phloem loading, generation of ROS, and so forth (Cakmak and Yazici, 2010). Mg is also involved in many critical biochemical and physiological processes in plants, thus, influencing plant growth and development (Cakmak and Yazici, 2010; Gransee and Führs, 2013).

Recently, the effect of interactions between Mg, ROS, and Ca^{2+} on root hairs has been reported. Low Mg availability resulted in longer and denser root hairs in *Arabidopsis* with greater concentrations of ROS and Ca^{2+} in the root tip and a stronger Ca^{2+} concentration gradient in the root hair tip (Niu *et al.*, 2014). However, when treated with diphenylene iodonium (DPI, an NADPH-oxidase inhibitor) or 1,2-bis(*o*-aminophenoxy)ethane-N,N,N′,N′,-tetraacetic acid (BAPTA, a Ca^{2+} chelator), the Ca^{2+} concentration gradient was eliminated and the growth of root hair was blocked in low-Mg treatment (Frahry and Schopfer, 1998; Yoshioka *et al.*, 2001; Kadota *et al.*, 2004). Instead, root hair growth was blocked in high-Mg treatment and the opposite was true when supplied with $CaCl_2$ or phenazine methosulfate (PMS, a ROS generator) in *Arabidopsis* under high-Mg treatment (Zhang *et al.*, 2009; Niu *et al.*, 2014). The study outlined by Niu *et al.* (2014) showed that NADPH oxidase in root was positive regulated by low Mg and was inhibited by high Mg levels, which is further evidence that Mg could control ROS to regulate root hair development. Therefore, the development of root hairs is closely related to Mg availability and to ROS and Ca^{2+} signaling.

Effect of ROS and Other Substances on Root Hairs

Phosphorus (P) is an important macronutrient in plants, not only as a constituent of key cellular molecules such as ATP, phospholipids, and nucleic acids, but also by playing a pivotal role in energy conservation and metabolic regulation (Marschner, 1995; Raghothama and Karthikeyan, 2005; Shin *et al.*, 2005). Phosphorus deficiency in the soil often induces bushy root hairs in plants such as *Arabidopsis* and citrus (Zhu *et al.*, 2005; Cao *et al.*, 2013). The response to P deprivation in roots was accompanied by an increase in ROS (Shin *et al.*, 2005). There is interaction between ROS and P in root hair growth and development.

Rho of plants (ROP), known as ras-related C3 botulinum toxin substrate (RAC) proteins, are Rho-related GTPases that participate in an array of physiological processes, including cytoskeleton organization and dynamics, vesicle trafficking, and hormone signaling (Zheng and Yang, 2000; Jones *et al.*, 2007; Chen *et al.*, 2012; Poraty-Gavra *et al.*, 2013). The *ROP* family genes play a vital role in ROS production in root hair cells (Molendijk *et al.*, 2001; Jones *et al.*, 2007; Wong *et al.*, 2007). In *Arabidopsis*, plants of both overexpressing wild-type *ROP2* and constitutively active *rop2* mutant could increase ROS production in root hair and thereby stimulated its growth (Jones *et al.*, 2002, 2007). However, dominant *rop2* mutants presented the opposite phenotype (Jones *et al.*, 2002, 2007). If active *ROP2* gene was expressed in the *rhd2* mutant (NADPH oxidase loss-of-function), ROS formation and root hair growth were blocked partially, suggesting that *RHD2* is required for *ROP2*-dependent ROS formation in root hair development (Jones *et al.*, 2002, 2007). With regard to other ROPs, *ROP11* disrupted the organization of the actin cytoskeleton and membrane cycling, resulting in root hair swelling and modulated pH fluctuations to regulate root hair tip growth (Bloch *et al.* 2005, 2011). *ROP11* interacted with the NADPH oxidase respiratory burst oxidase homolog F in the N-terminal cytoplasmic region to regulate ROS production in root hair (Yan *et al.*, 2016). This suggests that ROPs modulate ROS production, thus, regulating root hair development.

In addition, tip growth of root hairs could be regulated by oscillations in extracellular pH and ROS (Monshausen *et al.*, 2007). In *Arabidopsis rhd2* mutants, which lack a functional version of the NADPH oxidase ATRBOH C, the root hair defective phenotype could be compensated by increased pH in the growth medium, in company with lower ROS levels and a lack of the oscillatory production of ROS at the root hair tip (Monshausen *et al.*, 2007). Thus, it may indicate that the NADPH oxidase AtRBOH C is necessary for oscillatory ROS production, and that pH and ROS interact to influence expansion of the growing root hair.

Adenosine triphosphate (ATP), an important modulator of NADPH oxidases in plant cells, has been found to modulate both the intracellular Ca^{2+} level and ROS levels in root hairs (Jeter *et al.*, 2004; Kim *et al.*, 2006; Song *et al.*, 2006; Cárdenas *et al.*, 2008). It has been proved that cellular ATP induced changes in ROS concentration in root hair of *Arabidopsis, Phaseolus vulgaris* and *Medicago* (Kim *et al.*, 2006; Song *et al.*, 2006; Cárdenas *et al.*, 2008; Roux and Steinebrunner, 2007).

Future Perspectives

Traditionally ROS production by plants has been associated with responses to physiological stresses and defense mechanisms against pathogen attack. In addition, ROS play an essential role in root hair growth. Root hair morphogenesis is driven by an amalgam of interacting processes controlled by complex signaling events. Some of the major factors involved in the signaling process during root hair development have been identified, such as NADPH oxidase/RHD2, ROS, Ca^{2+}, Mg^{2+}, P, ROPs, and so forth. It is not clear how these signaling components regulate root hairs at the molecular and cellular level, or how these signals interact with auxin, ethylene, and abscisic acid to regulate root hair development. On the other hand, the evidence of the role of a Ca^{2+} gradient in root hair tip-growing system is scarce.

It has been noted that *SCN1* controls the site of root hair outgrowth by spatially controlling ROS production. Nevertheless, we cannot completely rule out the possibility that *SCN1* spatially regulates the production of ROS by other mechanisms, for example, microfilament organization, endocytosis, or ROPs.

More work is needed to clone the genes of additional root hair mutants and elucidate their roles, as well as undertaking reverse genetics and mutant complementation studies to add to our current knowledge of the signaling networks involving ROS-regulated root hair initiation and tip growth.

Acknowledgments

This study was supported by the Plan in Scientific and Technological Innovation Team of Outstanding Young, Hubei Provincial Department of Education (T201604), and the Open Fund of Institute of Root Biology, Yangtze University (R201401).

References

Apel K, Hirt H. (2004) Reactive oxygen species: metabolism, oxidative stress, and signal transduction. *Ann Rev Plant Biol* 55: 374–389.

Bassil E, Nijo G, Baluska F, Brown PH. (2005) Boron deficiency rescues the *Arabidopsis thaliana rhd2* root hair phenotype: are reactive oxygen species involved? *Eur J Cell Biol* 84(suppl.): 63–63.
Bibikova TN, Zhigilei A, Gilroy S. (1997) Root hair growth in *Arabidopsis thaliana* is directed by calcium and an endogenous polarity. *Planta* 203: 495–505.
Bloch D, Lavy M, Efrat Y, *et al.* (2005) Ectopic expression of an activated RAC in *Arabidopsis* disrupts membrane cycling. *Mol Biol Cell* 16: 1913–1927.
Bloch D, Monshausen G, Singer M, Gilroy S, Yalovsky S. (2011) Nitrogen source interacts with ROP signalling in root hair tip-growth. *Plant Cell Environ* 34: 76–88.
Bush DS. (1995) Calcium regulation in plant cells and its role in signaling. *Annu Rev Plant Biol* 46: 95–122.
Cakmak I, Yazici AM. (2010) Magnesium: a forgotten element in crop production. *Better Crops* 94: 23–25.
Cao X, Chen CL, Zhang DJ, Shu B, Xiao J, Xia RX. (2013) Influence of nutrient deficiency on root architecture and root hair morphology of trifoliata orange (*Poncirus trifoliata* L. Raf.) seedlings under sand culture. *Sci Hortic* 162: 100–105.
Cárdenas L, Martínez A, Sánchez F, Quinto C. (2008) Fast, transient and specific intracellular ROS changes in living root hair cells responding to Nod factors (NFs). *Plant J* 56: 802–813.
Carol R J, Takeda S, Linstead P, *et al.* (2005) A RhoGDP dissociation inhibitor spatially regulates growth in root hair cells. *Nature* 438: 1013–1016.
Carol RJ, Dolan L. (2001) Building a hair: tip growth in *Arabidopsis thaliana* root hairs. *Phil Trans R Soc B* 357: 815–821.
Carol RJ, Dolan L. (2006) The role of reactive oxygen species in cell growth: lessons from root hairs. *J Exp Bot* 57: 1829–1834.
Chen X, Naramoto S, Robert S, *et al.* (2012) *ABP1* and *ROP6* GTPase signaling regulate clathrin-mediated endocytosis in *Arabidopsis* roots. *Curr Biol* 22: 1326–1332.
Cormack RGH. (1947) A comparative study of developing epidermal cells in white mustard and tomato roots. *Am J Bot* 34: 310–314.
Costa A, Drago I, Behera S, *et al.* (2010) H_2O_2 in plant peroxisomes: an *in vivo* analysis uncovers a Ca^{2+}-dependent scavenging system. *Plant J* 62: 760–772.
Felle HH, Hepler PK. (1997) The cytosolic Ca^{2+}-concentration gradient of *Sinapis alba* root hairs as revealed by Ca^{2+}-selective microelectrode tests and fura-dextran ratio imaging. *Plant Physiol* 114: 39–45.
Foreman J, Demidchik V, Bothwell JH, *et al.* (2003) Reactive oxygen species produced by NADPH oxidase regulate plant cell growth. *Nature* 422: 442–446.
Frahry G, Schopfer P. (1998) Inhibition of O_2-reducing activity of horseradish peroxidase by diphenyleneiodonium. *Phytochemistry* 48: 223–227.
Gahoonia TS, Care D, Nielsen NE. (1997) Root hairs and phosphorus acquisition of wheat and barley cultivars. *Plant Soil* 191: 181–188.
Galway ME, Heckman JW, Schiefelbein JW. (1997) Growth and ultrastructure of *Arabidopsis* root hairs: the *rhd3* mutation alters vacuole enlargement and tip growth. *Planta* 201: 209–218.
Gill SS, Tuteja N. (2010) Reactive oxygen species and antioxidant machinery in abiotic stress tolerance in crop plants. *Plant Physiol Biochem* 48: 909–930.
Gransee A, Führs H. (2013) Magnesium mobility in soils as a challenge for soil and plant analysis, magnesium fertilization and root uptake under adverse growth conditions. *Plant Soil* 368: 5–21.
Grierson C, Schiefelbein J. (2002) Root hairs. *The Arabidopsis Book* 1: e0060.
Hepler PK, Vidali L, Cheung AY. (2001) Polarized cell growth in higher plants. *Annu Rev Cell Dev Biol* 17: 159–187.
Herrmann A, Felle HH. (1995) Tip growth in root hair cells of *Sinapis alba* L.: signifcance of internal and external Ca^{2+} and pH. *New Phytol* 129: 523–533.

Jeter CR, Tang W, Henaff E, Butterfield T, Roux SJ. (2004) Evidence of a novel cell signaling role for extracellular adenosine triphosphates and diphosphates in *Arabidopsis*. *Plant Cell* 16: 2652–2664.

Jones DL, Sha JE, Kochian LV. (1995) Role of calcium and other ions in directing root hair tip growth in *Limnobium stoloniferum*. Inhibition of tip growth by aluminum. *Planta* 197: 672–680.

Jones MA, Shen JJ, Fu Y, Li H, Yang Z, Grierson CS. (2002) The *Arabidopsis ROP2* GTPase is a positive regulator of both root hair initiation and tip growth. *Plant Cell* 14: 763–776.

Jones MA, Raymond MJ, Yang ZB, Smirnoff N. (2007) NADPH oxidase-dependent reactive oxygen species formation required for root hair growth depends on ROP GTPase. *J Exp Bot* 58: 1261–1270.

Joo JH, Bae YS, Lee JS. (2001) Role of auxin-induced reactive oxygen species in root gravitropism. *Plant Physiol* 126: 1055–1060.

Juárez SPD, Mangano S, Estevez JM. (2015) Improved ROS measurement in root hair cells. In: Estevez JM (ed.), *Plant Cell Expansion: Methods and Protocols*. Springer Science + Business Media, New York, pp. 67–71.

Kadota Y, Goh T, Tomatsu H, *et al.* (2004) Cryptogein-induced initial events in tobacco BY-2 cells: pharmacological characterization of molecular relationship among cytosolic Ca^{2+} transients, anion efflux and production of reactive oxygen species. *Plant Cell Physiol* 45: 160–170.

Kim MC, Chung WS, Yun D, Cho MJ. (2009) Calcium and calmodulin-mediated regulation of gene expression in plants. *Mol Plant* 2: 13–21.

Kim SY, Sivaguru M, Stacey G. (2006) Extracellular ATP in plants. Visualization, localization, and analysis of physiological significance in growth and signaling. *Plant Physiol* 142: 984–992.

Lee SH, Cho HT. (2009) Auxin and root hair morphogenesis. *Plant Cell Monographs* 12: 45–64.

Li S, Yu JL, Zhu M, Zhao F, Luan S. (2012) Cadmium impairs ion homeostasis by altering K^+, and Ca^{2+}, channel activities in rice root hair cells. *Plant Cell Environ* 35: 1998–2013.

Li T, Lin G, Zhang X, Chen Y, Zhang S, Chen B. (2014) Relative importance of an arbuscular mycorrhizal fungus (*Rhizophagus intraradices*) and root hairs in plant drought tolerance. *Mycorrhiza* 24: 595–602.

Liszkay A, Zalme EVD, Schopfer P. (2004) Production of reactive oxygen intermediates ($O_2^{\cdot-}$, H_2O_2, and $^{\cdot}OH$) by maize roots and their role in wall loosening and elongation growth. *Plant Physiol* 136: 3114–3123.

Lohar DP, Haridas S, Gantt JS, Vandenbosch KA. (2007) A transient decrease in reactive oxygen species in roots leads to root hair deformation in the legume–rhizobia symbiosis. *New Phytol* 173: 39–49.

Ma YX. (1999) *The Root of Wheat*. China Agriculture Press, Beijing.

Marschner A. (1995) *Mineral Nutrition of Higher Plants*. Academic Press, San Diego, CA.

Miller DD, Callaham DA, Gross DJ, Hepler PK. (1992) Free Ca^{2+} gradient in growing pollen tubes of *Lilium*. *J Cell Sci* 101: 7–12.

Miller G, Suzuki N, Ciftci-Yilmaz S, Mittler R. (2009) Reactive oxygen species homeostasis and signalling during drought and salinity stresses. *Plant Cell Environ* 33: 453–467.

Mittler R, Vanderauwera S, Suzuki N, *et al.* (2011) ROS signaling: the new wave? *Trends Plant Sci* 16: 300–309.

Molendijk AJ, Bischoff F, Rajendrakumar CSV, *et al.* (2001) *Arabidopsis thaliana* Rop GTPases are localized to tips of root hairs and control polar growth. *EMBO J* 20: 2779–2788.

Møller IM, Sweetlove LJ. (2010) ROS signalling – specificity is required. *Trends Plant Sci* 15: 370–374.

Monshausen GB, Bibikova TN, Messerli MA, Shi C, Gilroy S. (2007) Oscillations in extracellular pH and reactive oxygen species modulate tip growth of *Arabidopsis* root hairs. *Proc Natl Acad Sci USA* 104: 20996–21001.

Nestler J, Liu S, Wen TJ, *et al.* (2014) *Roothairless5*, which functions in maize (*Zea mays* L.) root hair initiation and elongation, encodes a monocot-specific NADPH oxidase. *Plant J* 79: 729–740.

Niu Y, Jin G, Yong SZ. (2014) Root development under control of magnesium availability. *Plant Signal Behav* 9: e29720.

O'Donnell BV, Tew DG, Jones OT, England PJ. (1993) Studies on the inhibitory mechanism of iodonium compounds with special reference to neutrophil NADPH oxidase. *Biochem J* 290: 41–49.

Parker JS, Cavell AC, Dolan L, Roberts K, Grierson CS. (2000) Genetic interactions during root hair morphogenesis in *Arabidopsis*. *Plant Cell* 12: 1961–1974.

Peret B, Clement M, Nussaume L, Desnos T. (2011) Root developmental adaptation to phosphate starvation: better safe than sorry. *Trends Plant Sci* 16: 442–450.

Peterson RL, Farquhar ML. (1996) Root hairs: specialized tubular cells extending root surfaces. *Bot Rev* 62: 1–40.

Pierson ES, Miller DD, Callaham DA, *et al.* (1994) Pollen tube growth is coupled to the extracellular calcium ion flux and the intracellular calcium gradient: effect of BAPTA-type buffers and hypertonic media. *Plant Cell* 6: 1815–1828.

Pierson ES, Miller DD, Callaham DA, van Aken J, Hackett G, Hepler PK. (1996) Tip-localized calcium entry fluctuates during pollen tube growth. *Dev Biol* 174: 160–173.

Poraty-Gavra L, Zimmermann P, Haigis S, *et al.* (2013) The *Arabidopsis* Rho of plants GTPase *AtROP6* functions in developmental and pathogen response pathways. *Plant Physiol* 161: 1172–1188.

Raghothama KG, Karthikeyan AS. (2005) Phosphate acquisition. *Plant Soil* 274: 37–49.

Renew S, Heyno E, Schopfer P, Liszkay A. (2005) Sensitive detection and localization of hydroxyl radical production in cucumber roots and *Arabidopsis* seedlings by spin trapping electron paramagnetic resonance spectroscopy. *Plant J* 44: 342–347.

Roux SJ, Steinebrunner I. (2007) Extracellular ATP: an unexpected role as a signaler in plants. *Trends Plant Sci* 12: 522–527.

Sanders D, Pelloux J, Brownlee C, Harper JF. (2002) Calcium at the crossroads of signaling. *Plant Cell* 14: S401–417.

Schiefelbein JW, Shipley A, Rowse P. (1992) Calcium influx at the tip of growing root-hair cells of *Arabidopsis thaliana*. *Planta* 187: 455–459.

Shin R, Berg RH, Schachtman DP. (2005) Reactive oxygen species and root hairs in *Arabidopsis* root response to nitrogen, phosphorus and potassium deficiency. *Plant Cell Physiol* 46: 1350–1357.

Shoresh M, Spivak M, Bernstein N. (2011) Involvement of calcium-mediated effects on ROS metabolism in the regulation of growth improvement under salinity. *Free Radical Bio Med* 51: 1221–1234.

Singh R, Singh S, Parihar P, *et al.* (2016) Reactive oxygen species (ROS): beneficial companions of plants' developmental processes. *Front Plant Sci* 7: 1299.

Singh S, Srivastava PK, Kumar D, Tripathi DK, Chauhan DK, Prasad SM. (2015) Morpho-anatomical and biochemical adapting strategies of maize (*Zea mays* L.) seedlings against lead and chromium stresses. *Biocatal Agric Biotechnol* 4: 286–295.

Singh S, Tripathi DK, Singh S, *et al.* (2017) Toxicity of aluminium on various levels of plant cells and organism: a review. *Environ Exp Bot* 137: 177–193.

Singh VP, Tripathi DK, Kumar D, Chauhan DK. (2011) Influence of exogenous silicon addition on aluminium tolerance in rice seedlings. *Biol Trace Elem Res* 144: 1260–1274.

Song CJ, Steinebrunner I, Wang X, Stout SC, Roux SJ. (2006) Extracellular ATP induces the accumulation of superoxide via NADPH oxidases in *Arabidopsis*. *Plant Physiol* 140: 1222–1232.

Steffens B. (2014) The role of ethylene and ROS in salinity, heavy metal, and flooding responses in rice. *Front Plant Sci* 5: 685.

Sundaravelpandian K, Chandrika NNP, Schmidt W. (2013a) *PFT1*, a transcriptional mediator complex subunit, controls root hair differentiation through reactive oxygen species (ROS) distribution in *Arabidopsis*. *New Phytol* 197: 151–161.

Sundaravelpandian K, Chandrika NNP, Tsai YH, Schmidt W. (2013b) *PFT1*-controlled ROS balance is critical for multiple stages of root hair development in *Arabidopsis*. *Plant Signal Behav* 8: e24066.

Takeda S, Gapper C, Kaya H, Bell E, Kuchitsu K, Dolan L. (2008) Local positive feedback regulation determines cell shape in root hair cells. *Science* 319: 1241–1244.

Tanaka N, Kato M, Tomioka R, *et al.* (2014) Characteristics of a root hair-less line of *Arabidopsis thaliana* under physiological stresses. *J Exp Bot* 65: 1497–1512.

Tripathi DK, Singh VP, Kumar D, Chauhan DK. (2012a) Impact of exogenous silicon addition on chromium uptake, growth, mineral elements, oxidative stress, antioxidant capacity, and leaf and root structures in rice seedlings exposed to hexavalent chromium. *Acta Physiol Plant* 34: 279–289.

Tripathi DK, Singh VP, Kumar D, Chauhan DK. (2012b) Rice seedlings under cadmium stress: effect of silicon on growth, cadmium uptake, oxidative stress, antioxidant capacity and root and leaf structures. *Chem Ecol* 28: 281–291.

Tripathi DK, Singh S, Singh VP, Prasad SM, Chauhan DK, Dubey NK. (2016a) Silicon nanoparticles more efficiently alleviate arsenate toxicity than silicon in maize cultivar and hybrid differing in arsenate tolerance. *Front Environ Sci* 4: 46.

Tripathi DK, Singh VP, Prasad SM, Dubey NK, Chauhan DK, Rai AK. (2016b) LIB spectroscopic and biochemical analysis to characterize lead toxicity alleviative nature of silicon in wheat (Triticum aestivum L.) seedlings. *J Photochem Photobiol B: Biol* 154: 89–98.

Tripathi DK, Singh S, Singh VP, Prasad SM, Dubey NK, Chauhan DK. (2017a) Silicon nanoparticles more effectively alleviated UV-B stress than silicon in wheat (Triticum aestivum) seedlings. *Plant Physiol Biochem* 110: 70–81.

Tripathi DK, Singh S, Singh S, *et al.* (2017b) Nitric oxide alleviates silver nanoparticles (AgNps)-induced phytotoxicity in Pisum sativum seedlings. *Plant Physiol Biochem* 110: 167–177.

Tripathi DK, Mishra RK, Singh S, *et al.* (2017c) Nitric oxide ameliorates zinc oxide nanoparticles phytotoxicity in wheat seedlings: implication of the ascorbate–glutathione cycle. *Front Plant Sci* 8: 1.

Tsukagoshi H, Busch W, Benfey PN. (2010) Transcriptional regulation of ros controls transition from proliferation to differentiation in the root. *Cell* 143: 606–616.

Weinl S, Kudla J. (2009) The CBL-CIPK Ca^{2+}-decoding signaling network: function and perspectives. *New Phytol* 184: 517–528.

White PJ, Broadley MR. (2003) Calcium in plants. *Ann Bot* 92: 487–511.

Wong HL, Pinontoan R, Hayashi K, *et al.* (2007) Regulation of rice NADPH oxidase by binding of Rac GTPase to its N-terminal extension. *Plant Cell* 19: 4022–4034.

Wu QS, Liu CY, Zhang DJ, Zou YN, He XH, Wu QH. (2016) Mycorrhiza alters the profile of root hairs in trifoliate orange. *Mycorrhiza* 26: 237–247.

Wymer CL, Bibikova TN, Gilroy S. (1997) Cytoplasmic free calcium distribution during the development of root hairs of *Arabidopsis thaliana*. *Plant J* 12: 427–439.

Xu R, Li Y. (2011) Control of final organ size by Mediator complex subunit 25 in *Arabidopsis thaliana*. *Development* 138: 4545–4554.

Yan M, Jing W, Xu N, Shen L, Zhang Q, Zhang WH. (2016) *Arabidopsis thaliana* constitutively active *ROP11* interacts with the NADPH oxidase respiratory burst oxidase homologue F to regulate reactive oxygen species production in root hairs. *Funct Plant Biol* 43: 221–231.

Yoshioka H, Sugie K, Park HJ, *et al.* (2001) Induction of plant gp91 *phox* homolog by fungal cell wall, arachidonic acid, and salicylic acid in potato. *Mol Plant Microbe Interact* 14: 725–736.
Yuan HM, Liu WC, Jin Y, Lu YT. (2013) Role of ros and auxin in plant response to metal-mediated stress. *Plant Signal Behav* 8: e24671.
Zhang DJ. (2015) The effects and mechanisms of auxin and ethylene on root hair development and growth in *Poncirus trifoliata* L. PhD thesis, Huazhong Agricultural University, Wuhan, China.
Zhang DJ, Xia RX, Cao X, Shu B, Chen CL. (2013) Root hair development of *Poncirus trifoliata*, grown in different growth cultures and treated with 3-indolebutyric acid and ethephon. *Sci Hortic* 160: 389–397.
Zhang YY, Zhu HY, Zhang Q, *et al.* (2009) Phospholipase Dα1 and phosphatidic acid regulate NADPH oxidase activity and production of reactive oxygen species in ABA-mediated stomatal closure in *Arabidopsis*. *Plant Cell* 21: 2357–2377.
Zheng ZL, Yang Z. (2000) The Rop GTPase: an emerging signaling switch in plants. *Plant Mol Biol* 44: 1–9.
Zhu C, Na Y, Guo Z, Meng Q, Gan L. (2016) An ethylene and ROS-dependent pathway is involved in low ammonium-induced root hair elongation in *Arabidopsis* seedlings. *Plant Physiol Bioch* 105: 37–44.
Zhu J, Kaeppler SM, Lynch JP. (2005) Mapping of QTL controlling root hair length in maize (*Zea mays* L.) under phosphorus deficiency. *Plant Soil* 270: 299–310.

17

Role of Reactive Oxygen Species Signaling in Cell Proliferation and Differentiation: An Overview

Abinaya Manivannan, Prabhakaran Soundararajan, and Byoung Ryong Jeong

Gyeongsang National University, Jinju, South Korea

Introduction

In aerobic organisms like plants, the metabolism involves a cascade of redox reactions that results in the generation of reactive oxygen species (ROS) via electron-transfer reactions with oxygen (Dickinson and Chang, 2011). Production of ROS is continuous, and elevated levels cause oxidative damage, therefore detoxification mechanisms, such as the antioxidant defense system, have evolved in plants (Raymond and Segrè, 2006; S. Singh *et al.*, 2015, 2017; R. Singh *et al.*, 2016; Tripathi *et al.*, 2016a,b, 2017a–c). Due to the activity of these antioxidant mechanisms the abundance of ROS in cells can be controlled. Several studies have revealed the necessity of ROS for the proper functioning of various physiological and developmental processes (Hancock *et al.*, 2001; S. Singh *et al.*, 2015, 2017; R. Singh *et al.*, 2016; Tripathi *et al.*, 2016a,b, 2017a–c). The reports evidence the existence of different mechanisms for the systematic regulation of ROS to employ them as signaling molecules(Hancock *et al.*, 2001). Moreover, various proteins are sensitive to the redox status of the cell and form the basis of ROS-sensing networks (Wang *et al.*, 2011; Cremers and Jakob., 2013). Initially, because of their unstable and highly reactive nature, ROS were considered only as unwanted intermediates of metabolic processes (D'Autréaux and Toledano, 2007). However, more recently the role of ROS as signaling molecules has been confirmed.

In plants, growth occurs in a continuous manner throughout the life cycle, as a post-embryonic process characterized by the active division and expansion of cells (Novikova *et al.*, 2013). Therefore, stringent networks of different molecular mechanisms are involved in the maintenance of proliferation and differentiation of cells. In order to maintain normal growth, the rate of proliferation by mitotic cell division and programmed cell death has to be balanced (Novikova *et al.*, 2013). Hence, perturbation in cellular proliferation results in deformities and improper cellular development (Novikova *et al.*, 2013). Since cellular proliferation and differentiation are controlled by various internal and external cues, a better understanding of the process is necessary. Generally, progression of the cell cycle has been attributed to complex regulatory mechanisms such as transcription (Berckmans and De Veylder, 2009), protein–protein interactions, posttranslational modification of proteins (Francis, 2007), and degradation of proteins (Jurado *et al.*, 2008). Cell division in plants is similar to that in other eukaryotic organisms and involves the replication and segregation of DNA (Francis, 2007). The cell cycle comprises a segregation (S) and mitotic phase (M) with two gaps (G1 and G2). The time interval

Reactive Oxygen Species in Plants: Boon Or Bane - Revisiting the Role of ROS, First Edition. Edited by Vijay Pratap Singh, Samiksha Singh, Durgesh Kumar Tripathi, Sheo Mohan Prasad, and Devendra Kumar Chauhan.

between the M and S phases is called G1, whereas the interval between the S and M phases is termed G2 (Francis, 2007). For the proper transmission of genetic material to the progeny cells, the transitions between G1-S and G2-M have to be maintained (Novikova *et al.*, 2013). The maintenance of cell cycle phases is attributed to several molecular regulatory networks or proteins such as cyclin dependent protein kinases (CDKs) (Takatsuka *et al.*, 2009), phytohormones (Novikova *et al.*, 2013), and the redox status of the cellular components (Dewitte and Murray, 2003; Gechev *et al.*, 2006). Hence the present chapter will deal with the basics of ROS and their impact on the proliferation and differentiation of plant cells.

Basics of Reactive Oxygen Species

Reactive oxygen species (ROS) can be defined as follows: "A species of oxygen which are [*sic*] in a more reactive state than the molecular oxygen and in which oxygen is reduced to varying degrees" (Hancock *et al.*, 2001). Molecular oxygen (O_2) acts as the vital terminal electron acceptor in aerobic organisms like plants. The reduction of O_2 results in the formation of highly harmful reactive intermediates such as superoxide radical ($O_2^{\cdot-}$) (Hancock *et al.*, 2001). The first step of O_2 reduction results in the formation of $O_2^{\cdot-}$, which is subsequently reduced to H_2O_2 by the action of superoxide dismutase (SOD) (Gechev *et al.*, 2006). Due to its longer half-life (1 ms), the H_2O_2 molecule has the ability to drift from the site of synthesis to adjacent cellular compartments and even to neighboring cells (Henzler and Steudle, 2000; Bienert *et al.*, 2006). The oxidizing power of $O_2^{\cdot-}$ and H_2O_2 is potentially toxic to cells because these radicals can inactivate the function of several key enzymes involved in metabolic reactions (Halliwell, 2006). In particular, $O_2^{\cdot-}$ has the ability to hinder the activity of enzymes containing Fe-S clusters, and it also modifies the catalytic active site of the enzymes (Van Breusegem *et al.*, 2001; Halliwell, 2006). Moreover, the protonated form of superoxide, HO_2^{-}, can migrate across biological membranes and induce the peroxidation of lipids by the removal of protons from polyunsaturated fatty acids (Gechev *et al.*, 2006). Further, dismutated $O_2^{\cdot-}$ in the form of H_2O_2 can oxidize the thiol group of vital enzymes and hinder the activity of the enzymes. In addition, the harmful effect of $O_2^{\cdot-}$ and H_2O_2 is enhanced by interaction with metal ions resulting in the formation of hydroxyl radicals ($\cdot OH$) by the Haber–Weiss reaction (Kehrer, 2000). Because of the lack of detoxification mechanism for $\cdot OH$ radicals, cells rely on the strategic mechanism that prevents the formation of $\cdot OH$ radicals (Gechev *et al.*, 2006). Apart from the interaction with H_2O_2, $O_2^{\cdot-}$ can also react with the nitric oxide radical ($\cdot NO$) to form peroxynitrite ($ONOO^-$), which can rapidly protonate to peroxynitrous acid (ONOOH). Peroxynitrous acid is a powerful oxidizing agent that can damage all biomolecules (Gechev *et al.*, 2006; Halliwell, 2006). Further, it also stimulate the generation of reactive and non-reactive radicals (Halliwell, 2006).

Generation of ROS in Cells

ROS are produced in all cellular environments since they are inevitable intermediates of aerobic metabolism. This ubiquitous nature of ROS increases their complexity. Generally, ROS are maintained at constant basal levels; however, in response to different physiological circumstances such as cellular differentiation, abiotic stress, developmental stages, and programmed cell death, the ROS levels tend to increase (Dat *et al.*, 2000). The major site of ROS synthesis in plants is the chloroplast (Asada, 2006). The photosynthetic electron transfer chain generates $O_2^{\cdot-}$, particularly under conditions resulting in over-energization of the electron transfer chain (Dat *et al.*, 2000). The $O_2^{\cdot-}$ is predominantly generated by electron leakage from Fe-S centers of photosystem I (PSI) or by the reduction of ferredoxin to O_2 via the Mehler reaction

(Gechev *et al.*, 2006) and subsequent dismutation to H_2O_2 by the enzyme superoxide dismutase (SOD). Apart from chloroplast, the two other major sites of ROS production are peroxisomes and glyoxysomes during the process of photorespiration and fatty acid oxidation, respectively (Del Rio *et al.*, 2003).

In addition, mitochondrial respiration also generates $O_2^{\cdot-}$ and H_2O_2. In mitochondria, the primary source of ROS is NADH dehydrogenase, ubiquinone radical, and complex III (Møller, 2001). Although, ROS production in mitochondria is less than that in chloroplasts, the ROS generated in mitochondria play key roles in several cellular process including adaptation to stressful environments (Robson and Vanlerberghe, 2002; Foyer and Noctor, 2003). In apoplast, the major part of ROS production depends on the activity of plasmalemma-bound NADPH oxidase and cell-wall associated peroxidases, which are regulated by developmental and environmental cues (Dat *et al.*, 2000; Sagi and Fluhr, 2006). The accumulation of ROS in apoplast participates in the hypersensitive response against pathogen attack; it also influences growth, development, and death of cells (Torres *et al.*, 2002; Foreman *et al.*, 2003; Gechev and Hille, 2005; Gapper and Dolan, 2006). In addition to the above mentioned sources, ROS are also generated during the catabolism of purines by xanthine oxidase and the synthesis of deoxyribonucleotides by ribonucleotide reductase enzyme (Dat *et al.*, 2000). Further, several oxidases induced by abiotic and biotic stress signals are also responsible for the production of ROS.

Impact of ROS as Signaling Molecules

In recent years, the potential role of ROS in cell signaling has become more evident in both plants and animals. ROS possess various desirable characteristics for acting as signaling molecules. According to Hancock *et al.* (2001) for a molecule to be considered as a possible signaling molecule it has to satisfy the following criteria: (i) the molecule should be synthesized by a cell upon stimulation; (ii) it should exert its function in the cell where it is produced or in a nearby cell; and (iii) the molecule can be seized or removed in order to reverse the signal (Hancock *et al.*, 2001). Accordingly, ROS molecules are small, they diffuse short distances, and can be generated as well as removed by various mechanisms in cells. Evidently, ROS are involved in several physiological processes such as induction or inhibition of cellular proliferation, activation or inhibition of apoptosis, and induce necrosis in higher concentrations (Dat *et al.*, 2000; Gechev and Hille, 2005; Gapper and Dolan, 2006). Moreover, a wide range of genes are regulated by ROS like H_2O_2; for instance, in *Arabidopsis thaliana* H_2O_2 induced the expression of genes related to antioxidant metabolism (Gechev and Hille, 2005; Gechev *et al.*, 2006).

Various reports suggest that the spatial and temporal fluctuations of ROS levels in cells can act as signals required for different physiological activities such as growth (Gechev and Hille, 2005), development (Foreman *et al.*, 2003), abiotic stress tolerance (Torres *et al.*, 2002), response to pathogens (Torres *et al.*, 2006) and apoptosis (Apel and Hirt, 2004; Gechev *et al.*, 2006). In addition, the specificity of ROS for the stimulation of different signaling processes depends on multiple factors such as chemical identity, signal intensity, site of generation, developmental stage of the plant, pre-existing stress responses of the cell, and interaction with other signaling messengers (Kwak *et al.*, 2006; Zaninotto *et al.*, 2006). Earlier studies attributed the role of ROS in cell signaling to H_2O_2, which was more evident from microarray-based transcriptional analysis to identify genes regulated by H_2O_2 (Levine *et al.*, 1994; Desikan *et al.*, 2001; Vandenabeele *et al.*, 2004; Gechev *et al.*, 2005;). However, several other studies also elucidated the distinct transcriptional responses of genes with respect to ROS (Vranova *et al.*, 2002; Demidchik *et al.*, 2003; op den Camp *et al.*, 2003). In many plant species, lower levels of H_2O_2 and $O_2^{\cdot-}$ stimulate alleviation against the oxidative burst induced by the abiotic stress thus triggering the protective mechanisms (Vranova *et al.*, 2002; Gechev *et al.*, 2003). Moreover, programmed cell death is another major cascade of reactions that can be initiated by higher levels of $O_2^{\cdot-}$ and H_2O_2 (Montillet *et al.*, 2005; Laloi *et al.*, 2006; Van Breusegem and Dat, 2006). Apart from the levels of ROS, the site of production also critically determines the signaling capacity (Mullineaux *et al.*, 2006). For instance, the NADPH oxidase-mediated generation of $O_2^{\cdot-}$ in root hair tip stimulates the divalent calcium ions necessary for growth of root hairs (Foreman *et al.*, 2003). The activity of NADPH oxidase is triggered by Rho-like GTPases, which also regulate the differentiation of tracheary elements through the site-specific generation of ROS (Nakanomyo *et al.*, 2002). Further, the spatial regulation of ROS depends highly on the mobility and interaction of the ROS molecules among different components of cells (Gechev *et al.*, 2006). As mentioned earlier, H_2O_2 can migrate from the site of production over quite a long distance and across biological membranes in comparison with $O_2^{\cdot-}$ and $\cdot OH$ molecules (Bienert *et al.*, 2006). However, the permeability of membranes to H_2O_2 is low and passage occurs through specialized aquaporin-like channels called peroxporins (Henzler and Steudle, 2000). The transportation can also be considered as a strategic mechanism for the fine tuning of the level of H_2O_2 in the cells. The association between the mobility and the cross-compartment communication of H_2O_2 can be illustrated by the amount of H_2O_2 generated in the cytosol upon the absence of ascorbate peroxidase (APX), which results in the inhibition of chloroplastic APX and leads to perturbation of the chloroplastic antioxidant system (Davletova *et al.*, 2005). In addition, the catalase enzyme of peroxisomes can scavenge the H_2O_2 molecules generated in peroxisomes, while the infiltration of catalase to the extracellular space of the leaves leads to the detoxification of H_2O_2 generated upon photorespiration (Willekens *et al.*, 1997). Thus, the ROS molecules are not completely toxic to the cells. The basal levels of ROS play vital roles in several signaling networks, and the balance among the ROS is important for the regulation of metabolic processes. In the upcoming sections the primary role of different types of ROS in the differentiation and proliferation of cells will be discussed in detail.

ROS in the Proliferation of Cells

Division of eukaryotic cells occurs once they reach a specific size, or it can be triggered by several internal and extracellular factors. As plants are sessile in nature, the pattern of growth is chiefly regulated by environmental conditions (Schmidt and Schippers, 2015). Although the components controlling the plant cell cycle have been determined, the influence of ROS signals on cellular proliferation and differentiation is hardly explored. However, the molecular rationale behind the ROS-mediated regulation of the cell cycle has been explored using genetic and biochemical approaches. Therefore, in this section the impact of ROS signals during the proliferation of cells in shoot, root, and leaves will be elaborated.

ROS in the Proliferation of Shoot Apical Meristem

In general, the aerial parts of plants are formed from shoot apical meristem (SAM). However, plant growth hormones such as cytokinins and auxins play a vital role in the maintenance of the stem cell pool and in organ formation respectively(Scofield *et al.*, 2013). Morphologically, the shoot meristem is organized in a central zone containing slowly dividing stem cells surrounded by a peripheral zone with rapidly dividing cells (Aichinger *et al.*, 2012). During differentiation, the progeny of the stem cells in the central zone pass into the peripheral zone and differentiate into leaves or flowers. Further, the stem cell zone is classified into three layers along with a meristem-organizing center (Schmidt and Schippers, 2015). The cells in the layers of SAM are upregulated by the homeodomain transcription factor WUSCHEL (WUS), particularly in the organizing center. In contrast, CLAVATA 3, a signaling peptide located in the central zone, represses the activity of stem cells (Schmidt and Schippers, 2015). The redox signaling mediated regulation of SAM has been demonstrated by mutational studies in *A. thaliana* (Yadav *et al.*, 2014). Interestingly, the expression levels of antioxidant genes such as *GSH1*, *CATALASE3* (*CAT3*), and *GLUTATHIONE REDUCTASE1* (*ATGR1*) are higher in the organizing center in comparison with other parts of the SAM, illustrating the elevated ROS levels (Lai *et al.*, 2012; Benina *et al.*, 2015). According to previous reports, several transcription factors (TFs) involved in the maintenance of cellular proliferation in the different zones of SAM are at least partly regulated by ROS (Schmidt and Schippers, 2015). For example, NPR1 TFs are among the several molecular regulators activated by the redox signals that play a prominent role in the formation of leaf primordia (Tada *et al.*, 2008). In addition, several Teosinte branched protein (TCP) TFs regulate the proliferation of cells in different types of tissues such as internodes, leaves, and flowers by indirectly controlling the cell cycle genes (Li *et al.*, 2005; Hervé *et al.*, 2009; Kieffer *et al.*, 2011; Li *et al.*, 2012). In most cases, the TCPs are sensitive to redox signals; TCP15 and TCP20 are two TFs belonging to class I TCPs sensitive to redox signals and directly involved in the regulation of cell cycle-related genes (Li *et al.*, 2005; Hervé *et al.*, 2009; Kieffer *et al.*, 2011). Among the ROS, H_2O_2 plays an important function in the control of DNA binding by the class I TCPs by oxidizing Cys residues present in the DNA-binding and dimerization domains (Viola *et al.*, 2013). Subsequently, the oxidized TCP15 and TCP20 result in the termination of the cell cycle (Vernoux *et al.*, 2000). Although various signals and factors are involved in the regulation of cellular proliferation, ROS-mediated signals appear to play a major role, one that is still relatively unexplored.

ROS in the Proliferation of Root Apical Meristem

Similar to the shoot, the root tissues originate from a root apical meristem (RAM) consisting of a proximal meristem and quiescent center. The cells in the proximal meristem give rise to the root tissue, and the organization of the meristem is maintained by the quiescent center (Aichinger *et al.*, 2012). In RAM, cells in the proximal meristem divide rapidly with lengthy proliferation times due to a slow G1/S transition period during the cell cycle (Kerk and Feldman, 1995). Interestingly, fluorescent staining of the RAM cells demonstrated ROS-mediated regulation of cell cycle activity; for instance, cells in the quiescent center were oxidized predominantly than the proximal meristem cells by H_2O_2 (Jiang *et al.*, 2003). In addition, the oxidation of ascorbic acid and glutathione is higher in the quiescent center cells on comparison with proximal meristem cells (Jiang et al., 2003; Noctor and Foyer, 1998). According to Jiang *et al.* (2003), the possible reason behind the increase in GSSG in quiescent center cells could be due to the least amount of NADPH levels which is essential for the reduction of GSH in the quiescent center cells. Thus, the lower levels of GSH could attribute to the cessation of cell cycle at G1 phase in the cells habituated in the quiescent center (Vernoux *et al.*, 2000, Vivancos *et al.*, 2010). Likewise, the redox associated regulation of the activity of RAM is evident in the *root meristemless1* mutant, which is devoid of the enzyme necessary for the biosynthesis of GSH (Vernoux *et al.*, 2000). The reduction in the abundance of GSH corresponded with the inability to form an active meristematic region in root; however, supplementation with GSH recovered the meristematic activity (Vernoux *et al.*, 2000).

Endogenous auxin levels in the roots are strongly correlated with the redox status of the cells in both the quiescent center and proximal meristem (Jiang *et al.*, 2003). Auxins are a class of phytohormones that play a vital role in the growth and development of roots; in minor concentrations they enhance the elongation of roots, whereas higher levels hinder root growth (Teale *et al.*, 2006). In a manner similar to ascorbic acid and glutathione, the oxidation of auxin occurs more predominantly in the quiescent center cells than in the proximal meristem cells (Schmidt and Schippers, 2015). However, decapitation of the root cap or treatment with a polar auxin transport inhibitor such as n-1-naphthylphthalamic acid modulates the redox status of the cells in the quiescent center to more reduced state resulting in a higher rate of cell division (Jiang *et al.*, 2003). In contrast, the cells in the proximal meristem region became more oxidized, which lowered the rate of cell division (Jiang *et al.*, 2003). However, the possible rationale behind the endogenous auxin-mediated synergistic regulation of cell cycle and redox status is yet to be elucidated. But the performance of cells in the quiescent center can be attributed to the network of auxin and the antioxidants such as ascorbic acid and glutathione (Schmidt and Schippers, 2015). Evidently in maize root tissue, auxin induced the biosynthesis and activity of the ascorbate oxidase enzyme responsible for the oxidation of ascorbic acid to dehydroascorbic acid, thereby stimulating the oxidative state in the quiescent center cells (Kerk and Feldman, 1995). In addition the ascorbate oxidase enzyme possesses the ability to degrade the auxin, which establishes a mechanism to fine-tune redox homeostasis within the quiescent center (Kerk *et al.*, 2000). Furthermore, the influx and efflux auxin transporters are prone to ROS-mediated posttranslation modifications, which affect auxin transportation (Fernández-Marcos *et al.*, 2013). Thus, the ROS levels act as a vital factor in the mediation of the cell cycle directly, or indirectly by modulating redox status, and regulate the proliferation of cells in RAM.

ROS in Differentiation of Cells

ROS are also involved in the switch from cellular proliferation to differentiation state. In general, the proliferation of cells ceases after several cycles and the cells enter into the transition state in which the cells elongate and differentiation occurs (Schmidt and Schippers, 2015). This process of transition has been reported to be regulated by the redox state. In *A. thaliana* the type III heme peroxidase in involved during the elongation of roots (Tsukagoshi *et al.*, 2010). Heme peroxidase has the ability to scavenge or generate the H_2O_2 in different chemical environments (Passardi *et al.*, 2004). Moreover, the expression of peroxidase genes is repressed by the UPB1 transcription factors, and the lack of UPB1 results in the enlargement of root meristem and triggers the elongation of cells (Tsukagoshi *et al.*, 2010). On the other hand, overexpression of UPB1 leads to early transition to the differentiation stage. In addition, lack of UPB1 results in increased levels of superoxide and decreased H_2O_2 abundance, and vice versa upon overexpression (Tsukagoshi *et al.*, 2010; Schmidt and Schippers, 2015). Thus, the balance between the ROS radicals is vital for the transition from the proliferative to the differentiation stage.

During the development of leaves, the synergistic processes of proliferation and cellular differentiation have to occur in order to reach the optimal size (Lu *et al.*, 2014). In leaves the cellular proliferation and expansion starts from the tip and progresses in a basipetal manner (Schmidt and Schippers, 2015). Among the factors influencing the transition state of cells, the maturation of chloroplasts is considered an important cue in leaves (Gonzalez *et al.*, 2012). The report by Kim and Apel (2013) suggested that the development of chloroplasts was hampered by the herbicide norflurazon, which repressed the expression of nuclear encoded chloroplast genes. Moreover, the impairment of carotenoid biosynthesis increased the content of $O_2^{\cdot-}$ by inducing the photosensitization of chlorophyll (Kim and Apel, 2013). In the same study the norflurazon treatment suspended the initiation of expansion of cells illustrating that the backward stimulus from the chloroplast can regulate the transition state in leaves (Li *et al.*, 2009; Kim and Apel, 2013). In addition a mutational study suggested that the backward signal attributed by the chloroplast is mediated by $O_2^{\cdot-}$ (Kim and Apel, 2013). Taken together, ROS play a vital role during the switch to cellular differentiation from the proliferation phase in different tissues. However, the in-depth mechanism behind ROS-dependent regulation remains to be elucidated.

Conclusions

Overall, on the basis of the existing literature, ROS-associated signaling acts as an important factor during the proliferation and differentiation of cells in plants. The existence of various types of ROS radicals and their diverse occurrence in cells gives them optimal signaling properties to participate in a complex cascade of events. However, only a small part of these ROS-mediated signaling networks has been elucidated to date. In future, deeper insights into ROS signaling in various cell types in different developmental stages have to be gained.

References

Aichinger E, Kornet N, Friedrich T, Laux T. (2012) Plant stem cell niches. *Ann Rev Plant Biol* 63: 615–636.

Apel, K. and Hirt, H., 2004. Reactive oxygen species: metabolism, oxidative stress, and signal transduction. *Annu Rev Plant Biol*, 55: 373–399.

Asada K. (2006) Production and scavenging of reactive oxygen species in chloroplasts and their functions. *Plant Physiol* 141: 391–396.

Bashandy T, Guilleminot J, Vernoux T, *et al.* (2010) Interplay between the NADP-linked thioredoxin and glutathione systems in Arabidopsis auxin signaling. *The Plant Cell* 22: 376–391.

Benina M, Ribeiro DM, Gechev TS, Muller-Robber B, Schippers JH. (2015) A cell type-specific view on the translation of mRNAs from ROS-responsive genes upon paraquat treatment of Arabidopsis thaliana leaves. *Plant Cell Environ* 38: 349–363.

Berckmans B, De Veylder L. (2009) Transcriptional control of the cell cycle. *Curr Opin Plant Biol* 12: 599–605.

Bienert GP, Schjoerring JK, Jahn TP. (2006) Membrane transport of hydrogen peroxide. *BBA Biomembranes* 1758: 994–1003.

Cremers CM, Jakob U. (2013) Oxidant sensing by reversible disulfide bond formation. *J Biol Chem* 288: 26489–26496.

Dat J, Vandenabeele S, Vranová E, Van Montagu M, Inzé D, Van Breusegem F. (2000) Dual action of the active oxygen species during plant stress responses. *Cell Mol Life Sci* 57: 779–795.

D'Autréaux B, Toledano MB. (2007) ROS as signalling molecules: mechanisms that generate specificity in ROS homeostasis. *Nat Rev Mol Cell Biol* 8: 813–824.

Davletova S, Rizhsky L, Liang H, *et al.* (2005) Cytosolic ascorbate peroxidase 1 is a central component of the reactive oxygen gene network of Arabidopsis. *The Plant Cell* 17: 268–281.

DellaPenna D, Pogson BJ. (2006) Vitamin synthesis in plants: tocopherols and carotenoids. *Annu Rev Plant Biol* 57: 711–738.

Delledonne M, Zeier J, Marocco A, Lamb C. (2001) Signal interactions between nitric oxide and reactive oxygen intermediates in the plant hypersensitive disease resistance response. *Proc Natl Acad Sci USA* 98: 13454–13459.

Del Río LA, Corpas FJ, Sandalio LM, Palma JM, Barroso JB. (2003) Plant peroxisomes, reactive oxygen metabolism and nitric oxide. *IUBMB Life* 55: 71–81.
Demidchik V, Shabala SN, Coutts KB, Tester MA, Davies JM. (2003) Free oxygen radicals regulate plasma membrane Ca^{2+}-and K^{+}-permeable channels in plant root cells. *J Cell Sci* 116: 81–88.
Desikan R, Soheila AH, Hancock JT, Neill SJ. (2001) Regulation of the Arabidopsis transcriptome by oxidative stress. *Plant Physiol* 127: 159–172.
Dewitte W, Murray JA. (2003) The plant cell cycle. *Ann Rev Plant Biol* 54: 235–264.
Dickinson BC, Chang CJ. (2011) Chemistry and biology of reactive oxygen species in signaling or stress responses. *Nat Chem Biol* 7: 504–511.
Fernández-Marcos M, Sanz L, Lewis DR, Muday GK, Lorenzo O. (2013) Control of auxin transport by reactive oxygen and nitrogen species. In: Chen R, Baluška F (eds), *Polar Auxin Transport*. Springer, Berlin, Heidelberg, pp. 103–117.
Foreman J, Demidchik V, Bothwell JH, *et al.* (2003) Reactive oxygen species produced by NADPH oxidase regulate plant cell growth. *Nature* 422: 442–446.
Foyer CH, Noctor G. (2003) Redox sensing and signalling associated with reactive oxygen in chloroplasts, peroxisomes and mitochondria. *Physiol Plantarum* 119: 355–364.
Francis D. (2007) The plant cell cycle – 15 years on. *New Phytol* 174: 261–278.
Gapper C, Dolan L. (2006) Control of plant development by reactive oxygen species. *Plant Physiol* 141: 341–345.
Gechev TS, Hille J. (2005) Hydrogen peroxide as a signal controlling plant programmed cell death. *J Cell Biol* 168: 17–20.
Gechev TS, Gadjev I, Van Breusegem F, *et al.* (2002) Hydrogen peroxide protects tobacco from oxidative stress by inducing a set of antioxidant enzymes. *Cell Mol Life Sci* 59: 708–714.
Gechev T, Minkov I, Hille J. (2005) Hydrogen peroxide-induced cell death in Arabidopsis: transcriptional and mutant analysis reveals a role of an oxoglutarate-dependent dioxygenase gene in the cell death process. *IUBMB Life* 57: 181–188.
Gechev TS, Van Breusegem F, Stone JM, Denev I, Laloi C. (2006) Reactive oxygen species as signals that modulate plant stress responses and programmed cell death. *Bioessays* 28: 1091–1101.
Gonzalez N, Vanhaeren H, Inzé D. (2012) Leaf size control: complex coordination of cell division and expansion. *Trends Plant Sci* 17: 332–340.
Halliwell B. (2006) Reactive species and antioxidants. Redox biology is a fundamental theme of aerobic life. *Plant Physiol* 141: 312–322.
Hancock JT, Desikan R, Neill SJ. (2001) Role of reactive oxygen species in cell signalling pathways. *Biochem Soc Trans* 29: 345–349.
Henzler T, Steudle E. (2000) Transport and metabolic degradation of hydrogen peroxide in Chara corallina: model calculations and measurements with the pressure probe suggest transport of H_2O_2 across water channels. *J Exp Bot* 51: 2053–2066.
Hervé C, Dabos P, Bardet C, *et al.* (2009) In vivo interference with AtTCP20 function induces severe plant growth alterations and deregulates the expression of many genes important for development. *Plant Physiol* 149: 1462–1477.
Jiang K, Meng YL, Feldman LJ. (2003) Quiescent center formation in maize roots is associated with an auxin-regulated oxidizing environment. *Development* 130: 1429–1438.
Jurado S, Diaz-Trivino S, Abraham Z, Manzano C, Gutierrez C, del Pozo JC. (2008) SKP2A protein, an F-box that regulates cell division, is degraded via the ubiquitin pathway. *Plant Signal Behav* 3: 810–812.
Kehrer JP. (2000) The Haber-Weiss reaction and mechanisms of toxicity. *Toxicology* 149: 43–50.
Kerk NM, Feldman NJ. (1995) A biochemical model for the initiation and maintenance of the quiescent center: implications for organization of root meristems. *Development* 121: 2825–2833.

Kerk NM, Jiang K, Feldman LJ. (2000) Auxin metabolism in the root apical meristem. *Plant Physiol* 122: 925–932.
Kieffer M, Master V, Waites R, Davies B. (2011) TCP14 and TCP15 affect internode length and leaf shape in Arabidopsis. *Plant J* 68: 147–158.
Kim C, Apel K. (2013) $^1O_2$-mediated and EXECUTER-dependent retrograde plastid-to-nucleus signaling in norflurazon-treated seedlings of Arabidopsis thaliana. *Mol Plant* 6: 1580–1591.
Kwak JM, Nguyen V, Schroeder JI. (2006) The role of reactive oxygen species in hormonal responses. *Plant Physiol* 141: 323–329.
Lai AG, Doherty CJ, Mueller-Roeber B, Kay SA, Schippers JH, Dijkwel PP. (2012) CIRCADIAN CLOCK-ASSOCIATED 1 regulates ROS homeostasis and oxidative stress responses. *Proc Natl Acad Sci USA* 109: 17129–17134.
Laloi C, Przybyla D, Apel K. (2006) A genetic approach towards elucidating the biological activity of different reactive oxygen species in Arabidopsis thaliana. *J Exp Bot* 57: 1719–1724.
Levine A, Tenhaken R, Dixon R, Lamb C. (1994) H_2O_2 from the oxidative burst orchestrates the plant hypersensitive disease resistance response. *Cell* 79: 583–593.
Li C, Potuschak T, Colón-Carmona A, Gutiérrez RA, Doerner P. (2005) Arabidopsis TCP20 links regulation of growth and cell division control pathways. *Proc Natl Acad Sci USA* 102: 12978–12983.
Li Z, Wakao S, Fischer BB, Niyogi KK. (2009) Sensing and responding to excess light. *Ann Rev Plant Biol* 60:239–260.
Li ZY, Li B, Dong AW. (2012) The Arabidopsis transcription factor AtTCP15 regulates endoreduplication by modulating expression of key cell-cycle genes. *Mol Plant* 5: 270–280.
Lu D, Wang T, Persson S, Mueller-Roeber B, Schippers JH. (2014) Transcriptional control of ROS homeostasis by KUODA1 regulates cell expansion during leaf development. *Nat Commun* 5: Art no. 3767.
Møller IM. (2001) Plant mitochondria and oxidative stress: electron transport, NADPH turnover, and metabolism of reactive oxygen species. *Ann Rev Plant Biol* 52: 561–591.
Montillet JL, Chamnongpol S, Rustérucci C, *et al.* (2005) Fatty acid hydroperoxides and H2O2 in the execution of hypersensitive cell death in tobacco leaves. *Plant Physiol* 138: 1516–1526.
Mullineaux PM, Karpinski S, Baker NR. (2006) Spatial dependence for hydrogen peroxide-directed signaling in light-stressed plants. *Plant Physiol* 141: 346–350.
Nakanomyo I, Kost B, Chua NH, Fukuda H. (2002) Preferential and asymmetrical accumulation of a Rac small GTPase mRNA in differentiating xylem cells of Zinnia elegans. *Plant Cell Physiol* 43: 1484–1492.
Noctor G, Foyer CH. (1998) Ascorbate and glutathione: keeping active oxygen under control. *Ann Rev Plant Biol* 49:249–279.
Novikova GV, Nosov AV, Stepanchenko NS, Fomenkov AA, Mamaeva AS, Moshkov IE. (2013) Plant cell proliferation and its regulators. *Russ J Plant Physiol* 60: 500–506.
op den Camp RG, Przybyla D, Ochsenbein C, *et al.* (2003) Rapid induction of distinct stress responses after the release of singlet oxygen in Arabidopsis. *The Plant Cell* 15: 2320–2332.
Passardi F, Penel C, Dunand C. (2004) Performing the paradoxical: how plant peroxidases modify the cell wall. *Trends Plant Sci* 9: 534–540.
Raymond J, Segrè D. (2006) The effect of oxygen on biochemical networks and the evolution of complex life. *Science* 311: 1764–1767.
Robson CA, Vanlerberghe GC. (2002) Transgenic plant cells lacking mitochondrial alternative oxidase have increased susceptibility to mitochondria-dependent and -independent pathways of programmed cell death. *Plant Physiol* 129: 1908–1920.
Sagi M, Fluhr R. (2006) Production of reactive oxygen species by plant NADPH oxidases. *Plant Physiol* 141: 336–340.

Sarkar D, Sharma S. (2010) Oxidative burst-mediated ROS signaling pathways regulating tuberization in potato. In: Dutta Gupta S (ed.), *Reactive Oxygen Species and Antioxidants in Higher Plants*. Science Publishers, Enfield, pp.79–100.

Schmidt R, Schippers JH. (2015) ROS-mediated redox signaling during cell differentiation in plants. *BBA Gen Subjects* 1850: 1497–1508.

Scofield S, Dewitte W, Nieuwland J, Murray JA. (2013) The Arabidopsis homeobox gene SHOOT MERISTEMLESS has cellular and meristem-organisational roles with differential requirements for cytokinin and CYCD3 activity. *Plant J* 75: 53–66.

Singh R, Singh S, Parihar P, *et al.* (2016) Reactive oxygen species (ROS): beneficial companions of plants' developmental processes. *Front Plant Sci* 7: 1299.

Singh S, Srivastava PK, Kumar D, Tripathi DK, Chauhan DK, Prasad SM. (2015) Morpho-anatomical and biochemical adapting strategies of maize (*Zea mays* L.) seedlings against lead and chromium stresses. *Biocatal Agric Biotechnol* 4: 286–295.

Singh S, Tripathi DK, Singh S, *et al.* (2017) Toxicity of aluminium on various levels of plant cells and organism: a review. *Environ Exp Bot* 137: 177–193.

Tada Y, Spoel SH, Pajerowska-Mukhtar K, *et al.* (2008) Plant immunity requires conformational charges of NPR1 via S-nitrosylation and thioredoxins. *Science* 321: 952–956.

Takatsuka H, Ohno R, Umeda M. (2009) The Arabidopsis cyclin-dependent kinase-activating kinase CDKF;1 is a major regulator of cell proliferation and cell expansion but is dispensable for CDKA activation. *Plant J* 59: 475–487.

Teale WD, Paponov IA, Palme K. (2006) Auxin in action: signalling, transport and the control of plant growth and development. *Nat Rev Mol Cell Biol* 7: 847–859.

Torres MA, Dangl JL, Jones JD. (2002) Arabidopsis gp91phox homologues AtrbohD and AtrbohF are required for accumulation of reactive oxygen intermediates in the plant defense response. *Proc Natl Acad Sci USA* 99: 517–522.

Torres MA, Jones JD,Dangl JL. (2006) Reactive oxygen species signaling in response to pathogens. *Plant physiol*, 141(2): 373–378.

Tripathi DK, Singh S, Singh VP, Prasad SM, Chauhan DK, Dubey NK. (2016a) Silicon nanoparticles more efficiently alleviate arsenate toxicity than silicon in maize cultivar and hybrid differing in arsenate tolerance. *Front Environ Sci* 4: 46.

Tripathi DK, Singh VP, Prasad SM, Dubey NK, Chauhan DK, Rai AK. (2016b) LIB spectroscopic and biochemical analysis to characterize lead toxicity alleviative nature of silicon in wheat (Triticum aestivum L.) seedlings. *J Photochem Photobiol B: Biol* 154: 89–98.

Tripathi DK, Singh S, Singh VP, Prasad SM, Dubey NK, Chauhan DK. (2017a) Silicon nanoparticles more effectively alleviated UV-B stress than silicon in wheat (Triticum aestivum) seedlings. *Plant Physiol Biochem* 110: 70–81.

Tripathi DK, Singh S, Singh S, *et al.* (2017b) Nitric oxide alleviates silver nanoparticles (AgNps)-induced phytotoxicity in Pisum sativum seedlings. *Plant Physiol Biochem* 110: 167–177.

Tsukagoshi H, Busch W, Benfey PN. (2010) Transcriptional regulation of ROS controls transition from proliferation to differentiation in the root. *Cell* 143: 606–616.

Van Breusegem F, Dat JF. (2006) Reactive oxygen species in plant cell death. *Plant Physiol* 141: 384–390.

Van Breusegem F, Vranová E, Dat JF, Inzé D. (2001) The role of active oxygen species in plant signal transduction. *Plant Sci* 161: 405–414.

Vandenabeele S, Van Der Kelen K, Dat J, *et al.* (2003) A comprehensive analysis of hydrogen peroxide-induced gene expression in tobacco. *Proc Natl Acad Sci USA* 100: 16113–16118.

Vandenabeele S, Vanderauwera S, Vuylsteke M, *et al.* (2004) Catalase deficiency drastically affects gene expression induced by high light in Arabidopsis thaliana. *Plant J* 39: 45–58.

Vanderauwera S, Zimmermann P, Rombauts S, *et al.* (2005) Genome-wide analysis of hydrogen peroxide-regulated gene expression in Arabidopsis reveals a high light-induced transcriptional cluster involved in anthocyanin biosynthesis. *Plant Physiol* 139: 806–821.

Vernoux T, Wilson RC, Seeley KA, *et al.* (2000) The ROOT MERISTEMLESS1/CADMIUM SENSITIVE2 gene defines a glutathione-dependent pathway involved in initiation and maintenance of cell division during postembryonic root development. *The Plant Cell* 12: 97–109.

Viola IL, Güttlein LN, Gonzalez DH. (2013) Redox modulation of plant developmental regulators from the class I TCP transcription factor family. *Plant Physiol* 162: 1434–1447.

Vivancos PD, Wolff T, Markovic J, Pallardó FV, Foyer CH. (2010) A nuclear glutathione cycle within the cell cycle. *Biochem J* 431: 169–178.

Vranová E, Inzé D, Van Breusegem F. (2002) Signal transduction during oxidative stress. *J Exp Bot,* 53(372): 1227–1236.

Wang H, Wang S, Lu Y, *et al.* (2011) Proteomic analysis of early-responsive redox-sensitive proteins in Arabidopsis. *J Proteome Res* 11: 412–424.

Willekens H, Chamnongpol S, Davey M, *et al.* (1997) Catalase is a sink for H_2O_2 and is indispensable for stress defence in C3 plants. *EMBO J* 16: 4806–4816.

Yadav RK, Tavakkoli M, Xie M, Girke T, Reddy GV. (2014) A high-resolution gene expression map of the Arabidopsis shoot meristem stem cell niche. *Development* 141: 2735–2744.

Zaninotto F, La Camera S, Polverari A, Delledonne M. (2006) Cross talk between reactive nitrogen and oxygen species during the hypersensitive disease resistance response. *Plant Physiol* 141: 379–383.

Index

Reactive Oxygen Species in Plants: Boon Or Bane - Revisiting the Role of ROS, First Edition. Edited by Vijay Pratap Singh, Samiksha Singh, Durgesh Kumar Tripathi, Sheo Mohan Prasad, and Devendra Kumar Chauhan.

d

e

q

r